STUDY GUIDE
AND
SOLUTIONS MANUAL

Paula Yurkanis Bruice

University of California, Santa Barbara

ORGANIC CHEMISTRY

SIXTH EDITION

Prentice Hall

Boston Columbus Indianapolis New York San Francisco Upper Saddle River
Amsterdam Cape Town Dubai London Madrid Milan Munich Paris Montréal Toronto
Delhi Mexico City São Paulo Sydney Hong Kong Seoul Singapore Taipei Tokyo

Publisher: Dan Kaveney
Marketing Manager: Erin Gardner
Associate Editor: Jessica Neumann
Managing Editor, Chemistry and Geosciences: Gina M. Cheselka
Project Manager: Ed Thomas
Operations Specialist: Maura Zaldivar
Supplement Cover Designer: Paul Gourhan

Cover Image: Motherwell, Robert (1915-1991) © VAGA, NY Beside the Sea. 1968. Synthetic polymer paint, 30 1/4 x 21 7/8" (77 x 55.7 cm). Gift of the artist. (1095. 1969) Location: The Museum of Modern Art, New York, NY, U.S.A. Digital Image © The Museum of Modern Art/Licensed by SCALA / Art Resource, NY.

Printed in the United States of America

10 9 8 7 6 5 4 3 2 1

ISBN-13:   978-0-321-67682-5
ISBN-10:   0-321-67682-3

Prentice Hall
is an imprint of

www.pearsonhighered.com

TABLE OF CONTENTS

to my students

I am enormously grateful to Paul Sampson of Kent State University, who reworked all the problems to make this book as error free as possible and who gave me many important suggestions along the way. I am solely responsible for any errors that may remain. If you find any, please email me so they can be fixed in a future printing.

Try to work as many problems as possible, so you can truly enjoy the wonderful world of organic chemistry.

Paula Yurkanis Bruice
pybruice@chem.ucsb.edu

CHAPTER 1
Electronic Structure and Bonding • Acids and Bases

Important Terms

acid
(Brønsted acid)
a substance that donates a proton.

acid-base reaction
a reaction in which an acid donates a proton to a base.

acid dissociation constant
a measure of the degree to which an acid dissociates.

acidity
a measure of how easily a compound gives up a proton.

antibonding molecular orbital
a molecular orbital that results when two atomic orbitals with opposite phases interact. Electrons in an antibonding orbital decrease bond strength.

atomic number
tells how many protons (or electrons) the neutral atom has.

atomic orbital
an orbital associated with an atom; the three-dimensional area around a nucleus where electrons are most likely to be found.

atomic weight
the average mass of the atoms in the naturally occurring element.

aufbau principle
states that an electron will always go into the available orbital with the lowest energy.

base
(Brønsted base)
a substance that accepts a proton.

basicity
describes the tendency of a compound to share its electrons with a proton.

bond dissociation energy
the amount of energy required to break a bond in a way that allows each of the atoms to retain one of the bonding electrons, or the amount of energy released when a bond is formed.

bonding molecular orbital
a molecular orbital that results when two atomic orbitals with the same phase interact. Electrons in a bonding orbital increase bond strength.

bond length
the internuclear distance between two atoms at minimum energy (maximum stability).

Brønsted acid
a substance that donates a proton.

Brønsted base
a substance that accepts a proton.

buffer solution
solution of a weak acid and its conjugate base.

carbanion
a species containing a negatively charged carbon.

carbocation a species containing a positively charged carbon.

condensed structure a structure that does not show some (or all) of the covalent bonds.

conjugate acid the species formed when a base accepts a proton.

conjugate base the species formed when an acid loses a proton.

core electrons electrons in filled shells.

covalent bond a bond created as a result of sharing electrons.

degenerate orbitals orbitals that have the same energy.

delocalized electrons electrons that do not belong to a single atom nor are they shared in a bond between two atoms.

dipole a separation of positive and negative charges.

dipole moment (μ) a measure of the separation of charge in a bond or in a molecule.

double bond composed of a sigma bond and a pi bond.

electronegative describes an element that readily acquires an electron.

electronegativity the tendency of an atom to pull electrons toward itself.

electropositive describes an element that readily loses an electron.

electrostatic attraction an attractive force between opposite charges.

electrostatic potential map a map that allows you to see how electrons are distributed in a molecule.

equilibrium constant the ratio of products to reactants at equilibrium.

excited-state electronic configuration the electronic configuration that results when an electron in the ground state has been moved to a higher energy orbital.

formal charge the number of valence electrons − (the number of nonbonding electrons + 1/2 the number of bonding electrons).

free radical (radical) a species with an unpaired electron.

ground-state electronic configuration a description of which orbitals the electrons of an atom occupy when they are all in their lowest available energy orbitals.

Heisenberg uncertainty principle states that both the precise location and the momentum of an atomic particle cannot be simultaneously determined.

Henderson-Hasselbalch equation	$pK_a = pH + \log[HA]/[A^-]$
Hund's rule	states that when there are degenerate orbitals, an electron will occupy an empty orbital before it will pair up with another electron.
hybrid orbital	an orbital formed by hybridizing (mixing) atomic orbitals.
hydride ion	a negatively charged hydrogen (a hydrogen atom with an extra electron).
hydrogen ion (proton)	a positively charged hydrogen (a hydrogen atom without its electron).
inductive electron withdrawal	the pull of electrons through sigma bonds by an atom or a group of atoms.
ionic bond	a bond formed as a result of the attraction of opposite charges.
ionic compound	a compound composed of a positive ion and negative ion.
ionization energy	the energy required to remove an electron from an atom.
isotopes	atoms with the same number of protons but a different number of neutrons.
Kekulé structure	a model that represents the bonds between atoms as lines.
Lewis acid	a substance that accepts an electron pair.
Lewis base	a substance that donates an electron pair.
Lewis structure	a model that represents the bonds between atoms as lines or dots and the lone-pair electrons as dots.
lone-pair electrons	valence electrons not used in bonding.
mass number	the number of protons plus the number of neutrons in an atom.
molecular orbital	an orbital associated with a molecule that results from the combination of atomic orbitals.
molecular orbital (MO) theory	describes a model in which the electrons occupy orbitals as they do in atoms but the orbitals extend over the entire molecule.
node	a region within an orbital where there is zero probability of finding an electron.
nonbonding electrons	valence electrons not used in bonding.
nonpolar covalent bond	a bond formed between two atoms that share the bonding electrons equally.

octet rule   states that an atom will give up, accept, or share electrons in order to achieve a filled outer shell (or an outer shell that contains eight electrons) and no electrons of higher energy. Because a filled second shell contains eight electrons, this is known as the octet rule.

orbital   the volume of space around the nucleus where an electron is most likely to be found.

orbital hybridization   mixing of atomic orbitals.

organic compound   a compound that contains carbon.

Pauli exclusion principle   states that no more than two electrons can occupy an orbital and that the two electrons must have opposite spin.

pH   the pH scale is used to describe the acidity of a solution ($pH = -\log [H^+]$).

pi (π) bond   a bond formed as a result of side-to-side overlap of p orbitals.

pK_a   describes the tendency of a compound to lose a proton ($pK_a = -\log K_a$, where K_a is the acid dissociation constant).

polar covalent bond   a bond formed between two atoms that do not share the bonding electrons equally.

potential map (electrostatic potential map)   a map that allows you to see how electrons are distributed in a molecule.

proton   a positively charged hydrogen ion.

proton transfer reaction   a reaction in which a proton is transferred from an acid to a base.

quantum mechanics   the use of mathematical equations to describe the behavior of electrons in atoms or molecules.

radical (free radical)   a species with an unpaired electron.

resonance   having delocalized electrons.

resonance contributors   structures with localized electrons that together approximate the true structure of a compound with delocalized electrons.

resonance hybrid   the actual structure of a compound with delocalized electrons.

sigma (σ) bond   a bond with a symmetrical distribution of electrons about the internuclear axis.

single bond   a single pair of electrons shared between two atoms.

tetrahedral bond angle   the bond angle (109.5°) formed by an sp^3 hybridized central atom.

tetrahedral carbon   an sp^3 hybridized carbon; a carbon that forms covalent bonds using four sp^3 hybrid orbitals.

trigonal planar carbon   an sp^2 hybridized carbon.

triple bond   composed of a sigma bond and two pi bonds.

valence electron   an electron in an outermost shell.

valence-shell electron-pair repulsion (VSEPR) model   a model for the prediction of molecular geometry based on the minimization of electron repulsion between regions of electron density (that is, between bonds and nonbonding electrons) around an atom.

wave equation   an equation that describes the behavior of each electron in an atom or a molecule.

wave functions   a series of solutions to a wave equation.

Solutions to Problems

1.   The atomic number = the number of protons.
      The mass number = the number of protons + the number of neutrons
      All isotopes have the same atomic number; in the case of oxygen it is 8. Therefore:

      The isotope of oxygen with a mass number of 16 has 8 protons and 8 neutrons.
      The isotope of oxygen with a mass number of 17 has 8 protons and 9 neutrons.
      The isotope of oxygen with a mass number of 18 has 8 protons and 10 neutrons.

2.   All four atoms have two core electrons (in their filled shell); the valence electrons are in the outer shell.
      a. 3   **b.** 5   **c.** 6   **d.** 7

3.   **a.** Use the aufbau principle (electrons go into available orbitals with the lowest energy) and the Pauli
            exclusion principle (no more than 2 electrons are in each atomic orbital), and remember that the
            relative energies of the atomic orbitals are:

$$1s < 2s < 2p < 3s < 3p < 4s < 3d < 4p < 5s < 4d < 5p$$

The question can be answered if you remember that there are one s atomic orbital and 3 degenerate p atomic orbitals in each shell. To write the electronic configurations for Br and I, you need to remember that there are 5 degenerate d atomic orbitals.

$$\text{Cl}\quad 1s^2\, 2s^2\, 2p^6\, 3s^2\, 3p^5$$
$$\text{Br}\quad 1s^2\, 2s^2\, 2p^6\, 3s^2\, 3p^6\, 4s^2\, 3d^{10}\, 4p^5$$
$$\text{I}\quad\ \ 1s^2\, 2s^2\, 2p^6\, 3s^2\, 3p^6\, 4s^2\, 3d^{10}\, 4p^6\, 5s^2\, 4d^{10}\, 5p^5$$

b. Because the three atoms are in the same column of the periodic table, they all have the same number of valence electrons (7), and the valence electrons of each of the three atoms are in similar orbitals (2 are in an s orbital and 5 are in p orbitals).

4.   The atomic numbers can be found in the periodic table on the last page of the text. Notice that elements in the same column of the periodic table have the same number of valence electrons and their valence electrons are in similar orbitals.

a. carbon (atomic number = 6; 2 core, 4 valence): $1s^2\, 2s^2\, 2p^2$
        silicon (atomic number = 14; 10 core, 4 valence): $1s^2\, 2s^2\, 2p^6\, 3s^2\, 3p^2$

b. oxygen (atomic number = 8; 2 core, 6 valence): $1s^2\, 2s^2\, 2p^4$
        sulfur (atomic number = 16; 10 core, 6 valence): $1s^2\, 2s^2\, 2p^6\, 3s^2\, 3p^4$

c. nitrogen (atomic number = 7; 2 core, 5 valence): $1s^2\, 2s^2\, 2p^3$
        phosphorus (atomic number = 15; 10 core, 5 valence): $1s^2\, 2s^2\, 2p^6\, 3s^2\, 3p^3$

d. magnesium (atomic number = 12; 10 core, 2 valence): $1s^2\, 2s^2\, 2p^6\, 3s^2$
        calcium (atomic number = 20; 18 core, 2 valence): $1s^2\, 2s^2\, 2p^6\, 3s^2\, 3p^6\, 4s^2$

5.   **a.** Potassium is in the first column of the periodic table; therefore, like lithium and sodium that are also in the first column, potassium has one valence electron.
      b. It occupies a 4 s orbital.

6. The polarity of a bond can be determined by the difference in the electronegativities (given in Table 1.3 on page 10 of the text) of the atoms sharing the bonding electrons. The greater the difference in electronegativity, the more polar the bond.

    a. Cl—CH$_3$ **b.** H—OH   **c.** H—F   **d.** Cl—CH$_3$

7.   **a.** KCl has the most polar bond because its two bonded atoms have the greatest difference in electronegativity. The electronegativity difference in the four listed compounds are as follows:

$$KCl: 3.0 - 0.8 = 2.2$$
$$LiBr: 2.8 - 1.0 = 1.8$$
$$NaI: 2.5 - 0.9 = 1.6$$
$$Cl_2: 3.0 - 3.0 = 0$$

    b. Cl$_2$ has the least polar bond because the two chlorine atoms share the bonding electrons equally.

8. Solved in the text.

9. The dipole moment is the magnitude of the charge times the distance between the charges. The charge on H and F in HF is sufficiently larger than the charge on H and Cl in HCl to more than make up for the fact that the H—Cl bond is longer than the H—F bond.

10. To answer this question, compare the electronegativities of the two atoms sharing the bonding electrons using Table 1.3 on page 10 of the text. (Note that if the atoms being compared are in the same row of the periodic table the atom farther to the right is the more electronegative; if the atoms being compared are in the same column, the one closer to the top of the column is the more electronegative.)

    a. $\overset{\delta-}{H}O\overset{\delta+}{-H}$ **c.** $H_3\overset{\delta+}{C}\overset{\delta-}{-}NH_2$ **e.** $\overset{\delta-}{H}O\overset{\delta+}{-}Br$ **g.** $\overset{\delta+}{I}\overset{\delta-}{-}Cl$

    b. $\overset{\delta-}{F}\overset{\delta+}{-}Br$ **d.** $H_3\overset{\delta+}{C}\overset{\delta-}{-}Cl$ **f.** $H_3\overset{\delta-}{C}\overset{\delta+}{-}MgBr$ **h.** $H_2\overset{\delta+}{N}\overset{\delta-}{-}OH$

11.   **a.** LiH and HF are polar (they have a red end and a blue end).
    b. A potential map marks the edges of the molecule's electron cloud. The electron cloud is largest around the H in LiH, because that H has more electrons around it than the Hs in the other molecules.
    c. Because the hydrogen of HF is blue, we know that this compound has the most positively charged hydrogen and, therefore, it will be most apt to attract a negatively charged species.

12. By answering this question you will see that a formal charge is a book-keeping device. It does *not necessarily* tell you which atom has the greatest electron density or is the most electron deficient.

    a. oxygen   **c.** oxygen
    b. oxygen (it is the more red)   **d.** hydrogen (it is the most blue)

Notice that in hydroxide ion, the atom with the formal negative charge **is** the atom with the greater electron density. In the hydronium ion, however, the atom with the formal positive charge **is not** the most electron deficient atom.

13.   formal charge = number of valence electrons
                        − (number of lone-pair electrons + ½ the number of bonding electrons)

In all four structures, every H is singly bonded and thus has a formal charge = 1 − (0 + 1) = 0.
Similarly, all CH_3 carbon atoms have four bonds and a formal charge = 4 − (0 + 4) = 0.
The formal charges on the remaining atoms follows:

a. $CH_3-\overset{+}{\overset{\cdot\cdot}{O}}-CH_3$
                |
                H

formal charge on O
6 − (2 + 3) = +1

b. $H-\overset{\cdot\cdot}{\underset{|}{C}}-H$
                |
                H

formal charge on C
4 − (2 + 3) = −1

c. CH_3
                |
        $CH_3-\overset{+}{N}-CH_3$
                |
                CH_3

formal charge on N
5 − (0 + 4) = +1

d.   H   H
            |+ |−
        $H-N-B-H$
            | |
            H   H

formal charge on
N: 5 − (0 + 4) = +1
B: 3 − (0 + 4) = −1

14.

a. $\overset{\cdot\cdot}{\underset{}{O}}:$
        $^-:\overset{\cdot\cdot}{O}:\overset{+}{N}:\overset{\cdot\cdot}{O}:^-$

or

        $\overset{\cdot\cdot}{\underset{}{O}}:$
        $‖$
        $:\overset{\cdot\cdot}{O}-\overset{+}{N}-\overset{\cdot\cdot}{O}:^-$

b. $:\overset{\cdot\cdot}{O}::\overset{+}{N}::\overset{\cdot\cdot}{O}:$

or

        $:\overset{\cdot\cdot}{O}=\overset{+}{N}=\overset{\cdot\cdot}{O}:$

c.   H H
        $H:\overset{|}{C}:\overset{|}{C}:^-$
            H H

or

        H   H
        | |
        $H-C-C:^-$
        | |
        H   H

d.   H
        $H:\overset{+}{C}:\overset{|}{C}:H$
              H H

or

            H
            |
        $H-C-\overset{+}{C}-H$
            | |
            H   H

e.   H H
        $H:\overset{|}{C}:\overset{+}{N}:H$
            H H

or

        H   H
        | |
        $H-C-\overset{+}{N}-H$
        | |
        H   H

f. Na^+ $^-:\overset{\cdot\cdot}{O}:H$

or

        Na^+ $^-:\overset{\cdot\cdot}{O}-H$

g. $\overset{\cdot\cdot}{\underset{}{O}}:$
        $^-:\overset{\cdot\cdot}{O}:\overset{\cdot\cdot}{C}:\overset{\cdot\cdot}{O}:H$

or

        $\overset{\cdot\cdot}{\underset{}{O}}:$
        $‖$
        $^-:\overset{\cdot\cdot}{O}-C-\overset{\cdot\cdot}{O}-H$

h. $\overset{\cdot\cdot}{\underset{}{O}}:$
        $H:\overset{:}{C}:H$

or

        $\overset{\cdot\cdot}{\underset{}{O}}:$
        $‖$
        $H-C-H$

15.

a.   H H   H H
        $H:\overset{|}{C}:\overset{|}{C}:\overset{\cdot\cdot}{O}:H$ and $H:\overset{|}{C}:\overset{\cdot\cdot}{O}:\overset{|}{C}:H$
        H H   H   H

or

        H   H   H   H
        | | | |
    $H-C-C-\overset{\cdot\cdot}{O}-H$ and $H-C-\overset{\cdot\cdot}{O}-C-H$
        | | | |
        H   H   H   H

b.   H:C̈:C̈:C̈:Ö:H   and   H:C̈:C̈:Ö:C̈:H   and   H:C̈:C̈:C̈:H
(with H's above and below carbons)   H:Ö:H
                                                                            H

or

$$H-\underset{\underset{H}{|}}{\overset{\overset{H}{|}}{C}}-\underset{\underset{H}{|}}{\overset{\overset{H}{|}}{C}}-\underset{\underset{H}{|}}{\overset{\overset{H}{|}}{C}}-\ddot{O}-H$$ and $$H-\underset{\underset{H}{|}}{\overset{\overset{H}{|}}{C}}-\underset{\underset{H}{|}}{\overset{\overset{H}{|}}{C}}-\ddot{O}-\underset{\underset{H}{|}}{\overset{\overset{H}{|}}{C}}-H$$ and $$H-\underset{\underset{H}{|}}{\overset{\overset{H}{|}}{C}}-\underset{\underset{:\ddot{O}:}{|}}{\overset{\overset{H}{|}}{C}}-\underset{\underset{}{|}}{\overset{\overset{H}{|}}{C}}-H$$
                                                                                                              H

16.   **a.** $CH_3CH_2\ddot{N}H_2$ **c.** $CH_3CH_2\ddot{O}H$ **e.** $CH_3CH_2\ddot{C}l:$

b. $CH_3\ddot{N}HCH_3$ **d.** $CH_3\ddot{O}CH_3$ **f.** $H\ddot{O}NH_2$

17.   **a.** $CH_3CH_2CH_2Cl$ **c.** $CH_3CH_2\overset{\overset{O}{\|}}{C}\underset{\underset{CH_3}{|}}{N}CH_2CH_3$

b. $CH_3\overset{\overset{O}{\|}}{C}OCH_2CH_3$ **d.** $CH_3CH_2C≡N$

18.   **a.** the (green) chlorine atom   **c.** the (blue) nitrogen atoms
b. the (red) oxygen atoms   **d.** the (black) carbon atoms and (gray) hydrogen atoms

19.   **a.** $$H-\underset{\underset{H}{|}}{\overset{\overset{H}{|}}{C}}-\underset{\underset{H}{|}}{\overset{\overset{}{}}{N}}-\underset{\underset{H}{|}}{\overset{\overset{H}{|}}{C}}-\underset{\underset{H}{|}}{\overset{\overset{H}{|}}{C}}-\underset{\underset{H}{|}}{\overset{\overset{H}{|}}{C}}-H$$ **c.**

b.

d.

20. He_2^+ has three electrons. Using Figure 1.4 on page 23 of the text, two electrons would be in a bonding molecular orbital and one electron would be in an antibonding molecular orbital. Because there are more electrons in the bonding molecular orbital than in the antibonding molecular orbital, He_2^+ exists.

21.   **a.** π^* This involves out-of-phase overlap of atomic orbitals (the overlapping orbitals have different colors) leading to an antibonding molecular orbital. Given that this example involves the side-to-side overlap of p orbitals, this results in a π^* antibonding molecular orbital.

b. σ^* This involves out-of-phase overlap of atomic orbitals (the overlapping orbitals have different colors) leading to an antibonding molecular orbital. Given that this example involves the end-on overlap of atomic orbitals, this results in a σ^* antibonding molecular orbital.

c. π This involves in-phase overlap of atomic orbitals (the overlapping orbitals have the same color) leading to a bonding molecular orbital. Given that this example involves the side-to-side overlap of p orbitals, this results in a π bonding molecular orbital.

d. σ This involves in-phase overlap of atomic orbitals (the overlapping orbitals have the same color) leading to a bonding molecular orbital. Given that this example involves the end-on overlap of atomic orbitals, this results in a σ bonding molecular orbital.

22.   The carbon-carbon bonds form as a result of sp^3—sp^3 overlap.
The carbon-hydrogen bonds form as a result of sp^3—s overlap.

23.   **a(1).**   Solved in the text.
b(1).   Solved in the text.

a(2).   The carbon forms fours bonds and each chlorine forms one bond.

$$\ddot{C}l-\overset{\displaystyle :\ddot{C}l:}{\underset{\displaystyle :\ddot{C}l:}{\overset{|}{\underset{|}{C}}}}-\ddot{C}l:$$

b(2).   The carbon uses sp^3 orbitals to form the bonds with the chlorine atoms so the bond angles are all 109.5°.

a(3).   The first attempt at drawing a Lewis structure shown below shows that carbon does not have a complete octet and does not form the needed number of bonds.

$$H-\overset{\displaystyle H}{\underset{\displaystyle H}{\overset{|}{\underset{|}{C}}}}-C-\ddot{O}-H$$

Placing a double bond between the carbon and oxygen and moving the H from O to C solves the problem.

$$H-\overset{\displaystyle H}{\underset{\displaystyle H}{\overset{|}{\underset{|}{C}}}}-\overset{\displaystyle \ddot{O}:}{\overset{\|}{C}}-H$$

b(3).   The sp^3 hybridized CH_3 carbon has 109.5° bond angles and the sp^2 hybridized $CH=O$ carbon has 120° bond angles.

$$:\ddot{O} \quad H \quad H$$

(structure: H₃C–C(=O)–H with 109.5° at CH₃ carbon and 120° at the CH=O carbon)

a(4).   In order to fill their octets and form the required number of bonds, carbon and nitrogen must form a triple bond.

$$H-C\equiv N:$$

b(4).   Because the carbon is sp hybridized, the bond angle is 180°.

$$\overset{180°}{H-C\equiv N:}$$

24.   The bond angles in H_3O^+ are greater than those in H_2O (104.5°) and less than those in CH_4 (109.5°).

The sp^3 C in CH_4 has four equivalent single bonds and no lone pairs, so it has ideal tetrahedral bond angles (109.5°). The sp^3 O in H_2O has two single bonds and two lone pairs, resulting in a decreased bond angle due to the greater electron-electron repulsion associated with the more diffuse lone pairs. The sp^3 O in H_3O^+ has three single bonds and one lone pair, so its bond angles are intermediate between those in CH_4 and H_2O.

25.   Because the carbon is sp^3 hybridized and has one lone pair, its bond angles will be similar to those in NH_3 (107.3°), another species with an sp^3 hybridized central atom with a lone pair.

26.   The nitrogen atom has the greatest electron density.
The hydrogens are the bluest atoms. Therefore, they have the least electron density. In other words, they have the most positive (least negative) electrostatic potential.

27.   Water is the most polar—it has a deep red area and the most intense blue area.
Methane is the least polar—it is all the same color with no red or blue areas.

28.   Electrons in atomic orbitals farther from the nucleus form **longer** bonds; they also form **weaker** bonds due to less electron density in the region of orbital overlap. Therefore:

a.   **relative lengths** of the bonds in the halogens are: $Br_2 > Cl_2$
relative strengths of the bonds are: $Cl_2 > Br_2$

b.   **relative lengths**: $CH_3—Br > CH_3—Cl > CH_3—F$
relative strengths: $CH_3—F > CH_3—Cl > CH_3—Br$

29.   **a.**   **longer:**   **1.**   C—I   **2.**   C—Cl   **3.**   H—Cl
b.   **stronger:**   **1.**   C—Cl   **2.**   C—C   **3.**   H—F

30.   We know that the σ bond is stronger than the π bond, because the σ bond in ethane has a bond dissociation energy of 90 kcal/mol, whereas the bond dissociation energy of the double bond ($\sigma + \pi$) in ethene is 174 kcal/mole, which is less than twice as strong.
Because the σ bond is stronger, we know that it has more effective orbital-orbital overlap.

31.   Because electrons in an *s* orbital are closer on average to the nucleus than those in a *p* orbital, the greater the *s* character in the interacting orbitals, the stronger (and shorter) will be the bond. Therefore, the carbon-carbon sigma bond formed by sp^2—sp^2 overlap is stronger (and shorter), because an sp^2 orbital has 33.3% *s* character, whereas an sp^3 orbital has 25% *s* character.

32.   **a.**

b.

33.   The bond angle depends on the central atom.

a. sp^3 nitrogen with no lone pair: 109.5°   **c.** sp^3 carbon with no lone pair: 109.5°

b. sp^3 nitrogen with a lone pair: 107.3°   **d.** sp^3 carbon with no lone pair: 109.5°

34.   **a.**   **BeH$_2$**

Beryllium does not have any unpaired electrons in its ground state. Therefore, it cannot form any bonds unless it promotes an electron. After promotion, hybridization of the two orbitals (an *s* orbital and a *p* orbital) that contain an unpaired electron results in two *sp* hybrid orbitals.

$$\underset{\underset{\uparrow\downarrow}{}}{1s} \quad \underset{\underset{\uparrow\downarrow}{}}{2s} \quad \underset{\underset{}{}}{2p_x} \quad \xrightarrow{\text{promotion}} \quad \underset{\underset{\uparrow\downarrow}{}}{1s} \quad \underset{\underset{\uparrow}{}}{2s} \quad \underset{\underset{\uparrow}{}}{2p_x} \quad \xrightarrow{\text{hybridization}} \quad \underset{\underset{\uparrow\downarrow}{}}{1s} \quad \underset{\underset{\uparrow}{}}{2sp} \quad \underset{\underset{\uparrow}{}}{2sp}$$

Each *sp* orbital of beryllium overlaps the *s* orbital of a hydrogen. The two *sp* orbitals orient themselves to get as far away from each other as possible, resulting in a bond angle of **180°**.

bond angle = 180°   H—Be—H

BeH$_2$

b. BH₃

Without promotion, boron could form only one bond because it has only one unpaired electron. Promotion gives it three unpaired electrons. When the three orbitals (one *s* orbital and two *p* orbitals) that contain an unpaired electron are hybridized, three *sp²* orbitals result.

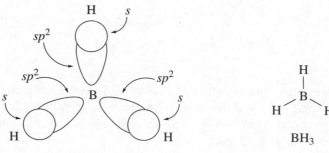

Each *sp²* orbital overlaps the *s* orbital of a hydrogen. The three *sp²* orbitals orient themselves to get as far away from each other as possible, resulting in bond angles of **120°**.

bond angles = 120°

c. CCl₄

The carbon in CCl₄ is bonded to four atoms, so it uses four *sp³* orbitals.
Each carbon-chlorine bond is formed by the overlap of an *sp³* orbital of carbon with an *sp³* orbital of chlorine. Because the four *sp³* orbitals of carbon orient themselves to get as far away from each other as possible, the bond angles are all **109.5°**.

bond angles = 109.5°

d. CO₂

The carbon in CO₂ is bonded to two atoms, so it uses two *sp* orbitals. Each carbon-oxygen bond is a double bond. One of the bonds of each double bond is formed by the overlap of an *sp* orbital of carbon with an *sp²* orbital of oxygen. The second bond of the double bond is formed as a result of side-to-side overlap of a *p* orbital of carbon with a *p* orbital of oxygen. Because carbon's two *sp* orbitals orient themselves to get as far away from each other as possible, the bond angle in CO₂ is **180°**.

O=C=O
bond angle = 180°

e. HCOOH

The double-bonded carbon and the double-bonded oxygen in HCOOH each use sp^2 orbitals; thus, the bonds around the double-bonded carbon are all 120°. The single-bonded oxygen forms bonds using sp^3 orbitals, while each hydrogen uses an s orbital. The predicted C—O—H bond angle is based on an sp^3 oxygen that has a reduced bond angle due to the greater electron-electron repulsion associated with the diffuse lone-pair electrons. The bond angle is predicted to be similar to that in H_2O (104.5°).

the σ bond is formed by sp^2-sp^2 overlap
the π bond is formed by p-p overlap

sp^2-s overlap

104.5°; all other bond angles are 120°

sp^3-s overlap

sp^2-sp^3 overlap

f. N_2

The triple bond consists of one σ bond and two π bonds. Each nitrogen has two sp orbitals; one is used to form the sigma bond and the other contains the lone pair. Each nitrogen has 2 p orbitals that are used to form the two π bonds. A bond angle is the angle formed by three atoms. Therefore, there are no bond angles in this compound.

:N≡N: the σ bond is formed by sp-sp overlap
each π bond is formed by p-p overlap

35.   The one with the greater electronegativity will have the smaller dipole moment because it will decrease the electron flow toward the F. Since CH_3F has a smaller dipole moment than CD_3F, we know that hydrogen is more electronegative than deuterium.

36.   The electrostatic potential map of ammonia is not symmetrical in shape because it is not symmetrical in the distribution of the charge—the nitrogen is more electron rich and therefore more red than the three hydrogens. The electrostatic potential map of the ammonium ion is symmetrical in shape because it is symmetrical in the distribution of the charge. Its symmetry results from the fact that nitrogen forms a bond with each of the four hydrogens and the four bond angles are all the same.

The nitrogen in the ammonium ion has significantly lower electron density than the nitrogen in ammonia due to the absence of the lone pair and an additional bond to hydrogen.

37.   **a, e, g,** and **h** have a dipole moment of zero because they are symmetrical molecules.

Because C is sp^3 hybridized, the bond angles in CH_2Cl_2 are 109.5°. Therefore, the compound is not symmetrical so it has a dipole moment.

Because Be is sp hybridized, the bond angle in $BeCl_2$ is 180°. Therefore, the compound is symmetrical, so it does not have a dipole moment.

Because B is sp^2 hybridized, the bond angles in BF_3 are 120°. Therefore, the compound is symmetrical so it does not have a dipole moment.

38.   The conjugate acid is obtained by adding an H^+ to the species.
  a. **(1)** $^+NH_4$ **(2)** HCl **(3)** H_2O **(4)** H_3O^+
  b. **(1)** $^-NH_2$ **(2)** Br^- **(3)** NO_3^- **(4)** HO^-

39.   **a.** The lower the pK_a, the stronger the acid, so the compound with $pK_a = 5.2$ is the stronger acid.
  b. The greater the dissociation constant, the stronger the acid, so the compound with a dissociation constant $= 3.4 \times 10^{-3}$ is the stronger acid.

40.
$$K_a = K_{eq}\left[H_2O\right]$$

$$K_{eq} = \frac{K_a}{\left[H_2O\right]} = \frac{4.53 \times 10^{-6}}{55.5} = 8.16 \times 10^{-8}$$

41. $K_a = 1.51 \times 10^{-5}$. It is a weaker acid than vitamin C whose K_a value was determined to be 6.76×10^{-5} in the Problem-Solving Strategy.

42.   **a.** $HO^- + H^+ \longrightarrow H_2O$

  b. $HCO_3^- + H^+ \rightleftharpoons H_2CO_3 \rightleftharpoons H_2O + CO_2$

  c. $CO_3^{2-} + 2\,H^+ \rightleftharpoons H_2CO_3 \rightleftharpoons H_2O + CO_2$

43.   If the pH is < 7, the body fluid is acidic; if the pH is > 7, the body fluid is basic.

  a. basic   **b.** acidic   **c.** basic

44.   Remember that a proton will be picked up by an atom that has one or more lone pairs. Notice that two oxygens have lone pairs in part **d**. To see which one gets the proton, see Table 1.8 on page 47 of the text.

  a. $CH_3CH_2\overset{+}{O}H_2$ **b.** CH_3CH_2OH **c.** $CH_3CH_2\overset{+}{N}H_3$ **d.** $CH_3CH_2-\overset{\overset{\displaystyle +OH}{\|}}{C}-OH$

45.   **a.** CH_3COO^- is the stronger base.
    Because CH_3COOH is the weaker acid, it has the stronger conjugate base.

  b. $^-NH_2$ is the stronger base.
    Because NH_3 is the weaker acid, it has the stronger conjugate base.

  c. H_2O is the stronger base.
    Because H_3O^+ is the weaker acid, it has the stronger conjugate base.

46.   The conjugate acids of the given bases have the following relative strengths:

$$CH_3\overset{+}{O}H_2 > CH_3\overset{\overset{\displaystyle O}{\|}}{C}OH > CH_3\overset{+}{N}H_3 > CH_3OH > CH_3NH_2$$

The bases, therefore, have the following relative strengths, since the weakest acid has the strongest conjugate base.

$$CH_3\overset{-}{N}H > CH_3O^- > CH_3NH_2 > CH_3\overset{\overset{\displaystyle O}{\|}}{C}O^- > CH_3OH$$

47.   In each of the following reactions, the position of equilibrium is established by analyzing the relative strengths of the acids on either side of the reaction—the equilibrium favors reaction of the stronger acid to form the weaker acid.

if the lone pairs are not shown:

a.   CH$_3$OH as an acid   CH$_3$OH + NH$_3$ $\rightleftharpoons$ CH$_3$O$^-$ + $\overset{+}{N}$H$_4$

CH$_3$OH as a base   CH$_3$OH + HCl $\rightleftharpoons$ CH$_3\overset{+}{O}$H + Cl$^-$
$\qquad\qquad\qquad\qquad\qquad\qquad\qquad\quad$ |
$\qquad\qquad\qquad\qquad\qquad\qquad\qquad\quad$ H

b.   NH$_3$ as an acid   NH$_3$ + CH$_3$O$^-$ $\rightleftharpoons$ $^-$NH$_2$ + CH$_3$OH

NH$_3$ as a base   NH$_3$ + HBr $\rightleftharpoons$ $\overset{+}{N}$H$_4$ + Br$^-$

if the lone pairs are shown:

a.   CH$_3$OH as an acid   CH$_3\ddot{O}$H + $\dot{\ddot{N}}$H$_3$ $\rightleftharpoons$ CH$_3\ddot{O}$:$^-$ + $\overset{+}{N}$H$_4$

CH$_3$OH as a base   CH$_3\ddot{O}$H + H$\overset{..}{\underset{..}{C}}$l: $\rightleftharpoons$ CH$_3\overset{+}{O}$H + :$\overset{..}{\underset{..}{C}}$l:$^-$
$\qquad\qquad\qquad\qquad\qquad\qquad\qquad\qquad\qquad$ |
$\qquad\qquad\qquad\qquad\qquad\qquad\qquad\qquad\qquad$ H

b.   NH$_3$ as an acid $\dot{\ddot{N}}$H$_3$ + CH$_3\ddot{O}$:$^-$ $\rightleftharpoons$ $^-$:$\ddot{N}$H$_2$ + CH$_3\ddot{O}$H

NH$_3$ as a base $\dot{\ddot{N}}$H$_3$ + H$\ddot{B}$r: $\rightleftharpoons$ $\overset{+}{N}$H$_4$ + :$\ddot{B}$r:$^-$

48.   Notice that in each case, the equilibrium favors reaction of the stronger acid to form the weaker acid.

a.   CH$_3$OH   +   HO$^-$ $\rightleftharpoons$ CH$_3$O$^-$ +   H$_2$O
$\quad$ p$K_a = 15.5$ $\qquad\qquad\qquad\qquad\qquad\qquad\qquad$ p$K_a = 15.7$

$\qquad$ CH$_3$OH   +   H$_3$O$^+$ $\rightleftharpoons$ CH$_3\overset{+}{O}$H$_2$ +   H$_2$O
$\qquad\qquad\qquad$ p$K_a = -1.7$ $\qquad\qquad$ p$K_a = -2.5$

$\qquad\qquad$ O $\qquad\qquad\qquad\qquad\qquad\qquad\qquad$ O
$\qquad\qquad$ || $\qquad\qquad\qquad\qquad\qquad\qquad\qquad$ ||
$\qquad\qquad$ C $\qquad\qquad\qquad\qquad\qquad\qquad\qquad$ C
$\quad$ CH$_3\qquad$ OH   +   HO$^-$ $\rightleftharpoons$ CH$_3\qquad$ O$^-$ +   H$_2$O
$\qquad$ p$K_a = 4.8$ $\qquad\qquad\qquad\qquad\qquad$ p$K_a = 15.7$

$\qquad\qquad$ O $\qquad\qquad\qquad\qquad\qquad\qquad\qquad$ $\overset{+}{O}$H
$\qquad\qquad$ || $\qquad\qquad\qquad\qquad\qquad\qquad\qquad$ ||
$\qquad\qquad$ C $\qquad$ + H$_3$O$^+$ $\rightleftharpoons$ $\qquad\qquad$ C $\qquad$ + H$_2$O
$\quad$ CH$_3\qquad$ OH $\qquad\qquad\qquad\qquad\qquad$ CH$_3\qquad$ OH
$\qquad\qquad\qquad$ p$K_a = -1.7$ $\qquad\qquad$ p$K_a = -6.1$

$\quad$ CH$_3$NH$_2$ +   HO$^-$ $\rightleftharpoons$ CH$_3\overset{-}{N}$H   +   H$_2$O
$\quad$ p$K_a = 40$ $\qquad\qquad\qquad\qquad\qquad\qquad\qquad$ p$K_a = 15.7$

$\quad$ CH$_3$NH$_2$ +   H$_3$O$^+$ $\rightleftharpoons$ CH$_3\overset{+}{N}$H$_3$ +   H$_2$O
$\qquad\qquad$ p$K_a = -1.7$ $\qquad\qquad$ p$K_a = 10.7$

b. HCl + H_2O $\rightleftharpoons$ H_3O^+ + Cl^-

   $pK_a = -7$ $pK_a = -1.7$

   NH_3 + H_2O $\rightleftharpoons$ $^+NH_4$ + HO^-

   $pK_a = 15.7$ $pK_a = 9.4$

49. **a.** $HC\equiv CH$ + HO^- $\rightleftharpoons$ $HC\equiv C^-$ + H_2O

   $pK_a = 25$ $pK_a = 15.7$

   b. $HC\equiv CH$ + $^-NH_2$ $\rightleftharpoons$ $HC\equiv C^-$ + NH_3

   $pK_a = 25$ $pK_a = 36$

   c. $^-NH_2$ would be a better base because the equilibrium favors the products.
   When HO^- removes a proton, the equilibrium favors the reactants.

50. $K_{eq} = \dfrac{K_a \text{ of the reactant acid}}{K_a \text{ of the product acid}}$

For **a** the reactant acid is HCl and the product acid is H_3O^+.

For **b** the reactant acid is CH_3COOH and the product acid is H_3O^+.

For **c** the reactant acid is H_2O and the product acid is $CH_3\overset{+}{N}H_3$.

For **d** the reactant acid is $CH_3\overset{+}{N}H_3$ and the product acid is H_3O^+.

a. $K_{eq} = \dfrac{10^7}{10^{1.7}} = 10^{7-1.7} = 10^{5.3} = 2.0 \times 10^5$

b. $K_{eq} = \dfrac{10^{-4.8}}{10^{1.7}} = 10^{-4.8-1.7} = 10^{-6.5} = 3.2 \times 10^{-7}$

c. $K_{eq} = \dfrac{10^{-15.7}}{10^{-10.7}} = 10^{-15.7+10.7} = 10^{-5} = 1.0 \times 10^{-5}$

d. $K_{eq} = \dfrac{10^{-10.7}}{10^{1.7}} = 10^{-10.7-1.7} = 10^{-12.4} = 4.0 \times 10^{-13}$

51. **a.** oxygen   **b.** H_2S **c.** CH_3SH

The size of an atom is more important than its electronegativity in determining stability. Therefore, even though oxygen is more electronegative than sulfur, H_2S is a stronger acid than H_2O and CH_3SH is a stronger acid than CH_3OH. Because the sulfur atom is larger, the electrons in its conjugate base are spread out over a greater volume, which stabilizes it. The more stable the base, the stronger its conjugate acid.

52. **a.** HBr **b.** $CH_3CH_2CH_2\overset{+}{O}H_2$ **c.** $CH_3CH_2CH_2OH$ **d.**

53.   **a.**   Because HF is the weakest conjugate acid, F^- is the strongest base.

      b.   Because HI is the strongest conjugate acid, I^- is the weakest base.

54.   Compare the acid strengths of the conjugate acids, recalling that the weaker acid has the stronger conjugate base.

      a. HO^- because H_2O is a weaker acid than H_3O^+.

      c. CH_3O^- because CH_3OH is a weaker acid than CH_3COOH.

      b. NH_3 because $^+NH_4$ is a weaker acid than H_3O^+.

      d. CH_3O^- because CH_3OH is a weaker acid than CH_3SH.

55.   **a.** $CH_3OCH_2CH_2OH$ because its conjugate base has the negative charge on oxygen stabilized by electron withdrawal by the CH_3O group.

      b. $CH_3CH_2CH_2\overset{+}{O}H_2$ because oxygen is more electronegative than nitrogen.

      c. $CH_3CH_2OCH_2CH_2OH$ because the electron-withdrawing oxygen is closer to the OH group.

      d. $CH_3CH_2\overset{\overset{\textstyle O}{\|}}{C}OH$ because the electron-withdrawing $C{=}O$ is closer to the OH group.

56. $\underset{\underset{\textstyle F}{|}}{CH_3CHCH_2OH} > \underset{\underset{\textstyle Cl}{|}}{CH_3CHCH_2OH} > \underset{\underset{\textstyle Cl}{|}}{CH_2CH_2CH_2OH} > CH_3CH_2CH_2OH$

The first two listed compounds are the most acidic because they have the electron-withdrawing substituent that stabilizes the conjugate base closest to the O—H bond. The first is more acidic than the second because fluorine is more electronegative than chlorine.

The second listed compound is a stronger acid than the third listed compound because the chlorine in the third compound is farther away from the O—H bond.

The compound on the far right does not have a substituent that withdraws electrons inductively, so it is the least acidic of the four compounds.

57.   The weaker acid has the stronger conjugate base.

      a. $\underset{\underset{\textstyle Br}{|}}{CH_3\overset{\overset{\textstyle O}{\|}}{C}HCO^-}$
      b. $\underset{\underset{\textstyle Cl}{|}}{CH_3\overset{\overset{\textstyle O}{\|}}{C}HCH_2CO^-}$
      c. $CH_3CH_2\overset{\overset{\textstyle O}{\|}}{C}O^-$
      d. $CH_3\overset{\overset{\textstyle O}{\|}}{C}CH_2CH_2O^-$

58.   Solved in the text.

59.   When a sulfonic acid loses a proton, the electrons left behind are shared by three oxygens.

In contrast, when a carboxylic acid loses a proton, the electrons left behind are shared by only two oxygens. The sulfonate ion, therefore, is more stable than the carboxylate ion.

The more stable the base, the stronger is its conjugate acid. Thus the sulfonic acid is a stronger acid than the carboxylic acid.

$$CH_3\overset{\overset{\textstyle \ddot{O}:}{\|}}{\underset{\underset{\textstyle \ddot{O}:}{\|}}{S}}-\ddot{O}:^- \longleftrightarrow CH_3\overset{\overset{\textstyle :\ddot{O}:^-}{|}}{\underset{\underset{\textstyle \ddot{O}:}{\|}}{S}}=\ddot{O}: \longleftrightarrow CH_3\overset{\overset{\textstyle \ddot{O}:}{\|}}{\underset{\underset{\textstyle :\ddot{O}:^-}{|}}{S}}=\ddot{O}:$$

a sulfonate ion

$$CH_3\overset{\overset{\textstyle \ddot{O}:}{\|}}{C}-\ddot{O}:^- \longleftrightarrow CH_3\overset{\overset{\textstyle :\ddot{O}:^-}{|}}{C}=\ddot{O}:$$

a carboxylate ion

60. **a.**

b.

61. **a.** $CH_3C\equiv\overset{+}{N}H$

b. CH_3CH_3

c. $F_3C\overset{\overset{\textstyle O}{\|}}{C}OH$

d. an sp^2 hybridized oxygen.

e. $CH_3C\equiv\overset{+}{N}H$ > $CH_3C\overset{\underset{\textstyle CH_3}{|}}{=}\overset{+}{N}HCH_3$ > $CH_3CH_2\overset{+}{N}H_3$

f. HNO_3 is more acidic than HNO_2. When the structure of the conjugate base of each acid is drawn out, you can see that NO_3 has a positively charged nitrogen and NO_2 does not. The positive charge decreases the electron density on the oxygen by inductive electron withdraw thereby stabilizing it. A stable base is a weak base. Therefore, it has the stronger conjugate acid.

62. **a.** CH_3COO^- **c.** H_2O **e.** $\overset{+}{N}H_4$ **g.** NO_2^-

b. $CH_3CH_2\overset{+}{N}H_3$ **d.** Br^- **f.** $HC\equiv N$ **h.** NO_3^-

63.

	a.	**b.**	**c.**
1.	neutral	charged	neutral
2.	neutral	charged	neutral
3.	charged	charged	neutral
4.	charged	charged	neutral
5.	charged	neutral	neutral
6.	charged	neutral	neutral

64.   **a.**   The COOH group will be in its basic form because the solution is more basic than its pK_a value.

The $^+NH_3$ group will be in its acidic form because the solution is more acidic than its pK_a value.

$$\underset{\underset{+NH_3}{|}}{CH_3\overset{\overset{\displaystyle O}{\|}}{C}HCO^-}$$

b.   No, alanine can never be without a charge. To be without a charge would require a group with a pK_a value of 9.69 to lose a proton before (be more acidic than) a group with a pK_a value of 2.34.

c.   As the pH becomes more basic than 2.34 the COOH group will become increasing more negatively charged.

As the pH becomes more acidic than 9.69, the $^+NH_3$ group will become increasing more positively charged.

Therefore, the amount of negative change will be the same as the amount of positive charge at the intersection of the two pK_a values.

$$\frac{2.34 + 9.69}{2} = 6.02$$

65.   **a.**   10.4 (two log units more basic than the pK_a)
   b.   2.7 (one log unit more acidic than the pK_a)
   c.   6.4 (two log units more acidic than the pK_a)
   d.   7.3 (when the pH is equal to the pK_a)
   e.   5.6 (one log unit more basic than the pK_a)

66.   **a.**   **1.**   pH = 4.9
         2.   pH = 10.7

b.   **1.**   pH > 6.9   Because the basic form is the form in which the compound is charged, the pH needs to be more than two units more basic than the pK_a.

   2.   pH < 8.7   Because the acidic form is the form in which the compound is charged, the pH needs to be more than two units more acidic than the pK_a.

67.   greater than pH 10.4
(As long as the pH is greater than the pK_a value of the compound, the majority of the compound will be in its basic form.)

68.   **a.** $CH_3COO^- + H^+ \rightleftharpoons CH_3COOH$

b. $CH_3COOH + HO^- \rightleftharpoons CH_3COO^-$

69.   **a.** $ZnCl_2 + CH_3\ddot{O}H \rightleftharpoons \underset{\underset{+}{\overset{|}{HOCH_3}}}{^-ZnCl_2}$

b. $FeBr_3$ + $:\overset{..}{\underset{..}{Br}}:^-$ $\rightleftharpoons$ $Br-\bar{F}eBr_3$

c. $AlCl_3$ + $:\overset{..}{\underset{..}{Cl}}:^-$ $\rightleftharpoons$ $Cl-\bar{A}lCl_3$

70.   **a, b, c,** and **h** are Brønsted acids (proton-donating acids). Therefore, they react with HO^- by donating a proton to it.

d, e, f, and **g** are Lewis acids. They react with HO^- by accepting a pair of electrons from it.

a. CH_3OH + HO^- $\rightleftharpoons$ CH_3O^- + H_2O

b. $\overset{+}{N}H_4$ + HO^- $\rightleftharpoons$ NH_3 + H_2O

c. $CH_3\overset{+}{N}H_3$ + HO^- $\rightleftharpoons$ CH_3NH_2 + H_2O

d. BF_3 + HO^- $\rightleftharpoons$ $HO-\bar{B}F_3$

e. $\overset{+}{C}H_3$ + HO^- $\rightleftharpoons$ CH_3OH

f. $FeBr_3$ + HO^- $\rightleftharpoons$ $HO-\bar{F}eBr_3$

g. $AlCl_3$ + HO^- $\rightleftharpoons$ $HO-\bar{A}lCl_3$

h. CH_3COOH + HO^- $\rightleftharpoons$ CH_3COO^- + H_2O

71.   **a.** $H:\overset{..}{\underset{..}{O}}:\overset{..}{\underset{..}{C}}:\overset{..}{\underset{..}{O}}:H$ **or** $H-\overset{..}{\underset{..}{O}}-\overset{\overset{\overset{..}{O}:}{\|}}{C}-\overset{..}{\underset{..}{O}}-H$

e. $H:\overset{H}{\underset{H}{C}}:\overset{..}{N}:H$ **or** $H-\overset{\overset{H}{|}}{\underset{\underset{H}{|}}{C}}-\overset{..}{N}-H$ (with H above)

b. $^-:\overset{..}{\underset{..}{O}}:\overset{..}{\underset{..}{C}}:\overset{..}{\underset{..}{O}}:^-$ **or** $^-:\overset{..}{\underset{..}{O}}-\overset{\overset{\overset{..}{O}:}{\|}}{C}-\overset{..}{\underset{..}{O}}:^-$

f. $H:\overset{H}{\underset{H}{C}}:N::N:$ **or** $H-\overset{\overset{H}{|}}{\underset{\underset{H}{|}}{C}}-\overset{+}{N}\equiv N:$

c. $H:\overset{..}{C}:H$ **or** $H-\overset{\overset{\overset{..}{O}:}{\|}}{C}-H$

g. $\overset{..}{\underset{..}{O}}::C::\overset{..}{\underset{..}{O}}$ **or** $\overset{..}{\underset{..}{O}}=C=\overset{..}{\underset{..}{O}}$

d. $H:\overset{..}{\underset{H}{N}}:\overset{..}{\underset{H}{N}}:H$ **or** $H-\overset{\overset{..}{}}{\underset{\underset{H}{|}}{N}}-\overset{\overset{..}{}}{\underset{\underset{H}{|}}{N}}-H$

h. $H:\overset{..}{\underset{H}{N}}:\overset{..}{\underset{..}{O}}:^-$ **or** $H-\overset{\overset{..}{}}{\underset{\underset{H}{|}}{N}}-\overset{..}{\underset{..}{O}}:^-$

72.   If the central atom is sp^3 hybridized, it will be tetrahedral; if it is sp^2 hybridized, it will be trigonal planar; and if it is sp hybridized, it will be linear.

a. sp^3, tetrahedral   **d.** sp^2, trigonal planar   **g.** sp, linear

b. sp^2, trigonal planar   **e.** sp^3, tetrahedral   **h.** sp^3, tetrahedral

c. sp^3, tetrahedral   **f.** sp^2, trigonal planar   **i.** sp^3, tetrahedral

73. **a.** $CH_3CH_2CH_3$ **b.** $CH_3CH{=}CH_2$ **c.** $CH_3C{\equiv}CCH_3$ **or** $CH_3CH_2C{\equiv}CH$

74. **a.** CCl_3CH_2OH > $CHCl_2CH_2OH$ > CH_2ClCH_2OH

b. The greater the number of electron-withdrawing chlorine atoms equidistant from the OH group, the stronger the acid.

75. The hybridization of the central atom determines the bond angle. If it is sp^3, the number of lone pairs on the central atom determines the bond angle.

a. 107.3°   **c.** 109.5°   **e.** 104.5°   **g.** 109.5°   **i.** 109.5°

b. 107.3°   **d.** 104.5°*   **f.** 120°(sp^2)   **h.** 180°(sp)

*104.5° is the correct prediction based on the bond angle in water.
However, the bond angle is actually somewhat larger (111.7°) because the bond opens up to minimize the interaction between the electron clouds of the relatively bulky CH_3 groups.

76.

	$1s$	$2s$	$2p_x$	$2p_y$	$2p_z$	$3s$	$3p_x$	$3p_y$	$3p_z$	$4s$	
a. Ca	⇅	⇅	⇅	⇅	⇅	⇅	⇅	⇅	⇅	⇅	$1s^2\,2s^2\,2p^6\,3s^2\,3p^6\,4s^2$
b. Ca²⁺	⇅	⇅	⇅	⇅	⇅	⇅	⇅	⇅	⇅		$1s^2\,2s^2\,2p^6\,3s^2\,3p^6$
c. Ar	⇅	⇅	⇅	⇅	⇅	⇅	⇅	⇅	⇅		$1s^2\,2s^2\,2p^6\,3s^2\,3p^6$
d. Mg²⁺	⇅	⇅	⇅	⇅	⇅						$1s^2\,2s^2\,2p^6$

77. **a.** C—F   >   C—O   >   C—N
b. C—Cl   >   C—Br   >   C—I
c. H—O   >   H—N   >   H—C
d. C—N   >   C—H   >   C—C

78. The stronger base has the weaker conjugate acid.

a. HO^- **b.** $CH_3\overline{N}H$ **c.** CH_3O^- **d.** Cl^-

79. **a.**

```
      H   O
      |   ||
  H — C — C — H
      |
      H
```

c.

```
      H   O
      |   ||
  H — C — C — O — H
      |
      H
```

b.

```
      H       H
      |       |
  H — C — O — C — H
      |       |
      H       H
```

d.

```
              H
              |
    H   H — C — H   H
    |       |       |
H — C ————— C ————— C — H
    |       |       |
    H       O       H
            |
            H
```

e.

$$H-\overset{\overset{\displaystyle H}{|}}{\underset{\underset{\displaystyle H}{|}}{C}}-\overset{\overset{\displaystyle H}{|}}{\underset{\underset{\displaystyle O}{|}}{C}}-\overset{\overset{\displaystyle H}{|}}{\underset{\underset{\displaystyle H}{|}}{C}}-C\equiv N$$

(with H below O)

f.

(branched hydrocarbon structure)

80. **a.** $\ddot{N}H_3$ + H—$\ddot{\underset{..}{C}}l$: $\rightleftharpoons$ $\overset{+}{N}H_4$ + :$\ddot{\underset{..}{C}}l$:$^-$

b. BF_3 + $CH_3\ddot{N}H_2$ $\rightleftharpoons$ $CH_3\overset{\overset{\displaystyle H}{|}}{\underset{\underset{\displaystyle H}{|}}{\overset{+}{N}}}{-}\overset{-}{B}F_3$

c.

$$\underset{H}{\overset{\ddot{O}:}{\underset{\diagdown}{\overset{\|}{C}}}}{\diagdown}_{OH} + H{-}\ddot{\underset{..}{C}}l: \rightleftharpoons \underset{H}{\overset{\overset{+}{\ddot{O}}H}{\underset{\diagdown}{\overset{\|}{C}}}}{\diagdown}_{OH} + :\ddot{\underset{..}{C}}l:^-$$

81. **a.** $\overset{\;+\;\rightarrow}{CH_3-Br}$ **c.** $\overset{\leftarrow\;+\;}{HO-NH_2}$ **e.** $\overset{\;+\;\rightarrow}{H_3C-OH}$

b. $\overset{\leftarrow\;+\;}{CH_3-Li}$ **d.** $\overset{\;+\;\rightarrow}{I-Br}$ **f.** $\overset{\leftarrow\;+\;}{(CH_3)_2N-H}$

82. **a.** $\overset{\overset{\displaystyle sp^2}{\Downarrow}}{CH_3CH=CH_2}$ **c.** $\overset{\overset{\displaystyle sp^3}{\Downarrow}}{CH_3CH_2OH}$ **e.** $\overset{\overset{\displaystyle sp^2}{\Downarrow}}{CH_3CH=NCH_3}$

b. $CH_3\overset{\overset{\displaystyle O \Leftarrow sp^2}{\|}}{C}CH_3$ **d.** $\overset{\overset{\displaystyle sp}{\Downarrow}}{CH_3C\equiv N}$ **f.** $\overset{\overset{\displaystyle sp^3}{\Downarrow}}{CH_3OCH_2CH_3}$

83.   formal charge = number of valence electrons
            − (number of lone-pair electrons + ½ the number of bonding electrons)

Each H has a single bond and is neutral; each C has four bonds and is neutral. Thus, the indicated formal charge is for O or N.

a. $H-\overset{\overset{\displaystyle H}{|}}{\underset{\underset{\displaystyle H}{|}}{C}}-\ddot{\underset{..}{O}}-H$ **b.** $H-\overset{\overset{\displaystyle H}{|}}{\underset{\underset{\displaystyle H}{|}}{C}}-\overset{+}{\ddot{\underset{..}{O}}}-H$ (with H below O) **c.** $H-\overset{\overset{\displaystyle H}{|}}{\underset{\underset{\displaystyle H}{|}}{C}}-\overset{-}{\ddot{\underset{..}{O}}}:$ **d.** $H-\overset{\overset{\displaystyle H}{|}}{\underset{\underset{\displaystyle H}{|}}{C}}-\dot{\underset{\underset{\displaystyle H}{|}}{N}}-H$

formal charge $6-(4+2)=0$ formal charge $6-(2+3)=+1$ formal charge $6-(6+1)=-1$ formal charge $5-(2+3)=0$

84.   **a.** $CH_3CH_2CHCOOH$ > CH_3CHCH_2COOH > $ClCH_2CH_2CH_2COOH$ > $CH_3CH_2CH_2COOH$
               | |
               Cl   Cl

   b.   An electron-withdrawing substituent makes the carboxylic acid more acidic, because it stabilizes its conjugate base by decreasing the electron density around the oxygen atom.

   c.   The closer the electron-withdrawing chloro substituent is to the acidic proton, the more it can decrease the electron density around the oxygen atom, so the more it stabilizes the conjugate base and therefore increases the acidity of its conjugate acid.

85.   **a.** $CH_3\overset{..}{\overset{+}{O}}-\overset{-}{B}F_3$ **b.** $CH_3\overset{..}{\overset{+}{O}}-H$ + Cl^- **c.** $CH_3\overset{+}{N}-\overset{-}{A}lCl_3$
                  | | |
                  CH_3 CH_3 with H above and H below N

86.   The open arrow in the structures below points to the shorter of the two indicated bonds in each compound.

   For **1**, **2**, and **3**:   A triple bond is shorter than a double bond, which is shorter than a single bond.

   For **4** and **5**:   Because an s orbital is closer to the nucleus than a p orbital, the greater the s character in the hybrid orbital, the shorter is the bond formed using that orbital. Therefore, the bond formed by a hydrogen and an sp carbon is shorter than the bond formed by a hydrogen and an sp^2 carbon, which is shorter than the bond formed by a hydrogen and an sp^3 carbon. (See Table 1.7 on page 38 of the text.)

   1. sp^3 sp^2 sp^2 ‖
         $CH_3CH=CHC\equiv CH$
                    sp sp

   4. $\overset{H}{\underset{H}{>}}C=CHC\equiv C-H$ with labels sp^2 sp^2 sp and ‖ above
         sp^2 sp^2 sp

   2. $\overset{O}{\underset{‖}{}}$ sp^2 sp^3
         CH_3CCH_2OH
         sp^3 sp^3
         sp^2

   5. $\overset{H}{\underset{H}{>}}C=CHC\equiv C-\overset{CH_3}{\underset{CH_3}{C}}-H$ with labels sp^2 sp^2 sp sp sp^3 and sp^3

   3.   ‖
         $CH_3NHCH_2CH_2N=CHCH_3$
         sp^3 sp^3 sp^3 sp^3 sp^2 sp^2 sp^3

87. CH_3CH_2SH > CH_3CH_2OH > $CH_3CH_2NH_2$ > $CH_3CH_2CH_3$

88.   If the solution is more acidic than the pK_a of the compound, the compound will be in its acidic form (with the proton). If the solution is more basic than the pK_a of the compound, the compound will be in its basic form (without the proton).

a. at pH = 3 $CH_3\overset{\overset{\displaystyle O}{\|}}{C}OH$

at pH = 6 $CH_3\overset{\overset{\displaystyle O}{\|}}{C}O^-$

at pH = 10 $CH_3\overset{\overset{\displaystyle O}{\|}}{C}O^-$

at pH = 14 $CH_3\overset{\overset{\displaystyle O}{\|}}{C}O^-$

b. at pH = 3 $CH_3CH_2\overset{+}{N}H_3$

at pH = 6 $CH_3CH_2\overset{+}{N}H_3$

at pH = 10 $CH_3CH_2\overset{+}{N}H_3$

at pH = 14 $CH_3CH_2NH_2$

c. at pH = 3 CF_3CH_2OH

at pH = 6 CF_3CH_2OH

at pH = 10 CF_3CH_2OH

at pH = 14 $CF_3CH_2O^-$

89.   In an alkene, six atoms are in the same plane: the two sp^2 hybridized carbons and the two atoms that are bonded to each of the two sp^2 hybridized carbons. The other atoms in the molecule will not be in the same plane with these six atoms.

If you put stars next to the six atoms that lie in a plane in each molecule, you might be able to see more clearly whether the indicated atoms lie in the same plane.

90.   In all four reactions, products are favored at equilibrium. (Recall that stronger acids reacts to form weaker acids.)

a. $CH_3COOH + CH_3O^- \rightleftharpoons CH_3COO^- + CH_3OH$

b. $CH_3CH_2OH + {}^-NH_2 \rightleftharpoons CH_3CH_2O^- + NH_3$

c. $CH_3COOH + CH_3NH_2 \rightleftharpoons CH_3COO^- + CH_3\overset{+}{N}H_3$

d. $CH_3CH_2OH + HCl \rightleftharpoons CH_3CH_2\overset{+}{O}H_2 + Cl^-$

91. **a.** $HC{\equiv}CCH_2OH$ > $CH_2{=}CHCH_2OH$ > $CH_3CH_2CH_2OH$

b. These three compounds differ only in the group that is attached to CH_2OH. The more electronegative the group attached to CH_2OH, the stronger the acid because inductive electron withdrawal stabilizes the conjugate base, and the more stable the base, the stronger its conjugate acid. An sp carbon is more electronegative than an sp^2 carbon, which is more electronegative than an sp^3 carbon.

92. If the central atom is sp^3 hybridized and it does not have a lone pair, the molecule will have tetrahedral bond angles (109.5°). Therefore, only $^+NH_4$ has tetrahedral bond angles. The following species are close to being perfectly tetrahedral: H_2O, H_3O^+, NH_3, $^-CH_3$. However, they all have bond angles slightly smaller than 109.5°.

93. **a.**

c.

The 3 carbons are all sp^3 hybridized.
All the bond angles are 109.5°.

b.

d.

The 4 carbons are all sp^2 hybridized.
All the bond angles are 120°.

94. The log of $10^{-4} = -4$, the log of $10^{-5} = -5$, the log of $10^{-6} = -6$, etc.
Because the $pK_a = -\log K_a$, the pK_a of an acid with a K_a of 10^{-4} is $-(-4) = 4$.
An acid with a K_a of 4.0×10^{-4} is a stronger acid than one with a K_a of 1.0×10^{-4}.
Therefore, the pK_a can be estimated as being between 3 and 4.

a. **1.** between 3 and 4
      2. between −2 and −1
      3. between 10 and 11
      4. between 9 and 10
      5. between 3 and 4

b. **1.** $pK_a = 3.4$
      2. $pK_a = -1.3$
      3. $pK_a = 10.2$
      4. $pK_a = 9.1$
      5. $pK_a = 3.7$

c. Nitric acid (HNO_3) is the strongest acid because it has the lowest pK_a value. (The lower the pK_a value, the stronger the acid.)

95.   The nitrogen in the upper left-hand corner is slightly red, indicating that it has a greater concentration of negative charge than the other ring nitrogen. At first glance, the bottom nitrogen appears to be an sp^3 nitrogen. However, it is not an sp^3 nitrogen, because its lone pair is delocalized. Delocalization of lone-pair electrons causes the left-hand nitrogen to have a greater concentration of negative charge (which causes it to be more basic) and the bottom nitrogen to have less negative charge (which causes it to be less basic). The nitrogen that is not in the ring is the most basic of the three nitrogens; that why it is the one that is most apt to be protonated. Protonating it puts a positive formal charge on the nitrogen, as indicated by the blue area.

96.   **a.** $CH_3CH_2CHCOOH$ > $CH_3CH_2CHCOOH$ > $CH_3CH_2CHCOOH$ > $CH_3CH_2CH_2COOH$
                   | | |
                   Cl   Br   OH

b.   The relative acidities show that Cl is a more electronegative substituent than OH, since the strongest acid has the substituent that is best at withdrawing electrons inductively.

97.   **a.**

c.

b.   **or**

d.   **or**

98.   The reaction with the most favorable equilibrium constant is the one with the greatest difference between the strengths of the reactant acid and the product acid. (In other words, the one that has the strongest reactant acid and the weakest product acid.)

a.   **1.** CH_3OH is a stronger reactant acid ($pK_a = 15.5$) than CH_3CH_2OH ($pK_a = 15.9$), and both reactions form the same product acid ($^+NH_4$). Therefore, the reaction of CH_3OH with NH_3 has the more favorable equilibrium constant.

2.   Both reactions have the same reactant acid (CH_3CH_2OH). The product acids are different: $^+NH_4$ is a stronger product acid ($pK_a = 9.4$) than $CH_3NH_3^+$ ($pK_a = 10.7$). Therefore, the reaction of CH_3CH_2OH with CH_3NH_2 has the more favorable equilibrium constant.

b.   Now we have to compare "apples" and "oranges" because the reaction with the most favorable equilibrium constant in **1** and the reaction with the most favorable equilibrium constant in **2** do not have any species in common: **1** has the stronger reactant acid, whereas **2** has the weaker product acid. Therefore, the equilibrium constants have to be calculated.
The reaction of CH_3OH with NH_3 has an equilibrium constant $= 7.9 \times 10^{-7}$

$$(K_{eq} = 10^{-15.5} / 10^{-9.4} = 10^{-6.1} = 7.9 \times 10^{-7}).$$

The reaction of CH_3CH_2OH with CH_3NH_2 has an equilibrium constant $= 6.3 \times 10^{-6}$

$$(K_{eq} = 10^{-15.9} / 10^{-10.7} = 10^{-5.2} = 6.3 \times 10^{-6}).$$

Therefore, the reaction of CH_3CH_2OH with CH_3NH_2 has the most favorable (greatest) equilibrium constant of the four reactions.

99. **a.** The first pK_a is lower than the pK_a of acetic acid because citric acid has additional oxygen-containing groups that withdraw electrons inductively.

b. The third pK_a is greater than the pK_a of acetic acid because loss of the third proton puts a third negative charge on the molecule. Increasing the number of charges on a species destablizes it.

100.

The dipole moment is 2.95 D because the two Cl's are withdrawing electrons in the same direction.

The dipole moment is 0 D because the two Cl's are withdrawing electrons in opposite direction.

101. The bond angles at the triple bonded carbons, when the bonding orbitals overlap maximally, are 180°. A 180° angle cannot fit into the ring structure. Therefore, the overlap between the sp orbital and the adjacent sp^3 orbital becomes distorted from the ideal end-on overlap. This poor overlap causes the compound to be unstable (compare with Figure 2.6 on page 101 of the text).

102.

$$K_a = \frac{[H^+][HO^-]}{[H_2O]}$$

Since $[H^+] = [HO^-]$, both must be 1×10^{-7} M

$$K_a = \frac{(1 \times 10^{-7})(1 \times 10^{-7})}{55.5}$$

$$K_a = 1.80 \times 10^{-16}$$

$$pK_a = -\log 1.80 \times 10^{-16}$$

$$pK_a = 15.7$$

The answer can also be obtained in the following way:

$$K_a = \frac{[H^+][HO^-]}{[H_2O]}$$

$$K_a[H_2O] = [H^+][HO^-]$$

take the log of both sides

$$\log K_a + \log[H_2O] = \log[H^+] + \log[HO^-]$$

multiply both sides by -1

$$-\log K_a - \log[H_2O] = -\log[H^+] - \log[HO^-]$$

$$pK_a - \log[H_2O] = pH + pOH$$

$$pK_a - \log[H_2O] = 14$$

$$pK_a = 14 + \log[H_2O]$$

$$pK_a = 14 + \log 55.5$$

$$pK_a = 14 + 1.7$$

$$pK_a = 15.7$$

103. From the following equilibria you can see that a carboxylic acid is neutral when it is in its acidic form and charged when it is in its basic form. An amine is charged when it is in its acidic form and neutral when it is in its basic form.

$$RCOOH \rightleftharpoons RCOO^- + H^+$$

$$R\overset{+}{N}H_3 \rightleftharpoons RNH_2 + H^+$$

Charged species will dissolve in water and neutral species will dissolve in ether.
When separating compounds, you want essentially all (100:1) of each compound in either its acidic form or its basic form.

From the Henderson-Hasselbalch equation it can be calculated that in order to obtain a 100:1 ratio of acidic form:basic form, the pH must be two pH units lower than the pK_a of the compound; and in order to obtain a 100:1 ratio of basic form:acidic form, the pH must be two pH units greater than the pK_a of the compound.

a. If both compounds are to dissolve in water, they both must be charged. Therefore, the carboxylic acid must be in its basic form, and the amine must be in its acidic form. To accomplish this, the pH will have to be at least two pH units greater than the pK_a of the carboxylic acid and at least two pH units less than the pK_a of the ammonium ion. In other words, it must be between pH 6.8 and pH 8.7.

b. For the carboxylic acid to dissolve in water, it must be charged (in its basic form), so the pH will have to be greater than 6.8. For the amine to dissolve in ether, it will have to be neutral (in its basic form), so the pH will have to be greater than 12.7 to have essentially all of it in the neutral form. Therefore, the pH of the water layer must be greater than 12.7.

c. To dissolve in ether, the carboxylic acid will have to be neutral, so the pH will have to be less than 2.8 to have essentially all the carboxylic acid in the acidic (neutral) form. To dissolve in water, the amine will have to be charged, so the pH will have to be less than 8.7 to have essentially all the amine in the acidic form. Therefore, the pH of the water layer must be less than 2.8.

104. Charged compounds will dissolve in water and uncharged compounds will dissolve in ether. The acidic forms of carboxylic acids and alcohols are neutral and the basic forms are charged. The acidic forms of amines are charged and the basic forms are neutral.

105. Using the following reaction as an example:

$$CH_3OH + NH_3 \rightleftharpoons CH_3O^- + \overset{+}{N}H_4$$

$$K_{eq} = \frac{[products]}{[reactants]}$$

$$K_{eq} = \frac{[CH_3O^-][^+NH_4]}{[CH_3OH][NH_3]}$$

$$K_a \text{ reactant acid} = \frac{[\text{H}^+][\text{CH}_3\text{O}^-]}{[\text{CH}_3\text{OH}]}$$

$$K_a \text{ product acid} = \frac{[\text{H}^+][\text{NH}_3]}{[^+\text{NH}_4]}$$

$$K_{eq} = \frac{K_a \text{ reactant acid}}{K_a \text{ product acid}} = \frac{\dfrac{[\text{H}^+][\text{CH}_3\text{O}^-]}{[\text{CH}_3\text{OH}]}}{\dfrac{[\text{H}^+][\text{NH}_3]}{[^+\text{NH}_4]}}$$

$$= \frac{\dfrac{[\text{CH}_3\text{O}^-]}{[\text{CH}_3\text{OH}]}}{\dfrac{[\text{NH}_3]}{[^+\text{NH}_4]}}$$

$$= \frac{[\text{CH}_3\text{O}^-][^+\text{NH}_4]}{[\text{NH}_3]\ [\text{CH}_3\text{OH}]}$$

106.   For a discussion of how to do problems such as Problems 106–108, see **Special Topic I** (pH, pK_a, and Buffers).

$$pK_a = pH + \log\frac{[\text{HA}]}{[\text{A}^-]}$$

The above equation, called the Henderson-Hasselbalch equation, shows that:

1.   When the value of the pH is equal to the value of the pK_a, the concentration of buffer in the acidic form [HA] equals the concentration of buffer in the basic form [A$^-$].

2.   When the solution is more acidic than the pK_a, more buffer species is in the acidic form than in the basic form

3.   When the solution is more basic than the pK_a, more buffer species is in the basic form than in the acidic form.

Because the pH of the blood (~7.3) is greater than the pK_a of the buffer (6.1), more buffer species is in the basic form than in the acidic form. Therefore, the buffer is better at neutralizing excess acid.

107.   **a.**   fraction present in the acidic form $= \dfrac{\text{amount in the acidic form}}{\text{amount in the acidic form} + \text{amount in the basic form}}$

$$= \frac{[\text{HA}]}{[\text{HA}]+[\text{A}^-]}$$

Because there are two unknowns, we must define one in terms of the other. By using the definition of the acid dissociation constant, we can define [A⁻] in terms of [HA], [H⁺], and K_a, so we have only one unknown.

$$K_a = \frac{[H^+][A^-]}{[HA]}$$

$$[A^-] = \frac{K_a[HA]}{[H^+]}$$

Substituting the definition of [A⁻] into the equation for the fraction present in the acidic form gives:

$$\frac{[HA]}{[HA]+[A^-]} = \frac{[HA]}{[HA]+\dfrac{K_a[HA]}{[H^+]}} = \frac{1}{1+\dfrac{K_a}{[H^+]}} = \frac{[H^+]}{[H^+]+K_a}$$

Therefore, the percent that is present in the acidic form is given by:

$$\frac{[H^+]}{[H^+]+K_a} \times 100$$

Because the pK_a of the acid is given as 5.3, we know that K_a is 5.0×10^{-6} (because $pK_a = -\log K_a$). Because the pH of the solution is given as 5.7, we know that [H⁺] is 2.0×10^{-6} (because $pH = -\log [H^+]$). Substituting into the equation for the percent present in the acidic form gives:

$$\frac{[H^+]}{[H^+]+K_a} = \frac{2.0 \times 10^{-6}}{2.0 \times 10^{-6} + 5.0 \times 10^{-6}} \times 100$$

$$\frac{2.0 \times 10^{-6}}{7.0 \times 10^{-6}} \times 100 = 29\%$$

b.   Fraction present in the acidic form $= \dfrac{[H^+]}{[H^+]+K_a} = 0.80$

$$[H^+] = 0.80 \left([H^+]+K_a\right)$$

$$[H^+] = 0.80\,[H^+] + 0.80\,K_a$$

$$0.20\,[H^+] = 0.80\,K_a$$

$$[H^+] = 4\,K_a$$

$$[H^+] = 4 \times 5.0 \times 10^{-6}$$

$$[H^+] = 20 \times 10^{-6}$$

$$pH = 4.7$$

108. In each problem, we define $[H^+] = x$. Then, $[A^-]$ is also x. In part **a**, since we have a 1.0 M solution, $[HA] = 1.0 - x$; in part **a**, since we have a 0.1 M solution, $[HA] = 0.1 - x$.

a.
$$K_a = \frac{[H^+][A^-]}{[HA]}$$
$$1.74 \times 10^{-5} = \frac{x^2}{1.0 - x}$$
$$1.74 \times 10^{-5} = x^2$$
$$x = 4.16 \times 10^{-3}$$
$$pH = 2.38$$

b.
$$K_a = \frac{[H^+][A^-]}{[HA]}$$
$$2.00 \times 10^{-11} = \frac{x^2}{0.1 - x}$$
$$2.00 \times 10^{-12} = x^2$$
$$x = 1.41 \times 10^{-6}$$
$$pH = 5.85$$

c. This question can be answered by plugging the given concentrations into the Henderson-Hasselbalch equation.

$$pK_a = pH + \log\frac{[acid]}{[base]}$$
$$3.76 = pH + \log\frac{0.3}{0.1}$$
$$3.76 = pH + \log 3$$
$$3.76 = pH + 0.48$$
$$pH = 3.76 - 0.48 = 3.28$$

Chapter 1 Practice Test

1.   Answer the following:

     a.   Which is a stronger acid, HCl or HBr?

     b.   Which is a stronger base, NH_3 or H_2O?

     c.   Which bond has a greater dipole moment, a carbon-oxygen bond or a carbon-fluorine bond?

     d.   Which has a dipole moment of zero, $CHCl_3$ or CCl_4?

2.   What is the hybridization of the carbon atom in each of the following compounds?

$$^+CH_3 \qquad\qquad {}^-CH_3 \qquad\qquad \cdot CH_3$$

3.   Draw the Lewis structure for $HCO_3{}^-$.

4.   The following compounds are drawn in their acidic forms, and their pK_a values are given.
       Draw the form in which each compound would predominantly exist at pH = 8.

$$
\begin{array}{cccc}
CH_3COOH & CH_3CH_2OH & \overset{\overset{\displaystyle H}{|}}{\underset{+}{CH_3OH}} & CH_3CH_2\overset{+}{N}H_3 \\[2em]
pK_a = 4.8 & pK_a = 15.9 & pK_a = -2.5 & pK_a = 11.2
\end{array}
$$

5.   Which compound has greater bond angles, H_3O^+ or $^+NH_4$?

6.   What is the conjugate base of NH_3?

7.   Give the structure of a compound that contains five carbons, two of which are sp^2 hybridized and three
       of which are sp^3 hybridized.

8.   a.   What products would be formed from the following reaction?

$$CH_3OH \quad + \quad \overset{+}{N}H_4 \quad \rightleftharpoons$$

     b.   Does the reaction favor reactants or products?

9.   a.   What orbitals do carbon's electrons occupy before promotion?

     b.   What orbitals do carbon's electrons occupy after promotion?

10.   Which of the following compounds is a stronger acid?

$$
\underset{\underset{\displaystyle Cl}{|}}{CH_3CHCH_2COOH} \quad \textbf{or} \quad \underset{\underset{\displaystyle Cl}{|}}{CH_3CH_2CHCOOH}
$$

11. For each of the following compounds, indicate the hybridization of the atom to which the arrow is pointing.

$$\text{O}=\text{C}=\text{O} \quad \overset{\overset{\text{O}}{\|}}{\text{HCOH}} \quad \text{HC}\equiv\text{N} \quad \text{CH}_3\text{OCH}_3 \quad \text{CH}_3\text{CH}=\text{CH}_2$$

12. Indicate whether each of the following statements is true or false.

a. A pi bond is stronger than a sigma bond.   T   F

b. A triple bond is shorter than a double bond.   T   F

c. The oxygen-hydrogen bonds in water are formed by the overlap of an sp^2 orbital of oxygen with an s orbital of hydrogen.   T   F

d. HO^- is a stronger base than $^-NH_2$.   T   F

e. A double bond is stronger than a single bond.   T   F

f. A tetrahedral carbon has bond angles of 107.5°.   T   F

g. A Lewis acid is a compound that accepts a share in a pair of electrons.   T   F

h. CH_3CH_3 is more acidic than $H_2C=CH_2$.   T   F

> ANSWERS TO ALL THE PRACTICE TESTS CAN BE FOUND AT THE END OF THE
> SOLUTIONS MANUAL.

SPECIAL TOPIC I

pH, pK_a, and Buffers

This is a continuation of the discussion on acids and bases found in Sections 1.16–1.26 on pages 41–61 of the text. First, we will see how the pH of solutions of acids and bases can be calculated. We will look at three different kinds of solutions.

1. A solution made by dissolving a strong acid or a strong base in water.

2. A solution made by dissolving a weak acid or a weak base in water.

3. A solution made by dissolving a weak acid and its conjugate base in water. Such a solution is known as a **buffer solution**.

Before we start, we need to review a few terms.

An acid is a compound that donates a proton, and a base is a compound that accepts a proton.

The degree to which an acid (HA) dissociates is described by its acid dissociation constant (K_a).

$$HA \rightleftharpoons H^+ + A^-$$

$$K_a = \frac{[H^+][A^-]}{[HA]}$$

The strength of an acid can be indicated by its acid dissociation constant (K_a) or by its pK_a value.

$$pK_a = -\log K_a$$

The stronger the acid, the **larger** its acid dissociation constant and the **smaller** its pK_a value.

For example, an acid with an acid dissociation constant of 1×10^{-2} (p$K_a = 2$) is a stronger acid than one with an acid dissociation constant of 1×10^{-4} (p$K_a = 4$).

While pK_a is used to describe the strength of an acid, pH is used to describe the acidity of a solution. In other words, pH describes the concentration of hydrogen ions in a solution.

$$pH = -\log [H^+]$$

The smaller the pH, the more acidic the solution:
acidic solutions have pH values <7; a neutral solution has a pH $= 7$; basic solutions have pH values >7.

A solution with a pH $= 2$ is more acidic than a solution with a pH $= 4$.
A solution with a pH $= 12$ is more basic than a solution with a pH $= 8$.

Determining the pH of a Solution

To determine the pH of a solution, the concentration of hydrogen ion [H$^+$] in the solution must be determined.

Strong Acids

A strong acid is one that dissociates completely in solution. Strong acids have pK_a values <1.

Because a strong acid dissociates completely, the concentration of hydrogen ions is the same as the concentration of the acid: a 1.0 M HCl solution contains 1.0 M [H$^+$]; a 1.5 M HCl solution contains 1.5 M [H$^+$]. Therefore, to determine the pH of a strong acid, the [H$^+$] value does not have to be calculated; it is the same as the molarity of the strong acid.

solution	[H$^+$]	pH
1.0 M HCl	1.0 M	0
1.0×10^{-2} M HCl	1.0×10^{-2} M	2.0
6.4×10^{-4} M HCl	6.4×10^{-4} M	3.2

Strong Bases

Strong bases are compounds such as NaOH or KOH that dissociate completely in water.

Because they dissociate completely, the [HO$^-$] is the same as the molarity of the strong base.

pOH describes the basicity of a solution. The smaller the pOH, the more basic the solution; just like the smaller the pH, the more acidic the solution.

$$pOH = - \log [HO^-]$$

[HO$^-$] and [H$^+$] are related by the ionization constant for water (K_w).

$$K_w = [H^+][HO^-] = 10^{-14}$$
$$pH + pOH = 14$$

solution	[HO$^-$]	pOH	pH
1.0 M NaOH	1.0 M	0	14.0 − 0 = 14.0
1.0×10^{-4} M NaOH	1.0×10^{-4} M	4.0	14.0 − 4.0 = 10.0
7.8×10^{-2} M NaOH	7.8×10^{-2} M	1.1	14.0 − 1.1 = 12.9

Weak Acids

A weak acid does not dissociate completely in solution. This means that [H$^+$] must be calculated before the pH can be determined.

Acetic acid (CH$_3$COOH) is an example of a weak acid. It has an acid dissociation constant of 1.74×10^{-5} (pK_a = 4.76). The pH of a 1.00 M solution of acetic acid can be calculated as follows:

$$CH_3COOH \rightleftharpoons H^+ + CH_3COO^-$$

$$K_a = \frac{[H^+][CH_3COO^-]}{[CH_3COOH]}$$

Each molecule of acetic acid that dissociates forms one proton and one molecule of acetate ion. Therefore, the concentration of protons in solution equals the concentration of acetate ions. Each has a concentration that can be represented by x. The concentration of acetic acid, therefore, is the concentration we started with minus x.

$$1.74 \times 10^{-5} = \frac{(x)(x)}{1.00 - x}$$

The denominator $(1.00 - x)$ can be simplified to 1.00 because 1.00 is much greater than x. (When we actually calculate the value of x, we see that it is 0.004. And $1.00 - 0.004 = 1.00$.)

$$1.74 \times 10^{-5} = \frac{x^2}{1.00}$$

$$x = 4.17 \times 10^{-3}$$

$$pH = -\log 4.17 \times 10^{-3}$$

$$pH = 2.38$$

Formic acid (HCOOH) has a pK_a value of 3.75. The pH of a 1.50 M solution of formic acid can be calculated as follows:

$$HCOOH \rightleftharpoons H^+ + HCOO^-$$

$$K_a = \frac{[H^+][HCOO^-]}{[HCOOH]}$$

A compound with a $pK_a = 3.75$ has an acid dissociation constant of 1.78×10^{-4}.

$$1.78 \times 10^{-4} = \frac{(x)(x)}{1.50 - x} = \frac{x^2}{1.50}$$

$$x^2 = 1.50\,(1.78 \times 10^{-4})$$

$$x^2 = 2.67 \times 10^{-4}$$

$$x = 1.63 \times 10^{-2}$$

$$pH = -\log(1.63 \times 10^{-2})$$

$$pH = 1.79$$

Weak Bases

When a weak base is dissolved in water, it accepts a proton from water, creating hydroxide ion.

Determining the concentration of hydroxide allows the pOH to be determined, and this in turn allows the pH to be determined.

The pH of a 1.20 M solution of sodium acetate can be calculated as follows:

$$CH_3COO^- + H_2O \; \rightleftharpoons \; CH_3COOH + HO^-$$

$$\frac{K_w}{K_a} = \frac{[HO^-][CH_3COOH]}{[CH_3COO^-]}$$

$$\frac{1.00 \times 10^{-14}}{1.74 \times 10^{-5}} = \frac{(x)(x)}{1.20 - x}$$

$$5.75 \times 10^{-10} = \frac{x^2}{1.20}$$

$$x^2 = 6.86 \times 10^{-10}$$

$$x = 2.62 \times 10^{-5} = [HO^-]$$

$$pOH = -\log 2.62 \times 10^{-5}$$

$$pOH = 4.58$$

$$pH = 14.00 - 4.58$$

$$pH = 9.42$$

Notice that by setting up the equation equal to K_w/K_a, we can avoid the introduction of a new term (K_b), because $K_w/K_a = K_b$.

Buffer Solutions

A buffer solution is a solution that maintains nearly constant pH in spite of the addition of small amounts of H^+ or HO^-. That is because a buffer solution contains both a weak acid and its conjugate base. The weak acid can donate a proton to any HO^- added to the solution, and the conjugate base can accept any H^+ that is added to the solution, so the addition of small amounts of HO^- or H^+ does not significantly change the pH of the solution.

A buffer can maintain nearly constant pH in a range of one pH unit on either side of the pK_a of the conjugate acid. For example, an acetic acid/sodium acetate mixture can be used as a buffer in the pH range 3.76–5.76 because acetic acid has a $pK_a = 4.76$; methylammonium ion/methylamine can be used as a buffer in the pH range 9.7–11.7 because the methylammonium ion has a $pK_a = 10.7$.

The pH of a buffer solution can be determined from the Henderson-Hasselbalch equation. This equation comes directly from the expression defining the acid dissociation constant. Its derivation is found on page 58 of the text.

Henderson-Hasselbalch equation

$$pK_a = pH + \log \frac{[HA]}{[A^-]}$$

The pH of an acetic acid/sodium acetate buffer solution (pK_a of acetic acid = 4.76) that is 1.00 M in acetic acid and 0.50 M in sodium acetate is calculated as follows:

$$pK_a = pH + \log \frac{[HA]}{[A^-]}$$

$$4.76 = pH + \log \frac{1.00}{0.50}$$

$$4.76 = pH + \log 2$$

$$4.76 = pH + 0.30$$

$$pH = 4.46$$

Remember from Section 1.25 that compounds exist primarily in their acidic forms in solutions that are more acidic than their pK_a values and primarily in their basic forms in solutions that are more basic than their pK_a values. Therefore, it could have been predicted that the above solution will have a pH less than the pK_a of acetic acid, because there is more conjugate acid than conjugate base present in the solution.

There are three ways a buffer solution can be prepared:

1. Weak Acid and Weak Base

A buffer solution can be prepared by mixing a solution of a weak acid with a solution of its conjugate base.

The pH of a formic acid/sodium formate buffer (pK_a of formic acid = 3.75) solution prepared by mixing 25 mL of 0.10 M formic acid and 15 mL of 0.20 M sodium formate is calculated as follows:

The following equation shows that the number of millimoles (mmol) of each of the buffer components can be determined by multiplying the number of milliliters (mL) by the molarity (M).

$$\textbf{M} = \textbf{molarity} = \frac{\textbf{moles}}{\textbf{liters}} = \frac{\textbf{millimoles}}{\textbf{milliliters}}$$

Therefore:

25 mL × 0.10 M = 2.5 mmol formic acid

15 mL × 0.20 M = 3.0 mmol sodium formate

$$pK_a = pH + \log \frac{[HA]}{[A^-]}$$

$$3.75 = pH + \log \frac{2.5}{3.0}$$

$$3.75 = pH + \log 0.83$$

$$3.75 = pH - 0.08$$

$$pH = 3.83$$

It could have been predicted that the above solution would have a pH greater than the pK_a of formic acid, because there is more conjugate base than conjugate acid present in the solution.

Notice that we use mmol rather than molarity (mmol/mL) because both the acid and the conjugate base are in the same solution so they have the same volume. Therefore, volumes will cancel in the equation.

2. Weak Acid and Strong Base

A buffer solution can be prepared by mixing a solution of a weak acid with a strong base such as NaOH. The NaOH reacts completely with the weak acid, thereby creating the conjugate base needed for the buffer solution. For example, if 20 mmol of a weak acid and 5 mmol of a strong base are added to a solution, the 5 mmol of strong base will react with 5 mmol of weak acid, creating 5 mmol of weak base and leaving behind 15 mmol of weak acid.

The pH of a solution prepared by mixing 10 mL of a 2.0 M solution of a weak acid with a pK_a of 5.86 with 5.0 mL of a 1.0 M solution of sodium hydroxide can be calculated as follows:

When the 20 mmol of HA and the 5.0 mmol of HO^- are mixed, the 5.0 mmol of strong base will react with 5.0 mmol of HA, with the result that 5.0 mmol of A^- will be formed and 15 mmol (20 mmol − 5.0 mmol) of HA will be left unreacted.

$$10 \, mL \times 2.0 \, M \; = \; 20 \, mmol \, HA \quad \longrightarrow \quad 15 \, mmol \, HA$$

$$5.0 \, mL \times 1.0 \, M \; = \; 5.0 \, mmol \, HO^- \quad \longrightarrow \quad 5.0 \, mmol \, A^-$$

$$pK_a \; = \; pH \; + \; \log \frac{[HA]}{[A^-]}$$

$$5.86 \; = \; pH \; + \; \log \frac{15}{5}$$

$$5.86 \; = \; pH \; + \; \log 3$$

$$5.86 \; = \; pH \; + \; 0.48$$

$$pH \; = \; 5.38$$

3. Weak Base and Strong Acid

A buffer solution can be prepared by mixing a solution of a weak base with a strong acid such as HCl. The strong acid will react completely with the weak base, thereby forming the conjugate acid needed for the buffer solution. The pH of an ethylammonium ion/ethylamine buffer (pK_a of $CH_3CH_2NH_3^+ = 11.0$) prepared by mixing 30 mL of 0.20 M ethylamine with 40 mL of 0.10 M HCl can be calculated as follows:

$$30 \, mL \times 0.20 \, M \; = \; 6.0 \, mmol \, RNH_2 \quad \longrightarrow \quad 2.0 \, mmol \, RNH_2$$

$$40 \, mL \times 0.10 \, M \; = \; 4.0 \, mmol \, H^+ \quad \longrightarrow \quad 4.0 \, mmol \, R\overset{+}{N}H_3$$

Notice that 4.0 mmol H^+ reacts with 4.0 mmol RNH_2, forming 4.0 mmol RNH_3^+, leaving behind 2.0 mmol RNH_2.

$$pK_a \; = \; pH \; + \; \log \frac{[HA]}{[A^-]}$$

$$11.0 \; = \; pH \; + \; \log \frac{4.0}{2.0}$$

$$11.0 \; = \; pH \; + \; \log 2.0$$

$$11.0 \; = \; pH \; + \; 0.30$$

$$pH \; = \; 10.7$$

Fraction Present in the Acidic or the Basic Form

A common question asked is what fraction of a buffer will be in a particular form at a given pH; either what fraction will be in the acidic form or what fraction will be in the basic form. This is an easy question to answer if you remember the following formulas that are derived starting on the bottom of this page:

$$\textbf{fraction present in the acidic form} = \frac{[H^+]}{K_a + [H^+]}$$

$$\textbf{fraction present in the basic form} = \frac{K_a}{K_a + [H^+]}$$

What fraction of an acetic acid/sodium acetate buffer (pK_a of acetic acid = 4.76; $K_a = 1.74 \times 10^{-5}$) is present in the acidic form at pH = 5.20; $[H^+] = 6.31 \times 10^{-6}$?

$$\frac{[H^+]}{K_a + [H^+]} = \frac{6.31 \times 10^{-6}}{(1.74 \times 10^{-5}) + (6.31 \times 10^{-6})}$$

$$= \frac{6.31 \times 10^{-6}}{(17.4 \times 10^{-6}) + (6.31 \times 10^{-6})}$$

$$= \frac{6.31 \times 10^{-6}}{23.7 \times 10^{-6}} = \frac{6.31}{23.7}$$

$$= 0.26$$

What fraction of a formic acid/sodium formate buffer (pK_a formic acid = 3.75; $K_a = 1.78 \times 10^{-4}$) is present in the basic form at pH = 3.90; $[H^+] = 1.26 \times 10^{-4}$?

$$\frac{K_a}{K_a + [H^+]} = \frac{1.78 \times 10^{-4}}{(1.78 \times 10^{-4}) + (1.26 \times 10^{-4})}$$

$$= \frac{1.78 \times 10^{-4}}{3.04 \times 10^{-4}} = \frac{1.78}{3.04}$$

$$= 0.586$$

$$= 0.59$$

The formulas describing the fraction present in the acidic or basic form are obtained from the definition of the acid dissociation constant.

$$K_a = \frac{[H^+][A^-]}{[HA]}$$

To derive the equation for the fraction present in the basic form, we need to define $[A^-]$ in terms of $[HA]$, so we can get rid of the $[A^-]$ term.

$$[A^-] = \frac{K_a[HA]}{[H^+]}$$

$$\text{fraction present in the acidic form} = \frac{[HA]}{[HA] + [A^-]} = \frac{[HA]}{[HA] + \dfrac{K_a[HA]}{[H^+]}} = \frac{1}{1 + \dfrac{K_a}{[H^+]}}$$

$$= \frac{[H^+]}{K_a + [H^+]}$$

To derive the equation for the fraction present in the basic form, we need to define [HA] in terms of [A$^-$], so we can get rid of the [HA] term.

$$K_a = \frac{[H^+][A^-]}{[HA]}$$

$$[HA] = \frac{[H^+][A^-]}{K_a}$$

$$\text{fraction present in the basic form} = \frac{[A^-]}{[HA] + [A^-]} = \frac{[A^-]}{[A^-] + \dfrac{[H^+][A^-]}{K_a}}$$

$$= \frac{1}{1 + \dfrac{[H^+]}{K_a}}$$

$$= \frac{K_a}{K_a + [H^+]}$$

Preparing Buffer Solutions

The type of calculations just shown can be used to determine how to make a buffer solution.

For example, how can 100 mL of a 1.00 M buffer solution with a pH = 4.24 be prepared if you have available to you 1.50 M solutions of acetic acid ($pK_a = 4.76$; $K_a = 1.74 \times 10^{-5}$), sodium acetate, HCl, and NaOH?

First we need to determine what fraction of the buffer will be present in the acidic form at pH = 4.24; [H$^+$] = 5.75×10^{-5}:

$$\frac{[H^+]}{K_a + [H^+]} = \frac{5.75 \times 10^{-5}}{(1.74 \times 10^{-5}) + (5.75 \times 10^{-5})}$$

$$= \frac{5.75 \times 10^{-5}}{7.49 \times 10^{-5}}$$

$$= 0.77$$

If a 1.00 M buffer solution is desired, the buffer must be 0.77 M in acetic acid and 0.23 M in sodium acetate.

There are three ways such a buffer solution can be prepared:

1. **By mixing the appropriate amounts of acetic acid and sodium acetate in water, and adding water to obtain a final volume of 100 mL.**

   The amount of acetic acid needed: $[CH_3COOH] = 0.77$ M

   $$M = \frac{mmol}{mL} = \frac{x\ mmol}{100\ mL} = 0.77\ M$$

   $$x = 77\ mmol$$

   Therefore, we need to have 77 mmol of acetic acid in the final solution.

   To obtain 77 mmol of acetic acid from a 1.50 M solution of acetic acid:

   $$\frac{77\ mmol}{y\ mL} = 1.50\ M$$

   $$y = 51.3\ mL$$

   Notice that the formula $M = mmol/mL$ was used twice. The first time it was used to determine the number of mmol of acetic acid that is needed in the final solution. The second time it was used to determine how that number of mmol can be obtained from an acetic acid solution of a known concentration.

   The amount of sodium acetate needed: $[CH_3COO^-] = 0.23$ M

   $$\frac{x\ mmol}{100\ mL} = 0.23$$

   $$x = 23\ mmol$$

   To obtain 23 mmol of sodium acetate from a 1.50 M solution of sodium acetate:

   $$\frac{23\ mmol}{y\ mL} = 1.50\ M$$

   $$y = 15.3\ mL$$

   The desired buffer solution can be prepared using: 51.3 mL 1.50 M acetic acid

   $$\underline{\begin{array}{l} 15.3\ mL\ 1.50\ M\ sodium\ acetate \\ 33.4\ mL\ H_2O \end{array}}$$

   100.0 mL

2. **By mixing the appropriate amounts of acetic acid and sodium hydroxide, and adding water to obtain a final volume of 100 mL.**

   Sodium hydroxide is used to convert some of the acetic acid into sodium acetate. This means that acetic acid will be the source of both acetic acid and sodium acetate.

   The concentrations needed are: $[CH_3COOH] = 1.00$ M

   $$[NaOH] = 0.23 \text{ M}$$

   The amount of acetic acid needed: $[CH_3COOH] = 1.00$ M

   $$\frac{x \text{ mmol}}{100 \text{ mL}} = 1.00 \text{ M}$$

   $$x = 100 \text{ mmol}$$

   To obtain 100 mmol of acetic acid from a 1.50 M solution of acetic acid:

   $$\frac{100 \text{ mmol}}{y \text{ mL}} = 1.50 \text{ M}$$

   $$y = 66.7 \text{ mL}$$

   The amount of sodium hydroxide needed: $[NaOH] = 0.23$ M

   $$\frac{x \text{ mmol}}{100 \text{ mL}} = 0.23 \text{ M}$$

   $$x = 23 \text{ mmol}$$

   To obtain 23 mmol of sodium hydroxide from a 1.50 M solution of NaOH:

   $$\frac{23 \text{ mmol}}{y \text{ mL}} = 1.50 \text{ M}$$

   $$y = 15.3 \text{ mL}$$

   The desired buffer solution can be prepared using: 66.7 mL 1.50 M acetic acid

   15.3 mL 1.50 M NaOH

   $$\underline{18.0 \text{ mL H}_2\text{O}}$$

   100.0 mL

3. **By mixing the appropriate amounts of sodium acetate and hydrochloric acid, and adding water to obtain a final volume of 100 mL.**

   Hydrochloric acid is used to convert some of the sodium acetate into acetic acid. This means that sodium acetate will be the source of both acetic acid and sodium acetate.

   The concentrations needed are: $[CH_3COONa] = 1.00$ M

   $$[HCl] = 0.77 \text{ M}$$

The amount of sodium acetate needed: $[CH_3COONa] = 1.00$ M

$$\frac{x \text{ mmol}}{100 \text{ mL}} = 1.00 \text{ M}$$

$$x = 100 \text{ mmol}$$

To obtain 100 mmol of sodium acetate from a 1.50 M solution of sodium acetate:

$$\frac{100 \text{ mmol}}{y \text{ mL}} = 1.50 \text{ M}$$

$$y = 66.7 \text{ mL}$$

The amount of hydrochloric acid needed:

$$[HCl] = 0.77 \text{ M}$$

$$\frac{x \text{ mmol}}{100 \text{ mL}} = 0.77 \text{ M}$$

$$x = 77 \text{ mmol}$$

To obtain 77 mmol of hydrochloric acid from a 1.50 M solution of HCl:

$$\frac{77 \text{ mmol}}{y \text{ mL}} = 1.50 \text{ M}$$

$$y = 51.3 \text{ mL}$$

100 mL of a 1.00 M acetic acid/acetate buffer cannot be made from these reagents, because the volumes needed (66.7 mL + 51.3 mL) add up to more than 100 mL. To make this buffer using sodium acetate and hydrochloric acid, you would need to use a more concentrated solution (>1.50 M) of sodium acetate or a more concentrated solution (>1.50 M) of HCl.

Problems on pH, pK_a, and Buffers

1.   Calculate the pH of each of the following solutions:

a. 1×10^{-3} M HCl

b. 0.60 M HCl

c. 1.40×10^{-2} M HCl

d. 1×10^{-3} M KOH

e. 3.70×10^{-4} M NaOH

f. a 1.20 M solution of an acid with a p$K_a = 4.23$

g. 1.60×10^{-2} M sodium acetate (pK_a of acetic acid = 4.76)

2.   Calculate the pH of each of the following buffer solutions:

    a.   A buffer prepared by mixing 20 mL of 0.10 M formic acid and 15 mL of 0.50 M sodium formate (pK_a of formic acid = 3.75).

    b.   A buffer prepared by mixing 10 mL of 0.50 M aniline and 15 mL of 0.10 M HCl (pK_a of the anilinium ion = 4.60).

    c.   A buffer prepared by mixing 15 mL of 1.00 M acetic acid and 10 mL of 0.50 M NaOH (pK_a of acetic acid = 4.76).

3.   What fraction of a carboxylic acid with $pK_a = 5.23$ would be ionized at pH = 4.98?

4.   What would be the concentration of formic acid and sodium formate in a 1.00 M buffer solution with a pH = 3.12 (pK_a of formic acid = 3.75)?

5.   You have found a bottle labeled 1.00 M RCOOH. You want to determine what carboxylic acid it is, so you decide to determine its pK_a value. How would you do this?

6.   **a.**   How would you prepare 100 mL of a buffer solution that is 0.30 M in acetic acid and 0.20 M in sodium acetate using a 1.00 M acetic acid solution and a 2.00 M sodium acetate solution?

    b.   The pK_a of acetic acid is 4.76. Would the pH of the above solution be greater or less than 4.76?

7.   You have 100 mL of a 1.50 M acetic acid/sodium acetate buffer solution that has a pH = 4.90. How could you change the pH of the solution to 4.50?

8.   You have 100 mL of a 1.00 M solution of an acid with a $pK_a = 5.62$ to which you add 10 mL of 1.00 M sodium hydroxide. What fraction of the acid will be in the acidic form? How much more sodium hydroxide will you need to add in order to have 40% of the acid in the acidic form?

9.   Describe three ways to make a 1.00 M acetic acid/sodium acetate buffer solution with a pH = 4.00.

10.   You have available to you 1.50 M solutions of acetic acid, sodium acetate, sodium hydroxide, and hydrochloric acid. How would you make 50 mL of each of the buffers described in the preceding problem?

11.   How would you make a 1.0 M buffer solution with a pH = 3.30?

12.   You are planning to carry out a reaction that will produce protons. In order for the reaction to take place at constant pH, it will be carried in a solution buffered at pH = 4.2. Would it be better to use a formic acid/formate buffer or an acetic acid/acetate buffer?

Solutions to Problems on pH, pK_a, and Buffers

1. **a.** pH $= -\log(1 \times 10^{-3})$
    pH $= 3$

    b. pH $= -\log 0.60$
    pH $= 0.22$

    c. pH $= -\log(1.40 \times 10^{-2})$
    pH $= 1.85$

    d. pOH $= -\log(1 \times 10^{-3})$
    pOH $= 3$
    pH $= 14 - 3 = 11$

    e. pOH $= -\log(3.70 \times 10^{-4})$

    pOH $= 3.43$
    pH $= 10.57$

    f. p$K_a = 4.23$, $K_a = 5.89 \times 10^{-5}$

$$K_a = \frac{[H^+][A^-]}{[HA]}$$

$$5.89 \times 10^{-5} = \frac{x^2}{1.20}$$

$$x^2 = 7.07 \times 10^{-5}$$

$$x = 8.41 \times 10^{-3}$$

$$\text{pH} = 2.08$$

    g. $\dfrac{K_w}{K_a} = \dfrac{[HO^-][HA]}{[A^-]}$ $(K_a = 10^{-4.76} = 1.74 \times 10^{-5})$

$$\frac{1.0 \times 10^{-14}}{1.74 \times 10^{-5}} = \frac{x^2}{1.60 \times 10^{-2}}$$

$$5.75 \times 10^{-10} = \frac{x^2}{1.60 \times 10^{-2}}$$

$$x^2 = 9.20 \times 10^{-12}$$

$$x = 3.03 \times 10^{-6}$$

$$\text{pOH} = 5.52$$

$$\text{pH} = 14.00 - 5.52 = 8.48$$

2.   a.   formic acid: 20 mL $\times$ 0.10 M = 2.0 mmol
sodium formate: 15 mL $\times$ 0.50 M = 7.5 mmol

$$pK_a = pH + \log \frac{[HA]}{[A^-]}$$

$$3.75 = pH + \log \frac{2.0}{7.5}$$

$$3.75 = pH + \log 0.27$$

$$3.75 = pH + (-0.57)$$

$$pH = 4.32$$

b. aniline: 10 mL $\times$ 0.50 M = 5.0 mmol $\longrightarrow$ 3.5 mmol aniline (RNH_2)
   HCl: 15 mL $\times$ 0.10 M = 1.5 mmol $\longrightarrow$ 1.5 mmol anilinium hydrochloride ($\overset{+}{R}NH_3$)

$$pK_a = pH + \log \frac{[HA]}{[A^-]}$$

$$4.60 = pH + \log \frac{1.5}{3.5}$$

$$4.60 = pH + \log 0.43$$

$$4.60 = pH + (-0.37)$$

$$pH = 4.97$$

c. acetic acid: 15 mL $\times$ 1.00 M = 15 mmol $\longrightarrow$ 10 mmol acetic acid
    NaOH: 10 mL $\times$ 0.50 M = 5.0 mmol $\longrightarrow$ 5.0 mmol sodium acetate

$$pK_a = pH + \log \frac{[HA]}{[A^-]}$$

$$4.76 = pH + \log \frac{10}{5.0}$$

$$4.76 = pH + \log 2$$

$$4.76 = pH + 0.30$$

$$pH = 4.46$$

3.   The ionized form is the basic form. Therefore, we need to use the equation that allows us to calculate the fraction present in the basic form.

$$\text{fraction present in basic from} = \frac{K_a}{K_a + [H^+]} = \frac{5.89 \times 10^{-6}}{(5.89 \times 10^{-6}) + (10.47 \times 10^{-6})} = \frac{5.89 \times 10^{-6}}{16.36 \times 10^{-6}}$$

$$= 0.36$$

4.   From the Henderson-Hasselbalch equation, we can calculate the ratio of the concentrations of formic acid and sodium formate at pH = 3.12.

$$pK_a = pH + \log \frac{[HA]}{[A^-]}$$

$$3.75 = 3.12 + \log \frac{[HA]}{[A^-]}$$

$$0.63 = \log \frac{[HA]}{[A^-]}$$

$$4.27 = \frac{[HA]}{[A^-]}$$

$$[HA] = 4.27\,[A^-]$$

Knowing the relative concentrations and knowing that the sum of the concentrations is 1.0, we can calculate the actual concentrations of formic acid and sodium formate.

$$[HA] + [A^-] = 1.0\ M$$

$$4.27\,[A^-] + [A^-] = 1.0\ M$$

$$5.27\,[A^-] = 1.0\ M$$

$$[A^-] = 0.19\ M$$

$$[\text{sodium formate}] = 0.19\ M$$
$$[\text{formic acid}] = 0.81\ M$$

5.   From the Henderson-Hasselbalch equation, we see that the pH of the solution is the same as the pK_a of the species in the acidic form when the concentration of the species in the acidic form is the same as the concentration of the species in the basic form.

$$pK_a = pH + \log \frac{[HA]}{[A^-]}$$

when $[HA] = [A^-]$,

$$pK_a = pH$$

Thus, in order to have a solution in which the pH will be the same as the pK_a, the number of mm of acid must equal the number of mm of conjugate base.

Preparing a solution of x mmol of RCOOH and $1/2\ x$ mmol NaOH will give a solution in which [RCOOH] = [RCOO$^-$].

For example: 20 mL of 1.00 M RCOOH = 20 mmol
                    10 mL of 1.00 M NaOH = 10 mmol

This will give a solution that contains 10 mmol RCOOH and 10 mmol RCOO$^-$.

The pH of this solution is the pK_a of RCOOH.

6.   **a.** $\dfrac{x\text{ mmol}}{100\text{ mL}} = 0.30\text{ M}$ $\dfrac{x\text{ mmol}}{100\text{ mL}} = 0.20\text{ M}$

                 $x = 30$ mmol of acetic acid $x = 20$ mmol of sodium acetate

       $\dfrac{30\text{ mmol}}{y\text{ mL}} = 1.00\text{ M}$ $\dfrac{20\text{ mmol}}{y\text{ mL}} = 2.00\text{ M}$

             $y = 30$ mL of 1.00 M acetic acid $y = 10$ mL of 2.00 M acetic acid

The buffer solution could be prepared by mixing: 30 mL of 1.00 M acetic acid
                                                          10 mL of 2.00 M sodium acetate
                                                         $\underline{\text{ 60 mL}}$ of water
                                                         100 mL

b.   Because the concentration of buffer in the acidic form (0.30 M) is greater than the concentration of buffer in the basic form (0.20 M), the pH of the solution will be less than 4.76.

7.   **original solution**

$$pK_a = pH + \log\frac{[HA]}{[A^-]}$$

$$4.76 = 4.90 + \log\frac{[HA]}{[A^-]}$$

$$-0.14 = \log\frac{[HA]}{[A^-]}$$

$$0.72 = \frac{[HA]}{[A^-]}$$

$$[HA] = 0.72\,[A^-]$$

$$[HA] + [A^-] = 1.50\text{ M}$$

$$0.72\,[A^-] + [A^-] = 1.50\text{ M}$$

$$1.72\,[A^-] = 1.50\text{ M}$$

$$[A^-] = 0.87\text{ M}$$

$$[HA] = 0.63\text{ M}$$

desired solution

$$pK_a = pH + \log\frac{[HA]}{[A^-]}$$

$$4.76 = 4.50 + \log\frac{[HA]}{[A^-]}$$

$$0.26 = \log\frac{[HA]}{[A^-]}$$

$$1.82 = \frac{[HA]}{[A^-]}$$

$$[HA] = 1.82[A^-]$$

$$[HA] + [A^-] = 1.50 \text{ M}$$

$$1.82[A^-] + [A^-] = 1.50 \text{ M}$$

$$2.82[A^-] = 1.50 \text{ M}$$

$$[A^-] = 0.53 \text{ M}$$

$$[HA] = 0.97 \text{ M}$$

The original solution contains 87 mmol of A^- (100 mL $\times$ 0.87 M).

The desired solution with a pH = 4.50 must contain 53 mmol of A^-.

Therefore, 34 mmol of A^- (87 − 53 = 34) must be converted to HA.

This can be done by adding 34 mmol of HCl to the original solution.

If you have a 1.00 M HCl solution, you will need to add 34 mL to the original solution in order to change its pH from 4.90 to 4.50.

$$\frac{34 \text{ mmol}}{x \text{ mL}} = 1.00 \text{ M}$$

$$x = 34 \text{ mL}$$

Note that after adding HCl to the original solution, it will no longer be a 1.50 M buffer; it will be more dilute (150 mmol/134 mL = 1.12 M).

The change in the concentration of the buffer solution will be less if a more concentrated solution of HCl is used to change the pH. For example, if you have a 2.00 M HCl solution:

$$\frac{34 \text{ mmol}}{x \text{ mL}} = 2.00 \text{ M}$$

$$x = 17 \text{ mL}$$

You will need to add 17 mL to the original solution, and the concentration of the buffer will be 1.28 M (150 mmol/117 mL = 1.28 M).

8.   acid: 100 mL $\times$ 1.00 = 100 mmol HA $\longrightarrow$ 90 mmol HA

NaOH: 10 mL $\times$ 1.00 = 10 mmol HO^- $\longrightarrow$ 10 mmol A^-

Therefore, 90% is in the acidic form.

For 40% to be in the acidic form you need:

40 mmol HA

60 mmol A^-

You need to have 60 mmol rather than 10 mmol in the basic form. To get the additional 50 mmol in the basic form, you would need to add 50 mL of 1.0 M NaOH.

9.

$$pK_a = pH + \log \frac{[HA]}{[A^-]}$$

$$4.76 = 4.00 + \log \frac{[HA]}{[A^-]}$$

$$0.76 = \log \frac{[HA]}{[A^-]}$$

$$5.75 = \frac{[HA]}{[A^-]}$$

$$[HA] = 5.75[A^-]$$

$$[HA] + [A^-] = 1.0 \, M$$

$$5.75[A^-] + [A^-] = 1.0 \, M$$

$$6.75[A^-] = 1.0 \, M$$

$$[A^-] = 0.15 \, M$$

$$[HA] = 0.85 \, M$$

a. [acetic acid] = 0.85 M
[sodium acetate] = 0.15 M

b. [acetic acid] = 1.00 M
[NaOH] = 0.15 M

c. [sodium acetate] = 1.00 M
[HCl] = 0.85 M

10. a.

$$\frac{x \, \text{mmol}}{50 \, \text{mL}} = 0.85 \, M$$

$$x = 42.5 \, \text{mmol of acetic acid}$$

$$\frac{42.4 \, \text{mmol}}{y \, \text{mL}} = 1.50 \, M$$

$$y = 28.3 \, \text{mL of 1.50 M acetic acid}$$

$$\frac{x \, \text{mmol}}{50 \, \text{mL}} = 0.15 \, M$$

$$x = 7.5 \, \text{mmol of sodium acetate}$$

$$\frac{7.5 \, \text{mmol}}{y \, \text{mL}} = 1.50 \, M$$

$$y = 5.0 \, \text{mL of 1.50 M sodium acetate}$$

28.3 mL of 1.50 M acetic acid

5.0 mL of 1.50 M sodium acetate

16.7 mL of H$_2$O
50.0 mL

b. $\dfrac{x\text{ mmol}}{50\text{ mL}} = 1.00\text{ M}$

$x = 50\text{ mmol of acetic acid}$

$\dfrac{50\text{ mmol}}{y\text{ mL}} = 1.50\text{ M}$

$y = 33.3\text{ mL of }1.50\text{ M acetic acid}$

$\dfrac{x\text{ mmol}}{50\text{ mL}} = 0.15\text{ M}$

$x = 7.5\text{ mmol of NaOH}$

$\dfrac{7.5\text{ mmol}}{y\text{ mL}} = 1.50\text{ M}$

$y = 5.0\text{ mL of }1.50\text{ M NaOH}$

33.3 mL of 1.50 M acetic acid

5.0 mL of 1.50 M NaOH

$\underline{\textbf{11.7 mL of }\mathbf{H_2O}}$
50.0 mL

c. $\dfrac{x\text{ mmol}}{50\text{ mL}} = 1.00\text{ M}$

$x = 50\text{ mmol of sodium acetate}$

$\dfrac{50\text{ mmol}}{y\text{ mL}} = 1.50\text{ M}$

$y = 33.3\text{ mL of }1.50\text{ M sodium acetate}$

$\dfrac{x\text{ mmol}}{50\text{ mL}} = 0.85\text{ M}$

$x = 42.5\text{ mmol of HCl}$

$\dfrac{42.5\text{ mmol}}{y\text{ mL}} = 1.5\text{ M}$

$y = 28.3\text{ mL of }1.5\text{ M HCl}$

We cannot make the required buffer with these solutions, because 33.3 mL + 28.3 mL > 50 mL.

11. Because formic acid has a $pK_a = 3.75$, a formic acid/formate buffer can be a buffer at pH = 3.30, since this pH is within one pH unit of the pK_a value.

$$pK_a = pH + \log\frac{[HA]}{[A^-]}$$

$$3.75 = 3.30 + \log\frac{[HA]}{[A^-]}$$

$$0.45 = \log \frac{[\text{HA}]}{[\text{A}^-]}$$

$$2.82 = \frac{[\text{HA}]}{[\text{A}^-]}$$

$$[\text{HA}] = 2.82\,[\text{A}^-]$$

$$[\text{HA}] + [\text{A}^-] = 1.0\ \text{M}$$

$$2.82\,[\text{A}^-] + [\text{A}^-] = 1.0\ \text{M}$$

$$3.82\,[\text{A}^-] = 1.0\ \text{M}$$

$$[\text{A}^-] = 0.26\ \text{M}$$

$$[\text{HA}] = 0.74\ \text{M}$$

The solution must have [formic acid] = 0.74 M and [sodium formate] = 0.26 M.

12.   The reaction to be carried out will generate protons that will react with the basic form of the buffer in order to keep the pH constant.

Therefore, the better buffer would be the one that has the larger percentage of the buffer in the basic form. So now we need to calculate the percentage that will be in the basic form at pH = 4.2 for each of the buffer systems.

From the following calculation we see that at pH = 4.20, 74% of the formate buffer will be in the basic form.

pH = 4.20, $[\text{H}^+] = 6.31 \times 10^{-5}$ Formic acid has a $pK_a = 3.75$; $K_a = 1.78 \times 10^{-4}$.

$$\begin{aligned}
\text{fraction of buffer in basic from} &= \frac{K_a}{K_a + [\text{H}^+]} = \frac{1.78 \times 10^{-4}}{(1.78 \times 10^{-4}) + (6.31 \times 10^{-5})} = \frac{1.78 \times 10^{-4}}{(1.78 \times 10^{-4}) + (0.63 \times 10^{-4})} \\
&= \frac{1.78 \times 10^{-4}}{2.41 \times 10^{-4}} \\
&= 0.74
\end{aligned}$$

From the following calculation we see that at pH = 4.20, 22% of the acetate buffer will be in the basic form.

pH = 4.20, $[\text{H}^+] = 6.31 \times 10^{-5}$ Acetic acid has a $pK_a = 4.76$; $K_a = 1.74 \times 10^{-5}$.

$$\begin{aligned}
\text{fraction of buffer in basic from} &= \frac{K_a}{K_a + [\text{H}^+]} = \frac{1.74 \times 10^{-5}}{(1.74 \times 10^{-5}) + (6.31 \times 10^{-5})} \\
&= \frac{1.78 \times 10^{-5}}{8.05 \times 10^{-5}} \\
&= 0.22
\end{aligned}$$

Therefore, the formate buffer is preferred, because it has a greater percentage of the buffer in the basic form.

CHAPTER 2
An Introduction to Organic Compounds:
Nomenclature, Physical Properties, and Representation of Structure

Important Terms

alcohol	a compound with an OH group in place of one of the hydrogens of an alkane (ROH).
alkane	a hydrocarbon that contains only single bonds.
alkyl halide	a compound with a halogen in place of one of the hydrogens of an alkane.
alkyl substituent	a substituent formed by removing a hydrogen from an alkane.
amine	a compound in which one or more of the hydrogens of NH_3 is replaced by an alkyl substituent (RNH_2, R_2NH, R_3N)
angle strain	the strain introduced into a molecule as a result of its bond angles being distorted from their ideal values.
anti conformer	the staggered conformer in which the largest substituents bonded to the two carbons are opposite each other. It is the most stable of the staggered conformers.
axial bond	a bond of the chair conformer of cyclohexane that is perpendicular to the plane in which the chair is drawn (an up-down bond).
banana bonds	the bonds in small rings that are slightly bent as a result of orbitals overlapping at an angle rather than overlapping head-on.
boat conformer	the conformer of cyclohexane that roughly resembles a boat.
boiling point	the temperature at which the vapor pressure of a liquid equals the atmospheric pressure.
chair conformer	the conformer of cyclohexane that roughly resembles a chair. It is the most stable conformation of cyclohexane.
cis fused	two rings fused together in such a way that if the second ring were considered to be two substituents of the first ring, the two substituents would be on the same side of the first ring.
cis isomer (for a cyclic compound)	the isomer with two substituents on the same side of the ring.
cis-trans stereoisomers (geometric isomers)	geometric (or *E,Z*) isomers.

common name   non-systematic nomenclature.

conformation   the three-dimensional shape of a molecule at a given instant.

conformational analysis   the investigation of various conformers of a molecule and their relative stabilities.

conformers   different conformations of a molecule.

constitutional isomers   molecules that have the same molecular formula but differ in the way the atoms are connected.
(structural isomers)

cycloalkane   an alkane with its carbon chain arranged in a closed ring.

1,3-diaxial interaction   the interaction between an axial substituent and the other two axial substituents on the same side of a cyclohexane ring.

dipole-dipole interaction   an interaction between the dipole of one molecule and the dipole of another.

eclipsed conformer   a conformer in which the bonds on adjacent carbons are parallel to each other when viewed looking down the carbon-carbon bond.

equatorial bond   a bond of the chair conformer of cyclohexane that juts out from the ring in approximately the same plane that contains the six ring carbon atoms.

ether   a compound in which an oxygen is bonded to two alkyl substituents (ROR).

flagpole hydrogens   the two hydrogens in the boat conformer of cyclohexane that are closest to each other; they are at the 1- and 4-positions of the ring.

functional group   the center of reactivity of a molecule.

gauche conformer   a staggered conformer in which the largest substituents bonded to the two carbons are gauche to each other; that is, their bonds have a dihedral angle of approximately 60°.

The substituents are gauche to each other.

gauche interaction   the interaction between two atoms or groups that are gauche to each other.

geometric isomers   cis-trans (or *E,Z*) isomers.
(cis-trans stereoisomers)

half-chair conformer   the least stable conformer of cyclohexane.

homolog   a member of a homologous series.

homologous series   a family of compounds in which each member differs from the next by one
                                    methylene group.

hydrocarbon   a compound that contains only carbon and hydrogen.

hydrogen bond   an unusually strong dipole-dipole attraction (5 kcal/mol) between a hydrogen
                                    bonded to O, N, or F and the lone pair of a different O, N, or F.

hyperconjugation   delocalization of electrons by the overlap of a σ orbital with an empty orbital.

induced dipole-induced   an interaction between a temporary dipole in one molecule and the dipole
dipole interaction   that the temporary dipole induces in another molecule.

IUPAC nomenclature   systematic nomenclature.

melting point   the temperature at which a solid becomes a liquid.

methylene group   a CH_2 group.

Newman projection   a way to represent the three-dimensional spatial relationships of atoms
                                    by looking down the length of a particular carbon-carbon bond.

packing   the property that determines how well individual molecules fit into a crystal
                                    lattice.

parent hydrocarbon   the longest continuous carbon chain in a molecule; if the molecule has a
                                    functional group, it is the longest continuous carbon chain that contains the
                                    functional group.

perspective formula   a way to represent the three-dimensional spatial relationships of atoms
                                    using two adjacent solid lines, one solid wedge and one hatched wedge.

polarizability   the ease with which an electron cloud of an atom can be distorted.

primary alcohol   an alcohol in which the OH group is bonded to a primary carbon.

primary alkyl halide   an alkyl halide in which the halogen is bonded to a primary carbon.

primary amine   an amine with one alkyl group bonded to the nitrogen.

primary carbon   a carbon bonded to only one other carbon.

primary hydrogen   a hydrogen bonded to a primary carbon.

quaternary ammonium   a nitrogen compound with four alkyl groups bonded to the nitrogen plus a
salt   negatively charged counter ion.

ring-flip (chair-chair interconversion)	the conversion of a chair conformer of cyclohexane into the other chair conformer. Bonds that are axial in one chair conformer are equatorial in the other chair conformer.
sawhorse projection	a way to represent the three-dimensional spatial relationships of atoms by looking at the carbon-carbon bond from an oblique angle.
secondary alcohol	an alcohol in which the OH group is bonded to a secondary carbon.
secondary alkyl halide	an alkyl halide in which the halogen is bonded to a secondary carbon.
secondary amine	an amine with two alkyl groups bonded to the nitrogen.
secondary carbon	a carbon bonded to two other carbons.
secondary hydrogen	a hydrogen bonded to a secondary carbon.
skeletal structure	a structure that shows the carbon-carbon bonds as lines and does not show the carbon-hydrogen bonds.
skew-boat conformer	one of the conformers of a cyclohexane ring.
solubility	the extent to which a compound dissolves in a solvent.
solvation	the interaction between a solvent and another molecule (or ion).
staggered conformer	a conformer in which the bonds on one carbon bisect the bond angles on the adjacent carbon when viewed looking down the carbon-carbon bond.

steric hindrance	hindrance due to groups occupying a volume of space.
steric strain	the repulsion between the electron cloud of an atom or group of atoms and the electron cloud of another atom or group of atoms.
straight-chain alkane	an alkane in which the carbons form a continuous chain with no branches.
structural isomers (constitutional isomers)	molecules that have the same molecular formula but differ in the way the atoms are connected.
symmetrical ether	an ether with two identical alkyl substituents bonded to the oxygen.
systematic nomenclature	a system of nomenclature based on rules such as IUPAC nomenclature.

tertiary alcohol   an alcohol in which the OH group is bonded to a tertiary carbon.

tertiary alkyl halide   an alkyl halide in which the halogen is bonded to a tertiary carbon.

tertiary amine   an amine with three alkyl groups bonded to the nitrogen.

tertiary carbon   a carbon bonded to three other carbons.

tertiary hydrogen   a hydrogen bonded to a tertiary carbon.

trans-fused   two rings fused together in such a way that if the second ring were considered to be two substituents of the first ring, the two substituents would be on opposite sides of the first ring.

trans isomer   the isomer with two substituents on opposite sides of the ring.
(for a cyclic compound)

twist-boat conformer   one of the conformers of a cyclohexane ring.

unsymmetrical ether   an ether with two different alkyl substituents bonded to the oxygen.

van der Waals forces   induced dipole-induced dipole interactions.

Solutions to Problems

1.   **a.** propyl alcohol   **b.** dimethyl ether   **c.** propylamine

2.   **a.** $CH_3CHCH_2CH_3$ **b.** CH_3CCH_3 (with CH_3 above and CH_3 below)

                 |
                CH_3

        2-methylbutane   2,2-dimethylpropane

3.   Notice that each carbon forms fours bonds and each hydrogen and bromine forms one bond.

        $CH_3CH_2CH_2CH_2Br$ $CH_3CHCH_2CH_3$ CH_3CHCH_2Br CH_3CCH_3 (with CH_3 above, Br below)

                                            Br CH_3

        n-butyl bromide   *sec*-butyl bromide   isobutyl bromide   *tert*-butyl bromide
            or
        butyl bromide

4.   "Dibromomethane does not have constitutional isomers" proves that carbon is tetrahedral.

    If carbon were not tetrahedral but flat, dibromomethane would have constitutional isomers because the two structures shown below would be different since the bromines would be 90° apart in one compound and 180° apart in the other compound. Only because carbon is tetrahedral are the two structures identical, giving dibromomethane no constitutional isomers.

                      H   Br
                      | |
                H—C—Br   H—C—H
                      | |
                      Br   Br

5.   **a.** CH_3CHOH (with CH_3 below)   **c.** CH_3CH_2CHI (with CH_3 below)   **e.** CH_3CNH_2 (with CH_3 above and CH_3 below)

    b. $CH_3CHCH_2CH_2F$ (with CH_3 below)   **d.** CH_3COH (with CH_3 above and CH_2CH_3 below)   **f.** $CH_3CH_2CH_2CH_2CH_2CH_2CH_2CH_2Br$

6.   **a.** ethyl methyl ether   **c.** *sec*-butylamine   **e.** isobutyl bromide
    b. methyl propyl ether   **d.** butyl alcohol or *n*-butyl alcohol   **f.** *sec*-butyl chloride

7.

a.

$$
\begin{array}{c}
\quad\quad\quad CH_3 \\
\quad\quad\quad | \\
CH_3CHCHCH_2CH_2CH_3 \\
\quad\quad | \\
\quad\quad CH_3
\end{array}
$$

b.

$$
\begin{array}{c}
CH_3 \quad CH_3 \quad CH_3 \\
| \quad\quad | \quad\quad | \\
CH_3CHCH_2C \!-\! CHCH_2CH_3 \\
\quad\quad\quad | \\
\quad\quad\quad CHCH_3 \\
\quad\quad\quad | \\
\quad\quad\quad CH_3
\end{array}
$$

c.

$$
\begin{array}{c}
\quad\quad\quad CH_2CH_3 \\
\quad\quad\quad | \\
CH_3CH_2CH_2CCH_2CH_2CH_2CH_2CH_2CH_3 \\
\quad\quad\quad | \\
\quad\quad\quad CH_2CH_3
\end{array}
$$

d.

$$
\begin{array}{c}
\quad CH_3 \\
\quad | \\
CH_3CCH_2CHCH_2CH_2CH_2CH_3 \\
\quad | \quad\quad | \\
\quad CH_3 \;\; CH_2CH_2CH_3
\end{array}
$$

e.

$$
\begin{array}{c}
CH_3 \quad\quad CH_3 \\
| \quad\quad\quad | \\
CH_3CHCH_2CHCHCH_2CH_2CH_3 \\
\quad\quad\quad | \\
\quad\quad\quad CH_2 \\
\quad\quad\quad | \\
\quad\quad\quad CHCH_3 \\
\quad\quad\quad | \\
\quad\quad\quad CH_3
\end{array}
$$

f.

$$
\begin{array}{c}
CH_3CH_2CH_2CHCH_2CH_2CH_2CH_3 \\
\quad\quad\quad | \\
\quad\quad\quad CH_3CCH_3 \\
\quad\quad\quad | \\
\quad\quad\quad CH_3
\end{array}
$$

8.

a.

#1 $CH_3CH_2CH_2CH_2CH_2CH_2CH_2CH_3$

octane

#2

$$
\begin{array}{c}
CH_3CHCH_2CH_2CH_2CH_2CH_3 \\
\quad\quad | \\
\quad\quad CH_3
\end{array}
$$

2-methylheptane

#3

$$
\begin{array}{c}
CH_3CH_2CHCH_2CH_2CH_2CH_3 \\
\quad\quad\quad | \\
\quad\quad\quad CH_3
\end{array}
$$

3-methylheptane

#4

$$
\begin{array}{c}
CH_3CH_2CH_2CHCH_2CH_2CH_3 \\
\quad\quad\quad\quad | \\
\quad\quad\quad\quad CH_3
\end{array}
$$

4-methylheptane

#5

$$
\begin{array}{c}
\quad CH_3 \\
\quad | \\
CH_3CCH_2CH_2CH_2CH_3 \\
\quad | \\
\quad CH_3
\end{array}
$$

2,2-dimethylhexane

#10

$$
\begin{array}{c}
CH_3 \quad CH_3 \\
| \quad\quad | \\
CH_3CH_2CH \!-\! CHCH_2CH_3
\end{array}
$$

3,4-dimethylhexane

#11

$$
\begin{array}{c}
CH_3 \quad CH_3 \\
| \quad\quad | \\
CH_3C \!-\! CHCH_2CH_3 \\
| \\
CH_3
\end{array}
$$

2,2,3-trimethylpentane

#12

$$
\begin{array}{c}
CH_3 \quad CH_3 \\
| \quad\quad | \\
CH_3CCH_2CHCH_3 \\
| \\
CH_3
\end{array}
$$

2,2,4-trimethylpentane

#13

$$
\begin{array}{c}
CH_3 \quad CH_3 \\
| \quad\quad | \\
CH_3CH \!-\! CCH_2CH_3 \\
\quad\quad | \\
\quad\quad CH_3
\end{array}
$$

2,3,3-trimethylpentane

#14

$$
\begin{array}{c}
CH_3 \quad CH_3 \quad CH_3 \\
| \quad\quad | \quad\quad | \\
CH_3CH \!-\! CH \!-\! CHCH_3
\end{array}
$$

2,3,4-trimethylpentane

#6 $CH_3CH_2CCH_2CH_2CH_3$
with CH_3 above and CH_3 below the C

3,3-dimethylhexane

#7 $CH_3CH-CHCH_2CH_2CH_3$
with CH_3 above each of the first two carbons

2,3-dimethylhexane

#8 $CH_3CHCH_2CHCH_2CH_3$
with CH_3 above the first and third carbons

2,4-dimethylhexane

#9 $CH_3CHCH_2CH_2CHCH_3$
with CH_3 above the first and fourth carbons

2,5-dimethylhexane

#15 CH_3C-CCH_3
with CH_3 CH_3 above and CH_3 CH_3 below

2,2,3,3-tetramethylbutane

#16 $CH_3CH_2CHCH_2CH_2CH_3$
with CH_2CH_3 below

3-ethylhexane

#17 $CH_3CH_2CHCHCH_3$
with CH_3 above and CH_2CH_3 below

3-ethyl-2-methylpentane

#18 $CH_3CH_2CCH_2CH_3$
with CH_3 above and CH_2CH_3 below

3-ethyl-3-methylpentane

b. The systematic name is under each structure.
c. Only #1 (*n*-octane) and #2 (isooctane) have common names.
d. #2, #7, #8, #9, #12, #13, #14, #17
e. #3, #8, #10, #11
f. #5, #11, #12, #15

9. **a.** 2,2,4-trimethylhexane
    b. 2,2-dimethylbutane
    c. 3-methyl-4-propylheptane
    d. 2,2,5-trimethylhexane
    e. 3,3-diethyl-4-methyl-5-propyloctane
    f. 5-ethyl-4,4-dimethyloctane
    g. 3,3-diethylhexane
    h. 4-isopropyloctane
    i. 2,5-dimethylheptane

10. **a.** $CH_3CH_2CH_2CH_2CH_3$

pentane

b. CH_3CCH_3 with CH_3 above and CH_3 below

2,2-dimethylpropane

c. $CH_3CHCH_2CH_3$ with CH_3 above

2-methylbutane

d. $CH_3CHCH_2CH_3$ with CH_3 above

2-methylbutane

11.

12. **a.**   **d.**

b.   **e.**

c.   **f.**

13. **a.**   **d.**   **g.**

b.   **e.**   **h.**

c.   **f.**   **i.**

14. **a.**
$$CH_3CHCHCHCH_2CH_2CH_3$$
with CH₃, CH₂CH₃ substituents above and CH₂CH₃ below

b.
$$CH_3CCH_2CH_2CHCH_3$$
with CH₃ substituents above and CH₃, CH₃ below

15. **a.** 1-ethyl-2-methylcyclopentane   **e.** 2-cyclopropylpentane
b. ethylcyclobutane   **f.** 1-ethyl-3-isobutylcyclohexane
c. 4-ethyl-1,2-dimethylcyclohexane   **g.** 5-isopropylnonane
d. 3,6-dimethyldecane   **h.** 1-*sec*-butyl-4-isopropylcyclohexane

16. **a.** *sec*-butyl chloride
2-chlorobutane
secondary

  c. isohexyl chloride
1-chloro-4-methylpentane
primary

  b. cyclohexyl bromide
bromocyclohexane
secondary

  d. isopropyl fluoride
2-fluoropropane
secondary

17. **a.** CH_2Cl Note that the name of a CH_2Cl sustituent is "chloromethyl," because a Cl is in place of one of the Hs of a methyl substituent

chloromethylcyclohexane

  b. Cl CH_3

1-chloro-1-methylcyclohexane

  c. CH_3 Cl

1-chloro-2-methylcyclohexane

CH_3

Cl

1-chloro-3-methylcylclohexane

CH_3

Cl

1-chloro-4-methylcyclohexane

18. **a.** **1.** methoxyethane
  2. ethoxyethane
  3. 4-methoxyoctane
  4. 1-isopropoxy-3-methylbutane
  5. 1-propoxybutane
  6. 2-isopropoxyhexane

  b. No.

  c. **1.** ethyl methyl ether
  2. diethyl ether
  3. no common name
  4. isopentyl isopropyl ether
  5. butyl propyl ether
  6. no common name

19. CH_3OH
common = methyl alcohol
systematic = methanol

$CH_3CH_2CH_2CH_2OH$
common = butyl alcohol or *n*-butyl alcohol
systematic = 1-butanol

CH_3CH_2OH
common = ethyl alcohol
systematic = ethanol

$CH_3CH_2CH_2CH_2CH_2OH$
common = pentyl alcohol or *n*-pentyl alcohol
systematic = 1-pentanol

$CH_3CH_2CH_2OH$
common = propyl alcohol or *n*-propyl alcohol
systematic = 1-propanol

$CH_3CH_2CH_2CH_2CH_2CH_2OH$
common = hexyl alcohol or *n*-hexyl alcohol
systematic = 1-hexanol

20. **a.** 1-pentanol
   primary

c. 5-methyl-3-hexanol
   secondary

b. 5-chloro-2-methyl-2-pentanol
   tertiary

d. 7-methyl-3,5-octanediol
   both alcohol groups are secondary

21.

CH_3
|
$CH_3CCH_2CH_2CH_3$
|
OH
2-methyl-2-pentanol

CH_3
|
$CH_3CH_2CCH_2CH_3$
|
OH
3-methyl-3-pentanol

CH_3 |
CH_3C—$CHCH_3$
| |
OH CH_3
2,3-methyl-2-butanol

22. **a.** 4-chloro-3-ethylcyclohexanol
   secondary

c. 1-bromo-5,5-dimethyl-3-heptanol
   secondary

b. 7,8-dimethyl-3-nonanol
   secondary

d. 4-methylcyclohexanol
   secondary

23. **a.** tertiary alkyl halide   **b.** tertiary alcohol   **c.** primary amine

24. **a.** hexylamine
   1-hexanamine
   primary

d. butylpropylamine
   N-propyl-1-butanamine
   secondary

b. *sec*-butylisobutylamine
   N-isobutyl-2-butanamine
   secondary

e. diethylpropylamine
   N,N-diethyl-1-propanamine
   tertiary

c. diethylmethylamine
   N-ethyl-N-methylethanamine
   tertiary

f. N-ethyl-3-methylcyclopentanamine
   it does not have a common name
   secondary

25. **a.** $CH_3CH_2CH_2NHCH_2CHCH_3$
                               |
                              CH_3

d. $CH_3CH_2CH_2NCH_2CH_2CH_3$
                          |
                         CH_3

b. $CH_3CH_2NHCH_2CH_3$

e. $CH_3CH_2CHCH_2CH_3$
                |
                N
           H_3C CH_3

c. $CH_3CHCH_2CH_2CH_2CH_2NH_2$
           |
          CH_3

f.

26.   **a.**   6-methyl-1-heptanamine
isooctylamine
primary

c.   4-methyl-*N*-propyl-1-pentanamine
isohexylpropylamine
secondary

b.   cyclohexanamine
cyclohexylamine
primary

d.   2,5-dimethylcyclohexanamine
no common name
primary

27.   **a.**   The bond angle is predicted to be similar to the bond angle in water (104.5°).
b.   The bond angle is predicted to be similar to the bond angle in ammonia (107.3°).
c.   The bond angle is predicted to be similar to the bond angle in water (104.5°).
d.   The bond angle is predicted to be similar to the bond angle in the ammonium ion (109.5°).

28.   pentane or 2-methylbutane

29.   **a.**   1, 4, and 5
b.   1, 2, 4, 5, and 6

30.   **a.**   Each water molecule has two hydrogens that can form hydrogen bonds, whereas each alcohol molecule has only one hydrogen that can form a hydrogen bond. Therefore, there are more hydrogen bonds between water molecules than between alcohol molecules.

b.   Each water molecule has two hydrogens that can form hydrogen bonds and two lone pairs that can accept hydrogen bonds, so each water molecule can form two hydrogen bonds. In contrast, ammonia has three hydrogens that can form hydrogen bonds but only one lone pair that can accept hydrogen bonds, so each ammonia molecule can form only one hydrogen bond, so it will have a lower boiling point than water.

c.   Each water molecule has two hydrogens that can form hydrogen bonds and two lone pairs that can accept hydrogen bonds, so each water molecule can form two hydrogen bonds. In contrast, HF has three lone pairs that can accept hydrogen bonds but only one hydrogen that can form a hydrogen bond, so each HF molecule can form only one hydrogen bond, so it will have a lower boiling point than water.

Notice that HF and ammonia can each form only one hydrogen bond, but HF has a higher boiling point because the hydrogen bond formed by HF is stronger since fluorine is more electronegative than nitrogen.

31.

32. **a.**

b.

c.

33. **a.**

b. $HOCH_2CH_2CH_2OH > CH_3CH_2CH_2OH > CH_3CH_2CH_2CH_2OH > CH_3CH_2CH_2CH_2Cl$

34.   Because cyclohexane is a nonpolar compound it will have the lowest solubility in the most polar solvent, which, of the solvents given, is ethanol.

$CH_3CH_2CH_2CH_2CH_2OH$ $CH_3CH_2OCH_2CH_3$ CH_3CH_2OH $CH_3CH_2CH_2CH_2CH_2CH_3$
   1-pentanol   diethyl ether   ethanol   hexane

35.   Starting with the least stable conformer and then obtaining the others by keeping the front carbon constant and rotating the back carbon clockwise.

a.

b.

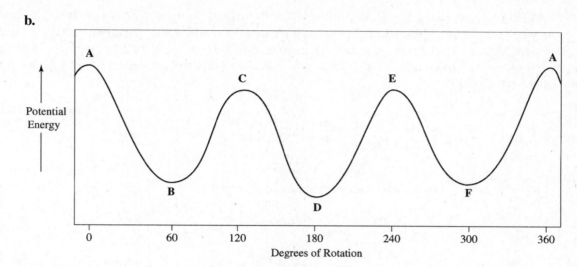

36. To draw the most stable conformer: have the largest group on the front carbon opposite the largest group on the back carbon.

a.

CH_3

H CH_3

H   H

CH_2CH_3

b.

CH_2CH_3

H   H

H CH_3

CH_2CH_3

c.

CH_2CH_3

H   H

CH_3 CH_3

CH_2CH_3

37. **a.** $180° - \dfrac{360°}{8}$

$180° - 45° = 135°$

b. $180° - \dfrac{360°}{9}$

$180° - 40° = 140°$

38. Hexethal would be expected to be the more effective sedative because it is less polar than barbital since hexethal has a hexyl group in place of the ethyl group of barbital. Being less polar, hexethal will be better able to penetrate the nonpolar membrane of the cell.

39. **a.**

Cl   Cl

Cl

Cl

Cl   Cl

b.

Cl   Cl   Cl

Cl   Cl   Cl

40. You can get the total strain energy of cycloheptane by subtracting the strainless heat of formation from the actual heat of formation:

The "strainless" heat of formation of cycloheptane is 7 (– 4.92) = – 34.4 kcal/mol.
The actual heat of formation of cycloheptane is – 28.2 kcal/mol (from Table 2.9 on page 104 of the text).

Therefore, the total strain energy of cycloheptane is – 28.2 – (– 34.4) = 6.2 kcal/mol.

41. Two 1,3-diaxial (gauche) interactions cause the chair conformer of fluorocyclohexane to be 0.25 kcal/mol less stable when the fluoro substituent is in the axial position than when it is in the equatorial position. The gauche conformer of 1-fluoropropane has one gauche interaction (see Figure 2.13 on page 106 of the text). Therefore, the gauche conformer is (0.25/2) = 0.13 kcal/mol less stable than the anti conformer that has no gauche interactions.

42.

$$K_{eq} = \frac{[\text{equatorial conformer}]}{[\text{axial conformer}]} = \frac{5.4}{1}$$

$$\% \text{ of equatorial conformer} = \frac{[\text{equatorial conformer}]}{[\text{equatorial conformer}] + [\text{axial conformer}]} \times 100$$

$$= \frac{5.4}{5.4 + 1} \times 100 = \frac{5.4}{6.4} \times 100 = 84\%$$

43.   **a.** cis   **b.** cis   **c.** cis   **d.** trans   **e.** trans   **f.** trans

44. Both *trans*-1,4-dimethylcyclohexane and *cis*-1-*tert*-butyl-3-methylcyclohexane have a conformer with two substituents in the equatorial position and a conformer with two substituents in the axial position. *cis*-1-*tert*-Butyl-3-methylcyclohexane will have a higher percentage of the diequatorial-substituted conformer because the bulky *tert*-butyl substituent will have a greater preference for the equatorial position than will a less bulky methyl substituent, since the larger substituent will have greater destabilizing 1,3-diaxial interactions when it is in an axial position.

45.   **a.**   **b.**

c. *trans*-1-Ethyl-2-methylcyclohexane is more stable because both substituents are in equatorial positions.

46.   **a.** one equatorial and one axial in each   **d.** one equatorial and one axial in each
  b. both equatorial in one and both axial in the other   **e.** one equatorial and one axial in each
  c. both equatorial in one and both axial in the other   **f.** both equatorial in one and both axial in the other

47.   **a.** One chair conformer of *trans*-1,4-dimethylcyclohexane has both substituents in equatorial positions, so it does not have any 1,3-diaxial interactions. The other chair conformer has both substituents in axial positions. When a substituent is in an axial position, it experiences two 1,3-diaxial interactions, so this chair conformer has a total of four 1,3-diaxial interactions. Because the 1,3-diaxial interaction between a methyl group and a hydrogen causes a strain energy of 0.9 kcal/mol, the chair conformer with both substituents in axial positions is 4 × 0.9 = 3.6 kcal/mol less stable than the chair conformer with both substituents in equatorial positions.

a 1,3-diaxial interaction

b. Each of the chair conformers of *cis*-1,4-dimethylcyclohexane has one substituent in an equatorial position and one in an axial position. Therefore, the two conformers have the same energy. (That is, the difference in energy between them is 0 kcal/mol.)

48.   Both condensed and skeletal structures are shown.

a.
$$CH_3$$
$$CH_3CH_2CHOCCH_3$$
$$CH_3 \; CH_3$$

b. $CH_3CHCH_2CH_2CH_2CH_2OH$
$$CH_3$$

c. $CH_3CH_2CHNH_2$
$$CH_3$$

d. $CH_3CHCH_2CH_2Br$
$$CH_3$$

e.
$$CH_3$$
$$CHCH_3$$
$$CH_3CH_2CH_2CHCHCH_2CH_2CH_2CH_3$$
$$CHCH_3$$
$$CH_3$$

f.
$$CH_2CH_3$$
$$CH_3CH_2N$$
$$CH_2CH_3$$

g. $CH_3CH_2CH_2CHCH_2CH_2CH_3$
$$CH_3CCH_3$$
$$CH_3$$

h.
$$Br$$
$$CH_3CHCH_2CH_2CCH_2CH_2CH_3$$
$$CH_3 \qquad Br$$

i.

$$CH_3$$

$$CH_3CHCHCH_2CH_2CH_3$$

$$OCH_2CH_3$$

j. $CH_3CH_2CH_2CH_2CHCH_2CH_2CH_2CH_3$

$$CHCH_3$$

$$CHCH_3$$

$$CH_3$$

k. $CH_3CH_2CHCHCH_2CH_2CH_2CH_3$

$$CH_3$$ (above)

$$CH_3$$ (below)

l. $CH_3CH_2CH_2CHCH_2CH_2CH_2CH_2CH_3$

$$CH_3CHCH_3$$

49.

50. **a.**

1. 2,2,6-trimethylheptane
2. 5-bromo-2-methyloctane
3. 5-methyl-3-hexanol
4. 3,3-diethylpentane
5. 5-bromo-*N*-ethyl-1-pentanamine
6. 2,3,5-trimethylhexane

7. 3-ethoxyheptane
8. 1,3-dimethoxypropane
9. *N,N*-dimethylcyclohexanamine
10. 3-ethylcyclohexanol
11. 1-bromo-4-methylcyclohexane

b.

1.

5. Br—[chain]—N/H—ethyl

9.

2. [structure with Br]
   Br

6. [branched structure]

10. [cyclohexane with ethyl and OH]
    OH

3. OH
   [structure]

7. [structure with O and ethyl]
   O

11. [cyclohexane with Br and methyl]
    Br

4. [quaternary carbon structure]

8. O—[chain]—O

51.   **C** and **D** are cis isomers. (In **C** both substituents are downward pointing; in **D** both substituents are upward pointing.)

52.   **a.**   **1.** 3   **2.** 4   **b.**   **1.** 6   **2.** 5   **c.**   **1.** 3   **3.** 4

53.   The first conformer is the most stable because the three substituents are more spread out, so its gauche interactions will not be as large—the Cl in the first conformer is between a CH_3 and an H, whereas the Cl in the other two conformers is between two CH_3 groups.

54.   **a.** [cyclohexane]

b. $\underset{\overset{|}{CH_3}\ \overset{|}{CH_3}}{\overset{\overset{|}{CH_3}\ \overset{|}{CH_3}}{CH_3C-CCH_3}}$

c. $\underset{\overset{|}{CH_3}\quad\ \overset{|}{CH_3}}{CH_3CHCH_2CHCH_3}$

55.   **a.**   2-butanamine
            sec-butylamine

        d.   1-ethoxypropane
                ethyl propyl ether

        b.   2-chlorobutane
                sec-butyl chloride

        e.   2-methylpentane
                isohexane

        c.   *N*-ethyl-2-butanamine
                sec-butylethylamine

        f.   2-propanamine
                isopropylamine

    g. 2-bromo-2-methylbutane
         tert-pentyl bromide

    i. bromocyclopentane
        cyclopentyl bromide

    h. 4-methyl-1-pentanol
         isohexyl alcohol

    j. cyclohexanol
        cyclohexyl alcohol

56.

a. 1-bromohexane
b. pentyl chloride
c. 1-butanol
d. 1-hexanol
e. hexane
f. 1-pentanol
g. 1-bromopentane
h. butyl alcohol

i. octane
j. isopentyl alcohol
    (The alcohol has the higher boiling point because it forms stronger hydrogen bonds. If you were asked to compare their solubilities in water, the amine is more soluble because it forms more hydrogen bonds with water.)
k. hexylamine (primary amines form stronger hydrogen bonds than do secondary amines)

57. **a.**

    b. The one on the right would predominate at equilibrium because it is more stable since both methyl groups are in equatorial positions.

    c.

    d. There would be equal amounts of each one at equilibrium, because they have the same stability since each conformer has one methyl group in an equatorial position and one methyl group in an axial position.

58. Ansaid is more soluble in water. It has a fluoro substituent that can hydrogen bond to water. Hydrogen bonding increases its solubility in water.

59. Note that each alcohol molecule has one hydrogen that can form a hydrogen bond and two lone pairs that can accept a hydrogen bond.

60. The student named only one compound correctly.

a. 2-bromo-3-pentanol
b. 4-ethyl-2,2-dimethylheptane
c. 3-methylcyclohexanol
d. 2,2-dimethylcyclohexanol
e. 5-(2-methylpropyl)nonane
f. 1-bromo-3-methylbutane

g. correct
h. 2,5-dimethylheptane
i. 5-bromo-2-pentanol
j. 3-ethyl-2-methyloctane
k. 2,3,3-trimethyloctane
l. 5-methyl-*N,N*-dimethyl-3-hexanamine

61. All three compounds are diaxial-substituted cyclohexanes. B has the highest energy. Only B has a 1,3-diaxial interaction between CH_3 and Cl, which will be greater than a 1,3-diaxial interaction between CH_3 and H or between Cl and H.

62. The only one is 2,2,3-trimethylbutane.

63. a.   c.   e.

b.   d.   f.

64. First draw the structure so you know what groups to put on the bonds in the Newman projections.

$$CH_3-CH-CH_2-CH_2-CH_2-CH_3$$
$$|$$
$$CH_3$$

a.

most stable

b.

least stable

c. Rotation can occur about all the C—C bonds. There are six carbon-carbon bonds in the compound, so there are five other carbon-carbon bonds, in addition to the $C_3 - C_4$ bond, about which rotation can occur.

d. Three of the carbon-carbon bonds have staggered conformers that are equally stable because each is bonded to a carbon with three identical substituents.

$$CH_3 \Downarrow CH - CH_2 - CH_2 - CH_2 \Downarrow CH_3$$
$$\Longrightarrow| \quad CH_3$$

65.

$CH_3CH_2CH_2CH_2CH_2Br$

   a. 1-bromopentane   primary alkyl halide
   b. pentyl bromide

$CH_3CH_2CH_2CHCH_3$
             |
             Br

   a. 2-bromopentane   secondary alkyl halide
   b. no common name

$CH_3CH_2CHCH_2CH_3$
          |
          Br

   a. 3-bromopentane   secondary alkyl halide
   b. no common name

      CH_3
      |
$CH_3CHCH_2CH_2Br$

   a. 1-bromo-3-methylbutane   primary alkyl halide
   b. isopentyl bromide

        CH_3
        |
$CH_3CH_2CHCH_2Br$

   a. 1-bromo-2-methylbutane   primary alkyl halide
   b. no common name

       Br
       |
$CH_3CH_2CCH_3$
       |
       CH_3

   a. 2-bromo-2-methylbutane   tertiary alkyl halide
   b. *tert*-pentyl bromide

      Br
      |
$CH_3CHCHCH_3$
        |
        CH_3

   a. 2-bromo-3-methylbutane   secondary alkyl halide
   b. no common name

      CH_3
      |
CH_3CCH_2Br
     |
     CH_3

   a. 1-bromo-2,2-dimethylpropane   primary alkyl halide
   b. no common name, but in older literature the common name neopentyl bromide is used.

   c. Four isomers do not have common names.
   d. Four isomers are primary alkyl halides.
   e. Three isomers are secondary alkyl halides.
   f. One isomer is a tertiary alkyl halide.

66.
   a. butane
   b. 1-propanol
   c. 5-propyldecane
   d. 4-propyl-1-nonanol
   e. 5-isopropyl-2-methyloctane or 2-methyl-5-(1-methylethyl)octane

   f. 6-chloro-4-ethyl-3-methyloctane
   g. 1-methoxy-5-methyl-3-propylhexane
   h. 6-isobutyl-2,3-dimethyldecane or 2,3-dimethyl-6-(2-methylpropyl)decane
   i. 8-methyl-4-decanamine
   j. 1-isobutyl-2-methylcyclohexane

67. **a.**

b.

c.

d.   **equally stable**

e.

f.

68. Alcohols with low molecular weights are more water soluble than alcohols with high molecular weights because, as a result of having fewer carbons, they have a smaller nonpolar component that has to be dragged into water.

69. **a.**

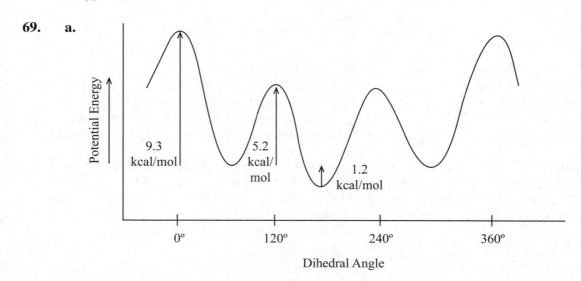

b.

c. 1.2 + 5.9 = 7.1 kcal/mol

d. 1.2 + 9.3 = 10.5 kcal/mol

70.   Six ethers have molecular formula = $C_5H_{12}O$.

$CH_3OCH_2CH_2CH_2CH_3$

1-methoxybutane
butyl methyl ether

$CH_3CHCH_2CH_3$
    |
    OCH_3

2-methoxybutane
sec-butyl methyl ether

$CH_3CH_2OCH_2CH_2CH_3$

1-ethoxypropane
ethyl propyl ether

CH_3CHCH_3
    |
    OCH_2CH_3

2-ethoxypropane
ethyl isopropyl ether

        CH_3
        |
CH_3COCH_3
        |
        CH_3

2-methoxy-2-methylpropane
tert-butyl methyl ether

$CH_3CHCH_2OCH_3$
    |
    CH_3

1-methoxy-2-methylpropane
isobutyl methyl ether

71.   The most stable conformer has two CH_3 groups in equatorial positions and one in an axial position. (The other conformer would have two CH_3 groups in axial positions and one in an equatorial position.)

72.   **a.** *N*-methyl-6-methyl-3-heptanamine
b. 3-ethyl-2,5-dimethylheptane
c. 1,2-dichloro-3-methylpentane

d. 2,3-dimethylpentane
e. 5-butyl-3,4-dimethylnonane
f. 5-butyl-3,3,9-trimethylundecane
        (undecane is an eleven carbon straight-chain
        hydrocarbon (Table 2.1 on page 69 of the text).

73.   One chair conformer of *trans*-1,2-dimethylcyclohexane has both substituents in equatorial positions, so it does not have any 1,3-diaxial interactions. However, the figure on top of page 109 of the text shows that the two methyl substituents are gauche to one another (as they would be in gauche butane; see Figure 2.13), giving it a strain energy of 0.9 kcal/mol.

a gauche
interaction

The other chair conformer of *trans*-1,2-dimethylcyclohexane has both substituents in axial positions. When a substituent is in an axial position, it experiences two 1,3-diaxial interactions. This chair conformer, therefore, has a total of four 1,3-diaxial interactions. Each diaxial interaction is between a CH₃ and an H, so each results in a strain energy of 0.9 kcal/mol. Therefore this chair conformer has a strain energy of 3.6 kcal/mol (4 × 0.9 = 3.6).

$$\text{H----CH}_3 \quad \text{a 1,3-diaxial interaction}$$

Thus, one conformer is 2.7 kcal/mole more stable than the other.

74.

$$\text{CH}_2\text{OH}$$

75.
a. 5-methyl-3-hexanol
b. 1-bromo-2-propylcyclopentane
c. 2-methyl-3-pentanol
d. 5-bromo-2-methyloctane

e. 1,5-hexanediol
f. 6-bromo-2-hexanol
g. 4-ethyl-3-methylcyclohexanol
h. 4-bromo-1-ethyl-2-methylcyclohexane

76.
a. 1-Hexanol has a higher boiling point than 3-hexanol because the alkyl group in 1-hexanol can better engage in van der Waals interactions. The OH group of 3-hexanol makes it more difficult for its 6 carbons to lie close to the 6 carbons of another molecule of 3-hexanol.

b. The floppy ethyl groups in diethyl ether interfere with the ability of the oxygen atom to engage in hydrogen bonding with water making it less soluble than tetrahydrofuran in which the alkyl groups are pinned back in a ring.

77. One of the chair conformers of *cis*-1,3-dimethylcyclohexane has both substituents in equatorial positions, so there are no unfavorable 1,3-diaxial interactions. The other chair conformer has three 1,3-diaxial interactions, two between a CH₃ and an H and one between two CH₃ groups. We know that a 1,3-diaxial interaction between a CH₃ and an H is 0.9 kcal/mol. Subtracting 1.8 from 5.4 (the energy difference between the two conformers) results in a value of 3.6 kcal/mol for the 1,3-diaxial interaction between the two CH₃ groups.

78.   Because bromine has a larger diameter than chlorine, one would expect bromine to have a greater preference for the equatorial position as would be indicated by a larger $\Delta G°$. However, Table 2.10 indicates that it has a smaller $\Delta G°$, indicating that it has less preference for the equatorial position than chlorine has. The C—Br bond is longer than the C—Cl bond, which causes bromine to be farther away than chlorine from the other axial substituents. Apparently, the longer bond more than offsets the larger diameter.

79.   **a.**   7-bromo-6-ethyl-5-decanol
   b.   5-chloro-3-ethyl-2,7-dimethylnonane
   c.   7,7-dimethyl-3-nonanol

80.   The conformer on the left has two 1,3-diaxial interactions between a CH_3 and an H (2×0.9 kcal/mole) for a total strain energy of 1.8 kcal/mol.

The conformer on the right has three 1,3-diaxial interactions, two between a CH_3 and an H (1.8 kcal/mole) and one between two CH_3 groups (3.6 kcal/mol; see Problem 77) for a total strain energy of 5.4 kcal/mol. Therefore, the conformer on the left will predominate at equilibrium.

81.   Problem 73 shows that the energy difference between the two chair conformers is 27 kcal/mol. To calculate the equilibrium constant needed to answer the question see Problem 23 on page 139.

$$\Delta G° = -2.7\,\text{kcal/mol}$$
$$\Delta G° = -RT \ln K_{eq}$$
$$-2.7\,\text{kcal/mol} = -1.986 \times 10^{-3}\,\text{kcal/mol K} \times 298\,\text{K} \times \ln\ K_{eq}$$
$$-2.7\,\text{kcal/mol} = -0.5918\,\text{kcal/mol} \ln K_{eq}$$
$$\ln K_{eq} = 4.56$$
$$K_{eq} = 95.6 = \frac{\text{both equatorial}}{\text{both axial}} = \frac{95.6}{1}$$

$$\begin{array}{l}\text{percentage of molecule} \\ \text{with both groups in} \\ \text{equatorial positions}\end{array} = \frac{\text{both equatorial}}{\text{both equatorial + both axial}} \times 100 = \frac{95.6}{95.6 + 1} \times 100 = 99\%$$

Chapter 2 Practice Test

1.   Name the following compound.

2.   Using Newman projections, draw the following conformers of hexane considering rotation about the C_3—C_4 bond.

a.   the most stable of all the conformers

b.   the least stable of all the conformers

c.   a gauche conformer

3.   Give both common and systematic names for each of the following compounds.

a. $CH_3CH_2CHCH_3$
             |
            Cl

b. $CH_3CHCH_2CH_2CH_2OH$
             |
            CH_3

c.   Br

4.   Label the three compounds in each set in order of decreasing boiling point.
       (Label the highest boiling compound #1, the next #2, and the lowest boiling #3.)

a. $CH_3CH_2CH_2CH_2CH_2Br$ $CH_3CH_2CH_2Br$ $CH_3CH_2CH_2CH_2Br$

b. $CH_3CH_2CH_2CH_2CH_3$ $CH_3CH_2CH_2CH_2OH$ $CH_3CH_2CH_2CH_2Cl$

c.
            CH_3 CH_3
             | |
     CH_3C——CCH_3 $CH_3CH_2CH_2CH_2CH_2CH_2CH_2CH_3$ $CH_3CHCH_2CH_2CH_2CH_2CH_3$
             | | |
            CH_3 CH_3 CH_3

5.   Give the systematic name for each of the following compounds.

a. $CH_3CHCH_2CH_2CHCH_2CH_3$
             | |
            CH_3 OH

c.   Br CH_3

b. $CH_3CH_2CHOCH_2CH_3$
             |
            $CH_2CH_2CH_2CH_3$

d.
            Cl
             |
     $CH_3CHCHCH_2CH_2CH_2Cl$
             |
            CH_2CH_3

6.   Draw the other chair conformer for the following compound.

7.   Draw the most stable conformer of *trans*-1-isopropyl-3-methylcyclohexane.

8.   Which of the following has:

    a.   the higher boiling point: diethyl ether or butyl alcohol?
    b.   the greater solubility in water: 1-butanol or 1-pentanol?
    c.   the higher boiling point: hexane or isohexane?
    d.   the higher boiling point: pentylamine or ethylmethylamine?
    e.   the greater solubility in water: ethyl alcohol or ethyl chloride?

9.   Give common and systematic names for each of the following compounds.

    a. $CH_3CHCH_2CH_2Br$ **b.** $CH_3CHCH_2CH_2OH$ **c.** $CH_3CHCH_2CH_2NH_2$
          CH_3 CH_3 CH_3

10.   Which is more stable:

    a.   A staggered conformer or an eclipsed conformer?

    b.   The chair conformer of methylcyclohexane with the methyl group in the axial position or the chair conformer of methylcyclohexane with the methyl group in the equatorial position?

    c.   Cyclohexane or cyclobutane?

11.   Give the structure of each of the following.

    a.   a secondary alkyl bromide that has three carbons

    b.   a secondary amine that has three carbons

    c.   an alkane with no secondary hydrogens

    d.   a constitutional isomer of butane

    e.   three compounds with molecular formula C_3H_8O

CHAPTER 3

Alkenes: Structure, Nomenclature, and an Introduction to Reactivity • Thermodynamics and Kinetics

Important Terms

acyclic	noncyclic.
addition reaction	a reaction in which atoms or groups are added to the reactant.
alkene	a hydrocarbon that contains a double bond.
allyl group	$CH_2 = CHCH_2—$
allylic carbon	an sp^3 carbon adjacent to a vinyl carbon.
Arrhenius equation	relates the rate constant of a reaction to the energy of activation and to the temperature at which the reaction is carried out $(k = Ae^{-E_a/RT})$.
catalyst	a species that increases the rate at which a reaction occurs without being consumed in the reaction.
cis isomer	the isomer with the hydrogens on the same side of the double bond.
degree of unsaturation	the sum of the number of π bonds and rings in a hydrocarbon.
E **isomer**	the isomer with the high-priority groups on opposite sides of the double bond.
electrophile	an electron deficient atom or molecule.
electrophilic addition reaction	an addition reaction in which the first species that adds to the reactant is an electrophile.
endergonic reaction	a reaction with a positive $\Delta G°$.
endothermic reaction	a reaction with a positive $\Delta H°$.
enthalpy	the heat given off $(-\Delta H°)$ or the heat absorbed $(+\Delta H°)$ during the course of a reaction.
entropy	a measure of the freedom of motion in a system.
exergonic reaction	a reaction with a negative $\Delta G°$.
exothermic reaction	a reaction with a negative $\Delta H°$.

**experimental energy of
activation**
$(E_a = \Delta H^{\ddagger} + RT)$

a measure of the approximate energy barrier to a reaction.
(It is approximate because it does not contain an entropy component.)

first-order rate constant

the rate constant of a first-order reaction.

**first-order reaction
(unimolecular reaction)**

a reaction whose rate is dependent on the concentration of one reactant.

**free energy of activation
$(\Delta G^{\ddagger})$**

the true energy barrier to a reaction.

functional group

the center of reactivity of a molecule.

**geometric isomers
(cis-trans stereoisomers)**

cis-trans (or *E,Z*) isomers.

**Gibbs free
energy change (ΔG°)**

the difference between the free energy content of the products and the free energy
content of the reactants at equilibrium under standard conditions (1M, 25 °C, 1 atm).

intermediate

a species formed during a reaction that is not the final product of the reaction.

kinetics

the field of chemistry that deals with the rates of chemical reactions.

kinetic stability

is indicated by $\Delta G^{\ddagger}$. If $\Delta G^{\ddagger}$ is large, the compound is kinetically stable (is not very
reactive). If $\Delta G^{\ddagger}$ is small, the compound is kinetically unstable (is very reactive).

Le Châtelier's principle

states that if an equilibrium is disturbed, the components of the equilibrium will
adjust in a way that will offset the disturbance.

mechanism of the reaction

a description of the step-by-step process by which reactants are changed into
products.

nucleophile

an electron-rich atom or molecule.

rate constant

the proportionality constant in the rate equation that quantifies the rate of a reaction.

**rate-determining step
or
rate-limiting step**

the step in a reaction that has the transition state with the highest energy.

**reaction coordinate
diagram**

describes the energy changes that take place during the course of a reaction.

saturated hydrocarbon

a hydrocarbon that is completely saturated with hydrogen (contains no double or
triple bonds).

second-order rate constant   the rate constant of a second-order reaction.

second-order reaction   a reaction whose rate is dependent on the concentration of two reactants, or on the square of the concentration of a single reactant.

solvation   the interaction between a solvent and another molecule (or ion).

thermodynamics   the field of chemistry that describes the properties of a system at equilibrium.

thermodynamic stability   is indicated by $\Delta G°$. If $\Delta G°$ is negative, the products are more stable than the reactants. If $\Delta G°$ is positive, the reactants are more stable than the products.

trans isomer   the isomer with the hydrogens on opposite sides of the double bond.

transition state   the energy maximum in a reaction step on a reaction coordinate diagram. In the transition state, bonds in the reactant that will break are partially broken and bonds in the product that will form are partially formed.

unsaturated hydrocarbon   a hydrocarbon that contains one or more double or triple bonds.

vinyl group $CH_2\!=\!CH-$

vinylic carbon   a carbon that is doubly bonded to another carbon.

Z isomer   the isomer with the high-priority groups on the same side of the double bond.

Solutions to Problems

1.   **a.** C_5H_8 **b.** C_4H_6 **c.** $C_{10}H_{16}$

2.   **a.** 3   **b.** 4   **c.** 1   **d.** 3   **e.** 13

3.   **a.** degree of unsaturation = 1   **b.** degree of unsaturation = 2   **c.** degree of unsaturation = 2

$CH_3CH{=}CH_2$

$CH_3C{\equiv}CH$

$CH_2{=}C{=}CH_2$

$HC{\equiv}CCH_2CH_3$
$CH_3C{\equiv}CCH_3$
$CH_2{=}CHCH{=}CH_2$
$CH_2{=}C{=}CHCH_3$

4.   **a.**

         $CH_3CH_2OCH{=}CH_2$ **c.**

       CH_3
       CH_3

       b. CH_3
       $CH_3C{=}CCH_2CH_2CH_2Br$ **d.** $CH_2{=}CHCH_2OH$
       CH_3

5.   **a.** 4-methyl-2-pentene   **d.** 1-bromo-4-methyl-3-hexene
       b. 2-chloro-3,4-dimethyl-3-hexene   **e.** 1,5-dimethylcyclohexene
       c. 1-bromocyclopentene   **f.** 1-butoxy-1-propene
                                           g. 8,8-dimethyl-1-nonene

6.   **a.** 5   **b.** 4   **c.** 4   **d.** 6

7.   **a.** 1 and 3

       b. 1. H_3C $CH_2CH_2CH_3$ H_3C H **3.** H_3C CH_3 H_3C H
                        $C{=}C$ $C{=}C$ $C{=}C$ $C{=}C$
                   H H H $CH_2CH_2CH_3$ H H H CH_3
                        cis   trans   cis   trans

8. **1.** [trans structure]
trans

[cis structure]
cis

2. [structure]

3. [trans structure]
trans

[cis structure]
cis

4. [structure]

9. $CH_3CH_2CH_2CH\!=\!CH_2$ $CH_3\overset{\overset{\displaystyle CH_3}{|}}{C}\!=\!CHCH_3$ $CH_3CH_2\overset{\overset{\displaystyle CH_3}{|}}{C}\!=\!CH_2$

10.   Only **C** has a dipole moment of zero, because the bond dipoles cancel since they are in opposite directions.

$$\underset{Cl}{\overset{H}{}}\!C\!=\!C\!\underset{H}{\overset{Cl}{}}$$

11.   **a.** $-I > -Br > -OH > -CH_3$
    b. $-OH > -CH_2Cl > -CH\!=\!CH_2 > -CH_2CH_2OH$

12.   **a.**
$\underset{H}{\overset{CH_3CH_2}{}}C\!=\!C\underset{H}{\overset{CH_3}{}}$ $\underset{H}{\overset{CH_3CH_2}{}}C\!=\!C\underset{CH_3}{\overset{H}{}}$

       Z   **E**

b.
$\underset{Cl}{\overset{CH_3CH_2}{}}C\!=\!C\underset{H}{\overset{CH_2CH_3}{}}$ $\underset{Cl}{\overset{CH_3CH_2}{}}C\!=\!C\underset{CH_2CH_3}{\overset{H}{}}$

       E   **Z**

c.
$\underset{CH_3CH_2}{\overset{CH_3CH_2CH_2CH_2}{}}C\!=\!C\underset{\underset{CH_3}{\overset{|}{CHCH_3}}}{\overset{CH_2Cl}{}}$ $\underset{CH_3CH_2}{\overset{CH_3CH_2CH_2CH_2}{}}C\!=\!C\underset{CH_2Cl}{\overset{\overset{CH_3}{\overset{|}{CHCH_3}}}{}}$

       Z   **E**

d.

HOCH₂CH₂ ... C(CH₃)₃

O=C ... C≡CH

H **Z**

HOCH₂CH₂ ... CH

O=C ... C(CH₃)₃

H **E**

13.   **a.**

c.

Cl

Cl

b.

Cl

Cl

d.   HO

O   H

HO

O   H

14.   **a.** (*E*)-2-heptene   **b.** (*Z*)-3,4-dimethyl-2-pentene   **c.** (*Z*)-1-chloro-3-ethyl-4-methyl-3-hexene

15.

CH₃

H₃C   CHCH₃

C=C

H   CH₂CH₂CH₂CH₃

16.   **a.**

(E)-2-methyl-2,4-hexadiene   (Z)-2-methyl-2,4-hexadiene

b.

(2E,4E)-2,4-heptadiene   (2E,4Z)-2,4-heptadiene

(2Z,4E)-2,4-heptadiene   (2Z,4Z)-2,4-heptadiene

c.

(E)-1,3-pentadiene   (Z)-1,3-pentadiene

b has four stereoisomers because each double bond can have either the *E* or the *Z* configuration.

a and **c** have only two stereoisomers because, in each case, there are two identical substituents bonded to one of the sp^2 carbons, so only one of the double bonds can have either the *E* or the *Z* configuration.

17.   **a.** $AlCl_3$ is the electrophile and NH_3 is the nucleophile.

b. The $H^{\delta+}$ of HBr is the electrophile and HO^- is the nucleophile.

18.   nucleophiles: H^- CH_3O^- $CH_3C{\equiv}CH$ NH_3

electrophiles: $CH_3\overset{+}{C}HCH_3$

19. 1.

Notice that drawing the arrows incorrectly leads to a bromine with
an incomplete octet and a positive charge and an oxygen with 10 valence
electrons and 2– charge.

Notice that drawing the arrows incorrectly leads to an oxygen with
an incomplete octet and a 2+ charge.

2. This one cannot be drawn because the arrow is supposed to show where the electrons move to,
but there are no electrons on the H to go anywhere.

3.

The product cannot be drawn because the destination of the
electrons in the breaking π bond is not clear.

4. This one cannot be drawn because the arrow is supposed to show where the electrons move to,
but there are no electrons on the C to go anywhere.

20. a.

electrophile   nucleophile

b.

nucleophile   electrophile

c.

nucleophile   electrophile

d.

electrophile   nucleophile

21.   See above.

22.   **a.**   Because the equilibrium constants for all the monosubstituted cyclohexanes in Table 2.10 on page 107 of the text are greater than 1, all of the equilibria have negative $\Delta G°$ values.

(Recall that $\Delta G° = -RT \ln K_{eq}$.)

b.   the *tert*-butylcyclohexane

c.   the *tert*-butylcyclohexane

d. $\Delta G° = -RT \times \ln K_{eq}$

$K_{eq} = 18$

$\Delta G° = -1.986 \times 10^{-3}$ kcal/mol K $\times$ 298 K $\times \ln 18$ (recall that $T = °C + 273$)

$\Delta G° = -0.59 \times \ln 18$ kcal/mol

$\Delta G° = -0.59 \times 2.89$ kcal/mol

$\Delta G° = -1.7$ kcal/mol

23.   **a.**   Solved in the text.

b. $$\Delta G° = -RT \ln K_{eq}$$

$$-2.1 = -1.986 \times 10^{-3} \times 298 \times \ln K_{eq}$$

$$\ln K_{eq} = 3.56$$

$$K_{eq} = 35$$

$$K_{eq} = \frac{[\text{isopropylcyclohexane}]_{equatorial}}{[\text{isopropylcyclohexane}]_{axial}} = \frac{35}{1}$$

$$\frac{\% \text{ of equatorial}}{\text{isopropylcyclohexane}} = \frac{[\text{isopropylcyclohexane}]_{equatorial}}{[\text{isopropylcyclohexane}]_{equatorial} + [\text{isopropylcyclohexane}]_{axial}} \times 100 = \frac{35}{35 + 1} \times 100$$

$$= \frac{35}{36} \times 100$$

$$= 97\%$$

c.   Isopropylcyclohexane has a greater percentage of the conformer with the substituent in the equatorial position because the isopropyl substituent is larger than the fluoro substituent. The larger the substituent, the less stable is the conformer in which the substituent is in the axial position because of the 1,3-diaxial interactions.

24. $\Delta S°$ is more significant in reactions in which the number of reactant molecules and the number of product molecules are not the same.

a.   **1.**   A + B $\rightleftharpoons$ C

2.   A + B $\rightleftharpoons$ C

   b. None of the four reactions has a positive $\Delta S°$.

   In order to have a positive $\Delta S°$, the products must have greater freedom of motion than the reactants. (In other words, there should be more molecules of products than molecules of reactant.)

25. **a.** **(1)**
$$\Delta G° = \Delta H° - T\Delta S°$$
(recall that $T = °C + 273$)
$$\Delta G° = -12 - (273 + 30)(.01)$$
$$\Delta G° = -12 - 3 = -15 \text{ kcal/mol}$$

$$\Delta G° = -RT \ln K_{eq}$$
$$\Delta G° = -(1.986 \times 10^{-3})(303) \ln K_{eq}$$
$$-15 = -0.60 \ln K_{eq}$$
$$\ln K_{eq} = 25$$
$$K_{eq} = 7.2 \times 10^{10}$$

(2)
$$\Delta G° = \Delta H° - T\Delta S°$$
$$\Delta G° = -12 - (273 + 150)(.01)$$
$$\Delta G° = -12 - 4 = -16 \text{ kcal/mol}$$

$$\Delta G° = -(1.986 \times 10^{-3})(423) \ln K_{eq}$$
$$-16 = -0.84 \ln K_{eq}$$
$$\ln K_{eq} = 19$$
$$K_{eq} = 1.8 \times 10^{8}$$

   b. For this reaction: the calculations show that increasing the temperature causes $\Delta G°$ to be more negative.

   c. For this reaction: the calculations show that increasing the temperature causes K_{eq} to be smaller, because $\ln K_{eq} = -\Delta G°/RT$.

26. **a.**

bonds broken		bonds formed		
π bond of ethene	62	C—H	101	
H—Cl	103	C—Cl	85	
	165 kcal/mol		186 kcal/mol	$\Delta H° = 165 - 186 = -21 \text{ kcal/mol}$

   b.

bonds broken		bonds formed		
π bond of ethene	62	C—H	101	
H—H	104	C—H	101	
	166 kcal/mol		202 kcal/mol	$\Delta H° = 166 - 202 = -36 \text{ kcal/mol}$

   c. Both are exothermic.

   d. Both reactions have sufficiently negative $\Delta H°$ values to expect that they would be exergonic as well.

27. **a.** **a** and **b** because the product is more stable than the reactant

   b. **b** because it has the smallest rate constant (greatest $\Delta G^{\ddagger}$) leading from the product to the transition state

   c. **c** because it has the largest rate constant (smallest $\Delta G^{\ddagger}$) leading from the product to the transition state

28. **a.** A thermodynamically **unstable** product is one that is less stable than the reactant.

   A kinetically **unstable** product is one that has a large rate constant (small $\Delta G^{\ddagger}$) for the reverse reaction.

b. A kinetically **stable** product is one that has a small rate constant (large $\Delta G^{\ddagger}$) for the reverse reaction.

29.   **a.** Solved in the text.

b. Decreasing the concentration of methyl chloride (by a factor of 10) will decrease **the rate of the reaction** (by a factor of 10) to 1×10^{-8} M s^{-1}.

c. Changing the concentration will not affect **the rate constant** (k) **of the reaction**, because rate constants do not depend on the concentration of the reactants.

30.   The rate constant for a reaction can be increased by **decreasing** the stability of the reactant (increasing its energy) or by **increasing** the stability of the transition state (decreasing its energy).

31.   Taking the logarithm of both sides of the Arrhenius equation gives the following equation (where k is the rate constant), which we can use to answer the questions:

$$\ln k \; = \; \ln A \, - \, \frac{E_a}{RT}$$

a. Increasing the experimental activation energy (E_a) will decrease the rate constant of a reaction (will cause the reaction to be slower).

b. Increasing the temperature (T) will increase the rate constant of a reaction (will cause the reaction to go faster).

32.   **a.** The first stated reaction has the greater equilibrium constant:

$$K_{eq} \; = \; \frac{1 \times 10^{-3}}{1 \times 10^{-5}} = 1 \times 10^{2} \qquad K_{eq} \; = \; \frac{1 \times 10^{-2}}{1 \times 10^{-3}} = 1 \times 10$$

b. Because both reactions start with the same concentration, the first stated reaction will form the most product because it has the greater equilibrium constant.

33.

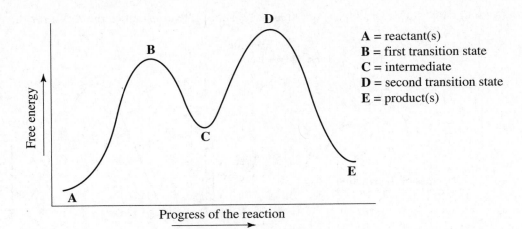

A = reactant(s)
B = first transition state
C = intermediate
D = second transition state
E = product(s)

34.   **a.** The first step in the forward direction (A → B) has the greatest free energy of activation.
  b. The first-formed intermediate is more apt to revert to reactants.
  c. The second step is the rate-determining step because it has the transition state with the highest energy.

Notice that the second step is rate-determining even though the first step has the greater energy of activation (steeper hill to climb). That is because it is easier for the intermediate that is formed in the first step to go back to starting material than to undergo the second step of the reaction. So, the second step is the rate-limiting step.

35.

  a. one (B)
  b. two
  c. the second step (k_2) (In this particular diagram, $k_2 > k_1$: if you had made the transition state for the second step a lot higher, you could have had a diagram in which $k_1 > k_2$.)
  d. the second step (k_{-1})   **f.**   B to C
  e. the second step in the reverse direction (k_{-1})   **g.**   C to B

36.   **a.** 3,8-dibromo-4-nonene   **c.** 1,5-dimethylcyclopentene   **e.** 4-methylcyclohexene
  b. (Z)-4-ethyl-3,7-dimethyl-3-octene   **d.** 3-ethyl-2-methyl-2-heptene   **f.** 4-ethyl-5-methylcyclohexene

37.   **a.** $CH_2{=}CHCH_2CH_2CH_2CH_3$ **b.** $CH_2{=}CHCHCH_2CH_3$ **c.** $CH_2{=}CHCCH_3$
with a CH_3 group on the b. structure and two CH_3 groups on the c. structure

38.

a.

$$
\underset{BrCH_2}{\overset{H}{}}C=C\underset{Br}{\overset{CH_2CH_2Br}{}}
$$

b.

$$
\underset{H}{\overset{H_3C}{}}C=C\underset{CH_3}{\overset{CH_2CH_2CH_2CH_3}{}}
$$

c.

$$
\underset{Br}{\overset{BrCH_2}{}}C=C\underset{CH_2CH_2CH_3}{\overset{\overset{CH_3}{|}{}CHCH_3}{}}
$$

d. $CH_2 = CHBr$

e.

f. $CH_2 = CHCH_2NHCH_2CH = CH_2$

39.

a.

b.

c.

d.

e.

f.

40.

a. $CH_2 = CHCH_2CH_2CH_2CH_3$
1-hexene

$CH_3CH = CHCH_2CH_2CH_3$
2-hexene

$CH_3CH_2CH = CHCH_2CH_3$
3-hexene

$$
CH_2 = \underset{\underset{CH_3}{|}}{C}CH_2CH_2CH_3
$$
2-methyl-1-pentene

$$
CH_2 = CH\underset{\underset{CH_3}{|}}{C}HCH_2CH_3
$$
3-methyl-1-pentene

$$
CH_2 = CHCH_2\underset{\underset{CH_3}{|}}{C}HCH_3
$$
4-methyl-1-pentene

$$
CH_3\underset{\underset{CH_3}{|}}{C} = CHCH_2CH_3
$$
2-methyl-2-pentene

$$
CH_3CH = \underset{\underset{CH_3}{|}}{C}CH_2CH_3
$$
3-methyl-2-pentene

$$
CH_3CH = CH\underset{\underset{CH_3}{|}}{C}HCH_3
$$
4-methyl-2-pentene

$$
CH_2 = \underset{\underset{CH_3}{|}}{\overset{\overset{CH_3}{|}}{C}}CHCH_3
$$
2,3-dimethyl-1-butene

$$
CH_3\underset{\underset{CH_3}{|}}{\overset{\overset{CH_3}{|}}{C}}CH = CH_2
$$
3,3-dimethyl-1-butene

$$
CH_3CH_2\underset{\underset{CH_2CH_3}{|}}{C} = CH_2
$$
2-ethyl-1-butene

$$
CH_3\underset{\underset{CH_3}{|}}{\overset{\overset{CH_3}{|}}{C}} = CCH_3
$$
2,3-dimethyl-2-butene

b. Of the compounds shown in part **a**, the following can have E and Z isomers:

2-hexene, 3-hexene, 3-methyl-2-pentene, 4-methyl-2-pentene

41. **a.** (E)-3-methyl-3-hexene

   b. *trans*-8-methyl-4-nonene or (E)-8-methyl-4-nonene

   c. *trans*-9-bromo-2-nonene or (E)-9-bromo-2-nonene

   d. 2,4-dimethyl-1-pentene

   e. 2-ethyl-1-pentene

   f. *cis*-2-pentene or (Z)-2-pentene

42.

43. Tamoxifen is an E isomer.

44.

45. $\Delta G^\circ = -RT \ln K_{eq}$

$\ln K_{eq} = -\Delta G^\circ / RT$

$\ln K_{eq} = -\Delta G^\circ / 0.59 \text{ kcal/mol}$

   a. $\ln K_{eq} = -2.72/0.59 \text{ kcal/mol}$

   $\ln K_{eq} = -4.6$

   $K_{eq} = [B]/[A] = 0.01$

   c. $\ln K_{eq} = 2.72/0.59 \text{ kcal/mol}$

   $\ln K_{eq} = 4.6$

   $K_{eq} = [B]/[A] = 100$

   b. $\ln K_{eq} = -0.65/0.59 \text{ kcal/mol}$

   $\ln K_{eq} = -1.10$

   $K_{eq} = [B]/[A] = 0.33$

   d. $\ln K_{eq} = 0.65/0.59 \text{ kcal/mol}$

   $\ln K_{eq} = 1.10$

   $K_{eq} = [B]/[A] = 3.0$

46. If the number of carbons is 40, $C_n H_{2n+2} = C_{40}H_{82}$. Therefore, a compound with molecular formula $C_{40}H_{56}$ is missing $(82 - 56)$ 26 hydrogens. This means that it has a total of 13 rings and π bonds, since each ring and π bond causes the compound to have two fewer hydrogens. Knowing that it has two rings and no triple bonds, we can conclude that it has 11 double bonds.

47. **a.** The C—Cl bond is a stronger bond because Cl uses a $3sp^3$ orbital to overlap the $2sp^3$ orbital of carbon, whereas Br uses a $4sp^3$ orbital. A $4sp^3$ has a greater volume than a $3sp^3$ and therefore has less electron density in the region of orbital-orbital overlap and therefore forms a weaker bond.

b. The Br—Br bond is a stronger bond because Br uses a $3sp^3$ orbital to overlap the $3sp^3$ orbital of the other bromine, whereas I uses a $5sp^3$ orbital. A $5sp^3$ has a greater volume than a $4sp^3$ and therefore has less electron density in the region of orbital-orbital overlap and therefore forms a weaker bond.

48.   **a.** *Z*   **b.** *E*   **c.** *E*   **d.** *Z*

49.   If the number of carbons is 30, $C_nH_{2n+2} = C_{30}H_{62}$. A compound with molecular formula $C_{30}H_{50}$ is missing 12 hydrogens. Because it has no rings, squalene has 6 π bonds (12/2 = 6).

50.   **a.** —CH=CH$_2$ > —CH(CH$_3$)$_2$ > —CH$_2$CH$_2$CH$_3$ > —CH$_3$

b. —OH > —NH$_2$ > —CH$_2$OH > —CH$_2$NH$_2$

c. —Cl > —C(=O)CH$_3$ > —C≡N > —CH=CH$_2$

51.   **a.**

(*E*)-2-methyl-2,4-hexadiene   (*Z*)-2-methyl-2,4-hexadiene

b.

(*E*)-1,5-heptadiene   (*Z*)-1,5-heptadiene

c.

1,4-pentadiene

52.   **a.** CH$_3$CH$_2$NH$_3^+$ **b.** (structure)   **c.** (structure)

53.   Only one name is correct.

a. 2-pentene   **d.** 3-heptene   **g.** 3-methyl-2-pentene
b. correct   **e.** 4-ethylcyclohexene   **h.** 2-methyl-1-hexene
c. 3-methyl-1-hexene   **f.** 2-chloro-3-hexene   **i.** 1-methylcyclopentene

54.

a.

$ClCH_2$ CH_3
    C=C
H H
            C=C
        H CH_3

c.

H CH_3
    C=C
CH_3CH_2 $CH_2CH_2CH_3$
            C=C
        H_3C H

b.

H CH_3
    C=C
CH_3CH_2 H
            C=C
        H $CH_2CH_2CH_3$

d.

Br
CH_3CH H
        C=C
    H CH_2CH_3
            C=C
        Br H

55.

a. 2   **d.** G   **g.** endergonic   **j.** G to F

b. B, D, F   **e.** A   **h.** exergonic

c. E to F   **f.** C   **i.** C

56. $\Delta G° = \Delta H° - T\Delta S°$

25 cal/mol = 25×10^{-3} kcal/mol

a. $\Delta G° = 20 - (298)(25 \times 10^{-3}) = 20 - 7.5$

$\Delta G° = 12.5$ kcal/mol

$\Delta G° = -RT \ln K_{eq}$

$12.5 = -1.986 \times 10^{-3} \times 298 \ln K_{eq}$

$12.5 = -0.59 \ln K_{eq}$

$-21.2 = \ln K_{eq}$

$6.2 \times 10^{-10} = K_{eq}$

b. $\Delta G° = 20 - (398)(25 \times 10^{-3}) = 20 - 10$

$\Delta G° = 10.0$ kcal/mol

$10.0 = -1.986 \times 10^{-3} \times 398 \ln K_{eq}$

$10.0 = -0.79 \ln K_{eq}$

$-12.7 = \ln K_{eq}$

$3.1 \times 10^{-6} = K_{eq}$

57.

(reaction mechanism scheme)

58.   **a.** $\Delta G° = - RT \ln K_{eq}$

$\Delta G° = - RT \ln 10$

$\Delta G° = -1.986 \times 10^{-3} \times 298 \times 2.3$

$\Delta G° = -1.36 \text{ kcal/mol}$

b. $\Delta G° = \Delta H° - 0$

$-1.36 = \Delta H° - 0$

$\Delta H° = -1.36 \text{ kcal/mol}$

c. $\Delta G° = 0 - T\Delta S°$

$-1.36 = 0 - 298\Delta S°$

$\Delta S° = 1.36/298 = 4.56 \times 10^{-3} \text{ kcal/(mol deg)}$

59. $\Delta G° = - RT \ln K_{eq}$

$\ln K_{eq} = - \Delta G°/RT$

$\ln K_{eq} = - \Delta G°/0.59 \text{ kcal/mol}$

$\ln K_{eq} = - 5.3/0.59 \text{ kcal/mol}$

$\ln K_{eq} = - 9.0$

$K_{eq} = [B]/[A] = 0.00013$

$[B]/[A] = 0.00013/1 = 0.13/1000$

The calculation shows that for every 1000 molecules in the chair conformation there is 0.13 molecule in a twist-boat conformation.

This agrees with the statement on page 105 of the text that for every thousand molecules of cyclohexane in a chair conformation, there is no more than one molecule in a twist-boat conformation.

60.   A step-by-step description of how to solve this problem is given in the box entitled "Calculating Kinetic Parameters" on page 153 of the text.

E_a can be determined from the Arrhenius equation ($\ln K = - E_a/RT$), because a plot of $\ln K$ versus $1/T$ gives a slope $= - E_a/R$.

$\ln 2.11 \times 10^{-5} = - 10.77$	$T = 304$	$1/T = 3.29 \times 10^{-3}$
$\ln 4.44 \times 10^{-5} = - 10.02$	$T = 313$	$1/T = 3.19 \times 10^{-3}$
$\ln 1.16 \times 10^{-4} = - 9.06$	$T = 324.5$	$1/T = 3.08 \times 10^{-3}$
$\ln 2.10 \times 10^{-4} = - 8.47$	$T = 332.8$	$1/T = 3.00 \times 10^{-3}$
$\ln 4.34 \times 10^{-4} = - 7.74$	$T = 342.2$	$1/T = 2.92 \times 10^{-3}$

$\text{slope} = - 8290$

$E_a = - (\text{slope}) R$

$E_a = - (-8290) \times 1.98 \times 10^{-3} \text{ kcal/mol}$

$E_a = 16.4 \text{ kcal/mol}$

To find $\Delta G°$:

$- \Delta G° = RT \ln kh/Tk_b$

> From the graph used to determine E_a, one can find the rate constant (k) at 30°.
> (It is 1.84×10^{-5} s^{-1}.)

$- \Delta G° = 1.98 \times 10^{-3} \times 303 \ \ln (1.84 \times 10^{-5} \times 6.625 \times 10^{-27})/(303 \times 1.3805 \times 10^{-10})$

$- \Delta G° = 1.98 \times 10^{-3} \times 303 \ \ln (1.22 \times 10^{-31})/(4.18 \times 10^{-8})$

$- \Delta G° = 1.98 \times 10^{-3} \times 303 \ \ln 2.92 \times 10^{-24}$

$- \Delta G° = 1.98 \times 10^{-3} \times 303 \times (- 54.2)$

$\Delta G° = 32.5$ kcal/mol

To find $\Delta H°$:

$\Delta H° = E_a - RT$

$\Delta H° = 16.4 - 1.98 \times 10^{-3} \times 303$

$\Delta H° = 16.4 - 0.6$

$\Delta H° = 15.8$ kcal/mol

To find $\Delta S°$:

$\Delta S° = (\Delta H° - \Delta G°)/T$

$\Delta S° = (15.8 - 32.5)/303$

$\Delta S° = (-16.7)/303$

$\Delta S° = 0.055$ kcal/(mol deg) $= 55$ cal/(mol deg)

Chapter 3 Practice Test

1.   Name each of the following compounds:

   a. $CH_3CH_2CHCH_2CH{=}CH_2$
   $\qquad\quad |$
   $\qquad\ \ CH_3$

   c. $CH_3CH_2CH{=}CHCH_2CH_2CHCH_3$
   $\qquad\qquad\qquad\qquad\qquad\quad |$
   $\qquad\qquad\qquad\qquad\qquad CH_2CH_3$

   b.

   d.

2.   Label the following substituents in order of decreasing priority in the *E,Z* system of nomenclature. Label the highest priority #1.

$$\underset{\displaystyle -\overset{\textstyle O}{\overset{\|}{C}}CH_3}{} \qquad -CH{=}CH_2 \qquad -Cl \qquad -C{\equiv}N$$

3.   Correct the incorrect names.

   a. 3-pentene

   c. 2-ethyl-2-butene

   b. 2-vinylpentane

   d. 2-methylcyclohexene

4.   Indicate whether each of the following statements is true or false:

   a. Increasing the energy of activation, increases the rate of the reaction.   T   F

   b. Decreasing the entropy of the products compared to the entropy of the reactants makes the equilibrium constant more favorable.   T   F

   c. An exergonic reaction is one with a $-\Delta G°$.   T   F

   d. An alkene is an electrophile.   T   F

   e. The higher the energy of activation, the more slowly the reaction will take place.   T   F

   f. Another name for *trans*-2-butene is (Z)-2-butene.   T   F

   g. *trans*-2-Butene has a dipole moment of zero.   T   F

   h. A reaction with a negative $\Delta G°$ has an equilibrium constant greater than one.   T   F

   i. Increasing the free energy of the reactants, increases the rate of the reaction.   T   F

   j. Increasing the free energy of the products, increases the rate of the reaction.   T   F

   k. The magnitude of a rate constant is not dependent on the concentration of the reactants.   T   F

5. Do the following compounds have the *E* or the *Z* configuration?

6. Draw structures for each of the following:

   a. allyl alcohol   **c.** *cis*-3-heptene

   b. 3-methylcyclohexene   **d.** vinyl bromide

7. What is the total number of π bonds and rings in a hydrocarbon with a molecular formula of C_8H_8?

8. Using curved arrows, show the movement of electrons in the following reaction mechanism:

9. Which of the following have cis-trans isomers?

   a. 1-pentene   **c.** 2-bromo-3-hexene

   b. 4-methyl-2-hexene   **d.** 2-methyl-2-hexene

10. Which of the following has a more favorable equilibrium constant (that is, which reaction favors products more)?

   a. A reaction with a $\Delta H°$ of 4 kcal/mol or a reaction with a $\Delta H°$ of 7 kcal/mol? (Assume a constant $\Delta S°$ value.)

   b. A reaction with a positive $\Delta S°$ value that takes place at 25 °C or the same reaction that takes place at 35 °C?

   c. A reaction in which two reactants form one product or a reaction in which one reactant forms two products? (Assume a constant $\Delta H°$ value.)

SPECIAL TOPIC II

An Exercise in Drawing Curved Arrows
"Pushing Electrons"

This is an extension of what you learned about drawing curved arrows in Section 3.6 on pages 132–136 of the text. Working through these problems will take only a little of your time. It will, however, be time well spent, because curved arrows will be used throughout the course and it is important that you are comfortable with this notation. (You will not encounter some of the reaction steps shown in this exercise for weeks or even months, so don't worry about why the chemical changes take place.)

Chemists use curved arrows to show how electrons move as covalent bonds break and/or new covalent bonds form. The tail of the arrow is positioned at the point where the electrons are in the reactant, and the head of the arrow points to where these same electrons end up in the product.

In the following reaction step, the bond between bromine and a carbon of the cyclohexane ring breaks and both electrons in the bond end up with bromine in the product. Thus, **the arrow starts at the electrons that carbon and bromine share in the reactant,** and **the head of the arrow points at bromine** because this is where the two electrons end up in the product.

Notice that the carbon of the cyclohexane ring is positively charged in the product, because it has lost the two electrons it was sharing with bromine. The bromine is negatively charged in the product, because it has gained the electrons that it shared with carbon in the reactant. The fact that two electrons move in this example is indicated by the two barbs on the arrowhead.

Notice that the arrow always starts at a bond or at a lone pair. It does not start at a negative charge.

In the following reaction step, a bond is being formed between the oxygen of water and a carbon of the other reactant. The arrow starts at one of the lone pairs of the oxygen and points at the atom (the carbon) that will share the electrons in the product. The oxygen in the product is positively charged, because the electrons that oxygen had to itself in the reactant are now being shared with carbon. The carbon that was positively charged in the reactant is not charged in the product, because it has gained a share in a pair of electrons.

Problem 1. Draw curved arrows to show the movement of the electrons in the following reaction steps: (You will find the answers immediately after Problem 10.)

a.

$$CH_3CH_2\overset{\overset{\displaystyle CH_3}{|}}{\underset{\underset{\displaystyle CH_3}{|}}{C}}-\ddot{B}\ddot{r}: \longrightarrow CH_3CH_2\overset{\overset{\displaystyle CH_3}{|}}{\underset{\underset{\displaystyle CH_3}{|}}{C}}{}^+ \quad + \quad :\ddot{B}\ddot{r}:^-$$

b.

$-\ddot{C}\ddot{l}: \longrightarrow$ $+$ $+$ $:\ddot{C}\ddot{l}:^-$

c.

$\overset{+}{\underset{\underset{\displaystyle H}{|}}{-\ddot{O}H}} \longrightarrow$ $+$ $+$ $H_2\ddot{O}:$

d. $CH_3CH_2-MgBr \longrightarrow CH_3\ddot{C}H_2 + {}^+MgBr$

e. $CH_3CH_2\underset{+}{C}HCH_3 + :\ddot{B}\ddot{r}:^- \longrightarrow CH_3CH_2\underset{\underset{\displaystyle :\ddot{B}r:}{|}}{C}HCH_3$

Frequently, chemists do not show the lone-pair electrons when they write reactions. The following are the same reaction steps you just saw except that the lone pairs are not shown.

Problem 2. Draw curved arrows to show the movement of the electrons:

a.

$$CH_3CH_2\overset{\overset{\displaystyle CH_3}{|}}{\underset{\underset{\displaystyle CH_3}{|}}{C}}-Br \longrightarrow CH_3CH_2\overset{\overset{\displaystyle CH_3}{|}}{\underset{\underset{\displaystyle CH_3}{|}}{C}}{}^+ \quad + \quad Br^-$$

b.

$-Cl \longrightarrow$ $+$ $+$ Cl^-

c.

$\overset{+}{\underset{\underset{\displaystyle H}{|}}{-\ddot{O}H}} \longrightarrow$ $+$ $+$ H_2O

d. $CH_3CH_2-MgBr \longrightarrow CH_3\bar{C}H_2 + {}^+MgBr$

The lone-pair electrons on Br^- in example **e** have to be shown in the reactant, because an arrow can start only at a bond or at a lone pair. Bromine's lone pairs do not have to be shown in the product.

e. $CH_3CH_2\underset{+}{C}HCH_3 + :\ddot{B}\ddot{r}:^- \longrightarrow CH_3CH_2\underset{\underset{\displaystyle Br}{|}}{C}HCH_3$

Many reaction steps involve both bond breaking and bond formation. In the following examples, the electrons in the bond that breaks are the same as the electrons in the bond that forms, so only one arrow is needed to show how the electrons move. As in the previous examples, the arrow starts at the point where the electrons are in the reactant, and the head of the arrow points to where these same electrons end up in the product (between the CH_3 carbon and the carbon that was previously positively charged in the first example, and between the carbons in the next example). Notice that the atom that loses a share in a pair of electrons (C in the first example, H in the second) ends up with a positive charge.

Frequently, the electrons in the bond that breaks are not the same as the electrons in the bond that forms. In such cases, two or more arrows are needed to show the movement of the electrons. In each of the following examples, look at the arrows that illustrate how the electrons move. Notice how the movement of the electrons allows you to determine both the structure of the products and the charges on the products.

$$CH_3CH_2 \overset{+}{\underset{\underset{H}{|}}{\ddot{O}}}H + H_2\ddot{O}: \longrightarrow CH_3CH_2 - \ddot{O}H + H_3\overset{+}{O}:$$

$$\underset{H_2C - CH_2}{\overset{+\cdot\cdot}{\ddot{B}r:}} + H\ddot{\ddot{O}}:^- \longrightarrow \underset{\underset{:OH}{|}}{\overset{:\ddot{B}r:}{\underset{}{|}}{H_2C - CH_2}}$$

$$:\ddot{B}r:^- + CH_3 \overset{H}{\underset{+}{\overset{|}{\ddot{O}}H}} \longrightarrow CH_3 - \ddot{B}r: + H_2O$$

$$\underset{\underset{H}{\overset{+}{\ddot{O}H}}}{\overset{:\ddot{O}H}{\underset{}{\underset{}{CH_3 - \overset{|}{\underset{|}{C}} - CH_3}}}} + H_2\ddot{O}: \longrightarrow \underset{:\ddot{O}H}{\overset{:\ddot{O}H}{\underset{}{CH_3 - \overset{|}{\underset{|}{C}} - CH_3}}} + H_3\overset{+}{O}:$$

Problem 3. Draw curved arrows to show the movement of the electrons that result in the formation of the given product(s).

a. $\underset{\underset{CH_3}{|}}{\overset{:\ddot{O}H}{\underset{}{CH_3 - \overset{|}{\underset{|}{C}} - \overset{+}{\underset{H}{\ddot{O}H}}}}} \longrightarrow \overset{+\ddot{O}H}{\underset{}{CH_3 - \overset{||}{C} - CH_3}} + H_2O$

b. $CH_3CH_2CH = CH_2 + H - Cl \longrightarrow CH_3CH_2\overset{+}{C}H - CH_3 + Cl^-$

c. $CH_3CH_2 - Br + \ddot{N}H_3 \longrightarrow CH_3CH_2 - \overset{+}{N}H_3 + Br^-$

d. $\underset{\underset{H}{|}}{\overset{CH_3}{\underset{}{CH_3\overset{|}{C} - \overset{+}{C}HCH_3}}} \longrightarrow \overset{CH_3}{\underset{+}{\underset{}{CH_3\overset{|}{C} - CH_2CH_3}}}$

Problem 4. Draw curved arrows to show the movement of the electrons that result in formation of the given products.

a. $CH_3CH=CHCH_3$ + $H-\overset{+}{\underset{H}{\overset{..}{O}}}-H$ ⟶ $CH_3\overset{+}{C}H-CH_2CH_3$ + $H_2\overset{..}{O}:$

b. $CH_3CH_2CH_2CH_2-\overset{..}{\underset{..}{C}l}:$ + $^-\overset{..}{C}\equiv N$ ⟶ $CH_3CH_2CH_2CH_2-C\equiv N$ + $:\overset{..}{\underset{..}{C}l}:^-$

c. $CH_3-\overset{:\overset{..}{O}H}{\underset{OH}{\overset{|}{\underset{|}{C}}}}-\overset{+..}{O}CH_3\overset{|}{H}$ ⟶ $CH_3-\overset{+\overset{..}{O}H}{\overset{||}{C}}-OH$ + $CH_3\overset{..}{O}H$

d. $CH_3-\overset{\overset{..}{O}:}{\overset{||}{C}}-H$ + CH_3-MgBr ⟶ $CH_3-\overset{:\overset{..}{O}:^-}{\underset{CH_3}{\overset{|}{\underset{|}{C}}}}-H$ + ^+MgBr

Problem 5. Draw curved arrows to show the movement of the electrons that result in formation of the given product(s).

a. $CH_3-\overset{O}{\overset{||}{C}}-CH_3$ + CH_3CH_2-MgBr ⟶ $CH_3-\overset{O^-}{\underset{CH_2CH_3}{\overset{|}{\underset{|}{C}}}}-CH_3$ + ^+MgBr

b. $CH_3CH_2CH_2-Br$ + $CH_3\overset{..}{\underset{..}{O}}:^-$ ⟶ $CH_3CH_2CH_2-\overset{..}{\underset{..}{O}}CH_3$ + Br^-

c.

d. $CH_3-\overset{:\overset{..}{O}:^-}{\underset{CH_3}{\overset{|}{\underset{|}{C}}}}-OCH_2CH_3$ ⟶ $CH_3-\overset{:\overset{..}{O}}{\overset{||}{C}}-CH_3$ + $CH_3CH_2O^-$

Problem 6. Draw curved arrows to show the movement of the electrons that result in formation of the given product(s).

a. $H\ddot{O}\!:^-$ + $CH_3CH\!-\!\overset{\overset{\displaystyle Br}{|}}{C}HCH_3$ $\longrightarrow$ $CH_3CH\!=\!CHCH_3$ + $H_2\ddot{O}\!:$ + Br^-
         $\underset{\displaystyle H}{|}$

b. $CH_3CH_2C\!\equiv\!C\!-\!H$ + $:\!\ddot{N}H_2$ $\longrightarrow$ $CH_3CH_2C\!\equiv\!C\!:^-$ + $\ddot{N}H_3$

c. $CH_3\overset{\overset{\displaystyle CH_3}{|}}{\underset{\underset{\displaystyle CH_3}{|}}{\overset{}{C}}}\!\!-\!CHCH_2CH_3$ $\longrightarrow$ $CH_3\overset{\overset{\displaystyle CH_3}{|}}{\underset{+}{C}}\!\!-\!\overset{}{\underset{\underset{\displaystyle CH_3}{|}}{C}}HCH_2CH_3$
         $\underset{+}{}$

d. $CH_2\!-\!\overset{\overset{\displaystyle CH_3}{|}}{\underset{+}{C}}CH_3$ + $H_2\ddot{O}\!:$ $\longrightarrow$ $CH_2\!=\!\overset{\overset{\displaystyle CH_3}{|}}{C}CH_3$ + $H_3\overset{+}{\ddot{O}}$
     $\underset{\displaystyle H}{|}$

Problem 7. Draw curved arrows to show the movement of the electrons that result in formation of the given products.

$CH_3CH_2\ddot{O}H$ + $H\!-\!\overset{\overset{\displaystyle H}{|}}{\underset{\underset{\displaystyle H}{|}}{\overset{+}{\ddot{O}}}}\!-\!H$ $\rightleftharpoons$ $CH_3CH_2\underset{+}{\overset{\overset{\displaystyle H}{|}}{\ddot{O}}}H$ + $H_2\ddot{O}\!:$

$CH_3\overset{+}{\ddot{N}}H_2$ + $H_2\underset{..}{O}\!:$ $\rightleftharpoons$ CH_3NH_2 + $H_3\overset{+}{\ddot{O}}\!:$
     $\underset{\displaystyle H}{|}$

Curved arrows are used to show the movement of electrons in each step of a reaction. For each step, you can tell how to draw the arrows by looking at the products that are formed in that step as a result of electron movement.

Problem 8. Draw curved arrows to show the movement of the electrons in each step of the following reaction sequences.

a. $CH_3CH\!=\!CH_2$ + $H\!-\!\ddot{B}r\!:$ $\longrightarrow$ $CH_3\overset{}{\underset{+}{C}H}\!-\!CH_3$ + $:\!\ddot{B}r\!:^-$ $\longrightarrow$ $CH_3\overset{}{\underset{\underset{\displaystyle :\ddot{B}r:}{|}}{C}H}\!-\!CH_3$

b. $CH_3\overset{\overset{\displaystyle CH_3}{|}}{\underset{\underset{\displaystyle CH_3}{|}}{C}}\!\!-\!Cl$ $\rightleftharpoons$ $CH_3\overset{\overset{\displaystyle CH_3}{|}}{\underset{\underset{\displaystyle CH_3 \, + \, Cl^-}{|}}{C}}+$ $\xrightarrow{\ddot{N}H_3}$ $CH_3\overset{\overset{\displaystyle CH_3}{|}}{\underset{\underset{\displaystyle CH_3}{|}}{C}}\!\!-\!\overset{+}{N}H_3$

c. $CH_3-\overset{\overset{\displaystyle :\ddot{O}:}{\|}}{C}-Cl$ + $H\ddot{O}:^-$ $\longrightarrow$ $CH_3-\overset{\overset{\displaystyle :\ddot{O}:^-}{|}}{\underset{\underset{\displaystyle :OH}{|}}{C}}-Cl$ $\longrightarrow$ $CH_3-\overset{\overset{\displaystyle \ddot{O}:}{\|}}{C}-OH$ + Cl^-

Problem 9. Draw curved arrows to show the movement of the electrons in each step of the following reaction sequences.

a.

b. $CH_3CH_2CH{=}CH_2$ $\xrightarrow{\overset{\overset{\displaystyle H}{|}}{\underset{\displaystyle CH_3\overset{+}{\ddot{O}}H}{}}}$ $CH_3CH_2\overset{+}{\underset{\displaystyle +}{C}}HCH_3$ $\xrightarrow[CH_3\ddot{O}H]{}$ $CH_3CH_2CHCH_3$

$CH_3CH_2CHCH_3$ with $\overset{\displaystyle :OCH_3}{\underset{\displaystyle +}{|}}$ and H

$\xrightarrow[CH_3\ddot{O}H|]{}$

$CH_3\overset{+}{\ddot{O}}H_2$ + $CH_3CH_2CHCH_3$ with OCH_3

c. $R-\overset{\overset{\displaystyle :\ddot{O}}{\|}}{C}-OCH_3$ + CH_3-MgBr $\longrightarrow$ $R-\overset{\overset{\displaystyle :\ddot{O}:^-}{|}}{\underset{\underset{\displaystyle CH_3}{|}}{C}}-OCH_3$ $\longrightarrow$ $R-\overset{\overset{\displaystyle :\ddot{O}}{\|}}{C}-CH_3$ + CH_3O^-

Problem 10. Use what the curved arrows tell you about electron movement to determine the product of each reaction step.

a. $CH_3CH_2\ddot{O}:^-$ + CH_3—$\ddot{B}r:$ ————→

b. CH_3—$\overset{\overset{\displaystyle +\ddot{O}H}{\|}}{C}$—$OCH_3$ + $H_2\ddot{O}:$ ————→

c. $H\ddot{O}:^-$ + CH_3CH_2CH—CH_2—Br ————→
     $\qquad\qquad\qquad\qquad$ |
     $\qquad\qquad\qquad\qquad$ H

d. CH_3CH_2—$\overset{\overset{\displaystyle :\ddot{O}:^-}{|}}{\underset{\underset{\displaystyle OH}{|}}{C}}$—$NH_2$ ————→

e. CH_3CH_2—$\overset{\overset{\displaystyle O}{\|}}{C}$—$H$ + CH_3—$MgBr$ ————→

f. CH_3—$\overset{\overset{\displaystyle CH_3}{|}}{\underset{\underset{\displaystyle CH_3}{|}}{C}}$—$\overset{+}{O}H$ ————→
     $\qquad\qquad\qquad\quad$ H

g. CH_3—$\overset{\overset{\displaystyle :\ddot{O}H}{|}}{\underset{\underset{\displaystyle OH}{|}}{C}}$—$\overset{+}{O}CH_3$ ————→
     $\qquad\qquad\qquad\quad$ H

Solutions to Problems on Electron Pushing

Problem 1

a. $CH_3CH_2\underset{CH_3}{\overset{CH_3}{C}}$—$\ddot{Br}:$ $\longrightarrow$ $CH_3CH_2\underset{CH_3}{\overset{CH_3}{C}}+$ + $:\ddot{Br}:^-$

b. $\longrightarrow$ +   + $:\ddot{C}l:^-$

c. $\overset{+}{\underset{H}{\ddot{O}H}}$ $\longrightarrow$ +   + $H_2\ddot{O}:$

d. $CH_3\overset{\frown}{C}H_2$—$MgBr$ $\longrightarrow$ $CH_3\overset{-}{\ddot{C}}H_2$ + ^+MgBr

e. $CH_3CH_2\underset{+}{C}HCH_3$ + $:\ddot{Br}:^-$ $\longrightarrow$ $CH_3CH_2\underset{:\ddot{Br}:}{C}HCH_3$

Problem 2

a. $CH_3CH_2\underset{CH_3}{\overset{CH_3}{C}}$—$Br$ $\longrightarrow$ $CH_3CH_2\underset{CH_3}{\overset{CH_3}{C}}+$ + Br^-

b. $\longrightarrow$ +   + Cl^-

c. $\underset{\overset{+|}{H}}{OH}$ $\longrightarrow$ +   + H_2O

d. $CH_3\overset{\frown}{C}H_2$—$MgBr$ $\longrightarrow$ $CH_3\overset{-}{C}H_2$ + ^+MgBr

e. $CH_3CH_2\underset{+}{C}HCH_3$ + $:\ddot{Br}:^-$ $\longrightarrow$ $CH_3CH_2\underset{Br}{C}HCH_3$

Problem 3

a. CH_3—$\underset{CH_3}{\overset{\overset{\displaystyle :\ddot{O}H}{|}}{C}}$—$\overset{+}{O}H$ $\longrightarrow$ CH_3—$\overset{\overset{\displaystyle ^+\ddot{O}H}{||}}{C}$—$CH_3$ + H_2O

b. $CH_3CH_2CH{=}CH_2$ + $H{-}Cl$ $\longrightarrow$ $CH_3CH_2\overset{+}{C}H{-}CH_3$ + Cl^-

c. $CH_3CH_2{-}Br$ + $\overset{..}{N}H_3$ $\longrightarrow$ $CH_3CH_2{-}\overset{+}{N}H_3$ + Br^-

d.

$$\underset{\underset{H}{|}}{CH_3\overset{\overset{CH_3}{|}}{C}{-}\overset{+}{C}HCH_3} \longrightarrow CH_3\overset{\overset{CH_3}{|}}{\underset{+}{C}}{-}CH_2CH_3$$

Problem 4

a. $CH_3CH{=}CHCH_3$ + $H{-}\underset{\underset{H}{|}}{\overset{+}{\overset{..}{O}}}{-}H$ $\longrightarrow$ $CH_3\overset{+}{C}H{-}CH_2CH_3$ + $H_2\overset{..}{O}{:}$

b. $CH_3CH_2CH_2CH_2{-}\overset{..}{\underset{..}{C}l}{:}$ + $^-\overset{..}{C}{\equiv}N$ $\longrightarrow$ $CH_3CH_2CH_2CH_2{-}C{\equiv}N$ + $:\overset{..}{\underset{..}{C}l}:^-$

c.

$$CH_3{-}\overset{\overset{:\overset{..}{O}H}{|}}{\underset{\underset{OH}{|}}{C}}{-}\overset{+}{\underset{\underset{H}{|}}{\overset{..}{O}}}CH_3 \longrightarrow CH_3{-}\overset{\overset{+\overset{..}{O}H}{\|}}{C}{-}OH + CH_3\overset{..}{O}H$$

d.

$$CH_3{-}\overset{\overset{\overset{..}{O}{:}}{\|}}{C}{-}H + CH_3{-}MgBr \longrightarrow CH_3{-}\underset{\underset{CH_3}{|}}{\overset{\overset{:\overset{..}{O}{:}^-}{|}}{C}}{-}H + {}^+MgBr$$

Problem 5

a.

$$CH_3{-}\overset{\overset{O}{\|}}{C}{-}CH_3 + CH_3CH_2{-}MgBr \longrightarrow CH_3{-}\underset{\underset{CH_2CH_3}{|}}{\overset{\overset{O^-}{|}}{C}}{-}CH_3 + {}^+MgBr$$

b. $CH_3CH_2CH_2{-}Br + CH_3\overset{..}{O}{:}^- \longrightarrow CH_3CH_2CH_2{-}OCH_3 + Br^-$

c.

d.

Problem 6

a.

b.

$$CH_3CH_2C\equiv C{-}H \quad + \quad :\overset{..}{N}H_2 \quad \longrightarrow \quad CH_3CH_2C\equiv \overset{..}{C}{^-} \quad + \quad :NH_3$$

c.

d.

Problem 7

$$CH_3\overset{+}{N}H_2 \quad + \quad H_2\overset{..}{O}: \quad \rightleftharpoons \quad CH_3NH_2 \quad + \quad H_3\overset{+}{O}:$$

Problem 8

a. $CH_3CH=CH_2$ + H—Br: $\longrightarrow$ $CH_3CH—CH_3$ + :Br:$^-$ $\longrightarrow$ $CH_3CH—CH_3$

:Br:

b. $CH_3C—Cl$ (with two CH_3 groups) $\rightleftharpoons$ CH_3C^+ (with two CH_3 groups) $+ Cl^-$ $\xrightarrow{:NH_3}$ $CH_3C—\overset{+}{N}H_3$ (with two CH_3 groups)

c. $CH_3—\overset{\overset{:O:}{\|}}{C}—Cl$ + HO:$^-$ $\longrightarrow$ $CH_3—\overset{\overset{:O:^-}{|}}{\underset{:OH}{C}}—Cl$ $\longrightarrow$ $CH_3—\overset{\overset{:O:}{\|}}{C}—OH$ + Cl^-

Problem 9

a. [cyclohexane with two CH₃ groups and OH, H] + H_2O^+ $\rightleftharpoons$ [cyclohexane with two CH₃, $\overset{+}{O}H$, H H + H_2O] $\rightleftharpoons$ [cyclohexane with two CH₃, $^+$ H] + $H_2\ddot{O}$:

[cyclohexane with two CH₃ and CH₃, H₃O⁺] $\xleftarrow{H_2\ddot{O}:}$ [cyclohexane with CH₃, $^+$, CH₃, H]

+ H_3O^+

b. $CH_3CH_2CH=CH_2$ $\xrightarrow{CH_3\overset{+}{O}H}$ $CH_3CH_2\overset{+}{C}HCH_3$ $\xrightarrow{CH_3\ddot{O}H}$ $CH_3CH_2CHCH_3$

$\overset{+}{:}OCH_3$

H

$CH_3\ddot{O}H$

$CH_3\overset{+}{O}H_2$ + $CH_3CH_2CHCH_3$

OCH_3

c. $R-\overset{:\overset{..}{O}:}{\underset{}{C}}-OCH_3$ + CH_3-MgBr ⟶ $R-\overset{:\overset{..}{O}:^-}{\underset{\underset{CH_3}{|}}{C}}-OCH_3$ ⟶ $R-\overset{:\overset{..}{O}}{\underset{}{C}}-CH_3$ + CH_3O^-

Problem 10

a. $CH_3CH_2\overset{..}{\underset{..}{O}}:^-$ + $CH_3\!-\!\overset{..}{\underset{..}{Br}}:$ ⟶ $CH_3CH_2OCH_3$ + Br^-

b. $CH_3-\overset{+\overset{..}{O}H}{\underset{}{C}}-OCH_3$ + $H_2\overset{..}{O}:$ ⟶ $CH_3-\overset{OH}{\underset{\underset{H}{\overset{+}{O}H}}{C}}-OCH_3$

c. $H\overset{..}{\underset{..}{O}}:^-$ + $CH_3CH_2\overset{|}{\underset{H}{CH}}\!-\!CH_2\!-\!Br$ ⟶ $CH_3CH_2CH{=}CH_2$ + H_2O + Br^-

d. $CH_3CH_2-\overset{:\overset{..}{O}:^-}{\underset{\underset{OH}{|}}{C}}-NH_2$ ⟶ $CH_3CH_2-\overset{O}{\underset{}{C}}-NH_2$ + HO^-

e. $CH_3CH_2-\overset{O}{\underset{}{C}}-H$ + CH_3-MgBr ⟶ $CH_3CH_2-\overset{O^-}{\underset{\underset{CH_3}{|}}{C}}-H$ + ^+MgBr

f. $CH_3-\overset{CH_3}{\underset{\underset{CH_3}{|}}{\overset{|}{C}}}-\overset{+}{\underset{H}{O}}H$ ⟶ $CH_3-\overset{CH_3}{\underset{\underset{CH_3}{|}}{\overset{|}{C}}}{}^+$ + H_2O

g. $CH_3-\overset{:\overset{..}{O}H}{\underset{\underset{OH}{|}}{\overset{|}{C}}}-\overset{+}{O}CH_3$ ⟶ $CH_3-\overset{+\overset{..}{O}H}{\underset{}{C}}-OH$ + CH_3OH

SPECIAL TOPIC III

Kinetics

These are problems based on the discussion on kinetics found in Section 3.7 of the text. The rate laws are derived in Appendix III on page A-6 of the text.

Problems

1. How long would it take for the reactant of a first-order reaction to decrease to one-half its initial concentration if the rate constant is 4.5×10^{-3} s^{-1} and the initial concentration of the reactant is:

    a. 1.0 M?　　　　　**b.** 0.50 M?

2. How long would it take for the reactants of a second-order reaction to decrease to one-half their initial concentration if the rate constant is 2.3×10^{-2} M^{-1} s^{-1} and the initial concentration of both reactants is:

    a. 1.0 M?　　　　　**b.** 0.50 M?

3. How many half-lives are required for a first-order reaction to reach >99% completion?

4. The initial concentration of a reactant undergoing a first-order reaction is 0.40 M. After 5 minutes, the concentration of the reactant is 0.27 M; after an additional 5 minutes, the concentration of the reactant is 0.18 M; and after an additional 5 minutes, the concentration of the reactant is 0.12 M.

    a. What is the average rate of the reaction during each five-minute interval?

    b. What is the rate constant of the reaction?

5. What percentage of a compound, undergoing a first-order reaction with a rate constant of 2.70×10^{-5} s^{-1}, would have reacted at the end of two hours?

6. How long would it take for a first-order reaction with a rate constant of 5.30×10^{-4} s^{-1} to reach 70% completion?

7. The following data were obtained in a study of the rate of inversion of sucrose at 25 °C. The initial concentration of sucrose was 1.00 M.

Time (minutes)	0	30	60	90	130	180
Sucrose inverted (M)	0	0.100	0.195	0.277	0.373	0.468

    a. What is the order of the reaction? (Assume that the reaction is either first order or second order.)

    b. What is the rate constant of the reaction?

8. Calculate the activation energy of a first-order reaction that is 20% complete in 15 minutes at 40 °C and is 20% complete in 3 minutes at 60 °C.

9. Analysis of a hydrolysis reaction occurring in a dilute aqueous solution of sucrose shows that 80 grams of the original 100 grams of sucrose remain after 10 hours. At this rate, how much sucrose would be left after 24 hours? (Assume a pseudo first-order reaction; see page A-8 of the text.)

Solutions to Problems on Kinetics

1. **a.** half-life of a first-order reaction $= t_{1/2} = \dfrac{\ln 2}{k_1}$ (See page A-7 of the text.)

$$= \frac{0.693}{k_1}$$

$$= \frac{0.693}{4.5 \times 10^{-3} \text{ s}^{-1}}$$

$$= 154 \text{ seconds}$$

$$= 2 \text{ minutes, } 34 \text{ seconds}$$

b. The half-life of a first-order reaction is independent of concentration, so the answer is the same as the answer for **a** (2 minutes, 34 seconds).

2. **a.** half-life of a second-order reaction $= t_{1/2} = \dfrac{1}{k_2 a}$ (See page A-8 of the text.)

$$= \frac{1}{2.3 \times 10^{-2} \text{ M}^{-1} \text{ s}^{-1} \, (1.00 \text{ M})}$$

$$= 43 \text{ seconds}$$

b. $t_{1/2} = \dfrac{1}{2.3 \times 10^{-2} \text{ M}^{-1} \text{ s}^{-1} \, (0.50 \text{ M})}$

$$= \frac{1}{1.5 \times 10^{-2} \text{ s}^{-1}}$$

$$= 87 \text{ seconds}$$

3. One-half of the compound reacts during the first half-life. One half of what is left after the first half-life reacts during the second half-life. One half of what is left after the second half-life reacts during the third half-life, etc. The following table shows that seven half-lives are required to reach >99% completion.

Number of half-lives	Percentage completion	
1	0.50 (100) = 50	50
2	0.50 (50) = 25	75
3	0.50 (25) = 12.5	87.5
4	0.50 (12.5) = 6.25	93.8
5	0.50 (6.25) = 3.125	96.9
6	0.50 (3.125) = 1.5625	98.4
7	0.50 (1.5625) = 0.78125	99.2

4.   **a.** $$\text{rate} = \frac{\text{change in concentration}}{\text{change in time}}$$

1ˢᵗ interval: $\text{rate} = \dfrac{0.40\,\text{M} - 0.27\,\text{M}}{5\,\text{min}} = \dfrac{0.13\,\text{M}}{5\,\text{min}} = 2.60 \times 10^{-2}\,\text{M min}^{-1}$

2ⁿᵈ interval: $\text{rate} = \dfrac{0.27\,\text{M} - 0.18\,\text{M}}{5\,\text{min}} = \dfrac{0.09\,\text{M}}{5\,\text{min}} = 1.80 \times 10^{-2}\,\text{M min}^{-1}$

3ʳᵈ interval: $\text{rate} = \dfrac{0.18\,\text{M} - 0.12\,\text{M}}{5\,\text{min}} = \dfrac{0.06\,\text{M}}{5\,\text{min}} = 1.20 \times 10^{-2}\,\text{M min}^{-1}$

Thus, we see that the **rate** of the reaction decreases with decreasing concentration.

b.   rate $= k$ [reactant], using the average concentration of the reactant during the 5-minute interval.

1ˢᵗ interval: $2.60 \times 10^{-2}\,\text{M min}^{-1} = k\,(0.335\,\text{M})$

$k = 7.8 \times 10^{-2}\,\text{min}^{-1}$

2ⁿᵈ interval: $1.80 \times 10^{-2}\,\text{M min}^{-1} = k\,(0.225\,\text{M})$

$k = 8.0 \times 10^{-2}\,\text{min}^{-1}$

3ʳᵈ interval: $1.20 \times 10^{-2}\,\text{M min}^{-1} = k\,(0.15\,\text{M})$

$k = 8.0 \times 10^{-2}\,\text{min}^{-1}$

The **rate constant**, as its name indicates, is constant during the course of the reaction.

5. a = the initial concentration (set at 100%)
x = the concentration that has reacted at time $= t$

$$\ln\frac{a}{a - x} = k_1 t \quad \text{(See page A-7 of the text.)}$$

$$\ln\frac{100}{100 - x} = 2.70 \times 10^{-5}\,\text{s}^{-1}\,(2\,\text{hours})$$

$$\ln\frac{100}{100 - x} = 2.70 \times 10^{-5}\,\text{s}^{-1}\,(7200\,\text{seconds})$$

$$\ln\frac{100}{100 - x} = 0.194$$

$$\frac{100}{100 - x} = 121$$

$$100 = 121 - 1.21x$$

$$1.21\,x = 21$$

$$x = 17.4\%$$

6.

$$\ln \frac{a}{a-x} = k_1 t$$

$$\ln \frac{100}{100-70} = 5.30 \times 10^{-4} \text{ s}^{-1} \, t$$

$$t = \frac{\ln 3.33}{5.30 \times 10^{-4} \text{ s}^{-1}}$$

$$t = \frac{1.20}{5.30 \times 10^{-4} \text{ s}^{-1}}$$

$$t = 2260 \text{ seconds}$$

$$t = 37.7 \text{ minutes}$$

7. To determine the order of the reaction, we will calculate the first-order and second-order rate constants based on the data provided. Since the rate constant for a reaction should be the same regardless of the set of data we use, the calculations will tell us the order of the reaction.

1st order

$$k_1 = \frac{-\ln \frac{a-x}{a}}{t}$$

$$k_1 = \frac{-\ln \frac{1.000 - 0.100}{1.000}}{30}$$

$$= \mathbf{3.51 \times 10^{-3}}$$

$$k_2 = \frac{-\ln \frac{1.000 - 0.195}{1.000}}{60}$$

$$= \mathbf{3.62 \times 10^{-3}}$$

$$k_1 = \frac{-\ln \frac{1.000 - 0.277}{1.000}}{90}$$

$$= \mathbf{3.60 \times 10^{-3}}$$

$$k_1 = \frac{-\ln \frac{1.000 - 0.373}{1.000}}{130}$$

$$= \mathbf{3.59 \times 10^{-3}}$$

$$k_1 = \frac{-\ln \frac{1.000 - 0.468}{1.000}}{180}$$

$$= \mathbf{3.51 \times 10^{-3}}$$

2nd order

$$k_2 = \frac{\frac{1}{a-x} - \frac{1}{a}}{t}$$

$$k_2 = \frac{\frac{1}{1.000 - 0.100} - \frac{1}{1.000}}{30}$$

$$= \frac{1.111 - 1.000}{30}$$

$$= \mathbf{3.70 \times 10^{-3}}$$

$$k_2 = \frac{\frac{1}{1.000 - 0.195} - 1.000}{60}$$

$$= \mathbf{4.03 \times 10^{-3}}$$

$$k_2 = \frac{\frac{1}{1.000 - 0.277} - 1.000}{90}$$

$$= \mathbf{4.26 \times 10^{-3}}$$

$$k_2 = \frac{\frac{1}{1.000 - 0.373} - 1.000}{130}$$

$$= \mathbf{4.57 \times 10^{-3}}$$

$$k_2 = \frac{\frac{1}{1.000 - 0.468} - 1.000}{180}$$

$$= \mathbf{4.89 \times 10^{-3}}$$

a. Because the calculated rate constants are relatively constant when the data are plugged into a first-order equation but vary considerably when the data are plugged into a second-order equation, one can conclude that the reaction is first-order.

b. $3.59 \times 10^{-3} \text{ min}^{-1}$

8. $k \text{ at } 40\,^{\circ}\text{C} = \dfrac{\ln \dfrac{1.00}{1.00-0.20}}{15 \text{ min}} = 1.49 \times 10^{-2} \text{ min}^{-1}$

$k \text{ at } 60\,^{\circ}\text{C} = \dfrac{\ln \dfrac{1.00}{1.00-0.20}}{3 \text{ min}} = 7.44 \times 10^{-2} \text{ min}^{-1}$

Using the equation for calculating kinetic parameters on page 153 of the text:

$$\ln k_2 - \ln k_1 = \frac{-E_a}{R}\left(\frac{1}{T_2} - \frac{1}{T_1}\right) (40\,^{\circ}\text{C} = 313 \text{ K}; 60\,^{\circ}\text{C} = 333 \text{ K})$$

$$\ln (7.44 \times 10^{-2}) - \ln (1.49 \times 10^{-2}) = \frac{-E_a}{1.986 \times 10^{-3} \text{ kcal}}\left(\frac{1}{333} - \frac{1}{313}\right)$$

$$-2.60 - (-4.21) = \frac{-E_a}{1.986 \times 10^{-3}}(0.00300 - 0.00319)$$

$$1.61 = \frac{-E_a}{1.986 \times 10^{-3}}(-0.00019)$$

$$E_a = \frac{1.61 \times 1.986 \times 10^{-3}}{0.00019}$$

$$E_a = 16.8 \text{ kcal/mol}$$

9.   Sucrose is hydrolyzed to form a mixture of glucose and fructose. Because there is excess water (it is the solvent), the reaction is a pseudo first-order reaction.

First, the rate constant of the reaction must be determined:

$$\ln \frac{a}{a-x} = k_1 t \quad \text{(See page A-7 of the text.)}$$

$$\ln \frac{100}{80} = k_1 \times 10 \text{ hours}$$

$$k_1 = 2.23 \times 10^{-2} \text{ hr}^{-1}$$

Now we can calculate the amount of sucrose that has reacted, and therefore, the amount that would be left.

$$\ln \frac{a}{a-x} = k_1 t$$

$$\ln \frac{100}{100-x} = 2.23 \times 10^{-2} \, (24 \, \text{hr})$$

$$\ln \frac{100}{100-x} = 0.535$$

$$\frac{100}{100-x} = 1.71$$

$$100 = 171 - 1.71x$$

$$1.71x = 71$$

$$x = 41.5 \, \text{g have reacted}$$

Therefore, 58.5 g would be left.

CHAPTER 4
The Reactions of Alkenes

Important Terms

acid-catalyzed reaction
a reaction catalyzed by an acid.

alkoxymercuration-reduction
addition of an alcohol to an alkene using mercuric acetate followed by sodium borohydride and sodium hydroxide.

carbocation rearrangement
the rearrangement of a carbocation to a more stable carbocation.

catalyst
a compound that increases the rate at which a reaction occurs without being consumed in the reaction. Because it does not change the equilibrium constant of the reaction, it does not change the amount of product that is formed.

catalytic hydrogenation
the addition of hydrogen to a double or a triple bond with the aid of a metal catalyst.

concerted reaction
a reaction in which all the bond-making and bond-breaking processes take place in a single step.

constitutional isomers (structural isomers)
molecules that have the same molecular formula but differ in the way the atoms are connected.

dimer
a molecule formed by joining together two identical molecules.

electrophilic addition reaction
an addition reaction in which the first species that adds to the reactant is an electrophile.

epoxide (oxirane)
an ether in which the oxygen is incorporated into a three-membered ring.

halohydrin
an organic molecule that contains a halogen atom and an OH group.

Hammond postulate
states that the transition state will be more similar in structure to the species (reactants or products) that it is closer to energetically.

heat of hydrogenation
the heat ($\Delta H°$) released in a hydrogenation reaction.

heterogeneous catalyst
a catalyst that is insoluble in the reaction mixture.

hormone
a compound that controls growth and other changes in tissues.

hydration
addition of water to a compound.

1,2-hydride shift
the movement of a hydride ion (a hydrogen with its pair of bonding electrons) from one carbon to an adjacent carbon.

hydroboration-oxidation
the addition of borane to an alkene (or to an alkyne) followed by reaction with hydrogen peroxide and hydroxide ion.

hydrogenation addition of hydrogen.

hyperconjugation delocalization of electrons by overlap of carbon-hydrogen or carbon-carbon σ bonds with an empty p orbital.

Markovnikov's rule the actual rule is: "When a hydrogen halide adds to an asymmetrical alkene, the addition occurs such that the halogen attaches itself to the carbon atom of the alkene bearing the least number of hydrogen atoms." Chemists use the rule as follows: The hydrogen adds to the sp^2 carbon that is bonded to the greater number of hydrogens.

A more general rule is: The electrophile adds to the sp^2 carbon that is bonded to the greater number of hydrogens.

mechanism of the reaction a description of the step-by-step process by which reactants are changed into products.

1,2-methyl shift the movement of a methyl group with its bonding electrons from one carbon to an adjacent carbon.

oxidation reaction a reaction that decreases the number of C—H bonds in the reactant or increases the number of C—O, C—N, or C—X bonds (X denotes a halogen).

oxymercuration-reduction addition of water to an alkene using a mercuric salt of a carboxylic acid as a catalyst, followed by reaction with sodium borohydride.

pheromone a chemical substance used for the purpose of communication.

primary carbocation a carbocation with the positive charge on a primary carbon.

reduction reaction a reaction that increases the number of C—H bonds in the reactant or decreases the number of C—O, C—N, or C—X bonds (X denotes a halogen).

regioselective reaction a reaction that leads to the preferential formation of one constitutional isomer over another.

secondary carbocation a carbocation with the positive charge on a secondary carbon.

steric effect an effect due to the space occupied by a substituent.

steric hindrance refers to bulky groups at the site of a reaction that make it difficult for the reactants to approach one another.

tertiary carbocation a carbocation with the positive charge on a tertiary carbon.

vicinal vicinal substituents are substituents on adjacent carbons.

Solutions to Problems

1. **a.** The σ bond orbitals of the carbon adjacent to the positively charged carbon are available for overlap with the vacant *p* orbital. Because the methyl cation does not have a carbon adjacent to the positively charged carbon, there are no σ bond orbitals available for overlap with the vacant *p* orbital.

b. An ethyl cation is more stable because the carbon adjacent to the positively charged carbon has three σ bond orbitals available for overlap with the vacant *p* orbital, whereas a methyl cation does not have any σ bond orbitals available for overlap with the vacant *p* orbital.

2. **a.** **1.** 3   **2.** 3   **3.** 6
b. *sec*-butyl cation

$$\underset{+}{CH_3CH_2\overset{\overset{\displaystyle CH_3}{|}}{C}CH_3} > \underset{+}{CH_3CH_2CHCH_3} > \underset{+}{CH_3CH_2CH_2CH_2}$$

3. **a.**

b. A halogen atom decreases the stability of the carbocation because, since it is an electronegative atom, it withdraws electrons away from the positively charged carbon. This increases the concentration of positive charge on the carbocation which makes it less stable.

$$\underset{\overset{|}{CH_3}}{\underset{+}{CH_3CHCH_2CH_2}} > \underset{\overset{|}{Cl}}{\underset{+}{CH_3CHCH_2CH_2}} > \underset{\overset{|}{F}}{\underset{+}{CH_3CHCH_2CH_2}}$$

Because fluorine is more electronegative than chlorine, the fluorine-substituted carbocation is less stable than the chlorine-substituted carbocation.

4. The transition state will resemble the one (reactants or products) that it is closer to on the reaction coordinate diagram; that is, the one that it is closer to in energy.
a. products   **b.** reactants   **c.** reactants   **d.** products

5. **a.** $CH_3CH_2\underset{\overset{|}{Br}}{CH}CH_3$ **c.** (cyclopentane with CH3 and Br)   **e.** (cyclohexane with CH3 and Br)

b. $CH_3CH_2\underset{\overset{|}{Br}}{\overset{\overset{\displaystyle CH_3}{|}}{C}}CH_3$ **d.** $CH_3\underset{\overset{|}{Br}}{\overset{\overset{\displaystyle CH_3}{|}}{C}}CH_2CH_2CH_3$ **f.** $CH_3CH_2\underset{\overset{|}{Br}}{CH}CH_3$

6. **a.** $CH_2{=}\underset{\overset{|}{}}{\overset{\overset{\displaystyle CH_3}{|}}{C}}CH_3$ **c.** (cyclohexyl)$-\overset{\overset{\displaystyle CH_3}{|}}{C}{=}CH_2$

b. (cyclohexyl)$-CH_2CH{=}CH_2$ **d.** (cyclohexylidene)${=}CHCH_3$ or (cyclohexenyl)$-CH_2CH_3$

(cyclohexyl)$-CH{=}CHCH_3$

Would not be a good choice, because it would lead to a mixture of two products.

7. a.

$$CH_3CH_2\underset{\underset{\displaystyle CH_3}{|}}{C}=CH_2$$

In both **a** and **b**, the compound that is more highly regioselective is the one where the choice is between forming a tertiary carbocation or a primary carbocation.

b.

In both **a** and **b**, the less regioselective compound is the one where the choice is between forming a tertiary carbocation or a secondary carbocation, because the difference in the stability of the two possible carbocations, and therefore the difference in the amount of product formed, is not as great as it is when the choice is between a tertiary and a primary carbocation.

8.   As long as the pH is greater than about −2.5 and less than about 15, more than 50% of 2-propanol would be in its neutral, nonprotonated form.

$$\overset{+}{ROH_2} \quad \underset{\xrightarrow{pK_a = -2.5}}{\rightleftharpoons} \quad ROH \quad \underset{\xrightarrow{pK_a = \sim 15}}{\rightleftharpoons} \quad RO^-$$

Recall that when the pH = pK_a, half the compound is in its acidic form and half is in its basic form. Therefore at a pH less than −2.5, more than half of the compound will be in its positively charged protonated form. At a pH greater than ~15, more than half of the compound will exist as the negatively charged anion.

Therefore, at a pH between −2.5 and ~15, more than half of the compound will exist in the neutral nonprotonated form.

9. a. three transition states   **b.** two intermediates

   c. The first step is the slowest step, so it has the smallest rate constant; the second step is fast because no bonds are being broken; the third step is fast because transfer of a proton from or to an O or an N is always a fast reaction.

10. a. $CH_3CH_2CH_2\underset{\underset{\displaystyle OH}{|}}{C}HCH_3$ **c.** $CH_3CH_2CH_2CH_2\underset{\underset{\displaystyle OH}{|}}{C}HCH_3$ and $CH_3CH_2CH_2\underset{\underset{\displaystyle OH}{|}}{C}HCH_2CH_3$

formed in equal amounts

b.

d.

11. a. 1. $CH_3\underset{\underset{\displaystyle Cl}{|}}{\overset{\overset{\displaystyle CH_3}{|}}{C}}CH_3$ **2.** $CH_3\underset{\underset{\displaystyle Br}{|}}{\overset{\overset{\displaystyle CH_3}{|}}{C}}CH_3$ **3.** $CH_3\underset{\underset{\displaystyle OH}{|}}{\overset{\overset{\displaystyle CH_3}{|}}{C}}CH_3$ **4.** $CH_3\underset{\underset{\displaystyle OCH_3}{|}}{\overset{\overset{\displaystyle CH_3}{|}}{C}}CH_3$

   b. The first step in all the reactions is addition of an electrophilic proton (H^+) to the carbon of the CH_2 group.

A *tert*-butyl carbocation is formed as an intermediate in each of the reactions.

c. The nucleophile that adds to the *tert*-butyl carbocation is different in each reaction. In reactions #3 and #4, there is a third step—a proton is lost from the group that acted as the nucleophile in the second step of the reaction.

3.

$$CH_3\underset{\underset{H_2\ddot{O}\ H}{\overset{+}{O}H}}{\overset{CH_3}{\underset{|}{C}}}CH_3 \longrightarrow CH_3\underset{\overset{|}{OH}}{\overset{CH_3}{\underset{|}{C}}}CH_3 + H_3O^+$$

4.

$$CH_3\underset{\underset{CH_3\ddot{O}H\ H}{\overset{+}{O}CH_3}}{\overset{CH_3}{\underset{|}{C}}}CH_3 \longrightarrow CH_3\underset{\overset{|}{OCH_3}}{\overset{CH_3}{\underset{|}{C}}}CH_3 + CH_3\overset{+}{O}H_2$$

12. Solved in the text.

13. a.

cyclohexene $+ CH_3OH \xrightarrow{H_2SO_4}$ cyclohexyl$-OCH_3$

b.

$$CH_2{=}\underset{\underset{CH_3}{|}}{C}CH_3 + CH_3OH \xrightarrow{H_2SO_4} CH_3O\underset{\overset{|}{CH_3}}{\overset{CH_3}{\underset{|}{C}}}CH_3$$

c.

cyclopentene $+ H_2O \xrightarrow{H_2SO_4}$ cyclopentanol$-OH$

d. $CH_3CH{=}CHCH_3 + H_2O \xrightarrow{H_2SO_4} CH_3\underset{\overset{|}{OH}}{\underset{|}{C}H}CH_2CH_3$

or

$CH_2{=}CHCH_2CH_3 + H_2O \xrightarrow{H_2SO_4} CH_3\underset{\overset{|}{OH}}{\underset{|}{C}H}CH_2CH_3$

14.

$CH_3\underset{\overset{|}{CH_3}}{C}{=}CH_2 \xrightarrow{H{-}OSO_3H} CH_3\overset{+}{\underset{\overset{|}{CH_3}}{C}}CH_3 \longrightarrow CH_3\underset{\overset{|}{CH_3}}{CH}CH_2CH_2\overset{+}{O}{-}\underset{\overset{|}{CH_3}}{\overset{CH_3}{\underset{|}{C}}}CH_3$

B: is any base that is present in the reaction mixture.

$HB^+ + CH_3\underset{\overset{|}{CH_3}}{CH}CH_2CH_2O\underset{\overset{|}{CH_3}}{\overset{CH_3}{\underset{|}{C}}}CH_3$

15. Solved in the text.

16.   **a.**

b.

c.

d.

e.

f.

1-Bromo-3-methylcyclohexane and 1-bromo-4-methylcyclohexane will be obtained in approximately equal amounts because, in each case, the intermediate is a secondary carbocation.

17.

CH₃
|
CH₃CCH₂CH₃
|
Br

Addition of H⁺ would form a carbocation that could rearrange.

Addition of Br⁺ forms a cyclic bromonium ion rather than a carbocation, so there is no rearrangement.

18.   **a.**   The first step in the reaction of propene with Br_2 forms a cyclic bromonium ion, whereas the first step in the reaction of propene with HBr forms a carbocation.

   b.   If the bromide ion were to attack the positively charged bromine atom, a highly unstable compound (with a negative charge on carbon and a positive charge on bromine) would be formed.

Notice that the electrostatic potential map of the cyclic bromonium ion on page 153 of the text shows that the ring carbons are the least electron dense (most blue) atoms in the intermediate and therefore the ones most susceptible to nucleophilic attack.

19.   The nucleophile that is present in greater concentration is more apt to collide with the intermediate. Therefore, if the solvent is a nucleophile, the major product will come from reaction of the solvent with the intermediate, because the concentration of the solvent is much greater than the concentration of the other nucleophile. (For example, in part **a** the concentration of CH_3OH is much greater than the concentration of Cl^-. In part **b**, the nucleophile comes from two equivalents of NaI and one equivalent of HBr.)

a.

CH₃   CH₃
| |
ClCH₂CCH₃   and   ClCH₂CCH₃
| |
OCH₃   Cl
major

c.

CH₃CH₂CHCH₃   and   CH₃CH₂CHCH₃
| |
OH   Cl
major

b.

CH₃CHCH₃   and   CH₃CHCH₃
| |
I   Br
major

d.

CH₃CH₂CHCH₃   and   CH₃CH₂CHCH₃
| |
OCH₃   Br
major

20.   Sodium and potassium achieve an outer shell of eight electrons by losing the single electron they have in the $3s$ (in the case of Na) or $4s$ (in the case of K) orbital, thereby becoming Na^+ and K^+. In order to form a covalent bond, they would have to regain electrons in these orbitals, thereby losing the stability associated with having an outer shell of eight electrons and no extra electrons.

21.   Because chlorine is more electronegative than iodine, iodine will be the electrophile. Therefore, it will end up attached to the sp^2 carbon that is bonded to the greater number of hydrogens.

22. **a.** CH$_2$CHCH$_2$CH$_3$ **b.** CH$_2$CHCH$_2$CH$_3$ **c.** CH$_2$CHCH$_2$CH$_3$ **d.** CH$_2$CHCH$_2$CH$_3$
        | | | | | | | |
        Br Br   Br OH   Br OCH$_2$CH$_3$ Br OCH$_3$

23. Look at the reagent (Cl$_2$) and remember what that reagent does when it reacts with an alkene (do not let the rest of the molecule confuse you): when Cl$_2$ adds to an alkene, it forms a cyclic chloronium ion intermediate. You know that the intermediate then reacts with a nucleophile. There are two nucleophiles in the solution that can react with the intermediate, a Cl$^-$ and the OH group at the end of the molecule. There is a greater probability that the OH group will attack the chloronium ion because, since it is attached to the reactant, it does not have to wander through the solution to find the chloronium ion as the Cl$^-$ has to. Loss of a proton gives the six-membered ring ether.

24. Notice that both the addition of water and the addition of an alcohol to an alkene can be carried out using either an acid catalyst (Section 4.5) or by mercuration/demercuration (Section 4.9).

a.

or

b.

or

c. $CH_3CH{=}CHCH_3$ $\xrightarrow[CH_3CH_2OH]{H_2SO_4}$ $CH_3CHCH_2CH_3$

$\qquad\qquad\qquad\qquad\qquad\qquad\qquad\qquad\qquad\quad |$

$\qquad\qquad\qquad\qquad\qquad\qquad\qquad\qquad\qquad OCH_2CH_3$

or

$\xrightarrow[\text{2. NaBH}_4]{\text{1. Hg(O}_2\text{CCF}_3)_2,\ CH_3CH_2OH}$

$\qquad\qquad\qquad CH_3 \qquad\qquad\qquad\qquad\qquad CH_3$

$\qquad\qquad\qquad |\qquad\qquad\qquad\qquad\qquad\qquad\ |$

d. $CH_2{=}CCH_2CH_3$ $\xrightarrow[CH_3OH]{H_2SO_4}$ $CH_3CCH_2CH_3$

$\qquad\qquad\qquad\qquad\qquad\qquad\qquad\qquad\qquad\quad |$

$\qquad\qquad\qquad\qquad\qquad\qquad\qquad\qquad\qquad OCH_3$

or

$\xrightarrow[\text{2. NaBH}_4]{\text{1. Hg(O}_2\text{CCF}_3)_2,\ CH_3OH}$

25. **a.**

$\qquad\qquad CH_3 \qquad\qquad\qquad CH_3 \qquad\ \ \text{1,2-hydride}\ \ \qquad\ CH_3 \qquad\qquad\qquad CH_3$

$CH_2{=}CHCHCH_3 \xrightarrow{H_2SO_4} CH_3\overset{+}{C}HCHCH_3 \xrightarrow{\text{shift}} CH_3CH_2\overset{+}{C}CH_3 \xrightarrow{H_2O} CH_3CH_2CCH_3$

$\qquad\qquad\qquad\qquad\qquad\qquad\qquad\qquad\qquad\qquad\qquad\qquad\qquad\qquad\qquad\qquad\qquad\qquad\qquad |$

$\qquad\qquad\qquad\qquad\qquad\qquad\qquad\qquad\qquad\qquad\qquad\qquad\qquad\qquad\qquad\qquad\qquad\qquad\qquad OH$

b.

$\qquad\qquad CH_3 \qquad\qquad\qquad\qquad\qquad\qquad\qquad\qquad\qquad CH_3$

$CH_2{=}CHCHCH_3 \xrightarrow[\text{2. NaBH}_4]{\text{1. Hg(O}_2\text{CCH}_3)_2,\ H_2O,\ THF} CH_3CHCHCH_3$

$\qquad\qquad\qquad\qquad\qquad\qquad\qquad\qquad\qquad\qquad\qquad\qquad\qquad\qquad |$

$\qquad\qquad\qquad\qquad\qquad\qquad\qquad\qquad\qquad\qquad\qquad\qquad\qquad OH$

In **a**, a carbocation rearrangement is required to get the desired product from the given starting material, so the product must be prepared using a reaction that forms a carbocation intermediate.

In **b,** the desired product must be obtained from a reaction that has an intermediate that will not rearrange. Thus, it cannot be prepared from a reaction that forms a carbocation intermediate.

26. **a.**

$CH_2CH_2CH_3$

c.

$H_3C \quad O \quad CH_3$
$H_3C \qquad\quad CH_3$

b.

d.

H_3C
$H_3C \qquad CH_2CH_3$

27. **a.** cyclohexene   **b.** 1-butene

28. Because one mole of BH_3 reacts with three moles of an alkene, one-third of a mole of BH_3 is needed to react with each mole of alkene. Therefore, two-thirds of a mole of BH_3 is needed to react with two moles of an alkene (in this case, 1-pentene).

29. **a.**

$\qquad\quad CH_3 \qquad\qquad\qquad\qquad\qquad\qquad CH_3$

$CH_3C{=}CHCH_3 \xrightarrow[\text{2. HO}^-,\ H_2O_2,\ H_2O]{\text{1. BH}_3/\text{THF}} CH_3CHCHCH_3$

$\qquad\qquad\qquad\qquad\qquad\qquad\qquad\qquad\qquad\qquad\qquad\qquad |$

$\qquad\qquad\qquad\qquad\qquad\qquad\qquad\qquad\qquad\qquad OH$

b.

$$\xrightarrow[\text{2. HO}^-, \text{H}_2\text{O}_2, \text{H}_2\text{O}]{\text{1. BH}_3/\text{THF}}$$

30.   **a.** CH_2=$CHCH_2CH_2CH_3$ **or** CH_3CH=$CHCH_2CH_3$

   b.

or   **or**

or   **or**

31.   **a.**   three alkenes: 1-butene, *cis*-2-butene, *trans*-2-butene

   b.   four alkenes: 3-methyl-1-pentene, (*E*)-3-methyl-2-pentene, (*Z*)-3-methyl-2-pentene, 2-ethyl-1-butene

   c.   five alkenes: 1-hexene, *cis*-2-hexene, *trans*-2-hexene, *cis*-3-hexene, *trans*-3-hexene

32.   Because alkene A has the smaller heat of hydrogenation, it is more stable.

33.   **a.**

This alkene is the most stable because it has the greatest number of alkyl substituents bonded to the sp^2 carbons.

   b.

This alkene is the least stable because it has the fewest number of alkyl substituents bonded to the sp^2 carbons.

   c.

This alkene has the smallest heat of hydrogenation because it is the most stable of the three alkenes.

34.

4 alkyl substituents   2 trans alkyl substituents   2 cis alkyl substituents   2 cis alkyl substituents that cause greater steric strain than those in *cis*-3-hexene

35.   Solved in the text.

36.   Hydroboration-oxidation of 3-methylcyclohexene would form both 2-methylcyclohexanol (the desired product) and 3-methylcyclohexanol, because borane could add to either the 1-position or the 2-position of the alkene, since the two transition states would have approximately the same stability. Therefore, less of the desired product would be formed from 3-methylcyclohexene than from 1-methylcyclohexene.

37.   **a.** $CH_3CH{=}CH_2$ $\xrightarrow[\text{CH}_3\text{OH}]{\text{H}_2\text{SO}_4}$ CH_3CHCH_3
                                                                    |
                                                                    OCH_3

   b. $CH_3CH_2CH{=}CHCH_3 + Br_2 \longrightarrow$ $CH_3CH_2\overset{\displaystyle Br}{\underset{\displaystyle Br}{CHCHCH_3}}$

   c.

   d.

   e.

   f.

38.   **a.**   **b.** $CH_3\overset{\displaystyle CH_3}{\underset{\displaystyle Br}{C}}CH_2CH_3$ **c.**   **d.** $CH_3CH_2\overset{\displaystyle CH_3}{\underset{\displaystyle Br}{C}}{-}\overset{}{\underset{\displaystyle CH_3}{C}HCH_3}$

39.   **a.**   electrophile   nucleophile

           $CH_3\overset{+}{C}HCH_3 + {:}\ddot{C}l{:}^{-} \longrightarrow CH_3CHCH_3$
                                                              |
                                                            ${:}\ddot{C}l{:}$

b.   nucleophile   electrophile

$$CH_3CH{=}CH_2 \; + \; H{-}Br \longrightarrow CH_3\overset{+}{C}H{-}CH_3 \; + \; Br^-$$

c.   electrophile
nucleophile

$$CH_3CH{=}CH_2 \longrightarrow CH_3CH_2{-}CH_2BH_2$$

$$+ \quad H{-}BH_2$$

nucleophile electrophile

40.

a.
$$\begin{array}{c} CH_3 \\ | \\ CH_3C{=}CHCH_3 \end{array} + \; HBr \longrightarrow \begin{array}{c} CH_3 \\ | \\ CH_3C{-}CH_2CH_3 \\ | \\ Br \end{array}$$

b.
$$\begin{array}{c} CH_3 \\ | \\ CH_3C{=}CHCH_3 \end{array} + \; \begin{array}{c} O \\ \| \\ RCOOH \end{array} \longrightarrow \begin{array}{c} H_3C \quad \overset{O}{\triangle} \\ H_3C \qquad CH_3 \end{array}$$

c.
$$\begin{array}{c} CH_3 \\ | \\ CH_3C{=}CHCH_3 \end{array} + \; HI \longrightarrow \begin{array}{c} CH_3 \\ | \\ CH_3C{-}CH_2CH_3 \\ | \\ I \end{array}$$

d.
$$\begin{array}{c} CH_3 \\ | \\ CH_3C{=}CHCH_3 \end{array} + \; Cl_2 \longrightarrow \begin{array}{c} CH_3 \\ | \\ CH_3C{-}CHCH_3 \\ | \quad | \\ Cl \quad Cl \end{array}$$

e.
$$\begin{array}{c} CH_3 \\ | \\ CH_3C{=}CHCH_3 \end{array} + \; ICl \longrightarrow \begin{array}{c} CH_3 \\ | \\ CH_3C{-}CHCH_3 \\ | \quad | \\ Cl \quad I \end{array}$$

f.
$$\begin{array}{c} CH_3 \\ | \\ CH_3C{=}CHCH_3 \end{array} \xrightarrow[Pd/C]{H_2} \begin{array}{c} CH_3 \\ | \\ CH_3CH{-}CH_2CH_3 \end{array}$$

g.
$$\begin{array}{c} CH_3 \\ | \\ CH_3C{=}CHCH_3 \end{array} + \; Br_2 \; + \; \text{excess } NaCl \longrightarrow \begin{array}{c} CH_3 \\ | \\ CH_3C{-}CHCH_3 \\ | \quad | \\ Cl \quad Br \end{array}$$

h.
$$\begin{array}{c} CH_3 \\ | \\ CH_3C{=}CHCH_3 \end{array} \xrightarrow[\text{2. } NaBH_4]{\text{1. } Hg(O_2CCH_3)_2,\ H_2O,\ THF} \begin{array}{c} CH_3 \\ | \\ CH_3C{-}CH_2CH_3 \\ | \\ OH \end{array}$$

i.

$$\underset{\substack{\displaystyle CH_3 \\ |}}{CH_3C}=CHCH_3 \ + \ H_2O \ \xrightarrow{\ H_2SO_4\ } \ \underset{\substack{| \\ \displaystyle OH}}{\overset{\substack{\displaystyle CH_3 \\ |}}{CH_3C}}-CH_2CH_3$$

j.

$$\underset{\substack{\displaystyle CH_3 \\ |}}{CH_3C}=CHCH_3 \ + \ Br_2 \ \xrightarrow{\ CH_2Cl_2\ } \ \underset{\substack{| \\ Br}}{\overset{\substack{\displaystyle CH_3 \\ |}}{CH_3C}}-\underset{\substack{| \\ Br}}{CHCH_3}$$

k.

$$\underset{\substack{\displaystyle CH_3 \\ |}}{CH_3C}=CHCH_3 \ + \ Br_2 \ \xrightarrow{\ H_2O\ } \ \underset{\substack{| \\ HO}}{\overset{\substack{\displaystyle CH_3 \\ |}}{CH_3C}}-\underset{\substack{| \\ Br}}{CHCH_3}$$

l.

$$\underset{\substack{\displaystyle CH_3 \\ |}}{CH_3C}=CHCH_3 \ + \ Br_2 \ \xrightarrow{\ CH_3OH\ } \ \underset{\substack{| \\ CH_3O}}{\overset{\substack{\displaystyle CH_3 \\ |}}{CH_3C}}-\underset{\substack{| \\ Br}}{CHCH_3}$$

m.

$$\underset{\substack{\displaystyle CH_3 \\ |}}{CH_3C}=CHCH_3 \ \xrightarrow[\text{2. } H_2O_2, \ HO^-, \ H_2O]{\text{1. } BH_3/THF} \ \underset{\substack{| \\ OH}}{\overset{\substack{\displaystyle CH_3 \\ |}}{CH_3CH}}-CHCH_3$$

n.

$$\underset{\substack{\displaystyle CH_3 \\ |}}{CH_3C}=CHCH_3 \ \xrightarrow[\text{2. } NaBH_4]{\text{1. } Hg(O_2CCF_3)_2, \ CH_3OH} \ \underset{\substack{| \\ OCH_3}}{\overset{\substack{\displaystyle CH_3 \\ |}}{CH_3C}}-CH_2CH_3$$

41.   **a.**   2,2-diethyl-3-isopropyloxirane or 3,4-epoxy-4-ethyl-2-methylhexane
      b.   3-ethyl-2,2-dimethyloxirane or 2,3-epoxy-2-methylpentane

42.   First draw the structures so you can see the number of alkyl groups bonded to the sp^2 carbons:

$$\underset{\substack{| \\ \displaystyle CH_3}}{\overset{\substack{\displaystyle CH_3 \\ |}}{CH_3CH}}=CCHCH_2CH_3 \qquad \underset{\substack{| \\ \displaystyle CH_3}}{\overset{\substack{\displaystyle CH_3 \\ |}}{CH_3C}}=CCH_2CH_2CH_3 \qquad \underset{\substack{| \\ \displaystyle CH_3}}{\overset{\substack{\displaystyle CH_3 \\ |}}{CH_3CH}}=CHCHCHCH_3$$

    3,4-dimethyl-2-hexene   2,3-dimethyl-2-hexene   4,5-dimethyl-2-hexene

    a.   2,3-Dimethyl-2-hexene is the most stable of the three alkenes because it has the greatest number alkyl substituents bonded to the sp^2 carbons.

    b.   4,5-Dimethyl-2-hexene has the fewest alkyl substituents bonded to the sp^2 carbons making it the least stable of the three alkenes. It, therefore, has the greatest heat of hydrogenation.

    c.   Because it is the most stable, 2,3-dimethyl-2-hexene has the smallest heat of hydrogenation.

43.

$$\xrightarrow[\text{H}_2\text{SO}_4]{\text{H}_2\text{O}}$$

forms a carbocation intermediate, so a carbocation rearrrangement can occur

$$\xrightarrow[\text{2. NaBH}_4]{\text{1. Hg(O}_2\text{CCH}_3)_2,\ \text{H}_2\text{O/THF}}$$

forms a cyclic mercurinium ion intermediate, so a carbocation rearrangement cannot occur

$$\xrightarrow[\text{2. H}_2\text{O}_2,\ \text{HO}^-,\ \text{H}_2\text{O}]{\text{1. BH}_3\text{/THF}}$$

BH_3 is the electrophile and H^- is the nucleophile.

44.

$$CH_3CHCH=CH_2\ +\ H-Br\ \longrightarrow\ CH_3\overset{H}{\underset{CH_3}{C}}-\overset{+}{C}HCH_3\ \xrightarrow[\text{shift}]{1,2\text{-hydride}}\ CH_3\overset{+}{C}CH_2CH_3$$

2-bromo-3-methylbutane   2-bromo-2-methylbutane

45.

a.
$$CH_3\overset{CH_3}{\underset{CH_3}{C}}=CCH_3$$

This compound is the most stable.
It has four alkyl substituents bonded to the sp^2 carbons.

b. $CH_2=CHCH_2CH_2CH_2CH_3$ $CH_2=CHCH_2CHCH_3$ with CH_3

$CH_2=CHCHCH_2CH_3$ with CH_3

$CH_3\overset{CH_3}{\underset{CH_3}{C}}CH=CH_2$

These compounds are the least stable.
Each has only one alkyl substituent bonded to the sp^2 carbons.

46.

a.
$$CH_3-\overset{:\overset{..}{O}:^-}{\underset{CH_3}{C}}-OCH_3\ \longrightarrow\ CH_3-\overset{:\overset{..}{O}}{\overset{\|}{C}}-CH_3\ +\ CH_3O^-$$

b. $CH_3C\equiv C-H\ +\ \ :\overset{..}{N}H_2\ \longrightarrow\ CH_3C\equiv C:^-\ +\ :NH_3$

c. $CH_3CH_2-Br\ +\ CH_3\overset{..}{\underset{..}{O}}:^-\ \longrightarrow\ CH_3CH_2-\overset{..}{\underset{..}{O}}CH_3\ +\ Br^-$

47.

48. **a.**   **c.**   **e.**   **g.**

b.   **d.**   **f.**   **h.**

49.   Diazomethane is a very reactive compound because the triple-bonded nitrogen has a strong propensity to depart from the carbon in order to form a very stable molecule of nitrogen gas. As it departs, the nucleophilic alkene attacks the electrophilic carbon in the same step, and also in the same step, the lone pair is the nucleophile that adds to the other sp^2 carbon of the alkene.

50. **a.** $\xrightarrow[\text{Pd/C}]{H_2}$

d. $CH_3CH_2CH{=}CHCH_2CH_3 \xrightarrow[H_2O]{Br_2}$

b. $CH_3CH_2CH_2CH{=}CH_2 \xrightarrow{HCl}$

e. $\xrightarrow[H_2O]{H_2SO_4}$

c. $\xrightarrow[\text{2. H}_2\text{O}_2,\ \text{HO}^-,\ \text{H}_2\text{O}]{\text{1. BH}_3/\text{THF}}$

f. $CH_3CH_2CH\!=\!CHCH_2CH_3 \xrightarrow[\substack{\text{NaBr}\\\text{excess}}]{\text{Cl}_2}$

or

$CH_3CH_2CH\!=\!CHCH_2CH_3 \xrightarrow[\substack{\text{NaCl}\\\text{excess}}]{\text{Br}_2}$

51.   a.   and

b.   (Note: D stands for deuterium, an isotope of hydrogen. DBr reacts in a manner very similar to HBr.) They would form different products:

and

52.   a. $\overset{\overset{\displaystyle CH_3}{|}}{CH_3\overset{+}{C}CH_3}$

c.   +

e. $\overset{\overset{\displaystyle CH_3}{|}}{CH_3C}\!=\!CHCH_2CH_3$

b. $CH_3\overset{+}{C}HCH_3$

d.

f.

53.   a.   To determine relative rates, the rate constant of each alkene is divided by the smallest rate constant of the series (3.51×10^{-8}).

relative rates

propene	$= (4.95 \times 10^{-8})/(3.51 \times 10^{-8}) = 1.41$
(Z)-2-butene	$= (8.32 \times 10^{-8})/(3.51 \times 10^{-8}) = 2.37$
(E)-2-butene	$= (3.51 \times 10^{-8})/(3.51 \times 10^{-8}) = 1$
2-methyl-2-butene	$= (2.15 \times 10^{-4})/(3.51 \times 10^{-8}) = 6.12 \times 10^3$
2,3-dimethyl-2-butene	$= (3.42 \times 10^{-4})/(3.51 \times 10^{-8}) = 9.74 \times 10^3$

b.   Both compounds form the same carbocation but, since (Z)-2-butene is less stable than (E)-2-butene, (Z)-2-butene has a smaller free energy of activation.

c.   2-Methyl-2-butene is more stable than (Z)-2-butene, and 2-methyl-2-butene forms a more stable carbocation intermediate (tertiary) and therefore a more stable transition state than does (Z)-2-butene (secondary). Knowing that 2-methyl-2-butene reacts faster tells us that the energy difference between the transition states is greater than the energy difference between the alkenes. This is what we would expect from the Hammond postulate, since the transition states look more like the carbocations than like the alkenes. Thus, we can conclude that 2-methyl-2-butene is the faster reacting compound because it forms the more stable carbocation.

d. 2,3-Dimethyl-2-butene is more stable than 2-methylbutene, and both compounds form a tertiary carbocation intermediate. On this basis, you would predict that 2,3-dimethyl-2-butene would react more slowly than 2-methylbutane. However, 2,3-dimethyl-2-butene has two sp^2 carbons that can react with a proton to form the tertiary carbocation, whereas 2-methyl-2-butene has only one sp^2 carbon that can react with a proton to form the tertiary carbocation. The fact that 2,3-dimethyl-2-butene reacts faster in spite of being more stable tells us that the more important factor is the greater number of collisions with the proper orientation that lead to a productive reaction in the case of 2,3-dimethyl-2-butene.

54. **a.**

has a greater dipole moment than

where the C—Cl dipoles oppose each other

b.

has a greater dipole moment than

Recall that an sp^2 carbon is more electronegative than an sp^3 carbon. Thus, the C—CH_3 and C—Cl dipoles enforce each other, leading to a higher dipole, in the compound on the left and oppose each other in the compound on the right.

55. No, he should not follow the his friend's advice. Adding the electrophile to the sp^2 carbon bonded to the greater number of hydrogens forms a secondary carbocation in preference to a primary carbocation. However, because of the electron-withdrawing fluoro substituents in this compound, the primary carbocation is more stable than the secondary carbocation, since in the secondary carbocation the positive charge is closer to the fluoro substituents. So the major product will be 1,1,1-trifluoro-3-iodopropane and not 1,1,1-trifluoro-2-iodopropane, the compound that would be predicted to be the major product by following the rule.

$$F_3CCH_2 \overset{+}{-}CH_2 \qquad\qquad F_3C\overset{+}{C}H-CH_3$$

more stable   less stable because of the nearby electron-withdrawing fluorines

56. **a.**

b. the first step (See the answer to Problem 9.)
c. H^+
d. 1-butene
e. the *sec*-butyl cation
f. methanol

57.   a.

b.

58.   Five- and six-membered rings are more stable than other size rings.

a.   Either a six-membered ring or a five-membered ring can be formed. However, because the more substituted three-membered ring carbon has the larger partial positive charge since it is secondary and the less substituted one is primary, the three-membered ring will break in the direction that forms the five-membered ring.

b.   A six-membered ring will form in preference to a less stable seven-membered ring. (In addition, formation of the six-membered ring involves the preferred attack on the more substituted ring carbon.)

59.   a.   Both 1-butene and 2-butene react with HCl to form 2-chlorobutane.

b.   Both alkenes form the same carbocation, and therefore have transition states that are close in energy, but because 2-butene is more stable than 1-butene, 2-butene has the greater free energy of activation.

c.   Because both alkenes form the same carbocation and since 1-butene is less stable (higher in energy) than 2-butene, one might expect that 1-butene would react more rapidly with HCl. However, as pointed out in Problem 53.d., there is also the consideration that 2-butene has two equivalent sites that can lead to the desired carbocation. Thus, if the greater number of collisions that can form the intermediate is the more important factor, 2-butene will react faster. In the absence of the actual rate data, it is impossible to know which (the difference in stability of the reactants or the difference in the number of productive collision) is the more important factor.

d. Both compounds form the same carbocation, and therefore have transition states that are close in energy, but since (Z)-2-butene is less stable, it will react more rapidly with HCl.

60.   a. [structure] and [structure]   **c.** [structure]

b. [structure]   **d.** [structure]

61.   a. 4 alkenes [structures]

b. 1-Methylcyclopentene is the most stable.

c. Because 1-methylcyclopentene is the most stable, it has the smallest heat of hydrogenation.

62.   a. [reaction mechanism showing H—Br addition, secondary and tertiary carbocations]

b. The initially formed carbocation is tertiary.

c. The rearranged carbocation is secondary; it undergoes another rearrangement to a more stable tertiary carbocation.

d. The initially formed carbocation rearranges in order to release the strain in the four-membered ring. (A tertiary carbocation with a strained four-membered ring is less stable than a secondary carbocation with an unstrained five-membered ring.)

63. It tells us that the first step of the mechanism is the slow step. If the first step is slow, the carbocation will react with water in a subsequent fast step, which means that the carbocation will not have time to lose a proton to reform the alkene, so all the deuterium atoms (D) will be retained in the unreacted alkene.

If the first step were not the slow step, an equilibrium would be set up between the alkene and the carbocation and, because the carbocation could lose either H^+ or D^+ when it reforms the alkene, all the deuterium atoms would not be retained in the unreacted alkene.

64.   **a.**   A proton adds to the alkene, forming a secondary carbocation, which undergoes a ring-expansion rearrangement to form a more stable tertiary carbocation.

65.   **a.**   The base removes a proton. Then do what is needed to get the known product of the reaction.

b.   Notice that this reaction has the same mechanism as that in part **a** except that instead of a base removing a proton and leaving behind its bonding electrons, heat causes CO_2 to be removed leaving behind its bonding electrons.

Chapter 4 Practice Test

1. Which member of each pair is more stable?

a. $CH_3\overset{+}{C}HCH_2CH_3$ or $CH_3\overset{+}{C}CH_3$ (with CH_3 substituent)

d. (cyclohexene with methyl) or (methylcyclohexene)

b. $CH_3CH_2CH_2\overset{+}{C}H_2$ or $CH_3CH_2\overset{+}{C}HCH_3$

c. $CH_3CH_2\overset{+}{C}H_2$ or $ClCH_2CH_2\overset{+}{C}H_2$

e.
$$\begin{matrix} H_3C \\ \end{matrix} C=C \begin{matrix} H \\ CH_3 \end{matrix}$$ or $$\begin{matrix} H_3C \\ H \end{matrix} C=C \begin{matrix} CH_3 \\ H \end{matrix}$$

2. Which would be a better compound to use as a starting material for the synthesis of 2-bromopentane?

$CH_3CH_2CH_2CH=CH_2$ or $CH_3CH_2CH=CHCH_3$

3. The addition of H_2 in the presence of Pd/C to alkenes **A** and **B** results in the formation of the same alkane. The addition of H_2 to alkene **A** has a heat of hydrogenation ($-\Delta H°$) of 29.7 kcal/mol, while the addition of H_2 to alkene **B** has a heat of hydrogenation of 27.3 kcal/mol. Which is the more stable alkene, **A** or **B**?

4. What is the major product of each of the following reactions?

a. $CH_2=\overset{\overset{\displaystyle CH_3}{|}}{C}CH_2CH_3$ +   HBr $\longrightarrow$

b. $CH_3\overset{\overset{\displaystyle CH_3}{|}}{C}HCH=CH_2$ +   HCl $\longrightarrow$

c. $CH_3CH_2CH=CH_2$ + Cl_2 $\xrightarrow{\text{H}_2\text{O}}$

d. $CH_3\overset{\overset{\displaystyle CH_3}{|}}{\underset{\underset{\displaystyle CH_3}{|}}{C}}CH=CH_2$ +   HBr $\longrightarrow$

5. What product would be obtained from the hydroboration-oxidation of the following alkenes?

a. 2-methyl-2-butene   **b.** 1-ethylcyclopentene

6.   Indicate how each of the following compounds could be synthesized using an alkene as one of the starting materials:

a. $\longrightarrow$
$$\underset{\overset{|}{CH_3}}{\overset{\overset{CH_3}{|}}{CH_3CCH_2CH_3}}$$

b. $\longrightarrow$

OH
Cl

c. $\longrightarrow$

CH$_3$
Br

d. $\longrightarrow$

CH$_3$
OH

7.   Indicate the carbocations that you would expect to rearrange to give a more stable carbocation.

$$\underset{\overset{|}{+}}{\overset{\overset{CH_3}{|}}{CH_3CH_2CHCHCH_3}}$$ $$\underset{+}{CH_3CH_2CHCH_3}$$

CH$_3$
$+$

CH$_3$
$+$

8.   Indicate whether each of the following statements is true or false:

a. The addition of Br$_2$ to 1-butene to form 1,2-dibromobutane is a concerted reaction.   T   F

b. The reaction of 1-butene with HCl will form 1-chlorobutane as the major product.   T   F

c. 2,3-Dimethyl-2-pentene is more stable than 3,4-dimethyl-2-pentene.   T   F

d. The reaction of HBr with 3-methylcyclohexene is more highly regioselective
than is the reaction of HBr with 1-methylcyclohexene.   T   F

e. The reaction of an alkene with a carboxylic acid forms an epoxide.   T   F

f. A catalyst increases the equilibrium constant of a reaction.   T   F

SPECIAL TOPIC IV

Exercise in Model Building

Do the following exercises using molecular models:

1. Build the enantiomers of 2-bromobutane.

   a. Try to superimpose them.

   b. Show that they are mirror images.

   c. Which one is (R)-2-bromobutane?

2. Build the erythro enantiomers of 3-bromo-2-butanol.

   a. Where are the Br and OH substituents (relative to each other) in the Fischer projection? (Recall that in a Fischer projection, the horizontal lines represent bonds that point out of the plane of the paper toward the viewer, and the vertical lines represent bonds that point back from the plane of the paper away from the viewer.)

   b. Where are the Br and OH substituents (relative to each other) in the most stable conformer considering rotation about the C-2—C-3 bond?

3. a. Build the compounds labeled "1" and "2" shown on the top of page 216 of the text.

   b. Build their mirror images.

   c. Show that "1" is superimposable on its mirror image but "2" is not.

4. Build the three stereoisomers of 2,3-dibromobutane.

5. Build the four stereoisomers of 2,3-dibromopentane. Why does 2,3-dibromopentane have four stereoisomers, whereas 2,3-dibromobutane has only three stereoisomers?

6. Build (S)-2-pentanol.

7. Build (2S,3R)-3-bromo-2-butanol. Rotate the model so its conformation corresponds to that shown in a Fischer projection. Compare its structure with the structure of (2S,3R)-3-bromo-2-butanol shown on page 221 of the text.

8. Rotate the model of (2S,3R)-3-bromo-2-butanol so it is in a perspective formula as a staggered conformer. Compare its structure with the structure of (2S,3R)-3-bromo-2-butanol shown on page 221 of the text.

9. Build the compounds shown in Problem 39 on page 223 of the text. Name the compounds.

10. Build (S)-1-bromo-2-methylbutane. Substitute the Br with an HO without breaking the bond to the asymmetric center to form 2-methyl-1-butanol. What is the configuration of your model of 2-methyl-1-butanol?

11. Build two models of *trans*-2-pentene. Add Br_2 to opposite sides of the double bond, forming the two enantiomers shown on page 236 of the text. Rotate them so they are Fischer projections. Are they erythro or threo enantiomers? Compare your answer with that given in the text.

12.   Build models of the molecules shown in Problem 99 on page 250 of the text. What is the configuration of the asymmetric center in each of the molecules?

13.   **a.**   Build *cis*-1-bromo-4-chlorocyclohexane. Build its mirror image. Are they superimposable?

      b.   Build *cis*-1-bromo-2-chlorocyclohexane. Build its mirror image. Are they superimposable?

14.   See the box on cyclic alkenes on page 234 of the text.

      a.   Build *cis*-cyclooctene.

      b.   Build *trans*-cyclooctene.

      c.   Build *cis*-cyclohexene.

      d.   Can you build *trans*-cyclohexene?

CHAPTER 5
Stereochemistry: The Arrangement of Atoms in Space:
The Stereochemistry of Addition Reactions

Important Terms

absolute configuration	the three-dimensional structure of a chiral compound. The configuration is designated by R or S. The configuration is called absolute to distinguish it from a relative configuration.
achiral	a molecule or object that contains an element (a plane or a point) of symmetry.
amine inversion	results when a compound containing an sp^3 hybridized nitrogen with a nonbonding pair of electrons rapidly inverts.
anti addition	an addition reaction in which the two added substituents add to opposite sides of the molecule.
asymmetric center	an atom that is bonded to four different substituents.
biochemistry	the chemistry associated with living organisms.
chiral	a chiral molecule has a nonsuperimposable mirror image.
chiral catalyst	a catalyst that will cause the synthesis of one enantiomer in great excess over the other.
chiral probe	something capable of distinguishing between enantiomers.
chromatography	a separation technique in which the mixture to be separated is dissolved in a solvent and the solution is passed through a column packed with an adsorbent stationary phase.
cis isomer	the isomer with substituents on the same side of a cyclic structure, or the isomer with the hydrogens on the same side of a double bond.
cis-trans isomers	result from not being able to rotate about a carbon-carbon bond. They are also called geometric isomers or E,Z isomers.
configuration	the three-dimensional structure of a chiral compound. The configuration is designated by R or S.
configurational isomers	stereoisomers that cannot interconvert unless a covalent bond is broken. Cis-trans isomers and isomers with asymmetric centers are configurational isomers.
constitutional isomers (structural isomers)	molecules that have the same molecular formula but differ in the way the atoms are connected.
dextrorotatory	the enantiomer that rotates the plane of polarization of plane-polarized light in a clockwise direction.

diastereomers	stereoisomers that are not enantiomers.
enantiomerically pure	only one enantiomer is present in an enantiomerically pure sample.
enantiomeric excess	how much excess of one enantiomer is present in a mixture of a pair of enantiomers, expressed as a percentage.
enantiomers	nonsuperimposable mirror-image molecules.
enzyme	a protein that catalyzes a biological reaction.
erythro enantiomers	the pair of enantiomers with similar groups on the same side (in the case of stereoisomers with adjacent asymmetric centers) when drawn in a Fischer projection.
Fischer projection	a method of representing the spatial arrangement of groups bonded to an asymmetric center. The asymmetric center is the point of intersection of two perpendicular lines; the horizontal lines represent bonds that project out of the plane of the paper toward the viewer, and the vertical lines represent bonds that project back from the plane of the paper away from the viewer.
isomers	nonidentical compounds with the same molecular formula.
levorotatory	the enantiomer that rotates the plane of polarization of plane-polarized light in a counterclockwise direction.
meso compound	a compound that possesses asymmetric centers and a plane of symmetry; it is achiral, because it has a plane of symmetry.
observed rotation	the amount of rotation observed in a polarimeter.
optically active	rotates the plane of polarization of plane-polarized light.
optically inactive	does not rotate the plane of polarization of plane-polarized light.
optical purity	how much excess of one enantiomer is present in a mixture of a pair of enantiomers.
perspective formula	a method of representing the spatial arrangement of groups bonded to an asymmetric center. Two bonds are drawn in the plane of the paper; a solid wedge is used to depict a bond that projects out of the plane of the paper toward the viewer, and a hatched wedge is used to represent a bond that projects back from the paper away from the viewer.
plane-polarized light	light that oscillates in a single plane passing through the direction the light travels.
plane of symmetry	an imaginary plane that bisects a molecule into pieces that are a pair of mirror images.
polarimeter	an instrument that measures the rotation of the plane of polarization of plane-polarized light.

polarized light	light that oscillates in only one plane.
racemic mixture (racemic modification, racemate)	a mixture of equal amounts of a pair of enantiomers.
***R* configuration**	after assigning relative priorities to the four groups bonded to an asymmetric center, if the lowest-priority group is on a vertical axis in a Fischer projection (or pointing away from the viewer in a perspective formula), an arrow drawn from the highest-priority group to the next highest-priority group goes in a clockwise direction.
receptor	a protein or other compound that binds a particular molecule.
regioselective	describes a reaction that leads to the preferential formation of one constitutional isomer over another.
relative configuration	the configuration of a compound relative to the configuration of another compound.
resolution of a racemic mixture	separation of a racemic mixture into the individual enantiomers.
***S* configuration**	after assigning relative priorities to the four groups bonded to an asymmetric center, if the lowest-priority group is on a vertical axis in a Fischer projection (or pointing away from the viewer in a perspective formula), an arrow drawn from the highest-priority group to the next highest-priority group goes in a counterclockwise direction.
specific rotation	the amount of rotation that will be observed for a compound with a concentration of 1.0 g/mL in a sample tube 1.0 decimeter long.
stereocenter (stereogenic center)	an atom at which the interchange of two groups produces a stereoisomer.
stereochemistry	the field of chemistry that deals with the structure of molecules in three dimensions.
stereoisomers	isomers that differ in the way the atoms are arranged in space.
stereoselective	describes a reaction that leads to the preferential formation of one stereoisomer over another.
stereospecific	describes a reaction in which the reactant can exist as stereoisomers and each stereoisomeric reactant leads to a different stereoisomeric product.
syn addition	an addition reaction in which the two added substituents add to the same side of the molecule.

threo enantiomers   the pair of enantiomers with similar groups on opposite sides (in the case of stereoisomers with adjacent asymmetric centers) when drawn in a Fischer projection.

trans isomer   the isomer with substituents on the opposite sides of a cyclic structure, or the isomer with the hydrogens on the opposite sides of a double bond.

Solutions to Problems

1. **a.** $CH_3CH_2CH_2OH$ CH_3CHOH $CH_3CH_2OCH_3$
                                                    |
                                                 CH_3

b. There are seven constitutional isomers with molecular formula $C_4H_{10}O$.

$CH_3CH_2CH_2CH_2OH$ CH_3CHCH_2OH CH_3COH $CH_3CHCH_2CH_3$
                                      | | |
                                   CH_3 CH_3 OH

with CH_3 above CH_3COH:
CH_3
|
CH_3COH
|
CH_3

$CH_3CH_2OCH_2CH_3$ $CH_3OCH_2CH_2CH_3$ CH_3OCHCH_3
                                                                        |
                                                                     CH_3

2. **a.**

b.

c. Br—⬡—Cl   Br—⬡····Cl   **d.**

3. **a**, **b**, **c**, **f**, and **h** are chiral.
      d, **e**, and **g** are each superimposable on its mirror image. These, therefore, are achiral.

4. **a.** F, G, J, L, N, P, Q, R, S, Z

      b. A, H, I, M, O, T, U, V, W, X, Y

5. **a**, **c**, and **f** have asymmetric centers.

6. Solved in the text.

7. **a**, **c**, and **f**, because in order to be able to exist as a pair of enantiomers, the compound must have an asymmetric center (except in the case of certain compounds with unusual structures). (See Problem 101.)

8.   Draw the first enantiomer with the groups in any order you want. Then draw the second enantiomer by drawing the mirror image of the first enantiomer. Your answer might not look exactly like the ones shown below because the first enantiomer can be drawn with the four groups on any of the four bonds. The next one is the mirror image of the first one.

a.   **1.**

$$CH_3\ \ \overset{C}{\underset{H}{\big|}}\ Br\cdots CH_2OH \qquad CH_3\ \ HOCH_2\cdots\overset{C}{\underset{H}{\big|}}\ Br$$

2.

$$CH_2CH_2Cl\ \ CH_3CH_2\cdots\overset{C}{\underset{H}{\big|}}CH_3 \qquad CH_2CH_2Cl\ \ H_3C\cdots\overset{C}{\underset{H}{\big|}}CH_2CH_3$$

3.

$$OH\ \ (CH_3)_2CH\cdots\overset{C}{\underset{H}{\big|}}CH_3 \qquad OH\ \ H_3C\cdots\overset{C}{\underset{H}{\big|}}CH(CH_3)_2$$

b.   **1.**

$$CH_3 \atop Br\text{—}\!\!\!\!\overset{\ }{\big|}\!\!\!\!\text{—}H \atop CH_2OH \qquad CH_3 \atop H\text{—}\!\!\!\!\overset{\ }{\big|}\!\!\!\!\text{—}Br \atop CH_2OH$$

2.

$$CH_2CH_2Cl \atop H\text{—}\!\!\!\!\overset{\ }{\big|}\!\!\!\!\text{—}CH_3 \atop CH_2CH_3 \qquad CH_2CH_2Cl \atop CH_3\text{—}\!\!\!\!\overset{\ }{\big|}\!\!\!\!\text{—}H \atop CH_2CH_3$$

3.

$$CH_3 \atop HO\text{—}\!\!\!\!\overset{\ }{\big|}\!\!\!\!\text{—}H \atop CH(CH_3)_2 \qquad CH_3 \atop H\text{—}\!\!\!\!\overset{\ }{\big|}\!\!\!\!\text{—}OH \atop CH(CH_3)_2$$

9.   Solved in the text.

10.   **A**, **B**, and **C** are identical.

11.   **a.** $—CH_2OH$ $—CH_3$ $—CH_2CH_2OH$ $—H$

　　　　　　① 　　　　③ 　　　　② 　　　　④

b. $—CH=O$ $—OH$ $—CH_3$ $—CH_2OH$

　　　　② 　　　① 　　　④ 　　　③

c. $—CH(CH_3)_2$ $—CH_2CH_2Br$ $—Cl$ $—CH_2CH_2CH_2Br$

　　　　② 　　　　③ 　　　① 　　　　④

d. $—CH=CH_2$ ② $—CH_2CH_3$ ③   [benzene ring] ① $—CH_3$ ④

C attached to 2 Cs   C attached to 3 Cs

12.   a. *R*   **b.** *R*

  c. To determine the configuration, first add the fourth bond to the asymmetric center. Remember that it cannot be drawn between the two solid bonds. (It can be drawn on either side of the solid wedge.)

  d. Draw the structure with the fourth bond to the asymmetric center. Switch a pair so that the H is on a hatched wedge. The configuration of the compound with the switched pair is *S*. Therefore, the configuration of the compound given in the question is *R*.

13.   a. *S*   **b.** *R*   **c.** *S*   **d.** *S*

14. The easiest way to determine whether two compounds are identical or enantiomers is to determine their configurations: If both are *R* (or both are *S*), they are identical. If one is *R* and the other is *S*, they are enantiomers.

  a. identical   **b.** enantiomers   **c.** enantiomers   **d.** enantiomers

15.   a.

$$CH_3CH_2 - \overset{\displaystyle Cl}{\underset{\displaystyle CH_3}{C}} \cdots H$$

  b.

$$CH_3CH_2 - \overset{\displaystyle Br}{\underset{\displaystyle CH_2Br}{C}} \cdots H$$

16.   a. levorotatory   **b.** dextrorotatory

17.

$$\text{specific rotation} = \frac{\text{observed rotation}}{\text{concentration} \times \text{length}}$$

$$[\alpha] = \frac{+13.4°}{\dfrac{2\ g}{50\ mL} \times 2\ dm} = \frac{+13.4°}{0.08} = +168$$

18.   a. −24   **b.** 0

19.   a. 50% of the mixture is excess (+)-mandelic acid

$$\text{optical purity} = 0.50 = \frac{\text{observed specific rotation}}{\text{specific rotation of the pure enantiomer}}$$

$$0.50 = \frac{\text{observed specific rotation}}{+158}$$

observed specific rotation = +79

  b. 0 (It is a racemic mixture.)

  c. 50% of the mixture is excess (−)-mandelic acid.
     observed specific rotation = −79 (For the calculation, see part **a.**)

20. **a.** From the data given, you cannot determine the configuration of naproxen.

b. 97% of the commercial preparation is (+)-naproxen; 3% is a racemic mixture
Therefore, the commercial preparation forms 98.5% (+)-naproxen and 1.5% (−)-naproxen.

21. Solved in the text.

22. As a result of the double bond, the compound has a cis isomer and a trans isomer. Because the compound also has an asymmetric center, the cis isomer can exist as a pair of enantiomers and the trans isomer can exist as a pair of enantiomers.

cis enantiomers

trans enantiomers

23. The statement is correct. Although there are many examples where there would be fewer than 2^n stereoisomers (for example, *trans*-2-butene has two stereocenters but only two stereoisomers, there are no situations where there will be more than 2^n stereoisomers).

24. **a.** enantiomers
b. identical compounds (Therefore, they are not stereoisomers.)
c. diastereomers

25. **a.** Find the sp^3 carbons that are bonded to four different substituents; these are the asymmetric centers. Cholesterol has eight asymmetric centers. They are indicated by arrows.

b. $2^8 = 256$

Only the stereoisomer shown above is found in nature.

26.   Your perspective formulas may not look exactly like the ones drawn here because you can draw the first one with the groups attached to any bonds you want. Just make certain that the second one is a mirror image of the first one.

Leucine has one asymmetric center, so it will have two stereoisomers.

Isoleucine has two asymmetric centers, so it has four stereoisomers. Again your perspective formulas may not look exactly like the ones drawn here. To make sure you have all four, determine the configuration of each of the asymmetric centers. You should have *R,R*, *S,S*, *R,S*, and *S,R*. Notice that the asymmetric centers in the mirror image have the opposite configurations.

27.   a.

b.

Br Cl
S R

Br Cl
R S

Br Cl
S S

Br Cl
R R

or

| CH₃ | |
CH₃
H—Br S   Br—H R
H—H   H—H
H—Cl R   Cl—H S
CH₂CH₃   CH₂CH₃

CH₃   CH₃
H—Br S   Br—H R
H—H   H—H
Cl—H S   H—Cl R
CH₂CH₃   CH₂CH₃

c.

R S   R S   S S   R R

or

CH₃   CH₃
Cl—H R   H—Cl S
Cl—H S   H—Cl R
CH₂CH₃   CH₂CH₃

CH₃   CH₃
H—Cl S   Cl—H R
Cl—H S   H—Cl R
CH₂CH₃   CH₂CH₃

d.

S   R   S   R

or

CH₂CH₂Br   CH₂CH₂Br
Br—H   H—Br
CH₂CH₃   CH₂CH₃
S   R

28.

29.

1-chloro-1-methylcyclooctane   *cis*-1-chloro-5-methylcyclooctane   *trans*-1-chloro-5-methylcyclooctane

30.   There is more than one diastereomer for **a, b,** and **d; c** has only one diastereomer.

To draw a diastereomer of **a, b,** or **d,** switch any one pair of substituents bonded to one of the asymmetric centers. Because any one pair can be switched, your diastereomer may not be the one drawn here, unless you happened to switch the same pair that I switched.

31.   **A** = identical   **B** = enantiomer   **C** = diastereomer   **D** = identical

32.   **B, D,** and **F** because each has two asymmetric centers and the same four groups bonded to each of the asymmetric centers.
A has two asymmetric centers but it does not have a stereoisomer that is a meso compound, because it does not have the same four groups bonded to each of the asymmetric centers.
C and **E** do not have a stereoisomer that is a meso compound, because they do not have asymmetric centers.

33.   Solved in the text.

34.   **a.**

b.

c.

CH$_3$
CH$_3$CHCH$_2$OH   no stereoisomers, because the compound does not have an asymmetric center

d.

CH$_2$CH$_3$
C
HOCH$_2$ Br   H
R

CH$_2$CH$_3$
C
H   CH$_2$OH
Br
S

or

CH$_2$OH
H———Br
CH$_2$CH$_3$
R

CH$_2$OH
Br———H
CH$_2$CH$_3$
S

e.   CH$_3$CH$_2$CCH$_2$CH$_3$ no stereoisomers, because the compound does not have an asymmetric center

CH$_3$
Cl

f.

H$_3$C   H
C——C——Br
HO   CH$_3$
H
R   S

H   CH$_3$
Br——C——C——OH
H$_3$C   H
R   S

CH$_3$
HO———H *R*
Br———H *S*
CH$_3$

CH$_3$
H———OH *S*
H———Br *R*
CH$_3$

or

H$_3$C   H
C——C——Br
H   CH$_3$
HO
S   S

H   CH$_3$
Br——C——C——H
H$_3$C   OH
R   R

CH$_3$
H———OH *S*
Br———H *S*
CH$_3$

CH$_3$
HO———H *R*
H———Br *R*
CH$_3$

g.

CH$_3$CH$_2$ Cl
H——C——C——H
Cl   CH$_2$CH$_3$
S   R
a meso compound

or

CH$_2$CH$_3$
H———Cl *S*
H———Cl *R*
CH$_2$CH$_3$
a meso compound

CH$_3$CH$_2$ H
H——C——C——Cl
Cl   CH$_2$CH$_3$
S   S

H   CHCH$_3$
Cl——C——C——H
CH$_3$CH$_2$ Cl
R   R

CH$_2$CH$_3$
H———Cl *S*
Cl———H *S*
CH$_2$CH$_3$

CH$_2$CH$_3$
Cl———H *R*
H———Cl *R*
CH$_2$CH$_3$

h.

Cl   Cl
| |
_/_/_
R   S
a meso compound

Cl   Cl   Cl   Cl
| | | |
_/_/_   _/_/_
R   R   S   S

or

CH₃		CH₃		CH₃	
H——Cl	*S*	H——Cl	*S*	Cl——H	*R*
H——H		H——H		H——H	
H——Cl	*R*	Cl——H	*S*	H——Cl	*R*
CH₃		CH₃		CH₃	

a meso compound

i.

Cl   Cl   Cl   Cl
| | | |
_/_/_/\   _/_/_/\
S   R   R   S

Cl   Cl   Cl   Cl
| | | |
_/_/_/\   _/_/_/\
S   S   R   R

CH₃		CH₃	
H——Cl	*S*	Cl——H	*R*
H——H		H——H	
H——Cl	*R*	Cl——H	*S*
CH₂CH₂CH₃		CH₂CH₂CH₃	

or

CH₃		CH₃	
H——Cl	*S*	Cl——H	*R*
H——H		H——H	
Cl——H	*S*	H——Cl	*R*
CH₂CH₂CH₃		CH₂CH₂CH₃	

j. The cis stereoisomer is a meso compound.

k. The cis stereoisomer is a meso compound.

l. This compound does not have any asymmetric centers, so it has only cis-trans isomers.

m.

n. This compound does not have any asymmetric centers, so it has only cis-trans isomers.

35.   **a.**

(2S,3R)-1,3-dichloro-2-pentanol   (2R,3S)-1,3-dichloro-2-pentanol

(2S,3S)-1,3-dichloro-2-pentanol   (2R,3R)-1,3-dichloro-2-pentanol

b.

(2S,3R)-1,3-dichloro-2-pentanol   (2R,3S)-1,3-dichloro-2-pentanol

(2S,3S)-1,3-dichloro-2-pentanol   (2R,3R)-1,3-dichloro-2-pentanol

36.

37.   Your answer might be correct yet not look like the answers shown here. If you can get the answer shown here by interchanging **two** pairs of groups bonded to an asymmetric center on the structure you drew, then your answer is correct. If you get the answer shown here by interchanging **one** pair of groups bonded to an asymmetric center, then your answer is not correct.

a.   **c.**

b.

Br   H
 \ |
  C — C — CH₂CH₃
 /| |\
H CH₃ Br

d.

     CH₂Br
      |
      C
     /|\
   Br H CH₂CH₃

38. The carbon of the COO⁻ group is C-1. The first structure is 2*R*,3*S*. Therefore, naturally occurring threonine, with a configuration of 2*S*,3*R*, is the mirror image of the first structure. Thus, the second structure is naturally occurring threonine.

39. **a.** (2*R*,3*R*)-2,3-dichloropentane
  b. (2*S*,3*S*)-2-bromo-3-chloropentane
  c. (1*R*,3*S*)-1,3-cyclopentanediol (naming the compound clockwise) or
     (1*S*,3*R*)-1,3-cyclopentanediol (naming the compound counterclockwise)
  d. (3*R*,4*S*)-3-chloro-4-methylhexane

40. The structures are named as drawn from left to right in Problem 28.

  (1*R*,2*S*)-2-methylcyclohexanol
  (1*S*,2*R*)-2-methylcyclohexanol
  (1*R*,2*R*)-2-methylcyclohexanol
  (1*S*,2*S*)-2-methylcyclohexanol

41. Solved in the text.

42. We see that the *R*-alkyl halide reacts with HO⁻ to form the *R*-alcohol. We are told that the product (the *R*-alcohol) is (−). We can, therefore, conclude that the (+) alcohol has the *S* configuration.

43. From the structures given on page 226 of the text, you can determine the configuration of the asymmetric center in each compound.

  a. *R*   **b.** *R*   **c.** *S*   **d.** *S*

44. Only **b** is true.

45. (+)-Limonene has the *R* configuration, so it is the stereoisomer found in oranges.

46. Compound **A** has two stereoisomers, because it has an asymmetric center (at N).
Compound **B** does not have stereoisomers, because it does not have an asymmetric center.
Compound **C** has an asymmetric center at N but, because of the lone pair, the two enantiomers rapidly interconvert, so it exists as a single compound.

47. **a.** No, because only one constitutional isomer can be formed as a product since 2-butene is a symmetrical alkene.
  b. No, because it forms a racemic mixture.
  c. No, because *cis*-butene and *trans*-butene form the same product.
  d. Yes, because two constitutional isomers are possible, but only one is formed.
  e. No, because it forms a racemic mixture.
  f. No, because *cis*-butene and *trans*-butene form the same product.

48. Only the stereoisomers of the major product of each reaction are shown.

a.

CH₂CH₂CH₃   CH₂CH₂CH₃   CH₃   CH₃
        C   C   **or**   H———Cl *S*   Cl———H *R*
H₃C⁄ ⁙H   H⁙ ⁄ CH₃   CH₂CH₂CH₃   CH₂CH₂CH₃
     Cl   Cl
     S   *R*

b.

CH₂CH₂CH₃   CH₂CH₂CH₃   CH₂CH₃   CH₂CH₃
        C   C   **or**   H———OH *S*   HO———H *R*
CH₃CH₂⁄ ⁙H   H⁙ ⁄ CH₂CH₃   CH₂CH₂CH₃   CH₂CH₂CH₃
     OH   HO
     S   *R*

c.

CH₃O—⟨cyclopentane⟩—CH₃

This compound does not have any stereoisomers, because it does not have any asymmetric centers.

d.

          CH₃
          │
CH₃CCH₂CH₃
          │
          Br

This compound does not have any stereoisomers, because it does not have an asymmetric center.

49. a.

⟨cyclopentane⟩   and   ⟨cyclopentane⟩

H₃C   CH₃   H₃C   ⁙CH₃

b.

H₃C   CH₃

Hydrogen will be less likely to add to the side of the ring where the methyl substituent is situated, because the methyl substituent provides steric hindrance to approach of the reactants to the surface of the catalyst. Addition of hydrogen to the bottom face of the alkene gives the *cis*-product that is shown.

50. a. 1. *trans*-3-heptene   **2.** *cis*-3-heptene

b. The enantiomer of each of the structures shown will also be formed, because the peroxyacid can approach both the top and the bottom of the plane defined by the double bond.

1.   O   **2.**   O
                                ╱ ╲   ╱ ╲
CH₃CH₂CH₂⁙⁙⁙C—C⁙⁙⁙H   H⁙⁙⁙C—C⁙⁙⁙H
              ⁄   ╲   ╱   ╲
             H   CH₂CH₃   CH₃CH₂CH₂   CH₂CH₃

51. a. ⟨cyclohexane⟩—OH

b. ⟨cyclohexane⟩   +   ⟨cyclohexane⟩

CH₃CH₂   OH   HO   CH₂CH₃

c.

R + S

or

HO ⎯ H R H ⎯ OH S

d.

52.

$$CH_2{=}CCH_2CH_2CH_3 \xrightarrow{Br_2} BrCH_2CCH_2CH_2CH_3$$

$$CH_2{=}CCH_2CH_2CH_3 \xrightarrow[Pd/C]{H_2} CH_3CHCH_2CH_2CH_3$$

$$CH_2{=}CCH_2CH_2CH_3 \xrightarrow[\text{2. } HO^-, H_2O_2, H_2O]{\text{1. } BH_3} HOCH_2CHCH_2CH_2CH_3$$

(* indicates an asymmetric center)

Each of the reactions forms a compound with one asymmetric center from a compound with no asymmetric centers. Therefore, each of the products will be a racemic mixture.

53.   You could identify the intermediate by determining the number of products that are formed. Because *trans*-2-butene forms a cyclic bromonium ion intermediate, it forms only the erythro enantiomers since only anti addition occurs. If the reaction formed a carbocation intermediate, both the erythro pair of enantiomers and the threo pair of enantiomers would be formed because both syn and anti addition could occur.

54.   a.

S + R **or**   H ⎯ Br S + Br ⎯ H R

b. $CH_3CH_2CCH_2CH_3$ This compound does not have an asymmetric center.

with CH_3 above and Br below the central carbon.

c.   Same as **b**.

d.

CH$_2$CH$_2$CH$_3$ ... CH$_3$CH$_2$ ... H ... Br   +   CH$_2$CH$_2$CH$_3$ ... H ... Br ... CH$_2$CH$_3$ **or**   H —— Br, CH$_2$CH$_3$ / CH$_2$CH$_2$CH$_3$ +   Br —— H, CH$_2$CH$_3$ / CH$_2$CH$_2$CH$_3$

S   *R*   *S*   *R*

e.

H$_3$C, H, C—C, Br, H, Br, CH$_2$CH$_3$ +   H, CH$_3$, Br—C—C, CH$_3$CH$_2$, H, Br   **or**   CH$_3$ / H—Br *S* / Br—H *S* / CH$_2$CH$_3$ +   CH$_3$ / Br—H *R* / H—Br *R* / CH$_2$CH$_3$

S S   *R R*

f.

CH$_2$CH$_2$CH$_2$CH$_3$ / C / BrCH$_2$, Br, H   +   CH$_2$CH$_2$CH$_2$CH$_3$ / C / H, Br, CH$_2$Br   **or**   H—Br, CH$_2$Br / CH$_2$CH$_2$CH$_2$CH$_3$ +   Br—H, CH$_2$Br / CH$_2$CH$_2$CH$_2$CH$_3$

R   *S*   *R*   *S*

55. Two different bromonium ions are formed because Br$_2$ can add to the double bond either from the top of the plane or from the bottom of the plane defined by the double bond, and the two bromonium ions are formed in equal amounts. Attacking the less hindered carbon of one bromonium ion forms one stereoisomer, whereas attacking the less hindered carbon of the other bromonium ion forms the other stereoisomer. Because Br$^-$ can attack the least sterically hindered carbon with equal ease from pathway **a** as from pathway **b**, equal amounts of the threo enantiomers will be obtained. Of course, some reaction will occur at the more hindered end of the bromonium ion, but it will occur to the same extent in both pathways.

pathway **a** / Br$_2$

pathway **b** / Br$_2$

H$_3$C, CH$_2$CH$_2$CH$_2$CH$_3$ / C=C / H, H

56. **a.**

and

The addition of Br and OH are anti, so in a cyclic compound these two substituents are trans to one another.

HO ... Br   Br ... OH

b. mechanism of the reaction

57.   a.   Only anti addition occurs. Because the reactant is trans, the product would be expected to be the erythro pair of enantiomers. However, in this case, each asymmetric carbon is attached to the same four groups, so the product is a meso compound. Thus, only one stereoisomer is obtained.

b.   Only syn addition occurs. Because the reactant is trans, the product is the threo pair of enantiomers.

c.   Only anti addition occurs. Because the reactant is cis, the product is the pair of enantiomers with the bromines on opposite sides of the ring.

d.   Only anti addition occurs. Because the reactant is cis, the product is the pair of enantiomers with the bromines on opposite sides of the ring.

e. Only syn addition occurs. Because the reactant is cis, the product would be the pair of enantiomers with the hydrogens on the same side of the ring, but in this case the product is a meso compound, so only one stereoisomer is obtained.

f. Only syn addition occurs. Because the reactant is cis, the product is the pair of enantiomers with the hydrogens on the same side of the ring.

58. **a.** CH₃CHCH₂Br
            |
            Cl

b. The *R* and *S* enantiomers will be formed in equal amounts.

59. Greater than 98% is excess of the *S* enantiomer; the remainder is a racemic mixture. Therefore, greater than 99% is the *S* enantiomer.

60. **a.** A racemic mixture of (*R*)-malate and (*S*)-malate would be obtained. (A product with one asymmetric center would be formed from a reactant with no asymmetric centers.)

b. A racemic mixture of (*R*)-malate and (*S*)-malate would again be obtained. In the absence of an enzyme, the reactions are neither stereoselective (part **a**) nor stereospecific (part **b**).

61. In (*R*)-thalidomide, the nitrogen would be attached to the asymmetric center by a solid wedge; a hydrogen would be attached by a hatched wedge.

62.

CH₃CH=CHCH₂CH₃
2 stereoisomers
[cis and trans]

CH₂=CHCH₂CH₂CH₃
no stereoisomers

CH₃C=CHCH₃ (with CH₃ above C)
no stereoisomers

CH₃CHCH=CH₂ (with CH₃)
no stereoisomers

CH₃CH₂C=CH₂ (with CH₃)
no stereoisomers

no stereoisomers

no stereoisomers

3 stereoisomers
[Cis is a meso compound.]
[Trans is a pair of enantiomers.]

no stereoisomers

no stereoisomers

63.   a.

b.   S   R   **or**   S   R

c.   a meso compound

d.

e.

f.   cis and trans

g.

h.

cis   trans

i. CH₃CH₂CCH₂CH₃ No isomers are possible for this compound, because it does not have an
   asymmetric center.

j.   or

k.

l.

cis and trans (no asymmetric centers)

64. Only the fourth one (CHFBrCl) has an asymmetric center.

65. **a.** (*R*)-3-methyl-1-pentene

   b. (*E*)-1-bromo-2-chloro-2-fluoro-1-iodoethene

   c. (2*R*,3*R*)-3-chloro-2-pentanol

   d. (*Z*)-2-bromo-1-chloro-1-fluoroethene

   e. 8-bromo-2-ethyl-1-octene

   f. (*E*)-1,3-dibromo-4,7-dimethyl-3-octene

   g. (*S*)-2-methyl-1,2,5-pentanetriol

   h. (2*S*,3*S*)-1,2-dibromo-2-methyl-3-pentanol

   i. (*E*)-4-(2-chloroethyl)-2,3-dimethyl-3-octene

66.   Mevacor has eight asymmetric centers.

67.   **a.** diastereomers   **e.** diastereomers
        b. enantiomers   **f.** identical
        c. constitutional isomers **g.** diastereomers
        d. diastereomers   **h.** identical

68.   **1.** Both *cis-* and *trans-*2-butene give these products; in each case a product with one asymmetric center is formed, so the product is a racemic mixture.

    2. Both *cis-* and *trans-*2-butene give these products; in each case a product with one asymmetric center is formed, so the product is a racemic mixture.

    3. *cis-*2-Butene forms a meso compound; the product has two asymmetric centers and only syn addition occurs.

        *trans-*2-Butene forms a pair of enantiomers; the product has two asymmetric centers and only syn addition occurs.

4. *cis*-2-Butene forms the threo pair of enantiomers; a product with two asymmetric centers is formed and only anti addition of Br_2 occurs.

   trans-2-Butene forms a meso compound; a product with two asymmetric centers is formed and only anti addition of Br_2 occurs.

5. *cis*-2-Butene forms the threo pair of enantiomers; a product with two asymmetric centers is formed and only anti addition of Br and OH occurs.

   trans-2-Butene forms the erythro pair of enantiomers; a product with two asymmetric centers is formed and only anti addition of Br and OH occurs.

6. Both *cis*- and *trans*-2-butene form this product; a product with no asymmetric centers is formed. $CH_3CH_2CH_2CH_3$

7. Both *cis*- and *trans*-2-butene form these products; a product with one asymmetric center is formed.

8. Both *cis*- and *trans*-2-butene form these products; a product with one asymmetric center is formed.

b. For the *cis*- and *trans*-2-butene to form different products, (1) the reaction must form two new asymmetric centers in the product and (2) either syn or anti addition must occur (but not both). Therefore, the cis and trans alkenes form different products when they react with a peroxyacid, with Br_2 in CH_2Cl_2, and with Br_2 in H_2O.

69. Only the compound on the far right is optically active, because it is the only one that has one or more asymmetric centers.

70. **a.** Because there are two asymmetric centers, there are four possible stereoisomers.

b.

71. Compound **c** does not have any asymmetric centers; it is, therefore, achiral but it does not have any stereoisomers.
Compounds **e** and **h** do not have any asymmetric centers; they have cis and trans stereosiomers.
Compounds **a**, **d**, **f**, **i**, and **j** have two asymmetric centers bonded to identical substituents. Thus, there are three stereoisomers, one of which is an achiral meso compound.
Compounds **b** and **j** each have two asymmetric centers bonded to different substituents. Thus, there are four stereoisomers, all of which are chiral.

72. **a.** diastereomers   **b.** enantiomers   **c.** identical   **d.** constitutional isomers

73. **a.**

b.

c.

(The H and OH are added to the same side.)

d.

(The two bromines are added to opposite sides
of the cyclohexane ring.)

74. **a.** (*R*)-citric acid   (OH has the highest priority; ^{14}C has a higher priority than ^{12}C.)

b. The reaction is catalyzed by an enzyme. Only one stereoisomer is typically formed in an enzyme-catalyzed reaction, because an enzyme has a chiral binding site that allows reagents to be delivered to only one side of the functional group of the reactant.

c. The product of the reaction will be achiral because, if it does not have a ^{14}C label, the two CH_2COOH groups will be identical and so it will not have an asymmetric center.

75. **a.**

b.

c.

d.

e.

H ''''' / CH₃ / H₃C ''''' Cl
H₃C   CH₃   Cl '''' H   +   Cl '''' H   +   H ''''' '''' CH₃   H₃C ''''' '''' H   +
S S   H₃C   CH₃   H₃C   Cl   Cl   CH₃
   cis   R   R   S   R   R   S
                                 trans

f.

H '''''   '''' H
D   S   R   D

g.   (CH₃)₂CCH₂CH₃   (CH₃)₂CCH₂CH₃   CH₂Br   CH₂Br
         | | | |
         C '''' H   C   **or**   H —|— Br   Br —|— H
        /   \   / \ | |
     BrCH₂   Br   H '''' CH₂Br   (CH₃)₂CCH₂CH₃   (CH₃)₂CCH₂CH₃
                        Br
         R   S   R   S

h.   CH₃CH₂   H   H   CH₂CH₃   CH₂CH₃   CH₂CH₃
         \   /   \   / | |
      H '''' C — C '''' CH₃   +   H₃CH '''' C — C '''' H   **or** H —|— CH₃ S   CH₃ —|— H   R
         /   \   /   \   CH₃ —|— H   S + H —|— CH₃ R
       H₃C   CH₂CH₂CH₃   CH₃CH₂CH₂   CH₃   CH₂CH₂CH₃   CH₂CH₂CH₃
         S   S   R   R

i.   CH₃CH₂   H   H   CH₂CH₃   CH₂CH₃   CH₂CH₃
         \   /   \   / | |
      H₃C '''' C — C '''' CH₃   +   H₃C '''' C — C '''' CH₃ **or** CH₃ —|— H R   H —|— CH₃ S
         /   \   /   \   CH₃ —|— H S + H —|— CH₃ R
        H   CH₂CH₂CH₃   CH₃CH₂CH₂   H   CH₂CH₂CH₃   CH₂CH₂CH₃
         R   S   R   S

j.

H '''''   '''' H   H '''''   '''' H
Cl S R CH₂CH₃   CH₃CH₂ Cl
                              S   R

76.   **a.** identical   **e.** constitutional isomers
          b. identical   **f.** diastereomers
          c. enantiomers   **g.** constitutional isomers
          d. constitutional isomers **h.** enantiomers

77.   optical purity = $\dfrac{+1.4}{+8.7}$ = 0.16 = 16% excess *R* enantiomer

100% − 16% = 84% is a racemic mixture

R enantiomer = 1/2 (84%) + 16% = 42% + 16% = 58%

78. **a.** *R*   **b.** *S*   **c.**

R

d.

S

e.

$$\begin{array}{c} Cl \\ H\!\!-\!\!\!\!\!\!\begin{array}{c}|\\|\end{array}\!\!\!\!\!\!-CH_3 \\ CH_2CH_3 \end{array}$$

R

f.

$$\begin{array}{c} CH_3 \\ H\!\!-\!\!\!\!\!\!\begin{array}{c}|\\|\end{array}\!\!\!\!\!\!-Cl \\ CH_2CH_3 \end{array}$$

S

79. $[\alpha] = \dfrac{\alpha}{l \times c} = \dfrac{-1.8°}{[2.0 \text{ dm}][0.15 \text{ g/mL}]} = -6.0$

80.   Butaclamol has four asymmetric centers; three of them are carbons and one is a nitrogen.

81.   The only way that *R* and *S* are related to (+) and (−) is that if one configuration (say, *R*) is (+), the other one is (−).

Because some compounds with the *R* configuration are (+) and some are (−), there is no way to determine whether a particular *R* enantiomer is (+) or (−) without putting the compound in a polarimeter or finding out whether someone else has previously determined how the compound rotates the plane of polarization of plane-polarized light.

82.   **a.**

b.

c.

CH$_2$CH$_3$—C(H$_3$C)(Cl)—H (S) + CH$_2$CH$_3$—C(CH$_3$)(Cl)—H (R)   **or**   H—C(CH$_3$)(CH$_2$CH$_3$)—Cl (S) + Cl—C(CH$_3$)(CH$_2$CH$_3$)—H (R)

d.

cyclohexane with CH$_3$ and Br substituents on one carbon

e.

CH$_3$CH$_2$—C(Br)(H)—C(H)(Br)(CH$_2$CH$_3$)   *R S*   a meso compound   **or**   CH$_2$CH$_3$ / H—Br *S* / H—Br *R* / CH$_2$CH$_3$

f.

CH$_3$CH$_2$—C(H)(Br)—C(H)(Br)(CH$_2$CH$_3$) *S S*   +   Br—C(H)(CH$_3$CH$_2$)—C(CH$_2$CH$_3$)(H)(Br) *R R*   **or**   H—Br *S* / Br—H *S* (CH$_2$CH$_3$) + Br—H *R* / H—Br *R* (CH$_2$CH$_3$)

g. The initially formed carbocation is secondary. It undergoes a 1,2-methyl shift to form a tertiary carbocation that gives the products shown below.

(CH$_3$)$_2$CH—C(Br)(CH$_3$)—CH$_2$CH$_3$ *R*   +   CH$_3$CH$_2$—C(Br)(CH$_3$)—CH(CH$_3$)$_2$ *S*   **or**   Br—C(CH$_2$CH$_3$)(CH$_3$)(CHCH$_3$—CH$_3$) *R* + CH$_3$—C(CH$_2$CH$_3$)(Br)(CHCH$_3$—CH$_3$) *S*

h.

CH$_2$CH$_3$—C(H$_3$C)(Br)—H *S*   +   CH$_2$CH$_3$—C(CH$_3$)(Br)—H *R*   **or**   H—C(CH$_3$)(CH$_2$CH$_3$)—Br *S* + Br—C(CH$_3$)(CH$_2$CH$_3$)—H *R*

i.

H$_3$C—C(Cl)(H)—C(H)(Cl)(CH$_3$)   *R S*   **or**   H—Cl *S* / H—Cl *R* (CH$_3$ ... CH$_3$)

a meso compound

j.

k.

a meso compound

l.

83.   **a.**   The compound has four stereoisomers.

b.   The first two stereoisomers are optically inactive because they are meso compounds. (They each have a plane of symmetry.)

84.   **a.**

b.

c.

85.   **a.** and **b.**

ethylcyclobutane   1,1-dimethylcyclobutane

cis-1,2-dimethylcyclobutane   *trans*-1,2-dimethylcyclobutane

cis-1,3-dimethylcyclobutane   *trans*-1,3-dimethylcyclobutane

c. **1.**   ethylcyclobutane
            1,1-dimethylcyclobutane
            1,2-dimethylcyclobutane
            1,3-dimethylcyclobutane

   2.   the three isomers of 1,2-dimethylcyclobutane
            the two isomers of 1,3-dimethylcyclobutane

   3.   *cis*- and *trans*-1,2-dimethylcyclobutane
            cis- and *trans*-1,3-dimethylcyclobutane

   4.   the two trans stereoisomers of 1,2-dimethylcyclobutane

   5.   all the isomers except the two trans stereoisomers of 1,2-dimethylcyclobutane

   6.   *cis*-1,2-dimethylcyclobutane
            (Note: *cis*-1,3-dimethylcyclobutane is not a meso compound, because it does not have any
            asymmetric centers.)

   7.   the two trans stereoisomers of 1,2-dimethylcyclobutane

   8.   *cis*-1,3-dimethylcyclobutane and *trans*-1,3-dimethylcyclobutane
            cis-1,2-dimethylcyclobutane and either of the enantiomers of
            trans-1,2-dimethylcyclobutane

86.

$$\text{observed specific rotation} = \frac{\text{observed rotation}}{\text{concentration} \times \text{length}}$$

$$\frac{-6.52°}{0.187 \text{ g/mL} \times 1 \text{ dm}} = -34.9$$

$$\% \text{ optical purity} = \frac{\text{observed specific rotation}}{\text{specific rotation of the pure enantiomer}} \times 100$$

$$= \frac{-34.9}{-39.0} \times 100$$

$$= 89.5\%$$

$$\% \text{ of the (+)-isomer} = \frac{100 - 89.5}{2} = 5.25\%$$

$$\% \text{ of the (−)-isomer} = 89.5 + 5.25 = 94.75\%$$

87. **a.** diastereomers   **c.** constitutional isomers
       b. identical   **d.** diastereomers

88. **a.**

b.

c.

d.

e.

89.   In the transition state for amine inversion, the nitrogen atom is sp^2 hybridized, which means it has bond angles of 120°. A nitrogen atom in a three-membered ring cannot achieve a 120° bond angle, so the amine inversion that would interconvert the enantiomers cannot occur. Therefore, the enantiomers can be separated.

90.   The first product would not be formed, because none of the bonds attached to C-2 were broken during the reaction. Therefore, the configuration at C-2 cannot change.

91.   The fact that the optical purity is 72% means that there is 72% enantiomeric excess of the S isomer and 28% racemic mixture. Therefore, the actual amount of the S isomer in the sample is 72% + 1/2(28%) = 86%. The amount of the R isomer in the sample is 1/2(28%) = 14%, (or 100% − 86% = 14%).

92.   **A** = a diastereomer   **C** = a diastereomer   **E** = a diastereomer
          B = a diastereomer   **D** = a diastereomer

93.

$\xrightarrow[\text{2. HO}^-,\ \text{H}_2\text{O}_2,\ \text{H}_2\text{O}]{\text{1. BH}_3/\text{THF}}$

$\xrightarrow[\text{H}_2\text{O}]{\text{Br}_2}$

94.   **a.**

b.

This is a pair of enantiomers because they are nonsuperimposable mirror images.

c.

This is the most stable isomer because, since the chloro substituents are all trans to each other, they can all be in the more stable equatorial position. (Recall that there is less steric strain when a substituent is in the equatorial position.)

95.

Tamiflu®

96.   Yes, as long as the Fischer projection is drawn with the #1 carbon at the top of the chain of carbons.

97.   Because fumarate is the trans isomer and it forms an erythro product, the enzyme must catalyze the anti addition of D_2O. (Recall: CIS-SYN-ERYTHRO)

98.   We can determine that $(-)$-1,4-dichloro-2-methylbutane has the S configuration because in the process of forming it from (S)-$(+)$-1-chloro-2-methylbutane, no bonds to the asymmetric center are broken. Therefore, the groups around the asymmetric center have the same relative configuration in the product as they had in the reactant. If the relative configuration is known, the absolute configuration can be determined.

$$CH_2Cl$$

S-configuration
(S)-$(+)$-1-chloro-2-methylbutane

$\xrightarrow{\ Cl_2\ }{hv}$

$$CH_2Cl$$

S-configuration
(S)-$(-)$-1,4-dichloro-2-methylbutane

99.   **a.**   **b.**   **c.**

100.   The trans compound exits as a pair of enantiomers.

As a result of ring-flip, each enantiomer has two chair conformations. In each case, the more stable conformation is the one with the larger group (the *tert*-butyl group) in the equatorial position.

101.   **a.**   The compounds do not have any asymmetric centers.

    b. **1.** is not chiral.

        2. is chiral. Because of its unusual geometry, it is a chiral molecule, even though it does not have any asymmetric centers, because it cannot be superimposed on its mirror image. This will be easier to understand if you build models.

mirror
images are superimposable

mirror
images are not superimposable

102.   The compound is not optically active because it has a center of symmetry.
A center of symmetry is a point, and if a line is drawn to this point from an atom or group and the line is then extended an equal distance beyond the point, the line would touch an identical atom or group.

point of symmetry

Chapter 5 Practice Test

1.   Are the following pairs of compounds identical or a pair of enantiomers?

a. CH_3CH_2———CH_2OH and H———CH_2CH_3 (with H on top, CH_3 on bottom for first; CH_3 on top, CH_2OH on bottom for second)

b.   and

2.   30 mL of a solution containing 2.4 g of a compound rotates the plane of polarized light −0.48° in a polarimeter with a 2-decimeter sample tube. What is the specific rotation of the compound?

3.   Which are meso compounds?

4.   Draw all the constitutional isomers with molecular formula C_4H_9Cl.

5.   Draw all the stereoisomers of each of the following:

a. 2,3-dibromobutane

b. 2,3-dibromopentane

6.   Draw all the products that would be obtained from each of the following reactions, indicating which isomers are formed:

a. 1-butene + HCl

c. *trans*-3-hexene + Br_2

b. 2-pentene + HBr

d. *trans*-3-heptene + Br_2

7.   *R*-(−)-2-methyl-1-butanol can be oxidized to (+)-2-methylbutanoic acid without breaking any of the bonds to the asymmetric center. What is the configuration of (−)-2-methylbutanoic acid?

(*R*)-(−)-2-methyl-1-butanol   (+)-2-methylbutanoic acid

8.   Draw the stereoisomers that would be obtained from each of the following reactions:

a.

c.

$$\xrightarrow[\text{Pd/C}]{H_2}$$

b.

d.

$$\xrightarrow[\text{Pd/C}]{H_2}$$

9.   Which of the following have the *R* configuration?

10.   Draw a diastereomer of the following compound:

11.   Indicate whether each of the following statements is true or false:

a.   Diastereomers have the same melting points.   T   F

b.   The addition of HBr to 3-methyl-2-pentene is a stereospecific reaction.   T   F

c.   The addition of HBr to 3-methyl-2-pentene is a stereoselective reaction.   T   F

d.   The addition of HBr to 3-methyl-2-pentene is a regioselective reaction.   T   F

e.   Meso compounds do not rotate the plane of polarization of
plane-polarized light.   T   F

f.   2,3-Dichloropentane has a stereoisomer that is a meso compound.   T   F

g.   All chiral compounds with the *R* configuration are dextrorotatory.   T   F

h.   A compound with three asymmetric centers can have a maximum of nine
stereoisomers.   T   F

CHAPTER 6
The Reactions of Alkynes • An Introduction to Multistep Synthesis

Important Terms

acetylide ion	the conjugate base of a terminal alkyne.

$$RC\equiv C^-$$ |
| **aldehyde** | a compound with a carbonyl group that is bonded to an alkyl group and to a hydrogen (or bonded to two hydrogens).

(structure: aldehyde forms) |
| **alkylation reaction** | a reaction that adds an alkyl group to a reactant. |
| **alkyne** | a hydrocarbon that contains a triple bond. |
| **carbonyl group** | a carbon doubly bonded to an oxygen.

(structure: carbonyl group) |
| **enol** | an alkene with an OH group bonded to one of the sp^2 carbons.

(structure: enol) |
| **geminal dihalide** | a compound with two halogen atoms bonded to the same carbon. |
| **internal alkyne** | an alkyne with its triple bond not at the end of the carbon chain. |
| **keto-enol tautomers** | a ketone and its isomeric enol. The keto and enol tautomers differ only in the location of a double bond and a hydrogen.

(structure: keto-enol tautomers) |
| **ketone** | a compound with a carbonyl group that is bonded to two alkyl groups.

(structure: ketone) |
| **π-complex** | a complex formed between an electrophile and a triple bond. |
| **polymer** | a large molecule made by covalently linking many small molecules called monomers. |

radical anion   a species with a negative charge and an unpaired electron.

retrosynthetic analysis   working backward (on paper) from a target molecule to available starting
or   materials.
retrosynthesis

tautomerization   interconversion of tautomers.

tautomers   constitutional isomers that are in rapid equilibrium; for example, keto and enol
                                         tautomers. The keto and enol tautomers differ only in the location of a double
                                         bond and a hydrogen.

terminal alkyne   an alkyne with its triple bond at the end of the carbon chain.

vinylic cation   a compound with a positive charge on a vinylic carbon.

vinylic radical   a compound with an unpaired electron on a vinylic carbon.

Solutions to Problems

1.   The molecular formula of a noncyclic hydrocarbon with 14 carbons and no π bonds is $C_{14}H_{30}(C_nH_{2n+2})$.

Because a compound has two fewer hydrogens for every ring and π bond, a compound with one ring and 4 π bonds (2 triple bonds) would have 10 fewer hydrogens than the C_nH_{2n+2} formula.
Thus, the molecular formula is **$C_{14}H_{20}$**.

2.   **a.** $ClCH_2CH_2C\equiv CCH_2CH_3$

c. $CH_3CHC\equiv CH$
$\quad\quad\quad\mid$
$\quad\quad\quad CH_3$

e. $HC\equiv CCH_2CCH_3$
with CH_3 above and CH_3 below

b. (octagon with one double bond / cyclooctyne ring)

d. $CH_3CH_2CHC\equiv CCH_2CHCH_3$
$\quad\quad\quad\quad\mid\quad\quad\quad\quad\mid$
$\quad\quad\quad\quad CH_3\quad\quad\quad CH_3$

f. $CH_3C\equiv CCH_3$

3. $HC\equiv CCH_2CH_2CH_2CH_3$
1-hexyne
butylacetylene

$CH_3C\equiv CCH_2CH_2CH_3$
2-hexyne
methylpropylacetylene

$CH_3CH_2C\equiv CCH_2CH_3$
3-hexyne
diethylacetylene

$CH_3CH_2CHC\equiv CH$
$\quad\quad\quad\mid$
$\quad\quad\quad CH_3$
3-methyl-1-pentyne
sec-butylacetylene

$CH_3CHCH_2C\equiv CH$
$\quad\quad\mid$
$\quad\quad CH_3$
4-methyl-1-pentyne
isobutylacetylene

$CH_3CHC\equiv CCH_3$
$\quad\quad\mid$
$\quad\quad CH_3$
4-methyl-2-pentyne
isopropylmethylacetylene

$\quad\quad CH_3$
$\quad\quad\mid$
$CH_3CC\equiv CH$
$\quad\quad\mid$
$\quad\quad CH_3$
3,3-dimethyl-1-butyne
tert-butylacetylene

4.   **a.** 5-bromo-2-pentyne
b. 6-bromo-2-chloro-4-octyne
c. 1-methoxy-2-pentyne
d. 3-ethyl-1-hexyne

5.   **a.** 6-methyl-2-octyne
b. 5-ethyl-4-methyl-1-heptyne
c. 2-bromo-4-octyne

6.   **a.** 1-hepten-4-yne
b. 4-methyl-1,4-hexadiene
c. 5-vinyl-5-octen-1-yne
d. 3-butyn-1-ol
e. 1,3,5-heptatriene
f. 2,4-dimethyl-4-hexen-1-ol

7.   **a.** (*E*)-1-bromo-1,3-pentadiene
b. 1-hepten-5-yne
c. (*E*)-4-hepten-1-yne

8.   pentane   1-pentene   1-pentyne

9. **a.** sp^2–sp^2 **d.** sp–sp^3 **g.** sp^2–sp^3
b. sp^2–sp^3 **e.** sp–sp **h.** sp–sp^3
c. sp–sp^2 **f.** sp^2–sp^2 **i.** sp^2–sp

10. The less stable reactant will be the more reactive reactant, if the less stable reactant leads to the more stable transition state, **or** if the less stable reactant leads to the less stable transition state **and** the difference in the stabilities of the reactants is greater than the difference in the stabilities of the transition states.

11. **a.** $CH_2\!=\!\underset{|}{C}CH_3$
           $\quad\; Br$

b. $CH_3\underset{|}{\overset{Br}{C}}CH_3$
         $\quad\;\; Br$

c. $CH_3\underset{|}{C}\!=\!CCH_3$
         $\quad Br \qquad\; Br$

d. $HC\!-\!CCH_3$ (with Br Br on each carbon)

e. $CH_3CH_2\underset{|}{\overset{Br}{C}}CH_3$
         $\qquad\quad Br$

f. $CH_3\underset{|}{\overset{Br}{C}}CH_2CH_2CH_3$ + $CH_3CH_2\underset{|}{\overset{Br}{C}}CH_2CH_3$

12. Only anti addition occurs, because the intermediate is a cyclic bromonium ion.
Therefore the product has the *E* configuration.

$$\underset{Br}{\overset{H_3C}{\diagdown}}C\!=\!C\underset{CH_3}{\overset{Br}{\diagup}}$$

13. Because the alkyne is not symmetrical, two ketones will be obtained.

$$\underset{O}{\overset{\;\;\;\;O}{\;}}$$
$CH_3CH_2\overset{\|}{C}CH_2CH_2CH_2CH_3$ and $CH_3CH_2CH_2\overset{\|}{C}CH_2CH_2CH_3$

14. **a.** $CH_3C\!\equiv\!CH$ **b.** $CH_3CH_2C\!\equiv\!CCH_2CH_3$ **c.** $HC\!\equiv\!C\!-\!\bigcirc$

The best answer for **b** is 3-hexyne, because it would form only the desired ketone. 2-Hexyne would form two different ketones, so only half of the product would be the desired ketone.

15. **a.** $CH_2\!=\!\underset{|}{C}CH_3$
           $\qquad\; OH$

Because the ketone has identical substituents bonded to the carbonyl carbon, it has only one enol tautomer.

b.

$$\underset{\text{OH}}{\text{CH}_3\text{CH}}\!=\!\text{CCH}_2\text{CH}_2\text{CH}_3 \quad \text{and} \quad \underset{\text{OH}}{\text{CH}_3\text{CH}_2\text{C}}\!=\!\text{CHCH}_2\text{CH}_3$$

E and *Z* isomers are possible for each of these enols.

$$\begin{array}{c} \text{H} \qquad\qquad \text{OH} \\ \diagdown\qquad\diagup \\ \text{C}\!=\!\text{C} \\ \diagup\qquad\diagdown \\ \text{H}_3\text{C} \qquad \text{CH}_2\text{CH}_2\text{CH}_3 \end{array}$$
E

$$\begin{array}{c} \text{H} \qquad\qquad \text{CH}_2\text{CH}_2\text{CH}_3 \\ \diagdown\qquad\diagup \\ \text{C}\!=\!\text{C} \\ \diagup\qquad\diagdown \\ \text{H}_3\text{C} \qquad \text{OH} \end{array}$$
Z

$$\begin{array}{c} \text{HO} \qquad\quad \text{H} \\ \diagdown\qquad\diagup \\ \text{C}\!=\!\text{C} \\ \diagup\qquad\diagdown \\ \text{CH}_3\text{CH}_2 \qquad \text{CH}_2\text{CH}_3 \end{array}$$
E

$$\begin{array}{c} \text{HO} \qquad\quad \text{CH}_2\text{CH}_3 \\ \diagdown\qquad\diagup \\ \text{C}\!=\!\text{C} \\ \diagup\qquad\diagdown \\ \text{CH}_3\text{CH}_2 \qquad \text{H} \end{array}$$
Z

c.

$$\underset{\text{OH}}{\text{CH}_2}\!=\!\overset{\displaystyle|}{\text{C}}\!-\!\bigcirc \quad \text{and} \quad \underset{\text{OH}}{\text{CH}_3}\overset{\displaystyle|}{\text{C}}\!=\!\bigcirc$$

Because each enol has identical groups bonded to one of its sp^2 carbons, *E* and *Z* isomers are not possible for either enol.

16. **a.** (1) $\underset{\text{O}}{\overset{\displaystyle\|}{\text{CH}_3\text{CH}_2\text{CCH}_3}}$ **b.** (1) $\underset{\text{O}}{\overset{\displaystyle\|}{\text{CH}_3\text{CH}_2\text{CCH}_3}}$ **c.** (1) $\underset{\text{O}}{\overset{\displaystyle\|}{\text{CH}_3\text{CH}_2\text{CH}_2\text{CCH}_3}}$ and $\underset{\text{O}}{\overset{\displaystyle\|}{\text{CH}_3\text{CH}_2\text{CCH}_2\text{CH}_3}}$

(2) $\underset{\text{O}}{\overset{\displaystyle\|}{\text{CH}_3\text{CH}_2\text{CH}_2\text{CH}}}$ (2) $\underset{\text{O}}{\overset{\displaystyle\|}{\text{CH}_3\text{CH}_2\text{CCH}_3}}$ (2) $\underset{\text{O}}{\overset{\displaystyle\|}{\text{CH}_3\text{CH}_2\text{CH}_2\text{CCH}_3}}$ and $\underset{\text{O}}{\overset{\displaystyle\|}{\text{CH}_3\text{CH}_2\text{CCH}_2\text{CH}_3}}$

major product because there is less steric hindrance to addition of BH_3 at the 2-position

17.   Ethyne (acetylene)

An alkyne can form an aldehyde only if the OH group adds to a terminal *sp* carbon. In the acid-catalyzed addition of water to a terminal alkyne, the proton adds to the terminal *sp* carbon. Therefore, the only way the OH group can add to a terminal *sp* carbon is if there are two terminal *sp* carbons in the alkyne. In other words, the alkyne must be ethyne.

18.   **a.** $\text{CH}_3\text{CH}_2\text{CH}_2\text{C}\!\equiv\!\text{CH}$ **or** $\text{CH}_3\text{CH}_2\text{C}\!\equiv\!\text{CCH}_3 \xrightarrow[\text{Pd/C}]{\text{H}_2} \text{CH}_3\text{CH}_2\text{CH}_2\text{CH}_2\text{CH}_3$

1-pentyne   2-pentyne

b. $\text{CH}_3\text{C}\!\equiv\!\text{CCH}_3 \xrightarrow[\substack{\text{Lindlar} \\ \text{catalyst}}]{\text{H}_2}$ $\begin{array}{c} \text{H}_3\text{C} \qquad\quad \text{CH}_3 \\ \diagdown\qquad\diagup \\ \text{C}\!=\!\text{C} \\ \diagup\qquad\diagdown \\ \text{H} \qquad\qquad \text{H} \end{array}$

2-butyne

c. $CH_3CH_2C{\equiv}CCH_3$ $\xrightarrow[\text{NH}_3\text{ (liq)}]{\text{Na}}$

2-pentyne   −78 °C

$$\begin{array}{c} CH_3CH_2 \qquad H \\ C{=}C \\ H \qquad CH_3 \end{array}$$

d. $CH_3CH_2CH_2CH_2C{\equiv}CH$ $\xrightarrow[\substack{\text{Lindlar} \\ \text{catalyst}}]{\text{H}_2}$ $CH_3CH_2CH_2CH_2CH{=}CH_2$

1-hexyne

19.   The base used to remove a proton must be stronger than the base that is formed as a result of proton removal. Therefore, the base used to remove a proton from a terminal alkyne must be a stronger base than the conjugate base of the terminal alkyne. A terminal alkyne has a $pK_a \sim 25$. In other words, any base whose conjugate acid has a pK_a greater than 25 can be used.

20.   The reaction of sodium amide with an alkane will not favor products because the carbanion that would be formed is a stronger base than the amide ion. (Recall that the equilibrium favors reaction of the strong acid (or strong base) and formation of the weak acid (or weak base); Section 1.19.)

$$CH_3CH_3 \;+\; {}^-NH_2 \;\rightleftharpoons\; CH_3\overset{-}{C}H_2 \;+\; NH_3$$

$pK_a > 60$ $pK_a = 36$

weaker acid   weaker base   stronger base   stronger acid

21. **a.** $CH_3CH_2CH_2\overset{-}{C}H_2 \;>\; CH_3CH_2CH{=}\overset{-}{C}H \;>\; CH_3CH_2C{\equiv}C^-$

b. ${}^-NH_2 \;>\; CH_3C{\equiv}C^- \;>\; CH_3CH_2O^- \;>\; F^-$

22. **a.** Solved in the text.

b. $H_2C{=}\overset{+}{C}H$

A triply bonded (sp) carbon is more electronegative than a doubly bonded (sp^2) carbon. Therefore, a triply bonded carbon with a positive charge would be less stable than a doubly bonded carbon with a positive charge. Thus, the vinyl cation is more stable.

23.   Solved in the text.

24. **a.** $HC{\equiv}CH$ $\xrightarrow[\text{2. CH}_3\text{CH}_2\text{CH}_2\text{Br}]{\text{1. NaNH}_2}$ $CH_3CH_2CH_2C{\equiv}CH$

b. $HC{\equiv}CH$ $\xrightarrow[\text{2. CH}_3\text{Br}]{\text{1. NaNH}_2}$ $CH_3C{\equiv}CH$ $\xrightarrow{\text{H}_2\text{/Lindlar catalyst}}$ $CH_3CH{=}CH_2$

c. $HC{\equiv}CH$ $\xrightarrow[\text{2. CH}_3\text{Br}]{\text{1. NaNH}_2}$ $CH_3C{\equiv}CH$ $\xrightarrow[\text{2. CH}_3\text{Br}]{\text{1. NaNH}_2}$ $CH_3C{\equiv}CCH_3$ $\xrightarrow[\substack{\text{Lindlar} \\ \text{catalyst}}]{\text{H}_2}$

$$\begin{array}{c} H_3C \qquad CH_3 \\ C{=}C \\ H \qquad H \end{array}$$

d. product of **a** $\xrightarrow[\text{2. HO}^-, \text{H}_2\text{O}_2, \text{H}_2\text{O}]{\text{1. BH}_3/\text{THF}}$ CH$_3$CH$_2$CH$_2$CH$_2$CH$\overset{\text{O}}{\parallel}$

e. product of **b** $\xrightarrow{\text{HBr}}$ CH$_3$CHCH$_3$ | Br

f. HC≡CH $\xrightarrow[\text{2. CH}_3\text{Br}]{\text{1. NaNH}_2}$ CH$_3$C≡CH $\xrightarrow{\text{excess HCl}}$ CH$_3$CCH$_3$ with Cl above and Cl below

25.

a. CH$_3$CH$_2$CCH$_3$ with Cl above and Cl below

b. CH$_3$CH$_2$CH$_2$CCH$_2$CH$_3$ with Cl above and Cl below

c. CH$_3$CH$_2$CH$_2$CCH$_2$CH$_2$CH$_3$ with Cl above and Cl below + CH$_3$CH$_2$CCH$_2$CH$_2$CH$_2$CH$_3$ with Cl above and Cl below

26.

a. CH$_3$C≡CCH$_2$CH$_2$CH$_3$

b. CH$_3$CH$_2$C≡CCHCH$_2$CH$_2$CH$_3$ | CH$_2$CH$_3$

c. CH$_3$C≡CH

d. CH$_2$=CHC≡CH

e. CH$_3$OC≡CH

f. CH$_3$CC≡CCHCH$_2$CH$_3$ with CH$_3$ above and CH$_3$ below first C, CH$_3$ above the CH

g. BrC≡CCH$_2$CH$_2$CH$_3$

h. cyclohexene ring with OH and CH$_3$ substituents

i. CH$_3$CH$_2$C≡CCH$_2$CH$_3$

j. CH$_3$CC≡CCCH$_3$ with CH$_3$ above and CH$_3$ below each quaternary carbon

k. cyclopentane—C≡CH

l. CH$_3$C≡CCH$_2$CHCHCH$_3$ with CH$_3$ above the CH and CH$_3$ below

27. CH$_3$CH$_2$$\overset{+}{\text{C}}$=CH$_2$ + :Cl:$^-$ ⟶ CH$_3$CH$_2$C=CH$_2$ | :Cl:

  electrophile   nucleophile

$$CH_3C\equiv CH \ + \ H{-}Br \longrightarrow CH_3\overset{+}{C}{=}CH_2 \ + \ Br^-$$

nucleophile   electrophile

$$CH_3C\equiv C{-}H \ + \ \ddot{:}NH_2^- \longrightarrow CH_3C\equiv C\overset{..}{:}^- \ + \ \ddot{N}H_3$$

electrophile   nucleophile

28. **a.** 5-bromo-2-hexyne   **c.** 5,5-dimethyl-2-hexyne   **e.** 1,5-cyclooctadiene
     b. 5-methyl-2-octyne   **d.** 6-chloro-2-methyl-3-heptyne   **f.** 1,6-dimethyl-1,3-cyclohexadiene

29.

30. The student named only one correctly.
     a. 4-methyl-2-hexyne   **c.** correct
     b. 7-bromo-3-heptyne   **d.** 2-pentyne

31. $HC\equiv CCH_2CH_2CH_2CH_2CH_3$

1-heptyne
pentylacetylene

$CH_3C\equiv CCH_2CH_2CH_2CH_3$

2-heptyne
butylmethylacetylene

$CH_3CH_2C\equiv CCH_2CH_2CH_3$

3-heptyne
ethylpropylacetylene

$CH_3CH_2CH_2CHC\equiv CH$
          |
          CH_3

3-methyl-1-hexyne

$CH_3CH_2CHCH_2C\equiv CH$
      |
      CH_3

4-methyl-1-hexyne

$CH_3CHCH_2CH_2C\equiv CH$
   |
   CH_3

5-methyl-1-hexyne
isopentylacetylene

$CH_3CH_2CHC\equiv CCH_3$
        |
        CH_3

4-methyl-2-hexyne
sec-butylmethylacetylene

$CH_3CHCH_2C\equiv CCH_3$
   |
   CH_3

5-methyl-2-hexyne
isobutylmethylacetylene

$CH_3CHC\equiv CCH_2CH_3$
   |
   CH_3

2-methyl-3-hexyne
ethylisopropylacetylene

$$CH_3CC \equiv CCH_3$$

with CH_3 substituents (structure)

4,4-dimethyl-2-pentyne
tert-butylmethylacetylene

$$CH_3CCH_2C \equiv CH$$

with CH_3 substituents (structure)

4,4-dimethyl-1-pentyne

$$CH_3CH_2CC \equiv CH$$

with CH_3 substituents (structure)

3,3-dimethyl-1-pentyne
tert-pentylacetylene

$$CH_3CH_2CHC \equiv CH$$

with CH_2CH_3 substituent (structure)

3-ethyl-1-pentyne

$$CH_3CHCHC \equiv CH$$

with CH_3 substituents (structure)

3,4-dimethyl-1-pentyne

32. Addition of Br_2 to the triple bond in the presence of H_2O forms an enol that immediately tautomerizes to a ketone.

$$CH_3CH_2CH_2C \equiv CH \xrightarrow[H_2O]{Br_2} CH_3CH_2CH_2C = CHBr \longrightarrow CH_3CH_2CH_2CCH_2Br$$

with OH on the vinyl carbon and O (ketone) in the product

33. a. $$CH_3CH_2CH_2C \equiv CH \xrightarrow[\text{2. HO}^-, H_2O_2, H_2O]{\text{1. BH}_3/\text{THF}} CH_3CH_2CH_2CH_2CH$$

with O (aldehyde)

b. $$CH_3CH_2CH = CH_2 \xrightarrow[\text{2. HO}^-, H_2O_2, H_2O]{\text{1. BH}_3} CH_3CH_2CH_2CH_2OH$$

c. $$CH_3CH_2CH_2C \equiv CCH_2CH_2CH_3 + H_2O \xrightarrow{H_2SO_4} CH_3CH_2CH_2CCH_2CH_2CH_2CH_3$$

with O (ketone)

This symmetrical alkyne will give the greatest yield of the desired ketone.
Because the reactant is not a terminal alkyne, the reaction can take place without the mercuric ion catalyst.

34. a. H_2 Lindlar catalyst   **b.** Na, NH_3 (liq), −78 °C   **c.** excess H_2, Pd/C

35. a. 1-octen-6-yne   **d.** 5-chloro-1,3-cyclohexadiene
b. *cis*-3-hexen-1-ol   **e.** 1-methyl-1,3,5-cycloheptatriene
c. 1,5-octadiyne

36. The molecular formula of the hydrocarbon is $C_{32}H_{56}$.
With one triple bond, two double bonds, and one ring, the degree of unsaturation is 5.
Therefore, the compound is missing 10 hydrogens from $C_nH_{2n+2} = C_{32}H_{66}$.

37.

a. CH₂=CCH₃ | Br

b. Br | CH₃CCH₃ | Br

c. (Br)(H)C=C(CH₃)(Br)

d. Br Br | | BrCHCCH₃ | Br

e. O ‖ CH₃CCH₃

f. O ‖ CH₃CH₂CH

g. CH₃CH₂CH₃

h. CH₃CH=CH₂

i. CH₃CH=CH₂

j. CH₃C≡C⁻

k. CH₃C≡CCH₂CH₂CH₃

38.

a. CH₃CH=CCH₃ | Br

b. Br | CH₃CH₂CCH₃ | Br

c. (Br)(H₃C)C=C(CH₃)(Br)

d. Br Br | | CH₃C—CCH₃ | | Br Br

e. O ‖ CH₃CCH₂CH₃

f. O ‖ CH₃CCH₂CH₃

g. CH₃CH₂CH₂CH₃

h. (H₃C)(H)C=C(CH₃)(H)

i. (H)(H₃C)C=C(CH₃)(H)

j. no reaction

k. no reaction

39.

a. Cl Cl / Cl-C structure with isopropyl, extra Cl

b. Cl / C=C with Cl and isobutyl

c. O ketone with isobutyl

d. HO— chain with methyl branch

e. pentane chain

f. cis pentene

g. trans branched alkene

40.   **a.**   **1.** $CH_3CHC\equiv CH$ $\xrightarrow[\text{Lindlar catalyst}]{H_2}$ $CH_3CHCH=CH_2$
                        | |
                        CH_3 CH_3

$\downarrow$ $\begin{array}{l}H_2O\\H_2SO_4\end{array}$

OH   H
| |
$CH_3CCH_2CH_3$ $\xleftarrow{H_2O}$ $CH_3\overset{+}{C}CH_2CH_3$ $\xleftarrow{}$ $CH_3\overset{+}{C}CHCH_3$
| | |
CH_3 CH_3 CH_3

2. $CH_3CHC\equiv CH$ $\xrightarrow[\text{Lindlar catalyst}]{H_2}$ $CH_3CHCH=CH_2$
                | |
                CH_3 CH_3

$\downarrow$ 1. BH$_3$/THF
2. H$_2$O$_2$, HO$^-$, H$_2$O

$CH_3CHCH_2CH_2OH$
|
CH_3

b.   3-Methyl-2-butanol will be a minor product obtained from both **1.** and **2.**

OH
|
$CH_3CHCH_2CHCH_3$
|
CH_3
3-methyl-2-butanol

3-Methyl-2-butanol will be obtained from **1.**, because water can attack the secondary carbocation before it has a chance to rearrange to the tertiary carbocation.

3-Methyl-2-butanol will be obtained from **2.**, because in the second step of the synthesis boron can also add to the other sp^2 carbon; it will be a minor product because the transition state for its formation is less stable than the transition state leading to the major product. Because a carbocation is not formed as an intermediate, a carbocation rearrangement cannot occur.

(The proton cannot add to the other sp^2 carbon in the second step of part **1.** because that would require the formation of a primary carbocation in the second step of the synthesis. Primary carbocations are so unstable that they can never be formed.)

41.   Three of the names are correct.
    a. 3-heptyne   **c.** correct   **e.** correct
    b. 5-methyl-3-heptyne   **d.** 6,7-dimethyl-3-octyne   **f.** correct

42.   Only **c** and **e** are keto-enol tautomers. Notice that an enol tautomer has an OH group bonded to an sp^2 carbon. The structures in **d** are not enol tautomers, because they do not have the oxygen on the same carbon.

43.   **a.** $HC\equiv CH$ $\xrightarrow[\text{HgSO}_4]{\text{H}_2\text{O, H}_2\text{SO}_4}$ $CH_2=CHOH$ $\rightleftharpoons$ $\overset{\displaystyle O}{\overset{\displaystyle \|}{CH_3CH}}$
                                                                        enol

b. $HC\equiv CH$ $\xrightarrow[\text{2. CH}_3\text{CH}_2\text{Br}]{\text{1. NaNH}_2}$ $CH_3CH_2C\equiv CH$ $\xrightarrow[\substack{\text{Lindlar}\\\text{catalyst}}]{\text{H}_2}$ $CH_3CH_2CH\!=\!CH_2$

$$\downarrow \text{Br}_2 \mid \text{CH}_2\text{Cl}_2$$

$$\underset{\underset{\displaystyle Br}{|}}{CH_3CH_2CHCH_2Br}$$

c. $HC\equiv CH$ $\xrightarrow[\text{2. CH}_3\text{Br}]{\text{1. NaNH}_2}$ $HC\equiv CCH_3$ $\xrightarrow[\text{HgSO}_4]{\text{H}_2\text{O, H}_2\text{SO}_4}$ $\overset{\displaystyle O}{\overset{\displaystyle \|}{CH_3CCH_3}}$

d. $HC\equiv CH$ $\xrightarrow[\text{2. CH}_3\text{Br}]{\text{1. NaNH}_2}$ $HC\equiv CCH_3$ $\xrightarrow[\text{2. CH}_3\text{CH}_2\text{Br}]{\text{1. NaNH}_2}$ $CH_3C\equiv CCH_2CH_3$

$$\downarrow \text{Na} \mid \text{NH}_3 \text{ (liq), } -78\,°\text{C}$$

e. $HC\equiv CH$ $\xrightarrow[\text{2. CH}_3\text{Br}]{\text{1. NaNH}_2}$ $HC\equiv CCH_3$ $\xrightarrow[\text{2. CH}_3\text{CH}_2\text{Br}]{\text{1. NaNH}_2}$ $CH_3C\equiv CCH_2CH_3$

$$\downarrow \text{H}_2 \mid \substack{\text{Lindlar}\\\text{catalyst}}$$

f. $HC\equiv CH$ $\xrightarrow[\text{2. CH}_3\text{Br}]{\text{1. NaNH}_2}$ $HC\equiv CCH_3$ $\xrightarrow[\text{2. CH}_3\text{CH}_2\text{Br}]{\text{1. NaNH}_2}$ $CH_3C\equiv CCH_2CH_3$

$$\downarrow \text{excess H}_2 \mid \text{Pd/C}$$

44. **a.** Syn addition of H_2 forms *cis*-2-butene; when Br_2 adds to *cis*-2-butene, the threo pair of enantiomers is formed.

b. Reaction with sodium and liquid ammonia forms *trans*-2-butene; when Br$_2$ adds to *trans*-2-butene, a meso compound is formed.

$$\begin{array}{c} CH_3 \\ H{-}\!\!\!-\!\!\!-Br \\ H{-}\!\!\!-\!\!\!-Br \\ CH_3 \end{array}$$

c. Anti addition of Cl$_2$ forms *trans*-2,3-dichloro-2-butene; when Br$_2$ adds to *trans*-2,3-dichloro-2-butene, a meso compound is formed.

$$\begin{array}{c} CH_3 \\ Cl{-}\!\!\!-\!\!\!-Br \\ Cl{-}\!\!\!-\!\!\!-Br \\ CH_3 \end{array}$$

45. **a.** $\overset{\displaystyle O}{\overset{\|}{CH_3CH_2CCH_3}}$ **b.** $\overset{\displaystyle O}{\overset{\|}{CH_3CH_2CH_2CCH_3}}$ **c.** cyclohexanone $=O$ **d.** cyclohexanecarbaldehyde $\overset{\displaystyle O}{\overset{\|}{CH}}$

46. **a.** $HC{\equiv}CH \xrightarrow[\text{2. } CH_3CH_2CH_2CH_2Br]{\text{1. } NaNH_2} CH_3CH_2CH_2CH_2C{\equiv}CH$

$$\xrightarrow[\substack{H_2SO_4}]{\substack{H_2O}}\Bigg|\ HgSO_4$$

$$\overset{\displaystyle O}{\overset{\|}{CH_3CH_2CH_2CH_2CCH_3}}$$

b. $HC{\equiv}CH \xrightarrow[\text{2. } CH_3CH_2Br]{\text{1. } NaNH_2} CH_3CH_2C{\equiv}CH \xrightarrow[\text{catalyst}]{\substack{H_2 \\ \text{Lindlar}}} CH_3CH_2CH{=}CH_2$

$$\Big|\ HBr$$

$$\underset{\displaystyle |}{CH_3CH_2CHCH_3}\ \ \text{Br}$$

c. $HC{\equiv}CH \xrightarrow[\text{2. } CH_3CH_2CH_2Br]{\text{1. } NaNH_2} CH_3CH_2CH_2C{\equiv}CH \xrightarrow[\text{catalyst}]{\substack{H_2 \\ \text{Lindlar}}} CH_3CH_2CH_2CH{=}CH_2$

$$\Bigg|\ \substack{H_2O \\ H_2SO_4}$$

$$\underset{\displaystyle |}{CH_3CH_2CH_2CHCH_3}\ \ \text{OH}$$

d.

e.

f.

47. The chemist can make 3-octyne by using 1-hexyne instead of 1-butyne. He would then need to use ethyl bromide (instead of butyl bromide) for the alkylation step:

$$CH_3CH_2CH_2CH_2C \equiv CH \xrightarrow[\text{2. } CH_3CH_2Br]{\text{1. } NaNH_2} CH_3CH_2CH_2CH_2C \equiv CCH_2CH_3$$

Or he could make the 1-butyne he needed by alkylating ethyne:

$$HC \equiv CH \xrightarrow[\text{2. } CH_3CH_2Br]{\text{1. } NaNH_2} CH_3CH_2C \equiv CH$$

$$\xrightarrow[\text{1. } NaNH_2]{\text{2. } CH_3CH_2CH_2CH_2Br}$$

$$CH_3CH_2CH_2CH_2C \equiv CCH_2CH_3$$

48. a. Only one product is obtained from the hydroboration-oxidation of 2-butyne, because the alkyne is symmetrical. Two different products can be obtained from hydroboration-oxidation of 2-pentyne, because the alkyne is not symmetrical.

$$CH_3C \equiv CCH_3 \qquad\qquad CH_3C \equiv CCH_2CH_3$$

1. BH₃/THF
2. HO⁻, H₂O₂, H₂O

1. BH₃/THF
2. HO⁻, H₂O₂, H₂O

$$CH_3CH_2CCH_3 \qquad CH_3CCH_2CH_3CH_3 \; + \; CH_3CH_2CCH_2CH_3$$

b. Only one product will be obtained from hydroboration-oxidation of any symmetrical alkyne, such as 3-hexyne or 4-octyne.

$$CH_3CH_2C \equiv CCH_2CH_3 \qquad\qquad CH_3CH_2CH_2C \equiv CCH_2CH_2CH_3$$

3-hexyne   4-octyne

49.   **a.**   The first step forms a trans alkene. Syn addition to a trans alkene forms the threo pair of enantiomers.

    b.   The first step forms a cis alkene. Syn addition to a cis alkene forms the erythro pair of enantiomers but, since each asymmetric carbon is bonded to the same four groups, the product here is a meso compound.

$$
\begin{array}{c}
\text{CH}_2\text{CH}_3 \\
\text{H}\!\!-\!\!|\!\!-\!\!\text{D} \\
\text{H}\!\!-\!\!|\!\!-\!\!\text{D} \\
\text{CH}_2\text{CH}_3
\end{array}
$$

50.   (3*E*,6*E*)-3,7,11-trimethyl-1,3,6,10-dodecatetraene

The configuration of the double bond at the 1-position and at the 10-position is not specified because isomers are not possible at those positions, since there are two hydrogens bonded to C-1 and two methyl groups bonded to C-11.

51.   **a.** $\text{HC}\!\equiv\!\text{CH} \xrightarrow{\text{NaNH}_2} \text{HC}\!\equiv\!\text{C}^- \xrightarrow{\text{CH}_3\text{Br}} \text{HC}\!\equiv\!\text{CCH}_3 \xrightarrow{\text{NaNH}_2} \ ^-\text{C}\!\equiv\!\text{CCH}_3$

$$\downarrow {\scriptstyle \text{CH}_3\text{CH}_2\text{CH}_2\text{CH}_2\text{CH}_2\text{Br}}$$

$$
\begin{array}{c}
\text{H}_3\text{C} \qquad\qquad \text{CH}_2\text{CH}_2\text{CH}_2\text{CH}_2\text{CH}_3 \\
\text{C}\!=\!\text{C} \\
\text{H} \qquad\qquad\qquad \text{H}
\end{array}
\xleftarrow[\text{Lindlar catalyst}]{\text{H}_2}
\text{CH}_3\text{CH}_2\text{CH}_2\text{CH}_2\text{CH}_2\text{C}\!\equiv\!\text{CCH}_3
$$

    b. $\text{HC}\!\equiv\!\text{CH} \xrightarrow{\text{NaNH}_2} \text{HC}\!\equiv\!\text{C}^- \xrightarrow{\text{CH}_3\text{CH}_2\text{Br}} \text{HC}\!\equiv\!\text{CCH}_2\text{CH}_3 \xrightarrow{\text{NaNH}_2} \ ^-\text{C}\!\equiv\!\text{CCH}_2\text{CH}_3$

$$\downarrow {\scriptstyle \text{CH}_3\text{CH}_2\text{CH}_2\text{Br}}$$

$$
\begin{array}{c}
\text{CH}_3\text{CH}_2 \qquad\qquad \text{H} \\
\text{C}\!=\!\text{C} \\
\text{H} \qquad\qquad \text{CH}_2\text{CH}_2\text{CH}_3
\end{array}
\xleftarrow[\text{NH}_3 \text{ (liq)}, -78\ ^\circ\text{C}]{\text{Na}}
\text{CH}_3\text{CH}_2\text{CH}_2\text{C}\!\equiv\!\text{CCH}_2\text{CH}_3
$$

c. $HC\equiv CH$ $\xrightarrow{NaNH_2}$ $HC\equiv C^-$ $\xrightarrow{CH_3CH_2Br}$ $HC\equiv CCH_2CH_3$ $\xrightarrow{NaNH_2}$ $^-C\equiv CCH_2CH_3$

CH_3CH_2Br ↓

$CH_3CH_2\underset{\underset{\displaystyle Br}{|}}{C}H\underset{\underset{\displaystyle OH}{|}}{C}HCH_2CH_3$ $\xleftarrow[H_2O]{Br_2}$ $\underset{H}{\overset{CH_3CH_2}{\diagdown}}C=C\underset{H}{\overset{CH_2CH_3}{\diagup}}$ $\xleftarrow[\text{Lindlar catalyst}]{H_2}$ $CH_3CH_2C\equiv CCH_2CH_3$

Because anti addition occurs in the last step, the threo enantiomers are formed.

If $Na/NH_3(liq)/-78\ °C$ is used instead of H_2/Lindlar catalyst in the fifth step, the trans alkene will be formed. Reaction of Br_2 and H_2O with the trans alkene will form the erythro enantiomers.

52. There are two reasons why hyperconjugation is less effective in stabilizing a positive charge on a vinylic cation than a positive charge on an alkyl cation.

First, an sp^2—s bond is stronger than an sp^3—s bond, so the sp^2 C—H bonds are less able to donate electron density to the adjacent positively charged carbon.

Second, an sp^2 carbon has 120° bond angles compared to the 109.5° bond angles of an sp^3 carbon, so the σ bonding orbitals of an sp^2 carbon are farther away from the positively charged carbon, which makes them less able to interact with that center.

$$\underset{\text{an alkyl cation}}{H-\underset{\underset{\displaystyle H}{|}}{\overset{\overset{\displaystyle H}{|}}{C}}-\overset{+}{C}\underset{\diagdown H}{\diagup^H}} \qquad \underset{\text{a vinylic cation}}{\overset{H}{\diagdown}\underset{H}{\diagup}C=\overset{+}{C}-H}$$

Chapter 6 Practice Test

1.   What reagents could be used to convert the given starting material into the desired product?

a.   [cyclohexyl]—C≡CH → [cyclohexyl]—CH₂CH=O

b.   [phenyl]—C≡CCH₃ → [phenyl with] C=C with H, H, and CH₃

2.   Draw the enol tautomer of the following ketone:

[cyclohexanone structure with O]

3.   Give the structure of each of the following:

a.   *sec*-butyl isobutyl acetylene   b.   2-methyl-1,3-cyclohexadiene

4.   Indicate whether each of the following statements is true or false:

a.   A terminal alkyne is more stable than an internal alkyne.   T   F

b.   Propyne is more reactive than propene toward reaction with HBr.   T   F

c.   1-Butyne is more acidic than 1-butene.   T   F

d.   An sp^2 carbon is more electronegative than an sp^3 carbon.   T   F

5.   Give the systematic name for each of the following compounds:

a.   CH₃CHC≡CCH₂CH₂Br   b.   CH₃CHC≡CCH₂CH₂OH
         | |
         CH₃   CH₃

6. What alkyne would be the best reagent to use for the synthesis of the following ketone?

$$\overset{\overset{\displaystyle O}{\displaystyle \|}}{CH_3CH_2CH_2CCH_3}$$

7. Rank the following compounds in order of decreasing acidity:
(Label the most acidic compound #1.)

NH_3 $CH_3C{\equiv}CH$ CH_3CH_3 H_2O $CH_3CH{=}CH_2$

8. Give an example of a ketone that has two enol tautomers.

9. Show how the target molecules could be prepared from the given starting materials.

a. $CH_3CH_2C{\equiv}CH$ $\longrightarrow$ $CH_3CH_2CH_2CH_2CH_2CH_3$

b. $CH_3CH_2C{\equiv}CH$ $\longrightarrow$ $CH_3CH_2\overset{\overset{\displaystyle |}{\underset{\displaystyle Br}{}}}{C}HCH_2CH_2CH_3$

c. $CH_3CH_2C{\equiv}CH$ $\longrightarrow$ $CH_3CH_2\overset{\overset{\displaystyle O}{\displaystyle \|}}{C}CH_2CH_2CH_3$

CHAPTER 7
**Delocalized Electrons and Their Effect on Stability, Reactivity, and pK_a •
More About Molecular Orbital Theory**

Important Terms

1,2-addition (direct addition)	addition to the 1- and 2-positions of a conjugated system.
1,4-addition (conjugate addition)	addition to the 1- and 4-positions of a conjugated system.
allene	a compound with two adjacent double bonds.
allylic carbon	a carbon adjacent to an sp^2 carbon of a carbon-carbon double bond.
allylic cation	a compound with a positive charge on an allylic carbon.
antibonding molecular orbital	the molecular orbital formed when out-of-phase orbitals overlap.
antisymmetric molecular orbital	a molecular orbital that does not have a plane of symmetry but would have if one half of the MO were turned upside down.
aromatic compound	cyclic compounds that are unusually stable because of large delocalization energies (see Section 15.1).
benzylic carbon	a carbon, joined to other atoms by single bonds, that is bonded to a benzene ring.
benzylic cation	a compound with a positive charge on a benzylic carbon.
bonding molecular orbital	the molecular orbital formed when in-phase orbitals overlap.
bridged bicyclic compound	a bicyclic compound in which the rings share two nonadjacent carbons.
common intermediate	an intermediate that reaction pathways have in common.
concerted reaction	a reaction in which all the bond making and breaking processes occur in a single step.
conjugate addition	addition to the 1- and 4-positions of a conjugated system.
conjugated diene	a compound with two conjugated double bonds.
conjugated double bonds	double bonds separated by one single bond.
contributing resonance structure	a structure with localized electrons that approximates the true structure of a compound with delocalized electrons.
cumulated double bonds	double bonds that are adjacent to one another.

cycloaddition reaction   a reaction in which two π-electron-containing reactants combine to form a single cyclic product.

[4 + 2] cycloaddition reaction   a cycloaddition reaction in which six π electrons participate in the transition state with four π electrons coming from one reactant and two π electrons coming from the other reactant.

delocalization energy (resonance energy)   the extra stability associated with a compound as a result of having delocalized electrons.

delocalized electrons   electrons that are not localized on a single atom or between two atoms.

Diels-Alder reaction   a [4 + 2] cycloaddition reaction.

diene   a hydrocarbon with two double bonds.

dienophile   an alkene that reacts with a diene in a Diels-Alder reaction.

direct addition (1,2-addition)   addition to the 1- and 2-positions of a conjugated system.

donation of electrons by resonance   donation of electrons through π bonds.

electron delocalization   the sharing of electrons by more than two atoms.

endo   a substituent is endo if it and the bridge are on opposite sides of a bicyclic compound.

equilibrium control   thermodynamic control.

exo   a substituent is exo if it and the bridge are on the same side of a bicyclic compound.

highest occupied molecular orbital (HOMO)   the highest energy molecular orbital that contains electrons.

isolated double bonds   a double bond separated from any other double bond by more than one single bond.

kinetic control   when a reaction is under kinetic control, the relative amounts of the products depend on the rates at which they are formed.

kinetic product   the product that is formed the fastest.

linear combination of molecular orbitals (LCAO)   the combination of atomic orbitals to produce molecular orbitals.

localized electrons   electrons that are restricted to a particular locality.

lowest unoccupied molecular orbital (LUMO)   the lowest energy molecular orbital that does not contain electrons.

pericyclic reaction   a reaction that takes place in one step as a result of a cyclic reorganization of electrons.

polyene   a compound that has several double bonds.

proximity effect   an effect caused by one species being close to another.

resonance   electron delocalization.

resonance contributor (resonance structure)   a structure with localized electrons that approximates the true structure of a compound with delocalized electrons.

resonance electron donation   donation of electrons through π bonds.

resonance electron withdrawal   withdrawal of electrons through π bonds.

resonance energy (delocalization energy) (resonance stabilization energy)   the extra stability a compound possesses as a result of having delocalized electrons.

resonance hybrid   the actual structure of a compound with delocalized electrons; it is represented by two or more resonance contributors with localized electrons.

s-cis-conformation   the conformation in which two double bonds of a conjugated diene are on the same side of a connecting single bond.

separated charges   a positive and a negative charge that can both be neutralized by the movement of electrons.

s-trans-conformation   the conformation in which two double bonds of a conjugated diene are on opposite sides of a connecting single bond.

symmetric molecular orbital   an orbital with a plane of symmetry so that one half is the mirror image of the other half.

thermodynamic control   when a reaction is under thermodynamic control, the relative amounts of the products depend on their stabilities.

thermodynamic product   the most stable product.

withdrawal of electrons by resonance   withdrawal of electrons through π bonds.

Solutions to Problems

1.   **a.**   **1.**   If stereoisomers are not included, three different monosubstituted compounds are possible.

$$BrC \equiv CC \equiv CCH_2CH_3 \qquad HC \equiv CC \equiv CCHCH_3 \qquad HC \equiv CC \equiv CCH_2CH_2Br$$
$$\underset{\displaystyle Br}{|}$$

If stereoisomers are included, four different monosubstituted compounds are possible, because the second listed compound has an asymmetric center so it can have both the R and S configuration.

2.   If stereoisomers are not included, two different monosubstituted compounds are possible.

$$BrCH = CHC \equiv CCH = CH_2 \qquad CH_2 = CC \equiv CCH = CH_2$$
$$\underset{\displaystyle Br}{|}$$

If stereoisomers are included, three different monosubstituted compounds are possible, because the first compound has a double bond that can have cis-trans isomers.

b.   **1.**   If stereoisomers are not included, five different disubstituted compounds are possible.

$$HC \equiv CC \equiv CCH_2CHBr_2 \qquad HC \equiv CC \equiv CCBr_2CH_3 \qquad HC \equiv CC \equiv CCHCH_2Br$$
$$\underset{\displaystyle Br}{|}$$

$$BrC \equiv CC \equiv CCH_2CH_2Br \qquad BrC \equiv CC \equiv CCHCH_3$$
$$\underset{\displaystyle Br}{|}$$

2.   If stereoisomers are not included, five different disubstituted compounds are possible.

$$CH_2 = CC \equiv CC = CH_2 \qquad CH_2 = CHC \equiv CC = CHBr \qquad CH_2 = CC \equiv CCH = CHBr$$
$$\underset{\displaystyle Br}{|} \ \underset{\displaystyle Br}{|} \qquad\qquad\qquad \underset{\displaystyle Br}{|} \qquad\qquad\qquad \underset{\displaystyle Br}{|}$$

$$CH_2 = CHC \equiv CCH = CBr_2 \qquad BrCH = CHC \equiv CCH = CHBr$$

c.   **1.**   If stereoisomers are included, seven different disubstituted compounds are possible, because two of the compounds have asymmetric centers so each can have either the R or S configuration.

2.   If stereoisomers are included, 10 different disubstituted compounds are possible, because the second and third compounds can have cis-trans isomers, and the fifth compound can have cis-cis, trans-trans, and cis-trans isomers.

2.   Ladenburg benzene is a better proposal. It would form one monosubstituted compound, three disubstituted compounds, and would not add Br_2, all in accordance with what early chemists knew about the structure of benzene.

Dewar benzene is not in accordance with what early chemists knew about the structure of benzene, because it would form two monosubstituted compounds, six disubstituted compounds, and it would add Br_2.

3.   a. **2, 4,** and **7** have delocalized electrons.
      1 does not have delocalized electrons, because π electrons cannot be moved toward an sp^3 carbon.
      3, 5, and **6** do not have delocalized electrons, because lone-pair electrons cannot be moved toward an sp^3 carbon.

4.   The resonance contributor that makes the greatest contribution to the hybrid is labeled **A**.
      B contributes less to the hybrid than **A**, and **C** contributes less to the hybrid than **B**.

      a. Solved in the text.

      b.

A is more stable than **B** because, unlike **B**, A does not have separated charges and does not have a positive charge on an oxygen.

c.   A is more stable than **B**, because the negative charge in **A** is on an oxygen, whereas that in **B** is on a less electronegative carbon.

B   **A**

d.   A is more stable than **B**, because **A** does not have separated charges. **B** is more stable than **C**, because the electronegative oxygen atom is closer to the positive charge in **C**. **C** is so unstable that it can be neglected.

A   **C**   **B**

e.   A is more stable than **B**, because the positive charge in **A** is on a less electronegative atom. (N is less electronegative than O.)

B   **A**

f. $\overset{+}{CH_3CH} - CH = CHCH_3 \longleftrightarrow CH_3CH = CH - \overset{+}{CH}CH_3$

equally stable

5.   **a.** $\overset{\delta+}{CH_3C} = \overset{\delta+}{CH} = CHCH_3$
          $|$
          CH_3

c.

e.

b.

d.

f. $\overset{\delta+}{CH_3CH} = CH = \overset{\delta+}{CHCH_3}$

6.   **a.**   All the carbon-oxygen bonds in the carbonate ion should be the same length, because each carbon-oxygen bond is represented in one resonance contributor by a double bond and in two resonance contributors by a single bond.

b.   Because the two negative charges are shared equally by three oxygens, each oxygen will have two-thirds of a negative charge.

7.   **a.**

Less stable, because the positive charge is shared by two primary carbons.

More stable, because the positive charge is shared by a primary and a secondary carbon.

b.

Less stable, because the positive charge is on a primary carbon.

More stable, because the positive charge is on a secondary carbon.

c.

Less stable, because the negative charge is not delocalized.

More stable, because the negative charge is delocalized.

d.

Less stable, because the positive charge is shared by an O and an N.

More stable, because the positive charge is shared by two nitrogens.

8. The second species has the greatest delocalization energy; it has three resonance contributors and none of them have separated charges. (See the answer to Problem 6.)
The first species has two resonance contributors and none of them have separated charges.
The third species has the smallest delocalization energy; it has two resonance contributors, one of which has separated charges.

9. The smaller the heat of hydrogenation (the absolute value of $\Delta H°$), the more stable the compound. Therefore, the relative stabilities of the dienes are:

conjugated diene > isolated diene > cumulated diene

10. $CH_2=CHCH_2CH=CH_2$ < $CH_3CH=CHCH=CH_2$ < $CH_3CH=CHCH=CHCH_3$ <

1,4-pentadiene   1,3-pentadiene   2,4-hexadiene   2,5-dimethyl-2,4-hexadiene

11. **a.** The compound with delocalized electrons is more stable than the compound in which all the electrons are localized.

Electrons are delocalized.   Electrons are localized.

b. Because nitrogen is less electronegative than oxygen, it is better able to share the positive charge.

$$CH_3\ddot{O}\text{—}\overset{+}{C}H_2 \quad \longleftrightarrow \quad CH_3\overset{+}{\ddot{O}}\text{=}CH_2 \qquad\qquad CH_3\ddot{N}H\text{—}\overset{+}{C}H_2 \quad \longleftrightarrow \quad CH_3\overset{+}{N}H\text{=}CH_2$$

more stable

c. In order for electron delocalization to occur, the atoms that share the (π) electrons must be in the same plane so their *p* orbitals can overlap. The two bulky *tert*-butyl groups do not allow enough room for the group with the positively charged carbon to be in the same plane as the benzene ring. Therefore, the carbocation cannot be stabilized by electron delocalization, since the *p* orbital of the positively charged carbon cannot overlap the *p* orbitals of the benzene ring since they are not in the same plane.

more stable

12.   The ψ_3 molecular orbital of 1,3-butadiene has three nodes (two vertical and one horizontal).

The ψ_4 molecular orbital of 1,3-butadiene has four nodes (three vertical and one horizontal).

13.   **a.** ψ_1 and ψ_2 are bonding molecular orbitals, and ψ_3 and ψ_4 are antibonding molecular orbitals.

b. ψ_1 and ψ_3 are symmetric molecular orbitals, and ψ_2 and ψ_4 are antisymmetric molecular orbitals.

c. ψ_2 is the HOMO and ψ_3 is the LUMO in the ground state.

d. ψ_3 is the HOMO and ψ_4 is the LUMO in the excited state.

e. If the HOMO is symmetric, the LUMO is antisymmetric, and vice versa.

14.   **a.** ψ_1, ψ_2, and ψ_3 are bonding molecular orbitals, and ψ_4, ψ_5, and ψ_6 are antibonding molecular orbitals.

b. ψ_1, ψ_3, and ψ_5 are symmetric molecular orbitals, and ψ_2, ψ_4, and ψ_6 are antisymmetric molecular orbitals.

c. ψ_3 is the HOMO and ψ_4 is the LUMO in the ground state.

d. ψ_4 is the HOMO and ψ_5 is the LUMO in the excited state.

e. If the HOMO is symmetric, the LUMO is antisymmetric, and vice versa.

15.   **a.** The ψ_1 molecular orbital of 1,3-butadiene has three bonding interactions, and the ψ_2 molecular orbital has two bonding interactions.

b. The ψ_1 molecular orbital of 1,3,5,7-octatetraene has seven bonding interactions, and the ψ_2 molecular orbital has 6 bonding interactions.

Notice that the ψ_1 molecular orbital has one bonding interaction between each of the overlapping *p* orbitals. Notice also that as the energy of the molecular orbital increases, the number of bonding interactions decreases.

16. In each case, the compound shown is the stronger acid because the negative charge that results when it loses a proton can be delocalized. Electron delocalization stabilizes the base and the more sable the base, the more acidic its conjugate acid. Electron delocalization is not possible for the other compound in each pair.

a.

b. $CH_3CH=CHOH \longrightarrow H^+ + CH_3CH=CHO^- \longleftrightarrow CH_3\bar{C}HCH=O$

c. $CH_3CH=CH\overset{+}{N}H_3 \longrightarrow H^+ + CH_3CH=CHNH_2 \longleftrightarrow CH_3\bar{C}HCH=\overset{+}{N}H_2$

17. a. Ethylamine is a stronger base, because when the lone pair on the nitrogen in aniline is protonated, it can no longer be delocalized into the benzene ring.

b. Ethoxide ion is a stronger base, because a negatively charged oxygen is a stronger base than a neutral nitrogen.

c. Ethoxide ion is a stronger base, because when the phenolate ion is protonated, the pair of electrons that is protonated can no longer be delocalized into the benzene ring.

18. The carboxylic acid is the most acidic because its conjugate base has greater delocalization energy than does the conjugate base of phenol. The alcohol is the least acidic because, unlike the negative charge on the conjugate base of phenol, the negative charge on the conjugate base of the alcohol cannot be delocalized.

19. Solved in the text.

20. Electron withdrawal by the methoxy group will make it a stronger acid (because it will make the conjugate base weaker).
Electron donation by the methoxy group will make it a weaker acid (because it will make the conjugate base stronger).

The pK_a values show that the methoxy substituted acid is a weaker acid (it has a larger pK_a value). Therefore, we know that resonance electron donation is a more important effect than inductive electron withdrawal.

21.

22.   **a.**   Solved in the text.

b.   The contributing resonance structures show that there are two sites that can be protonated.

site of protonation

site of protonation

23.   **a.**   The more reactive double bond is the one that forms a tertiary carbocation.

$$CH_2=CHCH_2CH_2CH_2\overset{CH_3}{\underset{:\overset{..}{Br}:}{\overset{|}{C}}}-CH_3 \longrightarrow CH_2=CHCH_2CH_2CH_2\overset{CH_3}{\underset{Br}{\overset{|}{C}}}-CH_3$$

Instead of Br⁻ being the nucleophile that adds to the tertiary carbon, the π bond can be the nucleophile. In that case, a stable six-membered ring cyclic will be formed. (See the Problem-Solving Strategy on page 175 of the text.)

$$CH_2=CHCH_2CH_2CH_2\overset{CH_3}{\underset{+}{\overset{|}{C}}}-CH_3 \longrightarrow$$

b.

$$HC\equiv CCH_2CH_2\overset{CH_2Cl}{\overset{|}{C}}\overset{\cdots H}{\underset{Cl}{}} \quad + \quad H\overset{CH_2Cl}{\underset{Cl}{\overset{|}{C}}}CH_2CH_2C\equiv CH$$

S   R

or

$$H-\overset{CH_2CH_2C\equiv CH}{\underset{CH_2Cl}{\overset{|}{\underset{|}{}}}}-Cl \quad + \quad Cl-\overset{CH_2CH_2C\equiv CH}{\underset{CH_2Cl}{\overset{|}{\underset{|}{}}}}-H$$

S   R

The double bond is more reactive than the triple bond. The reaction forms a new asymmetric center so a pair of enantiomers is formed.

c.   The more reactive double bond is the one that forms a tertiary carbocation.

24.   The indicated double bond is the most reactive in an electrophilic addition reaction with HBr because addition of an electrophile to this double bond forms the most stable carbocation (a tertiary allylic carbocation).

25. **a.** $CH_3CH{=}CH{-}CH{=}CHCH_3$

$\downarrow Cl_2$

$CH_3CH{-}CH{-}CH{=}CHCH_3$ $\longleftrightarrow$ $CH_3CH{-}CH{=}CH{-}\overset{+}{C}HCH_3$

$\downarrow Cl^-$ $\qquad\qquad\qquad\qquad$ $\downarrow Cl^-$

$CH_3CHCHCH{=}CHCH_3$ (Cl, Cl) $\quad + \quad$ $CH_3CHCH{=}CHCHCH_3$ (Cl, Cl)

1,2- addition product $\qquad\qquad$ **1,4- addition product**

b. $CH_3CH_2\underset{CH_3}{\overset{Br}{C}}{-}\underset{CH_3}{C}{=}CHCH_3$ $\quad + \quad$ $CH_3CH_2\underset{CH_3}{\overset{Br}{C}}{=}\underset{CH_3}{C}CHCH_3$

1, 2-addition product $\qquad\qquad$ **1, 4-addition product**

c. $\quad + \quad$

1, 2-addition product $\qquad\qquad$ **1, 4-addition product**

26.

product of the first reaction

the product has one asymmetric center;
two stereoisomers are obtained (a pair of enantiomers)

$CH_3CHCH_2CH_2CHCH_3$ (Br, Br)

product of the second reaction

the product has two asymmetric centers;
three stereoisomers are obtained (the first one is a meso compound)

27. CH$_2$=CH—CH=CH—CH=CH$_2$ $\xrightarrow{\text{HBr}}$ CH$_3$—CH—CH=CH—CH=CH$_2$
  |
  Br

CH$_3$—CH=CH—CH—CH=CH$_2$
  |
  Br

CH$_3$—CH=CH—CH=CH—CH$_2$
  |
  Br

28. **a.** The chlorine adds so that the positive charge in the carbocation is on a secondary carbon. (If the chlorine had added to the other end of the conjugated system, the positive charge would be on a primary carbon.)

b. The proton adds so that the positive charge in the carbocation is shared by a tertiary and a secondary carbon. (If the proton had added to the other end of the conjugated system, the positive charge would be shared by two secondary carbons.)

29. **a.** Addition at C-1 forms the more stable carbocation, because the positive charge is shared by two secondary allylic carbons. If the deuterium had added to C-4, the positive charge would have been shared by a secondary and a primary carbon.

b. DCl was used to cause the 1,2- and 1,4-products to be different. If HCl had been used, the 1,2- and 1,4-products would have been the same.

30.   **a.** The rate-determining step is formation of the carbocation

b. The product-determining step is reaction of the carbocation with the nucleophile.

31.   **a.** Solved in the text.

b.

kinetic product

thermodynamic product

c.

kinetic product   thermodynamic product

Notice that the 1,2-product is always the kinetic product.
The thermodynamic product is the product with the most highly substituted double bond.

32.   In order for a Diels-Alder reaction to occur, the overlapping orbitals of the reactants must have the same color (the same symmetry). In other words, they must both be symmetric or both be antisymmetric. In a [2+2] cycloaddition reaction at room temperature (both reactants will be in the ground state electronic configuration), the HOMO of one of the reactants will be symmetric and the LUMO of the other will be antisymmetric (see Figure 7.11 on page 316 of the text). Thus, they will not have the same symmetry and the reaction will not occur.

In contrast, a [2+2] cycloaddition reaction does occur under photochemical conditions. Under photochemical conditions, one of the alkenes will be in an excited state. Therefore, its HOMO will be antisymmetric and will be able to overlap with the antisymmetric LUMO of the other alkene.

33.   **a.**

c.

b.

d.

34. First draw the resonance contributors to determine where the charges are on the reactants. The major product is obtained by joining the negatively charged carbon of the diene with the positively charged carbon of the dienophile.

Because the reaction creates an asymmetric center, the product will be a racemic mixture.

35. The resonance contributors show that if the electron-donating substituent is at the end of the conjugated system, it and the electron-withdrawing substituent of the dienophile will be adjacent to one another in the product of the Diels-Alder reaction.

If the electron-donating substituent is not at the end of the conjugated system, it and the electron-withdrawing substituent of the dienophile will be opposite to one another in the product of the Diels-Alder reaction.

36. **a.**

b.

37.   **A** and **C** will not react, because they are both locked in an *s*-trans conformation.

       D and **E** will react, because they are both locked in an *s*-cis conformation.

       B and **F** will react, because they can rotate into an *s*-cis conformation.

38.   Solved in the text.

39.   **a.**   It is not optically active, because it is a meso compound.
              (It has 2 asymmetric centers and a plane of symmetry.)

       b.   It is not optically active, because it is a racemic mixture.
              (Identical amounts of the enantiomers will be obtained.)

40.   **a.**

       b.

       c.

       d.

       e.

       f.

41.   **a, b, d, e, k, l, m, n**

42.   **a.** and **b.**

       1.

                                         More stable, because the
                                         negative charge is on nitrogen
                                         rather than on carbon.

2. :N≡N—Ö:⁻ ⟷ ⁻:N̈=N=Ö:

More stable, because the
negative charge is on oxygen
rather than on nitrogen.

3. ⁻:Ö—N̈=Ö: ⟷ :O=N̈—Ö:⁻

Both are equally stable.

Additional resonance structures could be drawn for each of these three species,
but they are relatively unstable because each has an incomplete octet at one atom.

43. **a.** **b.**

44. **a.**

b.

c.

d.

45.

46.   **a.** different compounds   **d.** resonance contributors
           b. different compounds   **e.** different compounds
           c. resonance contributors

Notice that in the structures that are different compounds, both atoms and electrons have changed their locations. In contrast, in the structures that are resonance contributors, only the electrons have moved.

47.   **a.** There are six linear dienes with molecular formula C_6H_{10}.

b. Two are conjugated dienes. $CH_2{=}CHCH{=}CHCH_2CH_3$
                                          $CH_3CH{=}CHCH{=}CHCH_3$

c. Two are isolated dienes. $CH_2{=}CHCH_2CH{=}CHCH_3$
                                        $CH_2{=}CHCH_2CH_2CH{=}CH_2$

There are also two cumulated dienes. $CH_2{=}C{=}CHCH_2CH_2CH_3$
                                        $CH_3CH{=}C{=}CHCH_2CH_3$

48.   **a. 1.** $CH_3\overset{\frown}{C}H{=}CH{-}\overset{\frown}{\ddot{O}}CH_3 \longleftrightarrow CH_3\overset{..}{\overset{-}{C}}H{-}CH{=}\overset{+}{\ddot{O}}CH_3$

                    major   minor

2. $\underset{}{}$ CH_2NH_2 $\longleftrightarrow$ CH_2NH_2

The two resonance contributors have the same stability and contribute equally to the resonance hybrid.

3.

major   minor   minor   minor   major

4. $CH_3 - \overset{+}{N} \overset{O}{\underset{\ddot{\overset{\cdot}{O}}}{}}$ $\longleftrightarrow$ $CH_3 - \overset{+}{N} \overset{O^-}{\underset{O}{}}$

The two resonance contributors have the same stability.

5. $CH_3\bar{C}H - \overset{+}{N} \overset{O}{\underset{\ddot{\overset{\cdot}{O}}}{}}$ $\longleftrightarrow$ $CH_3\bar{C}H - \overset{+}{N} \overset{O^-}{\underset{O}{}}$ $\longleftrightarrow$ $CH_3CH = \overset{+}{N} \overset{O^-}{\underset{O^-}{}}$

minor   minor   major

6. $CH_3CH = CH - \overset{+}{C}H_2$ $\longleftrightarrow$ $CH_3\overset{+}{C}H - CH = CH_2$

minor   major

7.

The five contributors are equally stable and contribute equally to the resonance hybrid.

8. The minor contributor has a positive charge on oxygen and charge separation.

$CH_3CH_2 \overset{O}{\underset{\ddot{\overset{\cdot}{O}}CH_2CH_3}{\overset{\|}{C}}}$ $\longleftrightarrow$ $CH_3CH_2 \overset{O^-}{\underset{\overset{+}{O}CH_2CH_3}{C}}$

major   minor

9. $CH_3CH = CHCH = \overset{+}{C}HCH_2$ $CH_3\overset{+}{C}HCH = CHCH = CH_2$

minor   major
(the positive   (the positive
charge is on   charge is on
a primary   a secondary
carbon)   carbon)

$CH_3CH = CH\overset{+}{C}HCH = CH_2$
major (the positive charge is on
a secondary carbon)

10. The minor contributor has a negative charge on carbon, which is not as stable as a negative charge on oxygen.

$\bar{C}H_2 \overset{O}{\underset{CH_2CH_3}{\overset{\|}{C}}}$ $\longleftrightarrow$ $CH_2 \overset{O^-}{\underset{CH_2CH_3}{C}}$

minor   major

11. $CH_3\bar{C}H - C \equiv N$ $\longleftrightarrow$ $CH_3CH = C = \bar{N}$

minor   major

12.

major   minor   minor   minor   major

13. The minor contributor has a positive charge on nitrogen and charge separation.

major

minor

14. The minor contributors have a negative charge on carbon, which is not as stable as a negative charge on oxygen.

minor   minor   major

15. The minor contributor has a negative charge on carbon, which is not as stable as a negative charge on oxygen. Notice that the electrons on the center carbon can be delocalized onto both of the carbonyl oxygens.

minor   major

major

b. **2, 4,** and **7**

49. Both compounds form the same product when they are hydrogenated, so the difference in heats of hydrogenation will depend only on the difference in the stabilities of the reactants. Because 1,2-pentadiene has cumulated double bonds and 1,4-pentadiene has isolated double bonds, 1,2-pentadiene is less stable and, therefore, will have a greater heat of hydrogenation (a more negative $\Delta H°$).

$$CH_2{=}C{=}CHCH_2CH_3 \xrightarrow[\text{Pd/C}]{H_2} CH_3CH_2CH_2CH_2CH_3$$

1,2-pentadiene

$$CH_2{=}CHCH_2CH{=}CH_2 \xrightarrow[\text{Pd/C}]{H_2} CH_3CH_2CH_2CH_2CH_3$$

1,4-pentadiene

50. **a.** CH₃ĊHCH=CH₂

This makes the greater contribution, because the positive charge is on a secondary carbon.

b.   This makes the greater contribution, because the negative charge is on an oxygen.

c. This makes the greater contribution, because the positive charge is on a tertiary carbon.

51. **a.** The resonance contributors show that the carbonyl oxygen has the greater electron density.

b. The compound on the right has the greater electron density on its nitrogen, because the compound on the left has a resonance contributor with a positive charge on the nitrogen as a result of electron delocalization.

c. The compound with the cyclohexane ring has the greater electron density on its oxygen, because the lone pair on the nitrogen can be delocalized only onto the oxygen.

There is less delocalization onto oxygen by the lone pair in the compound with the benzene ring (path **a**) because the lone pair can also be delocalized away from the oxygen into the benzene ring (path **b**).

52. Remember that an sp^2 nitrogen is more electronegative than an sp^3 nitrogen, and the more electronegative the atom to which a hydrogen is attached, the stronger the acid. The stronger the acid, the weaker its conjugate base.

The lone pair is localized and the nitrogen is sp^2 hybridized, which is not as strong a base as an sp^3 hybridized nitrogen.

These are weak bases because the lone-pair electrons are delocalized.

The strongest base because the lone pair is localized and the nitrogen is sp^3 hybridized.

53. The carbocation is stable because the positive charge is shared by 10 carbons (the central carbon and three carbons of each of the three benzene rings) as a result of electron delocalization.

54. The structure shown below is the stronger acid because it has the weaker conjugate base. When this compound loses a proton, the electrons left behind can be delocalized onto six different carbons. The electrons left behind on the conjugate base of the other compound can be delocalized onto only three different carbons; they cannot be delocalized into the second benzene ring.

55. The methyl group on benzene can lose a proton easier than the methyl group on cyclohexane because the electrons left behind on the carbon in the former can be delocalized into the benzene ring. In contrast, the electrons left behind in the other compound cannot be delocalized.

56.

CH_3CH_2O — C(O$^-$) — $CH-C\equiv N$
#1 most stable because the negative charge is on oxygen

CH_3CH_2O — C(O) — $CH=C=N^-$
#2 because the negative charge is on nitrogen

CH_3CH_2O — C(O) — $CH-C\equiv N$
#3 because the negative charge is on carbon

$CH_3CH_2\overset{+}{O}$ — C(O$^-$) — $CH-C\equiv N$
#4 the least stable because the negative charge is on carbon and it has separated charges

57.   The more the electrons that are left behind when the proton is removed can be delocalized, the greater the stability of the base. The more stable the base, the more acidic its conjugate acid.
The negative charge on the base in the first compound can be delocalized onto two other carbons; the negative charge on the base in the second compound can be delocalized onto one other carbon; the negative charge on the base in the last compound cannot be delocalized.

Therefore, the compounds have the relative acidities shown below:

58.   **a.** CH₃—C(=O)—O⁻   The negative charge is shared by two oxygens.

   b. CH₃CH₂C̈H—C(=O)—CH₃   The negative charge is shared by a carbon and an oxygen.

   c. CH₃—C(=O)—CH—C(=O)—CH₃   The negative charge is shared by a carbon and two oxygens.

   d. <ring structure with N:⁻>   The negative charge is shared by a nitrogen and two oxygens.

59.   The stronger base is the less stable base of each pair in Problem 58.

   a. CH₃—C(=O)—CH₂O⁻   Less stable because the negative charge cannot be delocalized.

   b. CH₃C̈HCH₂—C(=O)—CH₃   Less stable because the negative charge cannot be delocalized.

   c. CH₃—C(=O)—CHCH₂—C(=O)—CH₃   Less stable because the negative charge can be delocalized onto only one oxygen.

d.

Less stable because the negative charge can be delocalized onto only one oxygen.

60. The resonance contributors of pyrrole are more stable because the positive charge is on nitrogen. In furan, the positive charge is on oxygen which, being more electronegative, is less stable with a positive charge.

61. **A** is the most acidic because the electrons left behind when the proton is removed can be delocalized onto two oxygens. **B** is the next most acidic because the electrons left behind when the proton is removed can be delocalized onto one oxygen. **C** is the least acidic because the electrons left behind when the proton is removed cannot be delocalized.

62. **a.** It has eight molecular orbitals.

b. ψ_1, ψ_2, ψ_3, and ψ_4 are bonding molecular orbitals; ψ_5, ψ_6, ψ_7 and ψ_8 are antibonding molecular orbitals.

c. ψ_1, ψ_3, ψ_5, and ψ_7 are symmetric molecular orbitals; ψ_2, ψ_4, ψ_6, and ψ_8 are antisymmetric molecular orbitals.

d. ψ_4 is the HOMO and ψ_5 is the LUMO in the ground state.

e. ψ_5 is the HOMO and ψ_6 is the LUMO in the excited state.

f. If HOMO is symmetric, the LUMO is antisymmetric, and vice versa.

g. It has seven nodes between the nuclei. It also has one node that passes through the nuclei.

63. The reaction of 1,3-cyclohexadiene with Br$_2$ forms 3,4-dibromocyclohexene as the 1,2-addition product and 3,6-dibromocyclohexene as the 1,4-addition product. (Recall that in naming the compounds, the double bond is at the 1,2-position.) The reaction of 1,3-cyclohexadiene with HBr forms 3-bromocyclohexene as both the 1,2-addition product and the 1,4-addition product.

3,4-dibromocyclohexene
1,2-addition product

3,6-dibromocyclohexene
1,4-addition product

1,3-cyclohexadiene

3-bromocyclohexene
1,2-addition product
1,4-addition product

64.   **a.** Only part **a** involves the reaction of two unsymmetrically substituted reactants. Therefore, only for part **a** do we need to look at the charge distribution in the reactants to determine the major product of the reaction.

We now need to join the negatively charged end of the diene and the positive end of the dienophile.

b.

c.

d.

65. **a.** **1.**

2.

3.

b. **1.**

2.

3.

66. **a.** $CH_2{=}CHCH{=}CHCH{=}CH_2$ $\xrightarrow{\text{HBr}}$ $CH_3CHCH{=}CHCH{=}CH_2$ +

1,3,5-hexatriene $\underset{\underset{\textstyle \text{Br}}{|}}{}$ **A**

$CH_3CH{=}CHCHCH{=}CH_2$ + $CH_3CH{=}CHCH{=}CHCH_2Br$

B $\overset{|}{\text{Br}}$ **C**

b. **A** will predominate if the reaction is under kinetic control because it is the 1,2-product and, therefore, will be the product formed most rapidly as a result of the proximity effect. In addition, **A** will be the 1,2-product regardless of which end of the conjugate system reacts with the electrophile.

c. **C** will predominate if the reaction is under thermodynamic control because it is the most stable diene. (It is the most substituted conjugated diene.)

67. The diene is the nucleophile, and the dienophile is the electrophile in a Diels-Alder reaction.

a. An electron-donating substituent in the diene would increase the rate of the reaction, because electron donation would increase its nucleophilicity.

b. An electron-donating substituent in the dienophile would decrease the rate of the reaction, because electron donation would decrease its electrophilicity.

c. An electron-withdrawing substituent in the diene would decrease the rate of the reaction, because electron withdrawal would decrease its nucleophilicity.

68. **a.** Addition of an electrophile to C-1 forms a carbocation with two resonance contributors, a *tertiary allylic carbocation* and a *secondary allylic carbocation*. Addition of an electrophile to C-4 forms a carbocation with two resonance contributors, a *tertiary allylic carbocation* and a *primary allylic carbocation*. Therefore, addition to C-1 results in formation of the more stable carbocation intermediate, and the more stable intermediate leads to the major products.

kinetic product   thermodynamic product

b. Addition of an electrophile to C-1 forms a carbocation with two resonance contributors; both are *tertiary allylic* carbocations. Addition of an electrophile to C-4 forms a carbocation with two resonance contributors, a *secondary allylic* carbocation and a *primary allylic* carbocation. Therefore, addition to C-1 results in formation of the more stable carbocation. Only one product is formed, because the carbocation is symmetrical.

This is the only product because
the carbocation is symmetrical.

69.   **a.** and **d.**

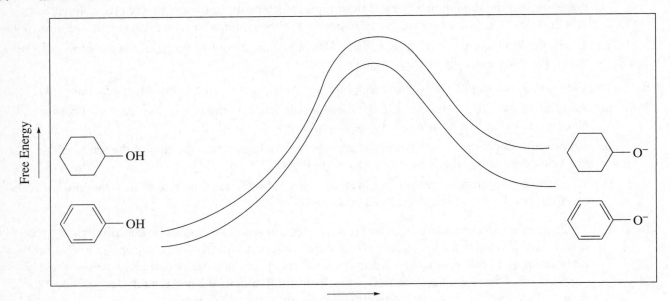

Progress of the Reaction

d.   The 2nd, 3rd, and 4th resonance contributors in **c** are more stable than those resonance contributors in **b**, because in **b** a positive charge is on the most electronegative atom (the oxygen) and there is charge separation. Therefore, the phenolate ion has greater electron delocalization than phenol. Thus, as shown in the energy diagram, the difference in energy between the phenolate ion and the cyclohexoxide ion is greater than the difference in energy between phenol and cyclohexanol.

e.   Because of greater electron delocalization in the phenolate ion compared to that in phenol, phenol has a larger K_a than cyclohexanol.

f.   Because it has a larger K_a (a lower pK_a), phenol is a stronger acid.

70.

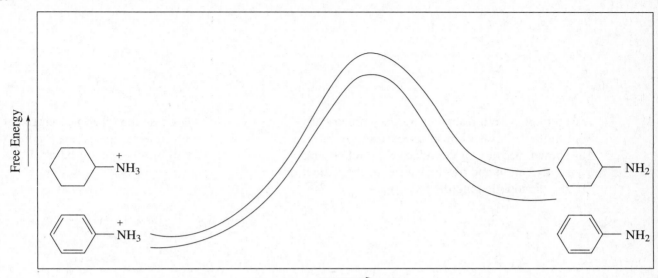

Progress of the Reaction

b.

c.

d. Aniline has greater electron delocalization than the anilinium ion. Thus in the energy diagram, the difference in energy between aniline and cyclohexylamine is greater than the difference in energy between the anilinium ion and the cyclohexylammonium ion.

e. Because of greater electron delocalization in aniline compared to that in the anilinium ion, the anilinium ion has a larger K_a than the cyclohexylammonium ion.

f. Because it has a larger K_a (a lower pK_a), the anilinium ion is a stronger acid than the cyclohexylammonium ion. Therefore, cyclohexylamine is a stronger base than aniline. (The stronger the acid, the weaker its conjugate base.)

71.   a.   Because the reaction creates an asymmetric center in the product, the product will be a racemic mixture.

b.

Even though both reactants are unsymmetrically substituted, they will be aligned primarily as shown, because of the relatively stable tertiary benzylic cation and delocalization of the π electrons of the dienophile onto the oxygen.

Because the reaction creates an asymmetric center, the product will be a racemic mixture.

c.

The products will be aligned primarily as shown because in the diene, a tertiary allylic carbocation is more stable than a secondary allylic carbocation. In the dienophile, a primary carbanion is more stable than a primary carbocation would be.

Because the reaction creates two asymmetric centers, the product will be a mixture of four stereoisomers.

72. The first pair is the preferred set of reagents because it has the more nucleophilic diene and the more electrophilic dienophile.

73. A Diels-Alder reaction is a reaction between a nucleophilic diene and an electrophilic dienophile.

a. The compound shown below is more reactive in both **1** and **2**, because electron delocalization increases the electrophilicity of the dienophile.

$$CH_2{=}CH{-}\overset{\overset{\displaystyle O}{\|}}{CH} \longleftrightarrow \overset{+}{CH_2}{-}CH{=}\overset{\overset{\displaystyle O^-}{|}}{CH}$$

b. The compound shown below is more reactive, because electron delocalization increases the nucleophilicity of the diene.

$$CH_2{=}CH{-}CH{=}CH{-}\overset{..}{\underset{..}{O}}CH_3 \longleftrightarrow {}^-CH_2{=}CH{-}CH{=}CH{-}\overset{+}{\underset{..}{O}}CH_3$$

74.

exo   endo

75. a.   +   **c.**   +

b.   +   **d.**   +

76.   The electrophile can add to either end of the conjugated system. Adding the electrophile to one end forms **A** and **B**; adding the electrophile to the other end forms **C** and **B**.

a. $\xrightarrow{Br_2}$

A   B   C

b.   A has two asymmetric centers but only two stereoisomers are obtained, because only anti addition of Br_2 can occur.

+

B has four stereoisomers because it has an asymmetric center and a double bond that can be in either the *E* or *Z* configuration.

+   +   +

C has two stereoisomers because it has one asymmetric center

77.   Nine of the compounds are shown below. Since each has one asymmetric center, each can have either the *R* or *S* configuration. Therefore, 18 different products can be obtained.

	1,3-Butadiene is the electrophile.	The 3,4-bond of 2-methyl-1,3-butadiene is the electrophile.	The 1,2-bond of 2-methyl-1,3-butadiene is the electrophile.
1,3-Butadiene is the nucleophile.			
2-Methyl-1,3-butadiene is the nucleophile. (the 1-position is on top)			
2-Methyl-1,3-butadiene is the nucleophile. (the 4-position is on top)			

78.   2-Methyl-1,3-pentadiene (with conjugated double bonds) is more stable than 2-methyl-1,4-pentadiene (with isolated double bonds). The rate-limiting step of the reaction is formation of the carbocation intermediate. 2-Methyl-1,3-pentadiene forms a more stable carbocation than does 2-methyl-1,4-pentadiene.

Since the more stable reactant forms the more stable carbocation, the relative free energies of activation for the rate-limiting steps of the two reactions depend on whether the difference in the stabilities of the reactants is greater or less than the difference in the stabilities of the transition states leading to formation of the carbocations (which depend on the difference in stabilities of the carbocations). Because the difference in the stabilities of the reactants is less than the difference in the stabilities of the transition states, the rate of reaction of HBr with 2-methyl-1,3-pentadiene is the faster reaction. (If the difference in the stabilities of the reactants had been greater than the difference in the stabilities of the transition states, the rate of reaction of HBr with 2-methyl-1,4-pentadiene would have been the faster reaction.)

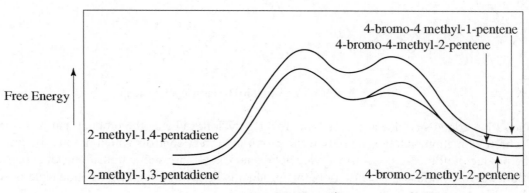

79. His recrystallization was not successful. Because maleic anhydride is a dienophile, it reacts with cyclopentadiene in a Diels-Alder reaction.

80. We saw in Problem 79 that maleic anhydride reacts with cyclopentadiene. The function of maleic anhydride in this reaction is to remove cyclopentadiene, since removal of a product drives the equilibrium toward products. (See Le Châtelier's principle on page 140 of the text.)

81.   The bridgehead carbon cannot have the 120° bond angle required for the sp^2 carbon of a double bond, because if it did, the compound would be too strained to exist.

82.   **a.**   Unless the reaction is being carried out under kinetic control, the amount of product obtained is not dependent on the rate at which the product is formed, so the relative amounts of products obtained will not tell you which product was formed faster.

   b.   In a thermodynamically controlled reaction, the product distribution depends on the relative stabilities of the products since the products come to equilibrium. Thus if the distribution of products that is obtained does not reflect the relative stabilities of the products, the reaction must have been kinetically controlled.

83.   She should follow her friend's advice. If she uses 2-methyl-1,3-cyclohexadiene, the product that is formed faster will be 3-chloro-3-methylcyclohexene, both if the proximity effect controls which product is formed faster *and* if the more stable transition state controls which product is formed faster, because this product is formed through a transition state in which the positive charge is primarily on a tertiary carbon. Thus the experiment will not be able to establish which of the two effects controls which product is formed faster.

3-chloro-3-methylcyclohexene

If she follows her friend's advice and uses 1-methyl-1,3-cyclohexadiene, the product that is formed faster will be 3-chloro-1-methylcyclohexene only if the proximity effect controls which product is formed faster. The product will be 3-chloro-3-methylcyclohexene if the more stable transition state controls which product is formed faster, because only this product is formed through a transition state in which the positive charge is primarily on a tertiary carbon.

3-chloro-1-methylcyclohexene   3-chloro-3-methylcyclohexene

Chapter 7 Practice Test

1.   For each of the following pairs of compounds, indicate the one that is the more stable:

a. $\overset{+}{-CH_2}$ or $\overset{+}{-CH_2}$

b. $CH_3\overset{-}{C}HCH_3$ or $CH_3\overset{-}{C}HC\equiv CH$

c. $CH_3\overset{-}{C}HCH_2\overset{O}{\overset{\|}{C}}CH_3$ or $CH_3\overset{-}{C}H\overset{O}{\overset{\|}{C}}CH_3$

d. $CH_2{=}CH\overset{+}{C}H_2$ or $CH_2{=}CHCH_2\overset{+}{C}H_2$

e. $\overset{+}{C}HCH_3$ or $\overset{+}{C}HCH_3$

2.   Draw resonance contributors for each of the following:

a. $CH_3CH{=}CH{-}\overset{..}{\underset{..}{O}}CH_3$

b. $CH_3CH{=}CH{-}CH{=}CH{-}\overset{+}{C}H_2$

c. $^-\overset{..}{C}H_2{-}CH{=}CH{-}\overset{O}{\overset{\|}{C}}H$

3.   Which compounds do not have delocalized electrons?

$CH_3CH_2NHCH{=}CHCH_3$ $CH_3\underset{+}{\overset{CH_3}{\overset{|}{C}}}CH_2CH{=}CH_2$ $CH_2{=}CHCH_2CH{=}CH_2$

$CH_2{=}CH\overset{O}{\overset{\|}{C}}CH_3$ $CH_3CH_2NHCH_2CH{=}CHCH_3$

4.   Give the products of each of the following reactions:

a.   + Br_2 $\longrightarrow$

b.   + $\underset{CH_2}{\overset{O}{\overset{\|}{\underset{\|}{CHCCH_3}}}}$ $\longrightarrow$

5.   Which of the following pairs are resonance contributors?

    a. CH_3CH_2OH and CH_3OCH_3

    b. $CH_3\overset{\overset{\displaystyle O}{\|}}{C}OH$ and $CH_3\overset{\overset{\displaystyle O}{\|}}{C}O^-$

    c. $CH_3\overset{\overset{\displaystyle O}{\|}}{C}OH$ and $CH_3\overset{\overset{\displaystyle O^-}{|}}{C}=\overset{+}{O}H$

    d. $CH_3CH_2\overset{\overset{\displaystyle O}{\|}}{C}H$ and $CH_3CH=\overset{\overset{\displaystyle OH}{|}}{C}H$

6.   Which of the following dienes can be used in a Diels-Alder reaction?

7.   Which is a stronger base?

    a. $\bigcirc$—NH_2 or $\bigcirc$—NH_2

    b. $\bigcirc$—CH_2O^- or $\bigcirc$—O^-

8.   Draw resonance contributors for each of the following:

    a. $\bigcirc$—$\ddot{N}H_2$

    b. $\bigcirc$—$\overset{+}{N}H_3$

    c. $\bigcirc$—$\ddot{\overset{..}{O}}{:}^-$

9.   Which resonance contributor makes a greater contribution to the resonance hybrid?

a.   or

b.   or

10.   Indicate whether each of the following statements are true or false:

a. A compound with four conjugated double bonds has four molecular orbitals.   T   F

b. ψ_1 and ψ_2 are symmetric molecular orbitals.   T   F

c. If ψ_3 is the HOMO in the ground state, ψ_4 will be the HOMO in the excited state.   T   F

d. If ψ_3 is the LUMO, ψ_4 will be the HOMO.   T   F

e. If the ground-state HOMO is symmetric, the ground-state LUMO will be
antisymmetric.   T   F

f. A conjugated diene is more stable than an isomeric isolated diene.   T   F

g. A single bond formed by an sp^2—sp^2 overlap is longer than a single bond
formed by an sp^2—sp^3 overlap.   T   F

h. The thermodynamically controlled product is the major product obtained
when the reaction is carried out under mild conditions.   T   F

i. 1,3-Hexadiene is more stable than 1,4-hexadiene.   T   F

11.   Give the four products that would be obtained from the following reaction. Ignore stereoisomers.

$$\underset{\underset{CH_3}{|}}{\overset{\overset{CH_3}{|}}{CH_2{=}C{-}C{=}CHCH_3}} \ + \ HBr \ \longrightarrow$$

12.   What reactants are necessary for the synthesis of the following compound via a Diels-Alder reaction?

13.   Rank the following carbocations in order of decreasing stability:

$$CH_3CH{=}CH\overset{+}{C}H_2 \quad CH_3CH{=}CH\overset{+}{C}HCH_3 \quad CH_3CH{=}CHCH_2\overset{+}{C}H_2 \quad CH_3CH{=}CH\overset{+}{C}CH_3$$
$$\underset{\qquad\qquad\qquad\qquad\qquad\qquad\qquad\qquad\qquad\qquad\qquad CH_3}{}$$

14.   Two 1,2-products and two 1,4-products can be obtained from the following reaction:

   a.   Which are the predominant 1,2- and 1,4-products?
   b.   Which of the above two products is the product of thermodynamic control?

15.   What reagents could be used to convert the given starting material into the given product?

16.   Give the product of the following reaction, showing its configuration:

17.   For each of the following reactions, give the major 1,2- and 1,4-products. Label the product of kinetic control and the product of thermodynamic control.

   a. $CH_2{=}CH{-}\underset{\underset{CH_3}{|}}{C}{=}CH_2 \quad + \quad HCl \longrightarrow$

   b.

Drawing Resonance Contributors

On pages 132–136 of the text and in Special Topic II of the *Study Guide and Solutions Manual*, we saw that chemists use curved arrows to show how electrons move when reactants are converted into products. Chemists also use curved arrows when they draw resonance contributors.

We have seen that delocalized electrons are electrons that are shared by more than two atoms. When electrons are shared by more than two atoms, we cannot use solid lines to represent the electrons. For example, in the carboxylate ion, a pair of electrons are shared by a carbon and two oxygens. Thus, the two oxygen atoms share the negative charge. We show the pair of delocalized electrons by a dotted line spread over the three atoms. We have seen that this structure is called a **resonance hybrid**.

carboxylate ion
resonance hybrid

Chemists do not like to use dotted lines when drawing structures because, unlike a solid line that represents two electrons, the dotted lines do not specify the number of electrons they represent. Therefore, chemists use structures with localized electrons (solid lines) to approximate the resonance hybrid that has delocalized electrons (dotted lines). These approximate structures are called **resonance contributors**. Curved arrows are used to show the movement of electrons in going from one resonance contributor to the next.

resonance contributors

Rules for Drawing Resonance Contributors

Resonance contributors are easy to draw if you remember these three simple rules for interconverting resonance contributors:

1. Only electrons move; atoms NEVER move.
2. The only electrons that can move are π electrons (electrons in π bonds) and lone-pair electrons.
3. Electrons are always moved toward an sp^2 or sp hybridized atom. An sp^2 carbon is a positively-charged carbon or a double-bonded carbon; an sp carbon is a triply-bonded carbon.
    (In Special Topic VI, we will see that electrons can also move toward a carbon with an unpaired electron, which is also an sp^2 carbon.)

Moving π electrons toward a positively-charged carbon.

In the following example, π electrons are moved toward a positively-charged carbon. Because the atom does not have a complete octet of electrons, it can accept the electrons. The carbon that is positively charged in the first resonance contributor is neutral in the second resonance contributor because it has received electrons. The carbon in the first resonance contributor that loses its share of the π electrons is positively charged in the second resonance contributor.

$$CH_3CH{=}CH{-}\overset{+}{C}HCH_3 \quad \longleftrightarrow \quad CH_3\overset{+}{C}H{-}CH{=}CHCH_3$$

We see that the following carbocation has three resonance contributors.

$$CH_2{=}CH{-}CH{=}CH{-}\overset{+}{C}HCH_3 \quad \longleftrightarrow \quad CH_2{=}CH{-}\overset{+}{C}H{-}CH{=}CHCH_3$$

$$\overset{+}{C}H_2{-}CH{=}CH{-}CH{=}CHCH_3$$

Notice that in going from one resonance contributor to the next, the total number of electrons in the structure does not change. Therefore, each of the resonance contributors must have the same net charge.

Problem 1. Draw the resonance contributors for the following carbocation:
(The answers can be found immediately after Problem 12.)

$$CH_3CH{=}CH{-}CH{=}CH{-}CH{=}CH{-}\overset{+}{C}H_2 \quad \longleftrightarrow$$

$$\longleftrightarrow$$

Problem 2. Draw the resonance contributors for the following carbocation:

$\longleftrightarrow \qquad \longleftrightarrow \qquad \longleftrightarrow \qquad \longleftrightarrow$

Moving π electrons toward a pair of atoms attached by a π bond.

In the following example, π electrons are moved toward a pair of atoms attached by a π bond. The atom to which the electrons are moved can accept them because the π bond can break.

Problem 3. Draw the resonance contributor for the following compound:

In the next example, the electrons can be moved equally easily to the left or to the right. When comparing the charges on the resonance contributors, we see that the charges on each of the end carbons cancel, so there is no charge on any of the carbons in the resonance hybrid.

$$\bar{\text{C}}\text{H}_2\text{—CH}=\text{CH—}\overset{+}{\text{C}}\text{H}_2 \quad \xleftrightarrow{\ a\ } \quad \text{CH}_2=\text{CH—CH}=\text{CH}_2 \quad \xleftrightarrow{\ b\ } \quad \overset{+}{\text{C}}\text{H}_2\text{—CH}=\text{CH—}\bar{\text{C}}\text{H}_2$$

$$\text{CH}_2=\!\!=\text{CH}=\!\!=\text{CH}=\!\!=\text{CH}_2$$

resonance hybrid

When electrons can be moved in either of two directions and there is a difference in the electronegativity of the atoms to which they can be moved, always move the electrons toward the more electronegative atom. For example, in the following example the electrons are moved toward oxygen, not toward carbon.

Notice that the first resonance contributor has a net charge of 0. Since the number of electrons in the molecule does not change, the other resonance contributor must have the same net charge. (A net charge of 0 does not mean that there is no charge on any of the atoms; a resonance contributor with a positive charge on one atom and a negative charge on another has a net charge of 0.)

Problem 4. Draw the resonance contributor for the following:

$$\text{CH}_2=\text{CH—C}\equiv\text{N} \quad \longleftrightarrow$$

Moving a lone pair toward a pair of atoms attached by a π bond.

In the following examples, lone-pair electrons are moved toward a pair of atoms attached by a π bond. Notice that the arrow starts at a pair of electrons, not at a negative charge. In the first example, each of the resonance contributors has a net charge of -1; in the second example, each of the resonance contributors has a net charge of 0.

The following species has three resonance contributors. Notice again that the arrow starts at a lone pair, not at a negative charge. The three oxygen atoms share the two negative charges. Therefore, each oxygen atom in the hybrid has 2/3 of a negative charge.

Problem 5. Draw the resonance contributor for the following:

$$
\underset{\underset{CH_3}{}\underset{NH_2}{}}{\overset{:\overset{..}{O}}{\underset{\|}{C}}} \longleftrightarrow
$$

Notice that in the next example the lone pair of electrons moves away from the most electronegative atom in the molecule. They do so because that is the only way electron delocalization can occur. The π electrons cannot move toward the oxygen because the oxygen atom has a complete octet (it is sp^3 hybridized). Recall that electrons can move only toward an sp^2 or sp hybridized atom.

$$
CH_3\overset{\frown}{CH}=CH\overset{\frown}{-}\overset{..}{\underset{..}{O}}CH_3 \longleftrightarrow CH_3\overset{-}{CH}-CH=\overset{+}{\underset{..}{O}}CH_3
$$

The compound in the next example has five resonance contributors. To get to the second resonance contributor, a lone pair on nitrogen is moved toward a pair of atoms attached by a π bond. Notice that the first and fifth resonance contributors are not the same; they are similar to the two resonance contributors of benzene. (See page 286 of the text.)

Problem 6. Draw the resonance contributors for the following:

The following species do not have delocalized electrons. Electrons cannot be moved toward an sp^3 hybridized atom because an sp^3 hybridized atom has a complete octet and, therefore, cannot accept any more electrons.

An sp^3 hybridized carbon cannot accept electrons.	An sp^3 hybridized carbon cannot accept electrons.	An sp^3 hybridized carbon cannot accept electrons.

$$
CH_2=CH\overset{\times}{-}CH_2-\overset{+}{C}H_2CH_3 \qquad CH_3CH=CH-CH_2\overset{\times}{-}\overset{..}{N}H_2 \qquad CH_3\overset{\overset{..}{O}:}{\underset{\|}{C}}-CH_2\overset{\times}{-}CH=CHCH_3
$$

Compare the above structures that do not have delocalized electrons with those that do. Notice that:

1. there can be only one single bond between the atom with the π bond and the atom with the positive or negative charge.
2. there can be only one single bond between the atom with the π bond and the atom with the lone pair.
3. there can be only one single bond between the atoms with the π bonds (that is, the double bonds must be conjugated).

Notice the difference in the resonance contributors for the four structures shown below. In the first example, a lone pair on the atom attached to the ring moves toward a pair of atoms attached by a π bond.

In the next example, first a π bond moves toward a pair of atoms attached by a π bond; the electron movement is toward the oxygen since it is more electronegative than carbon. Then to draw the other resonance contributors, a π bond moves to a positive charge.

In the next two examples, the atom attached to the ring has neither a lone pair nor a π bond. Therefore, the substituent can neither donate electrons into the ring nor accept electrons from the ring. Thus, these compounds have only two resonance contributors—the ones that are similar to the two resonance contributors of benzene.

Problem 7. Which of the following species have delocalized electrons?

$$CH_3CH{=}CHCH_2CH{=}CH_2$$
A

$$CH_3CH{=}CHCH{=}CH_2$$
B

$$CH_3CH{=}CHCH_2\overset{\cdot\cdot}{N}HCH_3$$
C

D $CH_3{-}\overset{\displaystyle :\overset{\cdot\cdot}{O}}{\underset{}{C}}{-}CH_2\overset{\cdot\cdot}{\underset{\cdot\cdot}{O}}CH_3$

E $CH_3{-}\overset{\displaystyle :\overset{\cdot\cdot}{O}}{\underset{}{C}}{-}\overset{\cdot\cdot}{N}HCH_3$

F $CH_3{-}\overset{\displaystyle :\overset{\cdot\cdot}{O}}{\underset{}{C}}{-}\overset{\cdot\cdot}{\underset{\cdot\cdot}{O}}H$

G $CH_3{-}\overset{\displaystyle :\overset{\cdot\cdot}{O}}{\underset{}{C}}{-}CH_2C{\equiv}N{:}$

H

I

J $\overset{\cdot\cdot}{\underset{\cdot\cdot}{O}}CH_3$

K $CH_2\overset{\cdot\cdot}{\underset{\cdot\cdot}{O}}CH_3$

L $\overset{\cdot\cdot}{\underset{\cdot\cdot}{O}}CH_3$

M $\overset{\cdot\cdot}{\underset{\cdot\cdot}{O}}CH_3$

N

O

Problem 8. Draw the resonance contributors for those compounds in Problem 7 that have delocalized electrons.

Problem 9. Draw the arrows to show how one resonance contributor leads to the next one.

a. [structure: $CH_2=CH-CH=CH-\ddot{O}-CH_2CH_3$ resonance series] ⟷ ⟷

b. [pyrrole resonance contributors] ⟷ ⟷ ⟷ ⟷

Problem 10. Draw the resonance contributors for the following species:

a. [cyclohexadienyl cation]

b. [1,3-cyclohexadiene]

c. [cyclohexa-2,5-dienone, $=\ddot{O}:$]

d. $CH_2=CH-CH=CH-\dot{N}H_2$

e. $CH_3-\overset{\displaystyle :\ddot{O}}{\overset{\|}{C}}-\overset{..}{\underset{..}{O}}-\overset{\displaystyle \ddot{O}:}{\overset{\|}{C}}-CH_3$

f. [$CH_2=CH-CH=CH-CH=CH-CH_3$ type cation, +]

Problem 11. Draw the resonance contributors for the following species:

a. $CH_3-\overset{\displaystyle :\ddot{O}}{\overset{\|}{C}}-\overset{..}{\underset{..}{C}}H^{-}-\overset{\displaystyle \ddot{O}:}{\overset{\|}{C}}-CH_3$

b. [phenyl group with $\ddot{O}CH_2CH_3$ substituent]

c.

d. [benzene ring with CH₂Cl̈: substituent]

e. [cyclopentadienyl cation, labeled +]

f. [cyclohexenone structure with :Ö]

Problem 12. Draw the resonance contributors for the following species:

a. [cyclopentadienyl anion]

b. CH₃—C(=Ö)—ÖCH₂CH₃

c. [CH₃CH₂—C(=Ö)—N̈H—CH₂CH₃]

d. [cyclohexene with N̈HCH₃ substituent]

e. [lactone structure with :Ö and Ö:]

f. [benzene ring with ĊHCH₃, labeled +]

Solutions to the Problems

Problem 1

$$CH_3CH=CH-CH=CH-CH=CH-\overset{+}{C}H_2 \longleftrightarrow CH_3CH=CH-CH=CH-\overset{+}{C}H-CH=CH_2$$

$$CH_3\overset{+}{C}H-CH=CH-CH=CH-CH=CH_2 \longleftrightarrow CH_3CH=CH-\overset{+}{C}H-CH=CH-CH=CH_2$$

Problem 2

Problem 3

Problem 4

$$CH_2=CH-C\equiv N: \longleftrightarrow \overset{+}{C}H_2-CH=C=\overset{..}{N}:^-$$

Problem 5

Problem 6

Problem 7

B, E, F, H, J, M, O

Problem 8

B $CH_3\bar{\ddot{C}}H-CH=CH-\overset{+}{C}H_2$ $\xleftarrow{a}$ $CH_3CH=CH-CH=CH_2$ $\xleftarrow{b}$ $CH_3\overset{+}{C}H-CH=CH-\bar{\ddot{C}}H_2$

E   [resonance structures of acetamide]

F   [resonance structures of acetic acid/protonated]

H   [resonance structures of cyclohexenyl cation/anion]

J   [resonance structures]

M   [resonance structures]

O   [resonance structures]

Problem 9

1. [resonance structures]

2. [resonance structures of pyrrole]

Problem 10

a.

b. Notice in the following example that the electrons can move either clockwise or counterclockwise:

c.

d. $CH_2=CH-\overset{\cdot\cdot}{C}H=CH-\overset{\cdot\cdot}{N}H_2 \longleftrightarrow \overset{}{C}H_2=CH-\overset{\cdot\cdot}{C}H-CH=\overset{+}{N}H_2$

$\updownarrow$

$^-\overset{}{C}H_2-CH=CH-CH=\overset{+}{N}H_2$

e. Notice in the following example that the electrons can be delocalized onto either of the two sp^2 oxygens:

f.

Problem 11

a.

b.

c.

d.

e.

f.

Problem 12

a.

b.

c.

d.

e.

f.

SPECIAL TOPIC VI

Molecular Orbital Theory

Chemists use models to describe such things as the bonding in molecules, the stability of molecules, and the reactions between molecules. The model used is generally the one that provides the best description of the molecule under consideration. One very powerful model is molecular orbital theory.

You were introduced to molecular orbital (MO) theory in Section 1.6 where you saw that electrons are assigned to a volume of space called an orbital. According to MO theory, covalent bonds are formed when atomic orbitals combine to form molecular orbitals. Let's review some important principles:

1. Orbitals are conserved. In other words, the number of molecular orbitals formed must equal the number of atomic orbitals combined. For example, when two atomic orbitals overlap, two molecular orbitals are formed—one lower in energy (a bonding MO) and one higher in energy (an antibonding MO, indicated by a *) than the overlapping orbitals.
2. Electrons always occupy the available *atomic orbitals* with the lowest energy, and no more than two electrons can occupy an *atomic orbital*. Similarly, electrons always occupy the available *molecular orbitals* with the lowest energy, and no more than two electrons can occupy a *molecular orbital*.
3. The relative energies of the molecular orbitals are: $\sigma < \pi < \pi^* < \sigma^*$.
4. The strongest covalent bonds are formed by electrons that occupy the molecular orbitals with the lowest energy. For example, the energy of a σ MO is lower than that of a π MO, and we have seen that a σ bond is stronger than a π bond (Section 1.14).

There are two ways atomic orbitals can overlap to form molecular orbitals; these are shown in Figures 1 and 2. In Figure 1, each of the overlapping atomic orbitals contributes one electron to the bond. In Figure 2, a filled atomic orbital (it has two electrons) overlaps an empty atomic orbital. In each case, we see that the electrons are stabilized (are lower in energy) as a result of orbital overlap.

Figure 1. Each of the overlapping atomic orbitals contributes one electron to the bond.

Figure 2. A bond is formed as a result of the overlap of a filled atomic orbital with an empty atomic orbital.

Using Molecular Orbital Theory to Describe Covalent Bond Formation

I. Each of the Overlapping Atomic Orbitals Contributes One Electron to the Bond

Take a look at Figure 1.4 on page 23 of the text. There you can see that an H—H bond is formed by the overlap of a 1s atomic orbital of a hydrogen atom with the 1s atomic orbital of another hydrogen atom; each of the atomic orbitals contributes one electron to the molecular orbital.

Figure 3 shows that the C—C bond in ethane is formed by the overlap of an sp^3 atomic orbital of carbon with an sp^3 atomic orbital of another carbon; again, each of the overlapping atomic orbitals contributes one electron to the bond. Figure 4 shows that the C—H bond of ethane is formed by the overlap of an sp^3 atomic orbital of carbon with the s atomic orbital of hydrogen. Because an s atomic orbital is more stable than an sp^3 atomic orbital (page 27 of the text), the MO formed by sp^3—s overlap is more stable than the MO formed by sp^3—sp^3 overlap. As a result, the C—H bond is stronger (and shorter) than the C—C bond.

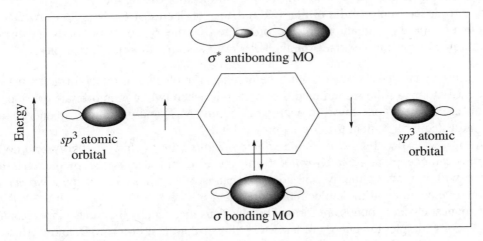

Figure 3. The C—C bond of ethane is formed by sp^3—sp^3 overlap.

Figure 4. A C—H bond of ethane is formed by sp^3—s overlap.

Figure 5 shows that the two sp^3 atomic orbitals that overlap to form the C—O bond of an alcohol or of an ether do not have the same energy. An electron is more stable in the atomic orbital of the more electronegative atom. Thus, the C—O bond is a little stronger and shorter than the C—C bond.

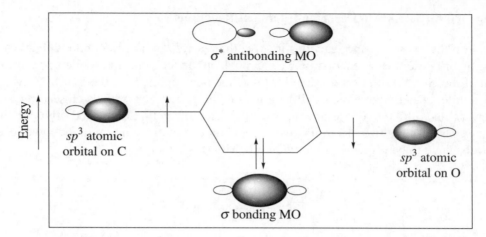

Figure 5. The C—O bond of an ether or of an alcohol is formed by sp^3—sp^3 overlap.

Figure 6 shows that the C—C σ bond of ethene is formed by the overlap of an sp^2 atomic orbital of carbon with an sp^2 atomic orbital of another carbon. The π bond of ethene is formed by the side-to-side overlap of two p orbitals (see Figure 1.5 on page 24 of the text). A π molecular orbital is less stable than a σ molecular orbital. The π bond, therefore, is weaker than the σ bond. Examine Figure 1.6 on page 24 of the text to see the molecular orbitals that are formed when the overlapping p orbitals do not each belong to the same kind of atom.

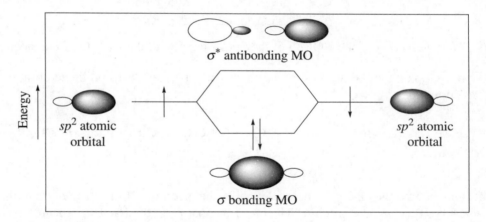

Figure 6. The C—C σ bond of ethene is formed by sp^2—sp^2 overlap.

II. A Filled Atomic Orbital Overlaps an Empty Atomic Orbital to Form a Bond

The overlap of a filled atomic orbital with an empty atomic orbital is the second way two atomic orbitals can overlap to form a bond (Figure 2).

For example, the bond formed between a Lewis base (such as NH_3) and a Lewis acid (such as BH_3) results from the base sharing a pair of electrons with the acid. Bond formation results from the overlap of a filled sp^3 orbital of nitrogen with an empty sp^2 orbital of boron. This reaction is discussed in Section 1.27 of the text.

Using Molecular Orbital Theory to Describe Chemical Reactions

We have seen that most organic reactions involve the reaction of a nucleophile with an electrophile. Molecular orbital theory describes a reaction between a nucleophile and an electrophile as occurring as the result of the interaction of the HOMO (highest occupied MO) of the nucleophile with the LUMO (lowest unoccupied MO) of the electrophile, because the most stabilizing interaction is between orbitals closest in energy. Notice that in these examples, a filled orbital overlaps an empty orbital. The interaction of a filled orbital with an empty orbital is stabilizing, because the two electrons involved in bond formation end up in the lower-energy bonding MO and no electrons have to be placed in an antibonding MO (Figure 7).

Figure 7. The interaction of the HOMO of the nucleophile with the LUMO of the electrophile.

We will start by looking at the reaction of 2-butene with HBr, an electrophilic addition reaction we first examined in Section 3.6.

In the first step of the reaction, the alkene is the nucleophile; the electrons of the π bond are in the π bonding MO; this is the HOMO. HBr is the electrophile. The electrons that form the H—Br bond are in a σ bonding MO. Therefore, the LUMO of HBr is the σ^* antibonding MO (Figure 8).

Figure 8. Interaction of the HOMO of the alkene with the LUMO of HBr.

Figure 9. Interaction of the HOMO of Br$^-$ with the LUMO of the carbocation.

In the second step of the reaction, Br^- is the nucleophile; its HOMO is a filled nonbonding sp^3 orbital. The carbocation is the electrophile; the LUMO is carbon's empty p orbital (Figure 9).

In Section 8.1 of the text, we used MO theory to explain why the alkyl halide in an S_N2 reaction undergoes back-side attack, resulting in inversion of configuration. The filled nonbonding orbital of the nucleophile is the HOMO. The LUMO of the electrophilic alkyl halide is the σ^* antibonding MO, which has its largest lobe at the back of the carbon in the C—Br bond, so this is where overlap is best. (See Figure 8.1 on page 336 of the text).

In Section 17.5 of the text, we used MO theory to explain how a nucleophile reacts with a carbonyl group to form a tetrahedral species. Once again the reaction results from the interaction of the HOMO of the nucleophile and the LUMO of the electrophile. In this case, a filled nonbonding orbital of the nucleophile is the HOMO and the LUMO is the π^* antibonding orbital of the carbonyl group. The π^* antibonding orbital is largest at the carbon atom, so this is where the nucleophile overlaps. (See Figure 17.3 on page 719 of the text.)

Molecular Orbital Theory and Delocalized Electrons

Molecular orbital theory is useful to describe compounds with delocalized electrons. For example, in Section 4.2 we saw that tertiary carbocations are more stable than secondary carbocations, which are more stable than primary carbocations, because the electrons in a filled σ bond are delocalized by overlapping the empty p orbital of the positively charged carbon. (See Figure 4.1 on page 158 of the text.) This kind of electron delocalization is known as hyperconjugation; it is one more example of the stabilization that results when a filled orbital overlaps an empty orbital.

Molecular orbital theory avoids having to use contributing resonance structures, because the electrons in the most stable MO are delocalized over the entire molecule. For example, in Section 7.8, we saw that 1,3-butadiene is more stable than 1,4-pentadiene, because the π electrons in 1,3-butadiene are delocalized over four sp^2 carbons, whereas the intervening methylene group in 1,4-pentadiene prevents the p orbitals of C2 and C4 from overlapping. Thus the π electrons of 1,4-pentadiene are localized, causing its molecular orbitals to have the same energy as those of ethene—another compound with localized electrons. (See pages 294–298 of the text.)

intervening
methylene group

1,3-butadiene   1,4-pentadiene

Let's now look at the molecular orbitals of the allyl cation, the allyl radical, and the allyl anion.

$$CH_2\!=\!CH\!-\!\overset{+}{C}H_2 \qquad CH_2\!=\!CH\!-\!\overset{\cdot}{C}H_2 \qquad CH_2\!=\!CH\!-\!\overset{..\,-}{C}H_2$$

the allyl cation   the allyl radical   the allyl anion

The three *p* atomic orbitals of the three allyl carbons combine to produce three π molecular orbitals (Figure 10).

Figure 10. The distribution of the electrons in the molecular orbitals of the allyl cation, the allyl radical, and the allyl anion.

The bonding MO (ψ_1) encompasses the three carbons. In an acyclic system, the number of bonding MOs always equals the number of antibonding MOs. Therefore, when there is an odd number of MOs, one of them must be a nonbonding molecular orbital; ψ_2 is the nonbonding MO. We have seen (Section 7.8) that as the energy of the MO increases, the number of nodes increases. Consequently ψ_2 must have a node—in addition to the one that ψ_1 has—that bisects the *p* orbitals. The only symmetrical position for a node to pass through in ψ_2 is through the middle carbon. You also know that it needs to pass through the middle carbon because that is the only way ψ_2 can be fully antisymmetric, which it must be since ψ_1 and ψ_3 are symmetric. (Recall that MOs alternate between being symmetric and antisymmetric; Section 7.8.)

     You can see why ψ_2 is called a nonbonding molecular orbital—there is no overlap between the *p* orbital on the middle carbon and the *p* orbital on either of the end carbons. Notice that a nonbonding MO has the same energy as the isolated *p* atomic orbitals. The third MO (ψ_3) is an antibonding MO.

     The two π electrons of the allyl cation are in the bonding MO, which means they are spread out over all three carbons. Consequently, the two carbon-carbon bonds are identical, with each having some double-bond character. The resonance contributors show that the positive charge is shared equally by the end carbon atoms, which is another way of showing that the stability of the allyl cation is due to electron delocalization.

$$CH_2\!=\!CH\!-\!\overset{+}{C}H_2 \quad\longleftrightarrow\quad \overset{+}{C}H_2\!-\!CH\!=\!CH_2$$

The contributing resonance structures show that when a nucleophile such as Br$^-$ reacts with an allyl cation, the Br$^-$ can bond to either of the end carbons—because they share the positive charge—but cannot bond to the middle carbon. Likewise, MO theory shows that only the end carbons have an empty orbital with which the filled orbital of Br$^-$ can overlap. The central carbon has a node so there can be no interaction with this carbon.

     The allyl radical has two electrons in the bonding MO, so these electrons are spread over all three carbon atoms. The third electron is in the nonbonding MO. The MO diagram shows that the third electron is shared equally by the end carbons with none of the electron density residing on the middle carbon. This agrees with what the resonance contributors show.

$$CH_2\!=\!CH\!-\!\overset{\cdot}{C}H_2 \quad\longleftrightarrow\quad \overset{\cdot}{C}H_2\!-\!CH\!=\!CH_2$$

Finally, the allyl anion has two electrons in the nonbonding MO. These two electrons are shared equally by the end carbons. This too agrees with what the resonance contributors show.

$$CH_2\!\!=\!\!CH\!\!-\!\!\overset{..}{\underset{}{C}}\overset{-}{H_2} \quad \longleftrightarrow \quad \overset{..}{\underset{}{C}}\overset{-}{H_2}\!\!-\!\!CH\!\!=\!\!CH_2$$

We have seen that both molecular orbital theory and contributing resonance structures can be used to explain electron delocalization. The choice is a matter of preference. Hyperconjugation is probably best shown by molecular orbital theory, because the contributing resonance structures that describe hyperconjugation would require breaking sigma bonds. Contributing resonance structures are sometimes preferred when one needs to see on which atoms charges reside.

CHAPTER 8
Substitution Reactions of Alkyl Halides

Important Terms

aprotic solvent	a solvent that does not have a hydrogen bonded to an oxygen or to a nitrogen; some aprotic solvents are polar, others are nonpolar.
backside attack	nucleophilic attack on the side of the carbon opposite to the side bonded to the leaving group.
base	a substance that accepts a proton.
basicity	describes the tendency of a compound to share its electrons with a proton.
bifunctional molecule	a molecule with two functional groups.
bimolecular	describes a reaction in which two molecules are involved in the transition state of the rate-determining step.
complete racemization	formation of a pair of enantiomers in equal amounts.
dielectric constant	a measure of how well a solvent can insulate opposite charges from one another.
elimination reaction	a reaction that removes atoms or groups from the reactant to form a π bond.
first-order reaction	a reaction whose rate is linearly dependent on the concentration of one reactant.
intermolecular reaction	a reaction that takes place between two molecules.
intimate ion pair	results when the covalent bond that joined the cation and anion has broken, but the cation and anion are still next to each other.
intramolecular reaction	a reaction that takes place within a molecule.
inversion of configuration	turning the carbon inside out like an umbrella so that the resulting product has a configuration opposite to that of the reactant.
ion-dipole interaction	the interaction between an ion and the dipole of a molecule.
kinetics	the field of chemistry that deals with the rates of chemical reactions.
leaving group	the group that is displaced in a substitution reaction.
nucleophile	an electron-rich atom or molecule.
nucleophilicity	a measure of how readily an atom or molecule with a lone pair attacks an atom.
nucleophilic substitution reaction	a reaction in which a nucleophile substitutes for an atom or group.
partial racemization	formation of a pair of enantiomers in unequal amounts.

protic solvent   a solvent that has a hydrogen bonded to an oxygen or to a nitrogen.

rate constant   the constant of proportionality in the rate law for a reaction; it describes how difficult it is to overcome the energy barrier of a reaction.

rate law   the equation that shows the relationship between the rate of a reaction and the concentration of the reactants.

second-order reaction   a reaction whose rate is linearly dependent on the concentration of two reactants.

S_N1 reaction   a unimolecular nucleophilic substitution reaction.

S_N2 reaction   a bimolecular nucleophilic substitution reaction.

solvent-separated ion pair   results when the cation and anion are separated by one or more solvent molecules.

solvolysis   reaction with the solvent.

steric effects   effects due to the fact that groups occupy a certain volume of space.

steric hindrance   refers to bulky groups at the site of a reaction that make it difficult for the reactants to approach each other.

substitution reaction   a reaction that changes one substituent of a reactant for another.

unimolecular   describes a reaction in which only one molecule is involved in the transition state of the rate-determining step.

Solutions to Problems

1. The rate of the reaction will be half of the original rate, since the concentration is half of the original concentration and the reaction rate is linearly dependent on the concentration of bromethane.

2. Increasing the height of the energy barrier decreases the magnitude of the rate constant; this causes the reaction to be slower.

3. The closer the methyl group is to the site of nucleophilic attack, the greater the steric hindrance to nucleophilic attack and the slower the rate of the reaction.

$$
\underset{}{CH_3CH_2CH_2CH_2CH_2Br} \; > \; \underset{\overset{|}{CH_3}}{CH_3CHCH_2CH_2Br} \; > \; \underset{\overset{|}{CH_3}}{CH_3CH_2CHCH_2Br} \; > \; \underset{\overset{|}{CH_3}}{CH_3CH_2\underset{\underset{CH_3}{|}}{C}Br}
$$

4.   a. Solved in the text.

   b. The S_N2 reaction of (R)-2-bromobutane with hydroxide ion will form (S)-2-butanol.

   c. The S_N2 reaction of (S)-3-chlorohexane and with hydroxide ion will form (R)-3-hexanol.

   d. The S_N2 reaction of 3-iodopentane (it does not have an asymmetric center) with hydroxide ion will form 3-pentanol (it does not have an asymmetric center).

5.   A protic solvent has a hydrogen bonded to an oxygen or to a nitrogen.

   a. aprotic   b. aprotic   c. protic   d. aprotic

6.   a. RO^- because ROH is weaker acid than RSH since the hydrogen is attached to a smaller atom.

   b. RS^- because it is less well solvated by water and sulfur is more polarizable than oxygen.

   c. RO^- because, although they differ in size, they are in an aprotic solvent. Remember that the stronger base is always the better nucleophile in an aprotic solvent.

7.   Remember that the stronger base is always the better nucleophile unless they differ in size AND they are in a protic solvent.

   a. They differ in size and because they are in a protic solvent, the larger one (Br^-) is the better nucleophile.

   b. They differ in size and because they are in an aprotic solvent, the stronger base (Cl^-) is the better nucleophile.

   c. Because the oxygen is negatively charged, CH_3O^- is the better nucleophile.

   d. Because the oxygen is negatively charged, CH_3O^- is the better nucleophile.

   e. Because H_2O is a stronger acid than NH_3, $^-NH_2$ is the stronger base and the better nucleophile.

   f. Because H_2O is a stronger acid than NH_3, $^-NH_2$ is the stronger base and the better nucleophile.

   g. They differ in size and because they are in a protic solvent, the larger one (I^-) is the better nucleophile.

   h. They differ in size and because they are in an aprotic solvent, the stronger base (Br^-) is the better nucleophile.

8.   Solved in the text.

9.   a. $CH_3CH_2Br + HO^-$

HO^- is a better nucleophile than H_2O.

b. $CH_3CHCH_2Br + HO^-$
$|$
CH_3

This alkyl halide has less steric hindrance toward nucleophilic attack.

c. $CH_3CH_2Cl + CH_3S^-$

CH_3S^- is a better nucleophile than CH_3O^- in a protic solvent (a solvent that can form hydrogen bonds).

d. $CH_3CH_2Br + I^-$

Br^- is a weaker base than $Cl,^-$ so Br^- is a better leaving group.

10.   In each reaction, the conjugate acid of the leaving group is stronger than the conjugate acid of the nucleophile. That means the leaving group is the weaker base (better leaving group).

H_2O	HCl	NH_3	HCl
$pK_a = 15.7$	$pK_a = -7.0$	$pK_a = 36$	$pK_a = -7.0$
H_2S	HBr	$RC\equiv CH$	HBr
$pK_a = 7.0$	$pK_a = -9.0$	$pK_a = 25$	$pK_a = -9.0$
ROH	HI	$HC\equiv N$	HI
$pK_a = 15.5$	$pK_a = -10.0$	$pK_a = 9.1$	$pK_a = -10.0$
RSH	HBr		
$pK_a = 10.5$	$pK_a = -9.0$		

11.   a. $CH_3CH_2OCH_2CH_2CH_3$

c. $CH_3CH_2\overset{+}{N}(CH_3)_3\ Br^-$

b. $CH_3CH_2C\equiv CCH_3$

d. $CH_3CH_2SCH_2CH_3$

12.   Solved in the text.

13.   Solved in the text.

14.   a. Reaction of an alkyl halide with ammonia gives a low yield of primary amine, because as soon as the primary amine is formed, it can react with another molecule of alkyl halide to form a secondary amine; the secondary amine can react with the alkyl halide to form a tertiary amine, which can then react with an alkyl halide to form a quaternary ammonium salt.

b. The alkyl azide is not treated with hydrogen until after all the alkyl halide has reacted with azide ion. Therefore, when the primary amine is formed, there is no alkyl halide for it to react with to form a secondary amine.

15.   The more stable the carbocation formed when the bromide ion departs, the faster the rate of the reaction:

$$\begin{array}{c} CH_3 \\ | \\ CH_3CBr \\ | \\ CH_3 \end{array} > \begin{array}{c} \\ CH_3CHBr \\ | \\ CH_3 \end{array} > CH_3CH_2CH_2Br > CH_3Br$$

16.   Because ethyl bromide and methyl bromide cannot dissociate to form a carbocation (the carbocation is too unstable to be formed), the alkyl halides have to react by an S_N2 pathway. The nucleophile (H_2O) is poor, so the S_N2 reactions are very slow.

17. Remember that a bromide ion is a better leaving group (a weaker base) than a chloride ion.

$$CH_3CH_2\overset{\overset{\textstyle CH_3}{|}}{\underset{\underset{\textstyle Br}{|}}{C}}CH_2CH_3 \;>\; CH_3\underset{\underset{\textstyle Br}{|}}{C}HCH_2CH_2CH_3 \;>\; CH_3\underset{\underset{\textstyle Cl}{|}}{C}HCH_2CH_2CH_3 \;>\; CH_3CH_2CH_2CH_2CH_2Cl$$

18. **A, B, C,** and **F** will form a substitution product in an S_N1 reaction that is different from the substitution product formed in an S_N2 reaction.

A The S_N2 and S_N1 reactions will form different products, because the secondary carbocation that is formed initially in the S_N1 reaction will undergo a 1,2-hydride shift to form a tertiary carbocation.

B The S_N2 and S_N1 reactions will form different products, because the secondary carbocation that is formed initially in the S_N1 reaction will undergo a 1,2-hydride shift to form a tertiary carbocation.

C The S_N2 and S_N1 reactions will form different products, because the secondary carbocation that is formed initially in the S_N1 reaction will undergo a 1,2-methyl shift to form a tertiary carbocation.

D The same substitution product will be obtained from S_N1 and S_N2 reactions. However, the S_N2 reaction will be too slow to measure because of steric hindrance.

E The same constitutional isomers will be obtained from both the S_N1 and S_N2 reactions. However, since the configuration of the asymmetric center is not specified in the reactant, the configuration of the asymmetric center in the product cannot be specified. If the configuration at C-4 were specified, one stereoisomer would be obtained from an S_N2 reaction (the product with the inverted configuration compared to the configuration of the reactant), and two products would be obtained from an S_N1 reaction (one product with the inverted configuration and one with the retained configuration).

F The S_N1 and S_N2 reactions will form different products because the secondary carbocation formed initially in the S_N1 reaction will undergo a 1,2-methyl shift to form a tertiary carbocation.

19.

3-chloro-3-methyl-1-butene

20.   a. 1. The S_N2 reaction proceeds with back-side attack.

2.

b. 1. The S_N1 reaction forms a carbocation; the nucleophile can add to either side of the carbocation.

2.

21.   *trans*-4-Bromo-2-hexene (the compound on the right) is more reactive in an S_N1 solvolysis reaction, because the carbocation that is formed is stabilized by electron delocalization. (It is a secondary allylic carbocation.) The other alkyl halide dissociates to form a secondary carbocation that cannot be stabilized by its π bond electrons.

22.   Most reactive in an S_N2 reaction:

a. Br^- is a better leaving group (a weaker base) than $Cl.^-$

b.   A primary carbon has less steric hindrance to back-side attack than a secondary carbon.

c.   A primary carbon has less steric hindrance to back-side attack than a secondary carbon.

d. S_N2 nucleophilic attack cannot occur on an sp^2 carbon.

e.   This compound has less steric hindrance to back-side attack, because the carbon undergoing back-side attack is bonded to a methyl group and an ethyl group, whereas the carbon undergoing back-side attack in the other compound is bonded to two ethyl groups.

f. S_N2 nucleophilic attack cannot occur on an sp^2 carbon.

23.   Most reactive in an S_N1 reaction:

a. Br^- is a better leaving group (a weaker base) than $Cl.^-$

b.   A secondary carbocation is more stable than a primary carbocation.

c.   A secondary carbocation is more stable than a primary carbocation.

d.   A benzyl carbocation is more stable than a vinyl carbocation.

e.   Both are equally reactive.

f.   An allyl carbocation is more stable than a vinyl carbocation.

24. **a.**

b.

25. **a.** An S_N2 reaction, because of the high concentration of a good nucleophile. Therefore, the product will have the inverted configuration compared to that of the reactant.

b. An S_N2 reaction, because a high concentration of a good nucleophile is used. Therefore, the product will have the inverted configuration compared to that of the reactant.

Even though ammonia is not negatively charged, it is a good nucleophile. Recall that an amine is a much stronger base than an alcohol. (The pK_a of a protonated alcohol is ~ −2.5. The pK_a of the ammonium ion is 9.4: see Appendix II on page A-4 of the text.)

c. An S_N2 reaction, because of the high concentration of a good nucleophile. Therefore, the product will have the inverted configuration compared to that of the reactant.

d. An S_N1 reaction, because a poor nucleophile is used. Notice that the initially formed secondary carbocation rearranges to a more stable tertiary carbocation. Only one product is formed because it does not have an asymmetric center.

   e. An S_N2 reaction, because of the high concentration of a good nucleophile. Therefore, the product will have the inverted configuration compared to that of the reactant.

$$CH_3 - \overset{\overset{\displaystyle CH_2CH_3}{|}}{\underset{\underset{\displaystyle OCH_3}{}}{C}} \cdots H$$

   f. An S_N1 reaction, because a poor nucleophile is used. Therefore, the product will have both the inverted and retained configurations compared to that of the reactant.

$$\underset{CH_3O}{H} \cdots \overset{\overset{\displaystyle CH_2CH_3}{|}}{C} - CH_3 \qquad CH_3 - \overset{\overset{\displaystyle CH_2CH_3}{|}}{\underset{\underset{\displaystyle OCH_3}{}}{C}} \cdots H$$

26.   The rate of an S_N1 reaction is not affected by increasing the concentration of the nucleophile, whereas the rate of an S_N2 reaction is increased when the concentration of the nucleophile is increased. Therefore, we first have to determine whether the reactions are S_N1 or S_N2 reactions.

   A is an S_N2 reaction, because the configuration of the product is inverted compared with that of the reactant.

   B is an S_N2 reaction, because the reactant is a primary alkyl halide.

   C is an S_N1 reaction, because the reactant is a tertiary alkyl halide.

   Because they are S_N2 reactions, **A** and **B** will go faster if the concentration of the nucleophile is increased. Because it is an S_N1 reaction, the rate of **C** will not change if the concentration of the nucleophile is increased.

27.   **a.** Solved in the text.

   b. $\displaystyle \underset{\text{by } S_N2}{\text{percentage}} = \frac{S_N2}{S_N2 + S_N1} \times 100 = \frac{k_2[\text{2-bromobutane}][\text{HO}^-]}{k_2[\text{2-bromobutane}][\text{HO}^-] + k_1[\text{2-bromobutane}]} \times 100$

$$\frac{S_N2}{S_N2 + S_N1} \times 100 = \frac{3.20 \times 10^{-5} \times 1 \times 10^{-3}}{3.20 \times 10^{-8} + 1.50 \times 10^{-6}} \times 100$$

$$= \frac{3.20 \times 10^{-8}}{3.20 \times 10^{-8} + 150 \times 10^{-8}} \times 100 = \frac{3.20}{153} \times 100 = 2.0\%$$

28.   Because both reactants in the rate-limiting step are neutral, the reaction will be faster if the polarity of the solvent is increased.

29.   **a.** Increasing the polarity will decrease the rate of the reaction because the concentration of charge on the reactants is greater (the reactants are charged) than the concentration of charge on the transition state.
   b. Increasing the polarity will decrease the rate of the reaction because the concentration of charge on the reactants is greater (the reactants are charged) than the concentration of charge on the transition state.
   c. Increasing the polarity will increase the rate of the reaction because the concentration of charge on the reactants is less (the reactants are not charged) than the concentration of charge on the transition state.

30. **a.** $CH_3Br + HO^- \longrightarrow CH_3OH + Br^-$

HO^- is better nucleophile than H_2O.

b. $CH_3I + HO^- \longrightarrow CH_3OH + I^-$

I^- is better leaving group than Cl^-.

c. $CH_3Br + NH_3 \longrightarrow CH_3\overset{+}{N}H_3 + Br^-$

NH_3 is a better nucleophile than H_2O.

d. $CH_3Br + HO^- \xrightarrow{\text{DMSO}} CH_3OH + Br^-$

Unlike ethyl alcohol, DMSO will not stabilize the nucleophile (and therefore decrease the rate of the reaction) by hydrogen bonding.

e. $CH_3Br + NH_3 \xrightarrow{\text{EtOH}} CH_3\overset{+}{N}H_3 + Br^-$

The more polar solvent will be more able to stabilize the transition state. (EtOH is ethanol.)

31. Solved in the text.

32. Acetate ion will be a more reactive nucleophile in dimethyl sulfoxide because dimethyl sulfoxide will not stabilize the nucleophile by hydrogen bonding, whereas methanol will stabilize it by hydrogen bonding.

33. Only an S_N1 reaction will give the product with retention of configuration. Since the S_N1 reaction is favored by a polar solvent, a greater percentage of the reaction will take place by an S_N1 pathway in 50% water/50% ethanol, the more polar of the two solvents.

34. **a.** HO〜〜〜Br   because it forms a six-membered ring, whereas the other compound would form a seven-membered ring. A seven-membered ring is more strained than a six-membered ring, so the six-membered ring is formed more easily (see Table 2.9 on page 104 of the text).

b. HO〜〜Br   because it forms a five-membered ring, whereas the other compound would form a four-membered ring. A four-membered ring is more strained than a five-membered ring, so the five-membered ring is formed more easily.

c. HO〜〜〜Br   because it forms a seven-membered ring, whereas the other compound would form an eight-membered ring. An eight-membered ring is more strained than a seven-membered ring, so the seven-membered ring is formed more easily; also, the Br and OH in the compound that leads to the eight-membered ring are less likely to be in the proper position relative to one another for reaction because there are more bonds around which rotation to an unfavorable conformation can occur. (See page 104 of the text.)

35. **a.** When hydroxide ion removes a proton from the OH group, the alkoxide ion cannot react in an intramolecular reaction with the alkyl chloride to form an epoxide, because it cannot reach the back side of the carbon attached to the chlorine. Thus, the only productive reaction is for hydroxide ion to react with the alkyl halide in an S_N2 reaction.

b. Hydroxide ion removes a proton from the OH group more rapidly than it attacks the alkyl chloride. Once the alkoxide ion is formed, it attacks the back side of the alkyl chloride, forming an epoxide.

c. Hydroxide ion removes a proton from the OH group and the alkoxide ion attacks the back side of the carbon attached to the bromine forming a six-membered ring ether.

$$BrCH_2CH_2CH_2CH_2CH_2OH \quad \xrightarrow{HO^-} \quad Br-CH_2CH_2CH_2CH_2CH_2\ddot{O}:^-$$

d. Hydroxide ion removes a proton from the OH group and the alkoxide ion attacks the back side of the carbon attached to the chlorine forming an epoxide.

e. $CH_3CH_2CH_2CH{=}CH_2 \xrightarrow[H_2O]{Cl_2} CH_3CH_2CH_2CH{-}CH_2Cl \xrightarrow{HO^-}$

36. **a.** CH_3OH **b.** CH_3NH_2 **c.** CH_3SCH_3 **d.** CH_3SH **e.** CH_3OCH_3 **f.** $CH_3\overset{+}{N}H_2CH_3$

(Notice that the product in part **c** is not protonated because its pK_a is $\sim$ −7; the product in "**f**" is protonated because its pK_a is ~11. In part **f**, $CH_3\overset{+}{N}H(CH_3)_2$ and $CH_3\overset{+}{N}(CH_3)_3$ can also be formed, depending on the concentration of bromomethane; see Problem 13 on page 346 of the text.)

37. **a.** **1.** $3° > 2° > 1°$

   2. An S_N1 reaction is not affected by the reactivity of the nucleophile, but a poor (weakly reactive) nucleophile favors an S_N1 by disfavoring an S_N2 reaction.

   3. An S_N1 reaction is not affected by the concentration of the nucleophile, but a low concentration of a nucleophile favors an S_N1 by disfavoring an S_N2 reaction.

   4. If the reactant is charged, an S_N1 reaction will be favored by the least polar solvent that will dissolve the reactant (generally an aprotic polar solvent).
   If the reactant is not charged, an S_N1 reaction will be favored by a protic polar solvent.

b. **1.** $1° > 2° > 3°$

   2. A highly reactive nucleophile favors an S_N2 reaction.

   3. A high concentration of a nucleophile favors an S_N2 reaction.

   4. If the reactant is charged, an S_N2 reaction will be favored by the least polar solvent that will dissolve the reactant (generally an aprotic polar solvent).
   If the reactant is not charged, an S_N2 reaction will be favored by a protic polar solvent.

38. If the atoms are in the same row, the stronger base is the better nucleophile. If the atoms are in the same column, the larger atom is the better nucleophile in the protic polar solvent because the solvent will form stronger hydrogen bonds with the smaller atom.

   a. HO^- **c.** H_2S **e.** I^-

   b. NH_3 **d.** HS^- **f.** Br^-

39. The weaker base is the better leaving group.

   a. H_2O **c.** H_2S **e.** I^-

   b. H_2O **d.** HS^- **f.** Br^-

40. **a.** HO^- **c.** CH_3NH_2 **e.** $CH_3CH_2S^-$ **g.** $^-C{\equiv}N$

   b. CH_3O^- **d.** HS^- **f.** $CH_3\overset{\overset{\text{O}}{\|}}{C}O^-$ **h.** $CH_3CH_2C{\equiv}C^-$

   In part **c**, a tertiary amine and a quaternary ammonium ion can also form unless a large excess of CH_3NH_2 is used; see Problem 13 on page 346 of the text.

41. **a.**

b.

c.

(A large excess of methylamine will have to be used in the second step in order to minimize the formation of a tertiary amine and a quarternary ammonium ion.)

42.   **a.** $CH_3CH_2S^- > CH_3CH_2O^- > CH_3\overset{\overset{\displaystyle O}{\|}}{C}O^-$ **c.** $NH_3 > H_2O$

   b. $-O^- >$ $-O^-$ **d.** $I^- > Br^- > Cl^-$

43.   The pK_a would increase (it would be a weaker acid) because of a decreased tendency to form a charged species in a less polar solvent. (See Problem 31.)

44.   **a.**

   b.   The products are obtained as a result of the nucleophiles reacting with the carbocation. 2-Bromo-2-methylpropane and 2-chloro-2-methylpropane form the same carbocation, so both alkyl halides form the same products.

45.   **a.**   The S_N2 reaction takes place with inversion of configuration.

(R)-2-bromopentane   (S)-2-methoxypentane

or

b. The S_N1 reaction takes place with racemization (either complete or partial).

(R)-2-bromopentane $\xrightarrow{CH_3OH}$ (S)-2-methoxypentane   +   (R)-2-methoxypentane

or

c. *trans*-1-Chloro-2-methylcyclohexane exists as a pair of enantiomers. The S_N2 reaction involves back-side attack.

$\xrightarrow[S_N2]{CH_3O^-}$

d. *trans*-1-Chloro-2-methylcyclohexane exists as a pair of enantiomers. The asymmetric center is lost when the carbocation that is formed in the S_N1 reaction undergoes a 1,2-hydride shift.

$\xrightarrow[S_N1]{CH_3OH}$ $\xrightarrow{\text{1,2-hydride shift}}$ $\xrightarrow{CH_3OH}$

e. 3-Bromo-2-methylpentane exists as a pair of enantiomers. The asymmetric center is lost when the carbocation that is formed in the S_N1 reaction undergoes a 1,2-hydride shift.

$$\underset{\underset{Br}{|}}{\overset{\overset{CH_3}{|}}{CH_3CHCHCH_2CH_3}} \longrightarrow \overset{\overset{CH_3}{|}}{CH_3\overset{+}{CH}CHCH_2CH_3} \longrightarrow \overset{\overset{CH_3}{|}}{CH_3\overset{+}{C}CH_2CH_2CH_3}$$

$\downarrow CH_3OH$

$$\underset{\underset{OCH_3}{|}}{\overset{\overset{CH_3}{|}}{CH_3CCH_2CH_2CH_3}}$$

f. 3-Bromo-3-methylpentane does not have an asymmetric center.

$$\underset{\underset{Br}{|}}{\overset{\overset{CH_3}{|}}{CH_3CH_2CCH_2CH_3}} \xrightarrow{CH_3OH} \underset{\underset{OCH_3}{|}}{\overset{\overset{CH_3}{|}}{CH_3CH_2CCH_2CH_3}}$$

46. **a.**

b.

c.

47.   Methoxide ion will be a better nucleophile in DMSO, because DMSO cannot stabilize the anion by hydrogen bonding.

48. **a.**

The nucleophile is less sterically hindered.

b.

The electron-withdrawing oxygen increases the electrophilicity of the carbon that the nucleophile attacks.

c.

Steric strain is decreased when the alkyl halide dissociates to form the carbocation in the rate-limiting step, since the hybridization of the carbon atom changes from sp^3 to sp^2, allowing the bond angle between the bulky groups to increase from 109.5° to 120°. The carbocation therefore is formed more rapidly.

d. $(CH_3)_3CBr$ $\xrightarrow{H_2O}$ $(CH_3)_3COH$ + HBr   Because the reactants are neutral the reaction will be faster in the more polar solvent.

49. CH_3CH_2Cl + K^+I^- $\underset{acetone}{\rightleftharpoons}$ CH_3CH_2I + K^+Cl^-

K^+Cl^- precipitates out in acetone, which drives the reaction to the right.

50.

51.   The slow step of the solvolysis reaction is formation of the carbocation. Since the products of the slow step are charged, they will be better stabilized by formic acid than by acetic acid, because formic acid has a greater dielectric constant. Because the products are more stabilized, the transition state leading to them will also be more stabilized so the reaction will be faster in the solvent with the greater dielectric constant. Therefore, the second reaction is faster.

52.   In parts **a–d**, the configuration of the asymmetric center attached to the Cl in the reactant will be inverted in the product.

a.

(2R,3S)

b. CH$_3$O—C(H)(CH$_3$)—C(H)(CH$_3$)—CH$_2$CH$_3$ **or**

(2R,3R)

c. CH$_3$, H, CH$_3$O—C—C—CH$_3$, CH$_2$CH$_3$ **or**

(2S,3S)

d. CH$_3$, H, CH$_3$O—C—C—CH$_2$CH$_3$, CH$_3$ **or**

(2S,3R)

e.

The alkyl halide dissociates to form a carbocation that undergoes a 1,2-methyl shift. Because the carbocation is planar, the migrating methyl group with its pair of electrons can add either to the top or the bottom of the carbocation. Therefore, the asymmetric center in the product can have either the *R* or *S* configuration.

f.

—CH$_2$OCH$_2$CH$_3$

53.   **a.**   We can predict that this is an S$_N$1 reaction, because acetate ion is a relatively poor nucleophile and the alkyl halide is sterically hindered.

b. We can predict that this is an S_N1 reaction, because acetate ion is a relatively poor nucleophile.

54.

a. $CH_3CH_2CH_2Br + HC\equiv C^- \longrightarrow CH_3CH_2CH_2C\equiv CH \xrightarrow[Pd/C]{H_2} CH_3CH_2CH_2CH_2CH_3$

b. $CH_3CH_2CH_2Br + HC\equiv C^- \longrightarrow CH_3CH_2CH_2C\equiv CH \xrightarrow[H_2O]{H_2SO_4,\ HgSO_4} CH_3CH_2CH_2\overset{\displaystyle O}{\overset{\|}{C}}CH_3$

c. $CH_3CH_2CH_2Br + HC\equiv C^- \longrightarrow CH_3CH_2CH_2C\equiv CH \xrightarrow[2.\ HO^-,\ H_2O_2,\ H_2O]{1.\ BH_3/THF}$

$CH_3CH_2CH_2CH_2\overset{\displaystyle O}{\overset{\|}{C}}H$

d. $CH_3CH_2CH_2Br + HC\equiv C^- \longrightarrow CH_3CH_2CH_2C\equiv CH \xrightarrow[\substack{Lindlar \\ catalyst}]{H_2} CH_3CH_2CH_2CH=CH_2$

$\downarrow \overset{\displaystyle O}{\overset{\|}{R}COOH}$

$CH_3CH_2CH_2\overset{O}{\overset{/\ \backslash}{CH}}{-}CH_2$

55. The equilbrium contstant is given by the relative stabilities of the products relative to the reactants. Therefore, any factor that stabilizes the products will increase the equilibrium constant.

$$K_{eq} = \frac{[\text{products}]}{[\text{reactants}]}$$

Ethanol will stabilize the charged products more than will diethyl ether because ethanol has a greater dielectric constant. Therefore the equilibrium will lie farther to the right (toward products) in ethanol.

56.

a. The reaction with quinuclidine had the larger rate constant because quinuclidine is less sterically hindered as a result of the substituents on the nitrogen being pulled back into a ring structure.

b. The reaction with quinuclidine had the larger rate constant for the same reason given in part **a.**

c. Isopropyl iodide has the larger difference in rate constants and therefore the larger ratio because, since it is more sterically hindered than methyl iodide, it is more affected by differences in the amount of steric hindrance in the nucleophile.

57. Because methanol is a poor nucleophile, the reaction will take place predominately via an S_N1 pathway.

A secondary benzylic carbocation is more stable and, therefore, is easier to form than a secondary carbocation, so the product will be the one shown.

$$OCH_3$$

58. I^- is a good nucleophile in a polar solvent such as methanol, so it reacts rapidly with methyl bromide, causing the concentration of I^- to decrease rapidly.

$$:\ddot{I}:^- + CH_3 \overset{\frown}{-} Br \xrightarrow{\text{fast}} CH_3 - I + :\ddot{B}r:^-$$

I^- is a good leaving group so methyl iodide undergoes an S_N2 reaction with methanol. Methanol is a poor nucleophile so the S_N2 reaction is slow. Therefore, iodide ion returns slowly to its original concentration.

$$CH_3\ddot{O}H + CH_3 \overset{\frown}{-} I \xrightarrow{\text{slow}} CH_3OCH_3 + :\ddot{I}:^- + H^+$$

59. The number of atoms in the ring is given by n. Three- and four-membered rings have strain, so they are harder to make than five- and six-membered rings. The three-membered ring is formed faster than the four-membered ring because the compound leading to the three-membered ring has one less carbon-carbon single bond that can rotate to give a conformer in which the reacting groups are positioned too far from one another for reaction.

Even though the compound that forms the five-membered ring has one more bond that can rotate to give a conformer in which the reacting groups are positioned too far from one another for reaction (compared to the compound that forms the four-membered ring), the five-membered ring is formed faster because it is relatively strain free. So lack of strain more than makes up for the lower probability of having the reacting groups in the proper position for reaction.

The rate of the ring-forming reaction gets slower as the size of the ring being formed gets larger, because the reactant has more bonds that can rotate to give conformers in which the reacting groups are positioned too far from one another for reaction.

60. The configuration of the asymmetric center attached to the Br in the reactant will be inverted in the product.

a.

CH$_3$O—C—C—CH$_3$ structure with H, CH$_3$CH$_2$, CH$_2$CH$_3$ **or**   Fischer projection: CH$_2$CH$_3$ / CH$_3$O—H / CH$_3$—H / CH$_2$CH$_3$ (3R,4S)

b.

CH$_3$O—C—C—CH$_2$CH$_3$ structure with H, CH$_3$CH$_2$, CH$_3$ **or**   Fischer projection: CH$_2$CH$_3$ / CH$_3$O—H / H—CH$_3$ / CH$_2$CH$_3$ (3R,4R)

c.

CH$_3$CH$_2$—C—C—CH$_2$CH$_3$ structure with H, CH$_3$O, CH$_3$ **or**   Fischer projection: CH$_2$CH$_3$ / H—OCH$_3$ / H—CH$_3$ / CH$_2$CH$_3$ (3S,4R)

d.

CH$_3$CH$_2$—C—C—CH$_3$ structure with H, CH$_3$O, CH$_2$CH$_3$ **or**   Fischer projection: CH$_2$CH$_3$ / H—OCH$_3$ / CH$_3$—H / CH$_2$CH$_3$ (3S,4S)

61. Tetrahydrofuran can solvate a charge better than diethyl ether can, because the floppy ethyl substituents of diethyl ether provide steric hindrance, making it difficult for the nonbonding electrons of the oxygen to approach the positive charge that is to be solvated.

62. **a.**

b.

The pair of enantiomers shown results from breaking the C2—C4 bond in the second step. Another pair of enantiomers could be formed by breaking the C2—C3 bond.

63. The cis isomer would be expected to react faster in an S_N1 reaction.

The rate-limiting step in an S_N1 reaction is formation of the carbocation intermediate. Both isomers form the same carbocation. The cis isomer (with one substituent in the equatorial position and one in the axial position) is less stable than the trans isomer (with both substituents in the equatorial position). Since the cis isomer is less stable, it has the smaller energy of activation and, therefore, the faster reaction rate.

cis-1-bromo-4-*tert*-butylcyclohexane   trans-1-bromo-4-*tert*-butylcyclohexane

64. A Diels-Alder reaction between hexachlorocyclopentadiene and 1,4-dichlorocyclopentene forms chlordane.

65. It will not undergo an S_N2 reaction, because of steric hindrance to back-side attack.
It will not undergo an S_N1 reaction, because the carbocation that would be formed is unstable, since the ring structure prevents it from achieving the 120° bond angles required for an sp^2 carbon.

Chapter 8 Practice Test

1. Which of the following is more reactive in an S_N1 reaction?

$$\overset{\displaystyle CH_3}{\underset{\displaystyle |}{}}$$

a. $CH_3CH_2CH_2CH_2CHBr$ **or** $CH_3CH_2CH_2\overset{\displaystyle CH_3}{\underset{|}{CH}}CH_2Br$

b. $CH_3CH\!\!=\!\!\overset{\displaystyle Br}{\underset{|}{C}}CH_3$ **or** $CH_3CH\!\!=\!\!CH\overset{\displaystyle Br}{\underset{|}{CH}}CH_3$

2. Which of the following is more reactive in an S_N2 reaction?

a. $CH_3CH_2\overset{\displaystyle CH_3}{\underset{|}{CH}}Br$ **or** $CH_3CH_2\overset{\displaystyle CH_2CH_3}{\underset{|}{CH}}Br$

b.   —Br   **or**   —CH_2Br

3. Which of the following alkyl halides forms a constitutional isomer as a result of an S_N1 reaction that is different from the isomer formed as a result of an S_N2 reaction?

$CH_3\overset{\displaystyle CH_3}{\underset{|}{CH}}\underset{\underset{\displaystyle Br}{|}}{CH}CH_3$ $CH_3\underset{\underset{\displaystyle Br}{|}}{CH}CH_2CH_3$

4. Indicate whether each of the following statements is true or false:

a. Increasing the concentration of the nucleophile favors an S_N1 reaction over an S_N2 reaction.   T   F

b. Ethyl iodide is more reactive than ethyl chloride in an S_N2 reaction.   T   F

c. In an S_N1 reaction, the product with the retained configuration is obtained in greater yield.   T   F

d. The rate of a substitution reaction in which none of the reactants is charged will increase if the polarity of the solvent is increased.   T   F

e. An S_N2 reaction is a two-step reaction.   T   F

f. The pK_a of a carboxylic acid is greater in water than it is in a mixture of dioxane and water.   T   F

g. 4-Bromo-1-butanol will form a cyclic ether faster than will 3-bromo-1-propanol.   T   F

5.   Answer the following:

    a. Which is a stronger base, CH_3O^- or CH_3S^-?

    b. Which is a better nucleophile in an aqueous solution, CH_3O^- or CH_3S^-?

6.   For each of the following pairs of S_N2 reactions, indicate the one that occurs with the greater rate constant:

    a. $CH_3CH_2CH_2Cl$ + HO^- **or** CH_3CHCH_3 + HO^-
                                      |
                                        Cl

    b. $CH_3CH_2CH_2Cl$ + HO^- **or** $CH_3CH_2CH_2I$ + HO^-

    c. $CH_3CH_2CH_2Br$ + HO^- **or** $CH_3CH_2CH_2Br$ + H_2O

    d. CH_3CHCH_3 $\xrightarrow[H_2O/CH_3OH]{CH_3O^-}$ **or** CH_3CHCH_3 $\xrightarrow[CH_3OH]{CH_3O^-}$
            |
          Br Br

    e. $BrCH_2CH_2CH_2CH_2NHCH_3$ **or** $BrCH_2CH_2CH_2NHCH_3$

7.   Circle the aprotic solvents.

    a. dimethyl sulfoxide   **b.** diethyl ether   **c.** ethanol   **d.** hexane

8.   How would increasing the polarity of the solvent affect the following?

    a. the rate of the S_N2 reaction of methylamine with 2-bromobutane

    b. the rate of the S_N1 reaction of methylamine with 2-bromobutane

    c. the rate of the S_N2 reaction of methoxide ion with 2-bromobutane

    d. the pK_a of acetic acid

    e. the pK_a of phenol

CHAPTER 9
Elimination Reactions of Alkyl Halides • Competition Between Substitution and Elimination

Important Terms

anti elimination	an elimination reaction in which the substituents being eliminated are removed from opposite sides of the molecule.
anti-periplanar	substituents are attached to parallel bonds on opposite sides of a molecule.
dehydrohalogenation	elimination of a proton and a halide ion.
deuterium kinetic isotope effect	ratio of the rate constant obtained for the reaction of a compound containing hydrogen and the rate constant obtained for the reaction of an analogous compound in which one or more of the hydrogens have been replaced by deuterium.
elimination reaction	a reaction that removes atoms or groups from the reactant to form a π bond.
β-elimination reaction or 1,2-elimination reaction	an elimination reaction where the groups being eliminated are bonded to adjacent carbons.
E1 reaction	a unimolecular elimination reaction.
E2 reaction	a bimolecualr elimination reaction.
regioselectivity	the preferential formation of a constitutional isomer.
sawhorse projection	a way to represent the three-dimensional spatial relationships of atoms by looking at the carbon-carbon bond from an oblique angle.
syn elimination	an elimination reaction in which substituents being eliminated are removed from the same side of the molecule.
syn-periplanar	substituents are attached to parallel bonds on the same side of a molecule.
target molecule	the desired end product of a synthesis.
Williamson ether synthesis	formation of an ether from the reaction of an alkoxide ion with an alkyl halide.
Zaitsev's rule	the rule that states that the more stable alkene product of an elimination reaction is obtained by removing a proton from the β-carbon that is bonded to the fewest hydrogens.

Solutions to Problems

1.   **a.** $\overset{\displaystyle CH_3}{\underset{|}{}}$ $CH_3C{=}CHCH_3$ **b.** $\overset{\displaystyle CH_3}{\underset{|}{}}$ $CH_3C{=}CHCH_3$ **c.** $\overset{\displaystyle CH_3}{}$ $CH_3CH_2\underset{\underset{\displaystyle CH_3}{|}}{\overset{|}{C}}CH{=}CH_2$

2.   **a.**   [structure with Br]   The alkene (2-butene) that is formed is more stable than the alkene (1-butene) that is formed from the other alkyl halide.

   b.   [cyclohexane with Br]   Br⁻ is a better leaving group (weaker base) than Cl⁻.

   c.   [structure with Br]   The alkene that is formed (2-methyl-2-pentene) is more stable than the alkene (4-methyl-2-pentene) that is formed from the other alkyl halide.

   d.   [structure with Cl]   The other alkyl halide cannot undergo an E2 reaction, because it does not have any β-hydrogens.

3.   The reaction of 2-bromo-2,3-dimethylbutane with sodium *tert*-butoxide forms two elimination products. Because of steric hindrance, the bulky base preferentially removes the most accessible β-hydrogen. Thus the less stable alkene (2,3-dimethyl-1-butene) is the one that has the lower "energy hill" that has to be climbed and therefore is the one that is easier to make. Remember that one of the factors that controls the rate of a reaction (the height of the "energy hill") is the number of effective collisions in a given period of time (page 144 of the text).

4. **a.** $CH_3CH\!=\!CHCH_3$ Removal of a hydrogen from the most substituted β-carbon forms the most stable "alkene-like" transition state.

b. CH₃

$CH_3C\!=\!CHCH_2CH_3$ Removal of a hydrogen from the most substituted β-carbon forms the most stable "alkene-like" transition state.

c. $CH_3CH\!=\!CHCH\!=\!CH_2$ The hydrogen is removed from the β-carbon that will lead to a conjugated alkene.

d. $CH_2\!=\!CHCH_2CH_3$ Removal of a hydrogen from the least substituted β-carbon forms the most stable "carbanion-like" transition state.

e.   The hydrogen is removed that will lead to a conjugated alkene.

f. CH₃

$CH_3CHCH\!=\!CHCH_3$ Removal of a hydrogen from the least β-carbon substituted carbon forms the most stable "carbanion-like" transition state.

5. **a.**   It forms the more stable alkene (the alkene with the greater number of substituents bonded to the sp^2 carbons).

b.   It forms the more stable alkene. (The new double bond is conjugated with the double bond that is already there.)

c.   It has four hydrogens that can be removed to form an alkene with two substituents on the sp^2 carbons, so it has a greater probability of an effective collision with the nucleophile than the other alkyl halide that has only two such hydrogens.

d.   It forms the more stable alkene. (The new double bond is conjugated with the phenyl substituent.)

6.

$$CH_3CH-\overset{\overset{\displaystyle CH_3}{|}}{\underset{\underset{\displaystyle Br}{|}}{C}}CH_2CH_3$$

3-bromo-2,3-dimethylpentane

$$CH_3\overset{\overset{\displaystyle CH_3}{|}}{C}=\underset{\underset{\displaystyle CH_3}{|}}{C}CH_2CH_3 \quad > \quad \underset{(CH_3)_2CH}{H_3C}\!\!\diagdown C=C\diagup\!\!\overset{CH_3}{\underset{H}{}} \quad > \quad \underset{(CH_3)_2CH}{H_3C}\!\!\diagdown C=C\diagup\!\!\overset{H}{\underset{CH_3}{}} \quad > \quad CH_3\underset{\underset{\displaystyle CH_3}{|}}{CH}\overset{\overset{\displaystyle CH_2}{\|}}{C}CH_2CH_3$$

| Four alkyl substituents are bonded to the sp^2 carbons. | Three alkyl substituents are bonded to the sp^2 carbons; the largest groups are on opposite sides of the double bond. | Three alkyl substituents are bonded to the sp^2 carbons; the largest groups are on the same side of the double bond. | Two alkyl substituents are bonded to the sp^2 carbons. |

7. The major product would be the one predicted by Zaitsev's rule, because the fluoride ion would dissociate in the first step, forming a carbocation. Loss of a proton from the carbocation would follow Zaitsev's rule, as it does in other E1 reactions.

8.   **a.**   B because it forms the more stable carbocation.   **c.**   B because it forms the more stable carbocation.

   b.   B because it forms the more stable alkene.   **d.**   A because it is less sterically hindered.

9.

10.   **a.**   E2 because a strong base is used.   **c.**   E1 because a weak base is used.

$$CH_3CH=CHCH_3$$

$$CH_3\overset{\overset{\displaystyle CH_3}{|}}{C}=CH_2$$

   b.   E1 because a weak base is used.   **d.**   E2 because a strong base is used.

$$CH_3CH=CHCH_3$$

$$CH_3\overset{\overset{\displaystyle CH_3}{|}}{C}=CH_2$$

e. E1 because a weak base is used. Notice that a 1,2-methyl shift occurs.

$$CH_3C{=}CCH_3$$

with CH_3 groups on the central carbons

f. E2 because a strong base is used.

$$CH_3\overset{\underset{\displaystyle CH_3}{|}}{C}CH{=}CH_2$$

with CH_3 below

Notice that even though the same alkyl halide is used in "**e**" and "**f**," different products are obtained because the carbocation that forms under E1 conditions rearranges.

11.   a. $$\dfrac{E2}{E2 + E1} = \dfrac{k_2[\textit{tert}\text{-butyl bromide}]\,[HO^-]}{k_2[\textit{tert}\text{-butyl bromide}]\,[HO^-] + k_1[\textit{tert}\text{-butyl bromide}]}$$

$$\dfrac{E2}{E2 + E1} = \dfrac{7.1 \times 10^{-5} \times 5.0}{7.1 \times 10^{-5} \times 5.0 + 1.5 \times 10^{-5}} = \dfrac{35.5 \times 10^{-5}}{35.5 \times 10^{-5} + 1.5 \times 10^{-5}} = \dfrac{35.5}{37} = 0.96$$

$$= 96\%$$

b. $$\dfrac{E2}{E2 + E1} = \dfrac{7.1 \times 10^{-5} \times 2.5 \times 10^{-3}}{7.1 \times 10^{-5} \times 2.5 \times 10^{-3} + 150 \times 10^{-7}} = \dfrac{1.78 \times 10^{-7}}{1.78 \times 10^{-7} + 150 \times 10^{-7}} = \dfrac{1.78}{152} = 1.2\%$$

12.   a.   1. $$CH_3CH_2CH{=}CCH_3$$ with CH_3 on the second sp2 carbon

No stereoisomers are possible.

2.

$$\underset{H}{\overset{CH_3CH_2}{\diagdown}} C{=}C \underset{CH{=}CH_2}{\overset{H}{\diagup}}$$

The major product is the conjugated diene with the bulkier group bonded to one sp^2 carbon on the opposite side of the double bond from the bulkier group bonded to the other sp^2 carbon.

3.

$$\underset{H}{\overset{CH_3CH_2}{\diagdown}} C{=}C \overset{H}{\diagup}\text{(phenyl)}$$

The major product is the conjugated alkene with the bulkier group bonded to one sp^2 carbon on the opposite side of the double bond from the bulkier group bonded to the other sp^2 carbon.

b. In none of the reactions is the major product dependent on whether you start with the R or S enantiomer of the reactant.

13.   a. Solved in the text.

c. $$\underset{CH_3CH_2CH_2}{\overset{H_3C}{\diagdown}} C{=}C \underset{H}{\overset{CH_2CH_3}{\diagup}}$$

d.

b. $$CH_3CH_2CH{=}CCH_3$$ with CH_3 on the second sp2 carbon

14.   E2 elimination reactions from six-membered rings occur only when the substituents to be eliminated are both in axial positions.

In the cis isomer, when Br is in an axial position, it has an axial hydrogen on each of the adjacent carbons. The one bonded to the same carbon as the ethyl group will be more apt to be the one eliminated with Br because the product formed is more stable and therefore more easily formed than the product formed when the other H is eliminated with Br. (Recall that when there is a choice, a hydrogen is removed from the β-carbon bonded to the fewest hydrogens.)

In the trans isomer, when Br is in an axial position, it has an axial hydrogen on only one adjacent carbon, and it is not the carbon that is bonded to the ethyl group. Therefore, a different product is formed. (Notice in this case there is no choice; there is only one hydrogen bonded to a β-carbon that can be removed.)

cis-1-bromo-2-ethylcyclohexane   trans-1-bromo-2-ethylcyclohexane

1-ethylcyclohexene   3-ethylcyclohexene

15.   In order for a six-membered ring to undergo an E2 reaction, the substituents that are to be eliminated must BOTH BE IN AXIAL POSITIONS.

When bromine and an adjacent hydrogen are both in axial positions, the large *tert*-butyl substituent is in an equatorial position in the cis isomer and in an axial position in the trans isomer. The rate constant for the reaction is $k'K_{eq}$.

Because a large substituent is more stable in an equatorial position than in an axial position, elimination of the cis isomer occurs through its more stable chair conformer (K_{eq} is large; see page 392 of the text), whereas elimination of the trans isomer has to occur through its less stable chair conformer (K_{eq} is small). The cis isomer, therefore, reacts more rapidly in an E2 reaction.

cis-1-bromo-4-*tert*-butylcyclohexane .   trans-1-bromo-4-*tert*-butylcyclohexane

16.   **a.**

(*S*)-2-chlorohexane

S_N2/E2 | CH_3O^−

(*R*)-2-methoxyhexane   major elimination product   minor elimination product
substitution product

b.

(*S*)-2-chlorohexane   (*R*)-2-methoxyhexane   (*S*)-2-methoxyhexane

substitution products

major elimination product   minor elimination product

c.

There are two stereoisomers of
trans-1-chloro-2-methylcyclohexane.

d.

1,2-hydride shift

CH_3OH

CH_3 CH_3

OCH_3 +

Notice that this is an E1 reaction. Thus, a carbocation forms and, in this case, it is a carbocation that rearranges.

e. CH_3CH_2

$\xrightarrow{CH_3O^-}$ $S_N2/E2$

major elimination product   minor elimination product

f. CH_3CH_2

$\xrightarrow{CH_3OH}$ $S_N1/E1$

major elimination product   minor elimination product

17.   Br$^-$ is the weakest base so it is the best leaving group, and F$^-$ is the strongest base and worst leaving group. It is easier to break a C—H bond than a C—D bond, so the more hydrogens on the β-carbons that are replaced by a deuterium, the slower the rate of the E2 reaction will be.

18.   The rate-limiting step in an E1 reaction is carbocation formation. Because the proton is removed in a subsequent fast step, the difference in the rate of removal of an H$^+$ versus a D$^+$ would not be reflected in the rate constant. Therefore, the deuterium kinetic isotope effect would be close to 1.

19. Because CH_3S^- is a better nucleophile and weaker base than CH_3O^-, the ratio of substitution (where Y^- reacts as a nucleophile) to elimination (where Y^- reacts as a base) will increase when the nucleophile is changed from CH_3O^- to CH_3S^-.

20. In order to undergo an elimination reaction under E2 conditions, the substituents that are to be eliminated (H and Br) must both be in axial positions. Drawing the compound in the chair conformation shows that when Br is in an axial position, neither of the adjacent β-carbons has a hydrogen in an axial position, so an elimination reaction cannot take place.

21. **a.** $CH_3CH_2CH_2Br$ .... This compound has less steric hindrance.

    b. .... I^- is a better leaving group (weaker base) than Br^-.

    c. $CH_3CH_2CH_2\overset{\overset{\displaystyle CH_3}{|}}{\underset{\underset{\displaystyle Br}{|}}{C}}CH_3$ .... A tertiary carbocation is easier to form than a secondary carbocation.

    d. .... This compound forms a conjugated double bond. The other compound cannot undergo an E2 reaction, because neither two adjacent double bonds nor a triple bond can be formed in a six-membered ring.

22. **a.**
  1. no reaction
  2. no reaction
  3. substitution and elimination
  4. substitution and elimination

    b.
  1. primarily substitution
  2. substitution and elimination (Because of the isopropyl group that slows down back-side attack and therefore formation of the substitution product, this primary alkyl halide will give more elimination product than will a primary alkyl halide such as 1-bromobutane.)
  3. substitution and elimination
  4. elimination

23.

$$CH_3\overset{\overset{\displaystyle CH_3}{|}}{\underset{\underset{\displaystyle CH_3}{|}}{C}}CH_2Br$$

1-bromo-2,2-dimethylpropane

    a. The bulky *tert*-butyl substituent blocks the back side of the carbon bonded to the bromine to nucleophilic attack, making an S_N2 reaction difficult. An S_N1 reaction is difficult because the carbocation formed when the bromide ion departs is an unstable primary carbocation.

b. It cannot undergo an E2 reaction, because the β-carbon is not bonded to a hydrogen.

It cannot undergo an E1 reaction, because that would require formation of a primary carbocation.

24. Because a strong base is used in the Williamson ether synthesis, the reaction is an S_N2 reaction, so a competing E2 reaction can also occur.

a. $CH_3CH_2CH_2Br + CH_3CH_2CH_2CH_2O^- \longrightarrow CH_3CH_2CH_2CH_2OCH_2CH_2CH_3$

butyl propyl ether

$+ CH_3CH=CH_2$

propene

b. $CH_3CH_2CH_2CH_2Br + CH_3CH_2CH_2O^- \longrightarrow CH_3CH_2CH_2CH_2OCH_2CH_2CH_3$

butyl propyl ether

$+ CH_3CH_2CH=CH_2$

1-butene

25. In order to maximize the amount of ether formed in the S_N2 reaction, make sure the less hindered group is provided by the alkyl halide.

a.

b.

c.

d.

Remember that an S_N2 reaction cannot be done on an aryl halide.

26.

27. a.

$$
\underset{\underset{Br}{|}}{\overset{\overset{CH_3}{|}}{CH_3CHCHCH_2CH_3}} \quad \xrightarrow[\text{S}_\text{N}2/\text{E2}]{HO^-} \quad \underset{\underset{OH}{|}}{\overset{\overset{CH_3}{|}}{CH_3CHCHCH_2CH_3}} \quad + \quad \overset{\overset{CH_3}{|}}{CH_3C\!=\!CHCH_2CH_3}
$$

The above S$_\text{N}$2 reaction will form the substitution product with the inverted configuration compared to the configuration of the starting material. But, because the configuration of the starting material is not specified (we do not know if it is the *R* isomer, the *S* isomer, or a mixture of *R* and *S*), we cannot specify the configuration of the S$_\text{N}$2 product.

$$
\underset{\underset{Br}{|}}{\overset{\overset{CH_3}{|}}{CH_3CHCHCH_2CH_3}} \quad \xrightarrow[\text{S}_\text{N}1/\text{E1}]{\overset{HO^-}{\text{dilute}}} \quad \underset{+}{\overset{\overset{CH_3}{|}}{CH_3CHCHCH_2CH_3}} \quad \xrightarrow{\substack{\text{1,2-hydride}\\\text{shift}}} \quad \underset{+}{\overset{\overset{CH_3}{|}}{CH_3CCH_2CH_2CH_3}}
$$

$$
\downarrow HO^-
$$

$$
\overset{\overset{CH_3}{|}}{CH_3C\!=\!CHCH_2CH_3} \quad + \quad \underset{\underset{OH}{|}}{\overset{\overset{CH_3}{|}}{CH_3CCH_2CH_2CH_3}}
$$

b.

$$
\underset{\underset{Br}{|}}{\overset{\overset{CH_3}{|}}{CH_3CH_2CCH_2CH_3}} \quad \xrightarrow[\text{S}_\text{N}2/\text{E2}]{HO^-}
$$

major   minor

A tertiary alkyl halide cannot undergo a substitution reaction under S$_\text{N}$2 conditions.

$$
\underset{\underset{Br}{|}}{\overset{\overset{CH_3}{|}}{CH_3CH_2CCH_2CH_3}} \quad \xrightarrow[\text{S}_\text{N}1/\text{E1}]{\overset{HO^-}{\text{dilute}}} \quad \underset{\underset{OH}{|}}{\overset{\overset{CH_3}{|}}{CH_3CH_2CCH_2CH_3}} \quad +
$$

minor

28. Because a cumulated diene is less stable than an alkyne, the transition state for its formation is less stable, so it is harder to make.

29. In the first elimination reaction, a hydrogen will be removed from the β-carbon bonded to the fewest hydrogens as expected. In the second elimination reaction, a hydrogen will be removed from the β-carbon that results in formation of a conjugated double bond.

$$
\underset{\underset{Cl}{|}\;\;\;\;\underset{Cl}{|}}{\overset{\overset{CH_3}{|}\;\;\;\;\overset{CH_3}{|}}{CH_3CHCHCH_2CHCHCH_3}} \xrightarrow{HO^-} \underset{\underset{Cl}{|}}{\overset{\overset{CH_3}{|}\;\;\;\;\;\;\;\overset{CH_3}{|}}{CH_3C\!=\!CHCH_2CHCHCH_3}} \xrightarrow{HO^-} \overset{\overset{CH_3}{|}\;\;\;\;\;\;\;\;\;\;\overset{CH_3}{|}}{CH_3C\!=\!CHCH\!=\!CHCHCH_3}
$$

major product = (*E*)-stereoisomer
minor product = (*Z*)-stereoisomer

30. $HOCH_2CH_2CH_2CH=CH_2$ $\xrightarrow{HBr}$ $HOCH_2CH_2CH_2\underset{\underset{Br}{|}}{C}HCH_3$ $\xrightarrow{NaH}$ CH_3

The synthesis shown on page 404 of the text will give a higher yield of the target molecule because the alkoxide ion attacks the back side of a primary alkyl halide.

In contrast, the alkoxide ion in the synthesis shown above attacks the back side of a secondary alkyl halide, which provides greater steric hindrance to back-side attack. Therefore, there will be greater competition from the E2 reaction that forms an elimination product, reducing the yield of the desired substitution product.

31.   **a.**

or

b.

c.

d.

cyclohexyl—CH=CH₂ →(Br₂)→ cyclohexyl—CHCH₂Br with Br →(excess NaNH₂)→ cyclohexyl—C≡CH

1. NaNH₂
2. CH₃CH₂Br

cyclohexyl—C≡CCH₂CH₃ →(H₂, Pd/C)→ cyclohexyl—CH₂CH₂CH₂CH₃

32.
a. H₃C and H on C=C, H and CH₃
c. H₃C and H on C=C, H and CH₃
e. H₃C and H on C=C, H and CH₂CH₃

b. cyclohexene
d. cyclohexane with =CH₂ (methylenecyclohexane)
f. cyclohexene with CH₃ (1-methylcyclohexene)

33.
a. H₃C and H on C=C, H and CH₃
c. H₃C and H on C=C, H and CH₃
e. H₃C and H on C=C, H and CH₂CH₃

b. cyclohexene
d. No reaction because a primary carbocation cannot be formed.
f. cyclohexene with CH₃

34.
a.
1. 3° > 2° > 1°
2. An E1 reaction is not affected by the strength of the base, but a weak base favors an E1 reaction by disfavoring an E2 reaction.
3. An E1 reaction is not affected by the concentration of the base, but a low concentration of a base favors an E1 reaction by disfavoring an E2 reaction.
4. If the reactant is charged, an E1 reaction will be favored by the least polar solvent that will dissolve the reactant (generally an aprotic polar solvent). If the reactant is not charged, an E1 reaction will be favored by a protic polar solvent.

b.
1. 3° > 2° > 1°
2. A strong base favors an E2 reaction.
3. A high concentration of a base favors an E2 reaction.
4. If the reactant is charged, an E2 reaction will be favored by the least polar solvent that will dissolve the reactant (generally an aprotic polar solvent). If the reactant is not charged, an E2 reaction will be favored by a protic polar solvent.

35. **a.** $CH_3CH_2CH_2\bar{C}H_2$

e. $CH_3CH_2C\!=\!CH_2$
         $\quad\quad\quad\quad\;\; |$
         $\quad\quad\quad\quad\;\;CH_3$

b. $CH_3\overset{+}{C}HCH_2CH_3$

f. $CH_3CH\bar{C}HCH_3$
         $\quad\quad\;\; |$
         $\quad\quad\;\;CH_3$

c. $CH_3\overset{+}{C}HCH\!=\!CH_2$
         stabilized by electron delocalization

g. $CH_3\overset{+}{C}CH_2CH_3$
         $\quad\quad\; |$
         $\quad\quad\;CH_3$

d. $CH_3\bar{C}HCH\!=\!CH_2$
         stabilized by electron delocalization

36. The predominant product is the elimination product because tertiary alkyl halides react with nucleophiles to form an elimination product and little substitution product.

$$
\begin{array}{ccc}
CH_3 & & CH_3\\
| & & |\\
CH_3\overset{|}{C}CH_3 + CH_3CH_2O^- & \longrightarrow & CH_3C\!=\!CH_2\\
| & & \\
Cl & & \text{predominant product}
\end{array}
$$

Rather than a tertiary alkyl halide and a primary alkoxide ion, he should have used a primary alkyl halide and a tertiary alkoxide ion.

$$
\begin{array}{ccc}
& CH_3 & CH_3\\
& | & |\\
CH_3CH_2Cl + CH_3CO^- & \longrightarrow & CH_3COCH_2CH_3 + Cl^-\\
& | & |\\
& CH_3 & CH_3
\end{array}
$$

37. **a.** $(CH_3)_3CBr \xrightarrow[\text{H}_2\text{O}]{\text{HO}^-}$
         $\begin{array}{c}H_3C\\ \diagdown\\ \quad\;C\!=\!CH_2 + I^-\\ \diagup\\ H_3C\end{array}$

Because Br⁻ is a better leaving group than Cl⁻.

b.

This compound is the only one that can undergo an E2 elimination reaction because the other compound does not have an axial hydrogen bonded to a β-carbon.

This compound will undergo an elimination reaction more slowly, because it can react only by an E1 pathway.

38. The very minor products that would be obtained from "anti-Zaitsev" elimination (that is, the less substituted alkenes) are not shown.

a.

$CH_3CH_2CH_2$ and H C=C H and CH_3 + $CH_3CH_2CH_2$ and CH_3 C=C H and H

             major   minor

b.

$CH_3CH_2CH_2$ and H C=C H and CH_3 + $CH_3CH_2CH_2$ and CH_3 C=C H and H

             major   minor

c. *trans*-1-Chloro-2-methylcyclohexane has two stereoisomers.

A hydrogen cannot be removed from the β-carbon bonded to the fewest hydrogens, because that hydrogen is not on an axial bond.

d. Both of the initially formed carbocations form the same carbocation as a result of a 1,2-hydride shift, so there is only one elimination product.

e.

CH_3CH_2 and H C=C H_3C and CH_3 + CH_3CH_2 and CH_3 C=C H_3C and H

             major   minor

f.

CH₃CH₂   H   CH₃CH₂   CH₃
    \   /   \   /
     C=C   +   C=C
    /   \   /   \
 H₃C   CH₃   H₃C   H

      major   minor

39.   **a.**   3-Bromocyclohexene forms 1,3-cyclohexadiene; 3-bromocyclohexane forms cyclohexene. 3-Bromocyclohexene reacts faster in an E2 reaction, because a conjugated double bond is more stable than an isolated double bond, so the transition state leading to formation of the conjugated double bond is more stable and, therefore, the conjugated double bond is easier to form.

b.   3-Bromocyclohexene also reacts faster in an E1 reaction, because it forms an allylic secondary carbocation which is more stable than the secondary carbocation formed by bromocyclohexane.

40.   **a.** $CH_3CHCH_2CH_3$ $\xrightarrow{CH_3Br}$ $CH_3CHCH_2CH_3$
                   | |
                   O^- OCH_3

b. $CH_3CH_2CH_2CH_2Br$ $\xrightarrow{CH_3O^-}$ $CH_3CH_2CH_2CH_2OCH_3$

c. $CH_3CH_2CH_2CH_2Br$ $\xrightarrow{CH_3NH_2}$ $CH_3CH_2CH_2CH_2\overset{+}{N}H_2CH_3$ $\xrightarrow{HO^-}$ $CH_3CH_2CH_2CH_2NHCH_3$

41.   **a.**   ethoxide ion, because elimination is favored by the bulkier base and *tert*-butoxide ion is bulkier than ethoxide ion

b.   ⁻SCN, because elimination is favored by the stronger base and ⁻OCN is a stronger base than ⁻SCN

c.   Br⁻, because elimination is favored by the stronger base and Cl⁻ is a stronger base than Br⁻

d. CH_3S^-, because elimination is favored by the stronger base and CH_3O^- is a stronger base than CH_3S^-

42.   The first compound has two axial hydrogens on β-carbons; the second compound has one axial hydrogen on a β-carbon but it cannot form the more substituted (more stable) alkene that can be formed by the first compound. The last compound cannot undergo an E2 reaction because it does not have an axial hydrogen on a β-carbon.

43.   These are all E2 reactions.

a.   The stereoisomer formed in greatest yield is the one with the largest groups on opposite sides of the double bond.

b.   No isomers are possible for this compound, because one of the sp^2 carbons is bonded to two hydrogens.

c.   No isomers are possible for this compound, because one of the sp^2 carbons is bonded to two methyl groups.

d.   Because it is an E2 reaction and there is only one hydrogen on the β-carbon, the stereoisomer formed in greater yield depends on the configuration of the reactant. The reactant can have four different configurations: *S,S; S,R; R,R;* and *R,S.* To determine the product of the reaction:

1.   Draw the reactant in a Fischer projection.

2.   From the Fischer projection, draw the Newman projection, remembering that the Fischer projection shows the compound in the eclipsed conformation.

3.   Rotate the Newman projection so the H and Br are anti.

4.   Put a double bond between the carbons bonded to the H and Br that are anti to each other; this allows you to determine the configuration of the alkene.

CH₂CH₃
CH₃ — Br
CH₃ — H
CH₂CH₃

(2S,3S)-3-bromo-
3,4-dimethylhexane

Et = CH₃CH₂

anti

$\xrightarrow{CH_3O^-}$

CH₃CH₂ / C=C \ CH₃ ... CH₃ ... CH₂CH₃

CH₂CH₃
CH₃ — Br
H — CH₃
CH₂CH₃

(2S,3R)-3-bromo-
3,4-dimethylhexane

anti

$\xrightarrow{CH_3O^-}$

CH₃CH₂ / C=C \ CH₂CH₃ ... CH₃ ... CH₃

CH₂CH₃
Br — CH₃
H — CH₃
CH₂CH₃

(2R,3R)-3-bromo-
3,4-dimethylhexane

$\xrightarrow{CH_3O^-}$

CH₃ / C=C \ CH₂CH₃ ... CH₃CH₂ ... H

CH₂CH₃
Br — CH₃
CH₃ — H
CH₂CH₃

(2R,3S)-3-bromo-
3,4-dimethylhexane

$\xrightarrow{CH_3O^-}$

CH₃CH₂ / C=C \ CH₃ ... CH₃ ... CH₃

44.

$$CH_3C-CHCH_3 \xrightarrow{\text{base}} CH_2=C-CHCH_3 + CH_3C=CCH_3$$

CH₃ CH₃ (with Br) → 2,3-dimethyl-1-butene   +   2,3-dimethyl-2-butene

CH$_2$CH$_3$
|
a. CH$_3$CH$_2$CO$^-$ Because it is the most sterically hindered base, **B** gives the highest
| percentage of the 1-alkene since the H that is most easily removed
CH$_2$CH$_3$ is the one that is most accessible to the base.

b. CH$_3$CH$_2$O$^-$ Because it is the least sterically hindered base, **C** gives the highest
                                         percentage of the more stable 2-alkene.

45.   **a.** These are S$_N$2/E2 reactions. The minor elimination products are not shown.

b. Only the substitution products are optically active; the elimination products do not have an
asymmetric center.

c.

Yes, all the products are optically active.

d. The cis enantiomers form the substitution products more rapidly, because there is less steric
hindrance from the adjacent substituent to back-side attack by the nucleophile.

e. The cis enantiomers form the elimination products more rapidly, because the alkenes formed from
the cis enantiomers are more substituted and, therefore, more stable. The more stable the alkene, the
lower the energy of the transition state leading to its formation and the more rapidly it is formed.

46.

47.   In an E2 reaction, both groups to be eliminated must be in axial positions.

When the bromine is in the axial position in the cis isomer, the *tert*-butyl substituent is in the more stable equatorial position.

When the bromine is in the axial position in the trans isomer, the *tert*-butyl substituent is in the less stable axial position.

Thus, elimination takes place via the most stable conformer in the cis isomer and via a less stable chair conformer in the trans isomer, so the cis isomer undergoes elimination more rapidly.

cis-1-bromo-4-*tert*-butylcyclohexane   *trans*-1-bromo-4-*tert*-butylcyclohexane

48.   In reactions **a–d**, the β-carbon from which the hydrogen is to be removed is bonded to only one hydrogen. Therefore, the configuration of the reactant determines the configuration of the E2 elimination product.

To determine the configuration of the product, see the instructions in Problem 43.

a. H₃C⟍ —C—C—CH₃ ... **or** ...

(2S,3S)-2-chloro-
3-methylpentane

anti

CH₃O⁻

E

b. H₃C⟍ —C—C—CH₂CH₃ ... **or** ...

(2S,3R)-2-chloro-
3-methylpentane

anti

CH₃O⁻

Z

c. Cl —C—C—CH₃ ... **or** ...

(2R,3S)-2-chloro-
3-methylpentane

CH₃O⁻

Z

d.

(2R,3R)-2-chloro-
3-methylpentane

e. This reactant has two hydrogens bonded to the β-carbon.

major product   minor product

49. Under E1 reaction conditions, both compounds would form the same product, because they would both form the same carbocation.

a.

b.

Under E2 reaction conditions, the two compounds would from different products, because the H and Br attached to the ring could only be eliminated in **a** where they are anti to one another. Because they are on the same side of the ring in **b**, a hydrogen would have to be removed from the other β-carbon.

a. $\xrightarrow{\text{E2}}$

b. $\xrightarrow{\text{E2}}$

50. In order to undergo an E2 reaction, a chlorine and a hydrogen bonded to an adjacent carbon must be trans to one another so they can both be in the required axial positions. Every Cl in the following compound has a Cl trans to it, so no Cl has a hydrogen trans to it. Thus, it is the least reactive of the isomers; it cannot undergo an E2 reaction.

51. The silver ion increases the ease of departure of the halogen atom.

$$CH_3CH=CHCH_2-\overset{..}{\underset{..}{Br}}: + Ag^+ \longrightarrow CH_3CH=CHCH_2-\overset{+}{\underset{..}{Br}}-Ag \longrightarrow CH_3CH=CH\overset{+}{C}H_2$$
$$+ \; AgBr$$

52. For a description of how to do this problem, see Problem 43.

a.   **or**

(3*S*,4*S*)-3-bromo-
4-methylhexane

$\downarrow CH_3O^-$

or   **+**

b.

(3R,4R)-3-bromo-
4-methylhexane

CH₃O⁻

c.

(3S,4R)-3-bromo-
4-methylhexane

CH₃O⁻

d.

(3R,4S)-3-bromo-
4-methylhexane

CH₃O⁻

53. **a.** $CH_3CH_2CD{=}CH_2$ and $CH_3CH_2CH{=}CH_2$

b. The deuterium-containing compound results from elimination of HBr, whereas the non-deuterium-containing compound results from elimination of DBr. The deuterium-containing compound will be obtained in greater yield, because a C—H bond is easier to break than a C—D bond.

54. **a.** $CH_3CH_2CH_2CH_2Br$ $\xrightarrow{\textit{tert}\text{-BuO}^-}$ $CH_3CH_2CH{=}CH_2$ $\xrightarrow{Br_2}$ $CH_3CH_2CHCH_2Br$ $\xrightarrow[]{\overset{\text{excess}}{{}^-NH_2}}$ $CH_3CH_2C{\equiv}CH$

$\underset{\displaystyle |}{} $

under that: Br

$\downarrow {}^-NH_2$

$\underset{H_2SO_4}{\overset{\displaystyle \overset{O}{\|}}{CH_3CH_2C-CH_2CH_2CH_3}} \xleftarrow[H_2SO_4]{H_2O} CH_3CH_2C{\equiv}CCH_2CH_3 \xleftarrow{CH_3CH_2Br} CH_3CH_2C{\equiv}C^-$

b. $BrCH_2CH_2CH_2CH_2Br \xrightarrow{\textit{tert}\text{-BuO}^-} CH_2{=}CHCH{=}CH_2 \xrightarrow[\substack{\text{(a Diels-Alder}\\\text{reaction)}}]{\Delta}$ (cyclohexene with CH=CH₂ substituent)

$\downarrow H_2 \mid Pd/C$

(cyclohexane with CH₂CH₃ substituent)

55. **a.** Both compounds form the same elimination product, because they both have hydrogens bonded to the same β-carbons that are anti to the bromine.

cis-4-bromocyclohexanol

trans-4-bromocyclohexanol

b. Only the trans isomer can undergo an intramolecular substitution reaction, because the S_N2 reaction requires back-side attack.

trans-4-bromocyclohexanol

The cis isomer can undergo only an intermolecular substitution reaction.

cis-4-bromocyclohexanol

c. The elimination reaction forms a pair of enantiomers, because the reaction creates an asymmetric center in the product. Both substitution reactions form a single stereoisomer, because the reaction does not create an asymmetric center in the product.

56. The given reaction conditions favor S_N1 and E1 reactions.

57.

There is only a substitution product.
The reactant does not undergo elimination because when Cl is in an axial position, neither of the β-carbons is bonded to an axial hydrogen.

CH₃O⁻
dilute
S_N1/E1

1,2-hydride
shift

58. **a.**

1S,2S

or

E2

b.

1S,2R

or

E2

Chapter 9 Practice Test

1.   Which of the following is more reactive in an E2 reaction?

   a.   [benzene ring]—CH_2CHCH_3 or   [benzene ring]—$CH_2CH_2CH_2Br$
                         |
                         Br

   b. $CH_3CH_2CHCH_3$ or CH_2=$CHCH_2CHCH_3$
                    | |
                    Br   Br

2.   Which of the following would give the greater amount of substitution product under conditions that favor S_N2/E2 reactions?

$$CH_3 \qquad\qquad CH_3$$
$$| \qquad\qquad\quad |$$
$$CH_3CBr \quad or \quad CH_3CHBr$$
$$|$$
$$CH_3$$

3.   What products are obtained when (R)-2-bromobutane reacts with CH_3O^-/ CH_3OH under conditions that favor S1/E1 reactions? Include the configuration of the products.

4.   What alkoxide ion and what alkyl bromide should be used to synthesize the following ethers?

                  CH_3
                  |
   a. $CH_3CH_2COCH_2CH_2CH_3$
                  |
                  CH_3

   b.   [cyclohexane ring]—O—[benzene ring]

   c.   [cyclohexane ring]—O—CH_3

5.   For each of the following pairs of E2 reactions, indicate the one that occurs with the greater rate constant:

   a. $CH_3CH_2CH_2Cl$ + HO^- or CH_3CHCH_3 + HO^-
                                                                    |
                                                                    Cl

   b. $CH_3CH_2CH_2Cl$ + HO^- or $CH_3CH_2CH_2I$ + HO^-

   c. $CH_3CH_2CH_2Br$ + HO^- or $CH_3CH_2CH_2Br$ + H_2O

d. CH_3CHCH_3 $\xrightarrow[\text{H}_2\text{O/CH}_3\text{OH}]{\text{CH}_3\text{O}^-}$ or CH_3CHCH_3 $\xrightarrow[\text{CH}_3\text{OH}]{\text{CH}_3\text{O}^-}$
      | |
      Br   Br

                                                   CH_3
                                                    |
e. CH_3CHCH_3 + HO^- or CH_3CCH_3 + HO^-
      | |
      Br   Br

6.   Give the major product of an E2 reaction of each of the following compounds with hydroxide ion:

a.

c.

b.

d.

7.   a. Which would be more reactive in an E2 reaction, *cis*-1-bromo-2-methylcyclohexane or *trans*-1-bromo-2-methylcyclohexane?

      b. Which would be more reactive in an E1 reaction, *cis*-1-bromo-2-methylcyclohexane or *trans*-1-bromo-2-methylcyclohexane?

CHAPTER 10
Reactions of Alcohols, Ethers, Epoxides, Amines, and Sulfur-Containing Compounds

Important Terms

alcohol	an organic compound with an OH functional group (ROH).
alkaloid	a natural product with a nitrogen heteroatom found in the leaves, bark, roots, or seeds of plants.
alkyl tosylate	an ester of *para*-toluenesulfonic acid.
alkyl triflate	an ester of trifluoromethanesulfonic acid.
antibiotic	a compound that interferes with the growth of a bacterial microorganism.
crown ether	a cyclic molecule that possesses several ether linkages around a central cavity.
dehydration	loss of water.
epoxide	an ether in which the oxygen is incorporated into a three-membered ring.
ether	a compound containing an oxygen bonded to two carbons (ROR).
inclusion compound	the complex formed when a crown ether specifically binds a metal ion or an organic molecule.
Hofmann elimination reaction	elimination of a proton and a tertiary amine from a quaternary ammonium hydroxide.
mercapto group	an SH group.
molecular recognition	the recognition of one molecule by another as a result of specific interactions.
phase-transfer catalysis	catalysis of a reaction by providing a way to bring a polar reagent into a nonpolar phase so that the reaction between a polar and a nonpolar compound can occur.
phase-transfer catalyst	a compound that carries a polar reagent into a nonpolar phase.
quaternary ammonium ion	a cation containing a nitrogen bonded to four alkyl groups (R_4N^+).
sulfide (thioether)	the sulfur analog of an ether (RSR).
sulfonate ester	the ester of a sulfonic acid (RSO_2OR).
sulfonium salt	$(R)_3S^+X^-$
thioether (sulfide)	the sulfur analog of an ether (RSR).
thiol (mercaptan)	the sulfur analog of an alcohol (RSH).

Solutions to Problems

1. They no longer have a lone pair of electrons.

2. **a.** Solved in the text.

   b. The conjugate acid of the leaving group of $CH_3OH_2^+$ is H_3O^+; its pK_a is −1.7.

   The conjugate acid of the leaving group of CH_3OH is H_2O; its pK_a is 15.5.

   Because H_3O^+ is a much stronger acid than H_2O, H_2O^+ is a much weaker base than HO^- and therefore is the better leaving group. Thus, $CH_3OH_2^+$ is more reactive than CH_3OH.

3. **a.** Solved in the text.

   b.

   c.

   Notice that in "**c**" a neutral amine ($CH_3CH_2NH_2$) is used instead of $CH_3CH_2NH^-$, because the weaker base ($CH_3CH_2NH_2$) favors the desired substitution reaction, whereas the stronger base ($CH_3CH_2NH^-$) would favor elimination at the expense of substitution (Section 9.8). The yield of the target molecule will still not be very high, since a tertiary amine and a quaternary ammonium ion will also be formed. (See Solved Problem 13 on page 346 of the text.)

   d. $CH_3CH_2CH_2CH_2OH$ $\xrightarrow[\text{2. }^-C\equiv N]{\text{1. HBr, }\Delta}$ $CH_3CH_2CH_2CH_2C\equiv N$

4. The relative reactivity would be:

$$\text{tertiary > primary > secondary}$$

   If secondary alcohols underwent an S_N2 reaction, they would be less reactive than primary alcohols, because they are more sterically hindered than primary alcohols at the carbon attached to the OH group.

5. All four alcohols undergo an S_N1 reaction, because they are either secondary or tertiary alcohols. Notice that in part **a**, arrows are shown in the first protonation step. However, you will often find that arrows are not shown in protonation steps as in parts **b**, **c**, and **d**.

   a.

b.

$+ \ H_2O$

The carbocation that is initially formed in **c** and the carbocation that is initially formed in **d** rearrange in order to form more stable carbocations.

c.

secondary
carbocation

1,2-methyl
shift

tertiary
carbocation

d.

+ H₂O
secondary
carbocation

1,2-hydride
shift

tertiary
carbocation

Cl⁻

6.   Solved in the text.

7.   **a.** $CH_3CH_2CH_2OH$ $\xrightarrow[\text{2. } CH_3CH_2S^-]{\text{1. TsCl/pyridine}}$ $CH_3CH_2CH_2SCH_2CH_3$

   b. $CH_3CH_2CH_2OH$ $\xrightarrow[\substack{\text{2. } CH_3CHCH_2O^- \\ | \\ CH_3}]{\text{1. TsCl/pyridine}}$ $CH_3CH_2CH_2OCH_2CHCH_3$

   with CH_3 branch

8.   **D** because it forms a very stable tertiary allylic carbocation intermediate.

9.   1-Bromo-1-methylcyclopentane would be formed because Br⁻ would react with the carbocation. If the carbocation were able to lose a proton before it reacts with Br⁻, HBr could add to the resulting alkene, thereby forming the same alkyl halide as would be formed by addition of Br⁻ to the carbocation.

10.   **a.** In order to synthesize an unsymmetrical ether (ROR′) by this method, two different alcohols (ROH and R′OH) would have to be heated with sulfuric acid. Therefore, three different ethers would be obtained as products. Consequently, the desired ether would account for considerably less than half of the total amount of ether that is synthesized.

$$\text{ROH} \ + \ \text{R'OH} \ \xrightarrow[\Delta]{\text{H}_2\text{SO}_4} \ \text{ROR} \ + \ \underset{\substack{\text{target} \\ \text{molecule}}}{\text{ROR'}} \ + \ \text{R'OR'}$$

b. It could be synthesized by a Williamson ether synthesis. (See Section 9.9 on page 400 of the text.)

$$\text{CH}_3\text{CH}_2\text{CH}_2\text{OH} \ \xrightarrow{\text{Na} \ \cdot} \ \text{CH}_3\text{CH}_2\text{CH}_2\text{O}^- \ \xrightarrow{\text{CH}_3\text{CH}_2\text{Br}} \ \text{CH}_3\text{CH}_2\text{CH}_2\text{OCH}_2\text{CH}_3 \ + \ \text{Br}^-$$

11.   **a.** $\underset{\overset{|}{\text{CH}_3}}{\text{CH}_3\text{CH}_2\text{CHCH}_2\text{OH}}$ **b.**   **c.** $\underset{\overset{|}{\text{OH}}}{\text{CH}_3\text{CH}_2\text{CHCH}_2\text{CH}_3}$ **d.**

12.   **a.**

b. Because of the difficulty of forming a primary carbocation, dehydration is an E2 reaction. The alkene that results is protonated and the proton that is removed is the one that results in formation of the most stable alkene.

c.

13. **a.**

b.

major product, because
it is formed by loss of
any one of 4 hydrogens

minor product,
because it is formed
by loss of any one
of 2 hydrogens

c.

d. In **d** and **f**, the reactant is a primary alcohol. Therefore, elimination of water takes place via an E2 reaction. Because the dehydration reaction is being carried out in an acidic solution, the alkene that is formed initially is protonated to form a carbocation. The proton that is then lost from the carbocation is the one that results in formation of the most stable alkene.

e.

1,2-methyl shift

f. $CH_2CH_2CH_2CH_2CH_2OH$

major product

14. **a.** $CH_3CH_2C{=}CCHCH_3$ with CH_3 above, CH_3 and CH_2CH_3 below

b.

$$H_3C, CH_3CH_2 \diagdown C{=}C \diagup CHCH_3 (CH_3), CH_2CH_3$$

15. **a.** $CH_3CH_2\overset{O}{\overset{\|}{C}}CH_2CH_3$

c. A tertiary alcohol is not oxidized to a carbonyl compound.

e. cyclohexanone (ring with =O)

b. $CH_3CH_2CH_2CH_2\overset{O}{\overset{\|}{C}}OH$

d. $CH_2\overset{O}{\overset{\|}{C}}CH_2\overset{O}{\overset{\|}{C}}CH_2CH_3$

f. $HO\overset{O}{\overset{\|}{C}}CH_2CH_2\overset{O}{\overset{\|}{C}}OH$

16. Chromic acid needs to be protonated to convert a poor leaving group ($^-$OH) into a good leaving group (H_2O). The alcohol displaces water in an S_N2 to form a protonated chromate ester, which loses a proton. The aldehyde is then formed in an E2 reaction.

$$HO{-}\overset{O}{\underset{O}{\overset{\|}{\underset{\|}{Cr}}}}{-}\ddot{O}H \;\rightleftarrows\; HO{-}\overset{O}{\underset{O}{\overset{\|}{\underset{\|}{Cr}}}}{-}\overset{H}{\overset{|}{\underset{+}{O}H}} \;\xrightarrow{S_N2}\; CH_3CH_2CH_2\overset{+}{O}{-}\overset{O}{\overset{\|}{Cr}}{-}OH$$

$CH_3CH_2CH_2\ddot{O}H$ H_2O $H_2\ddot{O}{:}$, H

$$CH_3CH_2CH{=}O \;+\; HCrO_3^- \;\xleftarrow{E_2}\; CH_3CH_2CH{-}O{-}\overset{O}{\overset{\|}{Cr}}{-}OH$$

$+\, H_3O^+$ $H_2\ddot{O}{:}$ H H_3O^+

17. Solved in the text.

18. We saw that HCl does not cleave ethers, because Cl$^-$ is not a strong enough nucleophile. F$^-$ is an even weaker nucleophile, so HF cannot cleave ethers. Therefore, ethers can be cleaved only with HBr or HI.

19. **a.** Solved in the text.

b. Cleavage occurs by an S_N1 pathway, because the benzyl cation that is formed is relatively stable; I$^-$ will react with the benzyl cation.

$$\text{C}_6\text{H}_5{-}CH_2{-}\overset{+}{\underset{H}{O}}{-}\text{C}_6\text{H}_5 \;\longrightarrow\; \text{C}_6\text{H}_5{-}\overset{+}{C}H_2 \;+\; HO{-}\text{C}_6\text{H}_5$$

$\downarrow$ I$^-$

$$\text{C}_6\text{H}_5{-}CH_2I$$

c. Cleavage occurs by an S_N1 pathway, because the benzyl cation that is formed is relatively stable; I^- will react with the benzyl cation.

d. Cleavage occurs by an S_N2 pathway, because a primary carbocation is too unstable to be formed.

$$HOCH_2CH_2CH_2CH_2CH_2I$$

e. Cleavage occurs by an S_N2 pathway, because a primary carbocation or a vinyl cation is too unstable to be formed.

enol   ketone

f. Cleavage will occur by an S_N1 pathway, because the tertiary carbocation that is formed is relatively stable; I^- will react with the tertiary carbocation.

20.

a.

b.

c.

d.

21. The reactivity of tetrahydrofuran is more similar to a noncyclic ether, because the five-membered ring does not have the strain that makes the epoxide reactive.

22. The stronger the base, the more difficult it is to displace. The leaving group of a protonated primary alcohol is water; its conjugate acid (H_3O^+) has a p$K_a = -1.7$. The leaving group of a protonated primary amine is ammonia; its conjugate acid ($^+NH_4$) has a p$K_a = 9.4$. Thus, ammonia is a much stronger base than water and far too basic to be displaced by Br^-.

$$Br^- \ + \ CH_3\overset{+}{N}H_3 \ \xrightarrow{\Delta} \ \text{no reaction}$$

23.   The major products come from removing a proton from the β-carbon that is bonded to the greater number of hydrogens.

$$\underset{\underset{CH_3}{|}}{\overset{\overset{CH_3}{|}}{CH_3CHCH_2N}} \quad + \quad CH_2{=}CHCH_3$$

24.   The minor products come from removing a proton from the β-carbon that is bonded to the fewest hydrogens.

$$\underset{\underset{CH_3}{|}}{CH_3C{=}CH_2} \quad + \quad \underset{\underset{CH_3}{|}}{\overset{\overset{CH_3}{|}}{NCH_2CH_2CH_3}}$$

25.   The only difference is the leaving group.

$$\underset{\underset{HO{:}^-}{H}}{\overset{\overset{CH_3 \quad CH_3}{| \qquad |+}}{CH_2{-}CH{-}NCH_3}} \qquad \underset{\underset{HO{:}^-}{H}}{\overset{\overset{CH_3}{|}}{CH_2{-}CH{-}Br}}$$

26.   In each case, a proton is removed from the β-carbon that is bonded to the greater number of hydrogens.

a. $\underset{\underset{CH_3}{|}}{CH_2{=}CHCH}\underset{\underset{CH_3}{|}}{CH_2CH_2NCH_3}$ **c.** $CH_3N\underset{\underset{CH_3}{|}}{CH_2}\underset{\underset{CH_3}{|}}{CH}CH_2CH{=}CH_2$

b.

$+ \quad N(CH_3)_3$

27.   **a.** Solved in the text.

b. $\underset{\underset{Br}{|}}{CH_3CH_2CH_2CHCH_3} \xrightarrow{(CH_3)_3N} \underset{\underset{CH_3}{|}}{\overset{}{CH_3CH_2CH_2CHCH_3}}$

$$\underset{\underset{CH_3}{|}}{CH_3{-}\overset{+}{N}{-}CH_3} \; Br^-$$

$$\downarrow \; Ag_2O \mid H_2O$$

$$(CH_3)_3N \quad + \quad CH_3CH_2CH_2CH{=}CH_2 \xleftarrow{\Delta} \underset{\underset{CH_3}{|}}{CH_3CH_2CH_2CHCH_3}$$

$$\underset{\underset{CH_3}{|}}{CH_3{-}\overset{+}{N}{-}CH_3} \; HO^-$$

c.

28. a. The primary amine is 2-methyl-3-pentanamine.

b. The secondary amine is 3-methylpyrrolidine.

29. a. The synthesis must be done with *tert*-butylthiol and a methyl halide. It cannot be done with methanethiol and *tert*-butyl bromide, because a tertiary alkyl halide cannot undergo an S_N2 reaction and so would form an elimination product rather than a substitution product.

b. This is the same problem as that of part **a**, using skeletal structures instead of condensed structures.

c. The highest yield of the target molecule is obtained by having the less substituted of the two R groups of the thioether be the alkyl halide and the more substituted be the thiol.

d. The synthesis must be done with these reagents because the sp^2 carbon of the benzene ring cannot undergo back-side attack.

30.   Melphalan is a good alkylating reagent because the sulfonate group is an excellent leaving group, because the electrons left behind when the group leaves can be delocalized onto two oxygens. In addition, Melphalan has two sulfonate groups so it can react with two nucleophiles.

31.   The first compound is too insoluble.

The second compound is used clinically; it has fewer carbons than the first compound, so it is more soluble in water.

The third compound is less reactive than the second compound, because the lone pair on the nitrogen can be delocalized into the benzene ring, so the lone pair is less apt to displace a chloride ion and form the three-membered ring that is needed for the compound to be an alkylating agent.

32.   Cyclophosphamide, Carmustine, and Chloroambucil are less reactive than Mechlorethamine because, unlike the lone-pair electrons on the nitrogen of Mechlorethamine, the lone-pair electrons on the nitrogen of the other three compounds can be delocalized and, therefore, these compounds are less apt to form the three-membered ring that is needed for the compound to be an alkylating agent.

33.

a. $CH_3CH_2CH_2O\overset{\text{O}}{\overset{\|}{C}}CH_3$

b. $CH_3CH_2CH_2CH_2Br$

c. $CH_3\underset{\underset{\text{CH}_3}{|}}{CH}CH_2CH_2O$—⬡

d. $CH_3CH_2\underset{\underset{\text{OH}}{|}}{CH}—\overset{\overset{\text{CH}_3}{|}}{\underset{\underset{\text{OCH}_3}{|}}{C}}CH_3$

e. $CH_3CH_2\underset{\underset{\text{CH}_3O}{|}}{CH}—\overset{\overset{\text{CH}_3}{|}}{\underset{\underset{\text{OH}}{|}}{C}}CH_3$

f. $CH_3CH=CHCH_3$

g. $CH_3\underset{\underset{\text{CH}_3}{|}}{CH}CH_2CH_2Cl$

34.   **a.**

The other alcohol cannot undergo dehydration to form an alkene, because its β-carbon is not bonded to a hydrogen.

b.

The rate-limiting step in a dehydration is carbocation formation. A secondary allylic cation is more stable than a secondary carbocation and therefore is formed more rapidly.

c.

A tertiary carbocation is more stable than a secondary carbocation.

d.

A secondary benzylic cation is more stable than a secondary carbocation.

e.

A secondary benzylic cation is more stable than a secondary carbocation.

35.

36. Only one S_N2 reaction takes place in reactions **a** and **b**, so the product has the inverted configuration compared to the configuration of the reactant.

a.

(*R*)-1-deuterio-1-propanol   (*S*)-1-deuterio-1-propanol

or

(*R*)-1-deuterio-1-propanol   (*S*)-1-deuterio-1-propanol

b.

(*R*)-1-deuterio-1-propanol   (*S*)-1-deuterio-1-methoxypropane

or

(*R*)-1-deuterio-1-propanol   (*S*)-1-deuterio-1-methoxypropane

c. Since the desired product has the same configuration as the starting material, it can be synthesized using reactions that don't break any of the bonds to the asymmetric center.

(*R*)-1-deuterio-1-propanol   (*R*)-1-deuterio-1-methoxypropane

It could also be synthesized using two consecutive S_N2 reactions. Since each one involves inversion of configuration, the final product will have the same configuration as the starting material. For example, treating the starting material with PBr_3 forms (S)-1-bromo-1-deuteriopropane. Treating (S)-1-bromo-1-deuteriopropane with methoxide ion forms (R)-1-deuterio-1-methoxypropane.

(R)-1-deuterio-1-propanol (S)-1-bromo-1-deuteriopropane (R)-1-deuterio-1-methoxypropane

37.   2-hexene and 3-hexene

$$CH_3CH_2CH_2CH_2CH_2CH_2OH \xrightarrow[\Delta]{H_2SO_4} CH_3CH_2CH_2CH_2CH=CH_2$$

$$H^+ + CH_3CH_2CH_2CH=CHCH_3 \rightleftharpoons CH_3CH_2CH_2CH_2\overset{+}{C}HCH_3$$

// 2-hexene

$$CH_3CH_2CH_2\overset{+}{C}HCH_2CH_3 \longrightarrow CH_3CH_2CH=CHCH_2CH_3 + CH_3CH_2CH_2CH=CHCH_3 + H^+$$

3-hexene   2-hexene
(cis and trans)   (cis and trans)

38.   **a.** $CH_3\underset{CH_3}{\overset{CH_3}{C}}Br + CH_3CH_2OH$

d.

g.

b. $CH_3\underset{CH_3}{C}HCH_2OH + CH_3I$

e.

h.

c.

f.

i. $CH_3\underset{CH_3}{C}HCH_2\underset{CH_3}{N}CH_3 + CH_2=CH_2$

39.   2,3-Dimethyl-2-butanol will dehydrate faster because it is a tertiary alcohol, whereas 3,3-dimethyl-2-butanol is a secondary alcohol.

2,3-dimethyl-2-butanol
a tertiary alcohol

3,3-dimethyl-2-butanol
a secondary alcohol

40.

unknown
alcohol

major
product

minor
product

1. BH$_3$/THF
2. H$_2$O$_2$, HO$^-$, H$_2$O

original unknown
alcohol

PCC

41.   The product of each reaction is an alkene.

Recall that alkenes undergo electrophilic addition reactions, and the first step in an electrophilic addition reaction is addition of an electrophile to the alkene. The acid-catalyzed dehydration reaction is reversible, because the electrophile (H$^+$) needed to react with the alkene in the first step of the reverse reaction is available.

The base-promoted elimination reaction of a hydrogen halide is not reversible, because an electrophile is not available to react with the alkene in the first step of the reverse reaction.

42.

43.

$$CH_3\underset{\underset{NH_2}{|}}{\overset{\overset{CH_3}{|}}{CH}}CHCH_3 + CH_3I \xrightarrow[\text{excess}]{K_2CO_3} CH_3\underset{\underset{\underset{\underset{CH_3}{|}}{CH_3\overset{+}{N}CH_3}}{|}}{\overset{\overset{CH_3}{|}}{CH}}CHCH_3 \quad I^-$$

primary amine

$\downarrow$ Ag$_2$O, H$_2$O

$$CH_2=CHCHCH_3 \xleftarrow{\Delta} CH_3\underset{\underset{\underset{\underset{CH_3 \quad HO^-}{|}}{CH_3\overset{+}{N}CH_3}}{|}}{\overset{\overset{CH_3}{|}}{CH}}CHCH_3$$

44. When (*S*)-2-butanol loses water, the asymmetric center in the reactant becomes a planar *sp*^{2} carbon. Therefore, the chirality is lost. When water attacks the carbocation, it can attack from either side of the planar carbocation, forming (*S*)-2-butanol and (*R*)-2-butanol with equal ease.

(*S*)-2-butanol

(*S*)-2-butanol   (*R*)-2-butanol

45. **a.**

b. $CH_3CH_2CH_2CH_2Br \xrightarrow{HO^-} CH_3CH_2CH_2CH_2OH \xrightarrow[\Delta]{H_2CrO_4} CH_3CH_2CH_2\overset{\overset{O}{\|}}{C}OH$

46.   *N*-Methylpiperidine forms 1,4-pentadiene.

2-Methylpiperidine forms 1,5-hexadiene.

β-carbon bonded
to the greatest number
of hydrogens

3-Methylpiperidine forms 2-methyl-1,4-pentadiene.

4-Methylpiperidine forms 3-methyl-1,4-pentadiene.

47. 3-Methyl-2-butanol is a secondary alcohol and, therefore, will undergo an S_N1 reaction. The carbocation intermediate that is formed rearranges to a more stable tertiary carbocation.

3-methyl-2-butanol

2-Methyl-1-propanol is a primary alcohol and, therefore, will undergo an S_N2 reaction. Because carbocations are not formed in S_N2 reactions, a carbocation rearrangement cannot occur.

2-methyl-1-propanol

48.

49.

50. Cyclopropane does not react with HO^-, because cyclopropane does not contain a leaving group; a carbanion is far too basic to serve as a leaving group. Ethylene oxide reacts with HO^-, because ethylene oxide contains an RO^- leaving group.

51. **Diethyl ether** is the ether that would be obtained in greatest yield, because it is a symmetrical ether. Since it is symmetrical, only one alcohol is used in its synthesis. Therefore, it is the only ether that would be formed. In contrast, the synthesis of an unsymmetrical ether requires two different alcohols. Therefore, the unsymmetrical ether is one of three different ethers that would be formed.

52. **a.**

b.

53.

54.

2-ethyloxirane

a. CH$_3$CH$_2$CHCH$_2$OH
          |
          OH

0.1 M HCl is a dilute solution
of HCl in water.

c. CH$_3$CH$_2$CHCH$_2$OH
          |
          OH

b. CH$_3$CH$_2$CHCH$_2$OH
          |
          OCH$_3$

d. CH$_3$CH$_2$CHCH$_2$OCH$_3$
          |
          OH

55. a.

b.

c. The six-membered ring is formed by attack on the more sterically hindered carbon of the epoxide. Attack on the less sterically hindered carbon is preferred.

56. Ethyl alcohol is not obtained as a product, because it reacts with the excess HI and forms ethyl iodide.

CH$_3$CH$_2$OCH$_2$CH$_3$ $\xrightarrow[\Delta]{HI}$ CH$_3$CH$_2$I + CH$_3$CH$_2$OH $\xrightarrow[\Delta]{HI}$ CH$_3$CH$_2$I + H$_2$O

57. B is any base in the solution (HSO$_4^-$, H$_2$O, ROH).

58.   **A** is the substitution product that is formed by methoxide ion attacking a carbon of the three-membered ring and eliminating the amino group, thereby opening the ring.

   B is the product of a Hofmann elimination reaction: methoxide ion removes a proton from a methyl group bonded to a ring carbon, eliminating the amino group. The red color disappears when Br_2 is added to **B**, because Br_2 adds to the double bond.

   When the aziridinium ion reacts with methanol, only **A**, the substitution product, is formed.

A   **B**

59.   a.

b.

c.

60.   **B** is the fastest reaction; **A** is the slowest reaction.

In order to form the epoxide, the alkoxide ion must attack the back-side of the carbon that is bonded to Br. This means that the OH and Br substituents must both be in axial positions. To be 1,2-diaxial, they must be trans to each other.

A does not form an epoxide, because the OH and Br substituents are cis to each other.
B and **C** can form epoxides because the OH and Br substituents are trans to each other.

The rate of formation of the epoxide is given by $k'K_{eq}$, where k' is the rate constant for the substitution reaction, and K_{eq} is the equilibrium constant for the equatorial/axial conformers.

When the OH and Br substituents are in the required diaxial position, the large *tert*-butyl substituent is in the equatorial position in **B** and in the axial position in **C**.

B   **C**

Because the more stable conformer is the one with the large *tert*-butyl group in the equatorial position, the OH and Br substituents are in the required diaxial position in the more stable conformer of **B**

(K_{eq} is large), whereas the OH and Br substituents are in the required diaxial position in the less stable conformer of **C** (K_{eq} is small). Therefore, **B** reacts faster than **C**.

B   **C**

61.   **a.**   The reaction of 2-chlorobutane with HO^- is an intermolecular reaction, so the two compounds have to find one another in the solution.

The following reaction takes place in two steps. The first is an intramolecular S_N2 reaction; the reaction is much faster than the above reaction since the two reactants are in the same molecule and can find one another relatively easily.

The second S_N2 reaction is fast because the strain of the three-membered ring and the positive charge on the nitrogen make the amine a very good leaving group.

   b.   The HO group is bonded to a different carbon, because HO^- attacks the least sterically hindered carbon of the three-membered ring.

62.

63. a.

b. Dehydration of the primary alcohol group cannot occur, because it cannot lose water via an E1 pathway, since a primary carbocation cannot be formed. It cannot lose water via an E2 pathway, because the β-carbon is not bonded to a hydrogen. However, dehydration of the tertiary alcohol group can occur. The product is an enol that tautomerizes to an aldehyde (see page 263 of the text).

64.

65. Cyclopentene oxide undergoes back-side attack by the nucleophile, so the two substituents in the products are trans to each other. Therefore, both the R,R-isomer and the S,S-isomer are obtained.

cyclopentene   dimethylamine R,R-isomer S,S-isomer
oxide

66.   Notice that the initially generated carbocation can undergo either a 1,2-methyl shift or a 1,2-alkyl shift.

a.

b.

67.   **a.**   A nitrogen is a stronger base than an oxygen, so unlike an epoxide that can be opened without the oxygen being protonated, the three-membered nitrogen-containing ring has to be protonated to improve the leaving propensity of the group.

b.   A nucleophile such as an NH_2 group on a chain of DNA can react with the three-membered ring. If a nucleophile on another chain of DNA reacts with another of the three-membered rings in this compound, the two DNA chains will be crosslinked.

Chapter 10 Practice Test

1. Which of the following reagents is the best one to use in order to convert methyl alcohol into methyl bromide?

    Br^- HBr Br_2 NaBr Br^+

2. Which of the following reagents is the best one to use in order to convert methyl alcohol into methyl chloride?

    Cl_2/CH_2Cl_2 Cl_2/hv Cl^- $SOCl_2$ NaCl

3. **a.** What would be the major product obtained from the reaction of the epoxide shown below in methanol containing 0.1 M HCl?

    b. What would be the major product obtained from the reaction of the epoxide in methanol containing 0.1 M $NaOCH_3$?

4. Give the major product(s) of each of the following reactions:

    a. $CH_3\overset{CH_3}{\underset{}{CH}}CH_2\overset{+}{\underset{\underset{CH_3}{|}}{N}}CH_2CH_2CH_2CH_3$ $\overset{\Delta}{\longrightarrow}$

    CH$_3$ HO$^-$

    b.

    c.

5. Give the major product that is obtained when each of the following alcohols is heated in the presence of H_2SO_4:

    a. $CH_3CH_2\overset{CH_3}{\underset{\underset{OH}{|}}{C}}\overset{}{\underset{\underset{CH_3}{|}}{CH}}CH_3$

    c. $CH_3CH_2CH_2CH_2CH_2OH$

$$CH_3$$
b. $CH_3CH_2CH_2CH\!-\!CCH_3$
$$\quad\quad\quad\quad\quad OH\quad CH_3$$

d.

6.   Indicate whether each of the following statements are true or false:

  a. Tertiary alcohols are easier to dehydrate than secondary alcohols.   T   F

  b. Alcohols are more acidic than thiols.   T   F

  c. Alcohols have higher boiling points than thiols.   T   F

  d. The acid-catalyzed dehydration of a primary alcohol is an S_N1 reaction.   T   F

  e. The Hofmann elimination reaction is an E2 reaction.   T   F

7.   What products would be obtained from heating the following ethers with one equivalent of HI?

$$CH_3$$
a. $CH_3CH_2COCH_3$
$$\quad\quad\quad CH_3$$

b.

8.   Give the major product of each of the following reactions:

$$CH_2CH_2$$
a. $CH_3CH_2CH_2NH_2$ + $CH_3CH_2\overset{+}{S}CH_2CH_3$ $\longrightarrow$

b. $CH_3CH_2CH_2OH$ + $SOCl_2$ $\xrightarrow{\text{pyridine}}$

c. $CH_3CH_2CH_2CH_2CH_2CH_2OH$ $\xrightarrow[\Delta]{H_2SO_4}$

d. $CH_3CH_2CH_2CH_2OH$ $\xrightarrow[CH_2Cl_2]{PCC}$

e. $CH_3CH_2CH_2CH_2OH$ $\xrightarrow{H_2CrO_4}$

CHAPTER 11
Organometallic Compounds

Important Terms

alkene metathesis (olefin metathesis)	breaks the double bond of an alkene (or the triple bond of an alkyne) and then rejoins the fragments to form a new double (or triple) bond between two carbons that were not previously bonded.
coupling reaction	a reaction that joins two carbon-hydrogen containing groups.
dialkyl lithium cuprate	$(R)_2CuLi$; prepared by treating an organolithium reagent with cuprous iodide.
Gilman reagent	a dialkyl lithium cuprate used to replace a halogen in an alkyl, aryl, or vinylic halide with an alkyl group.
Grignard reagent	the compound that results when magnesium is inserted between the carbon and halogen of an alkyl halide (RMgBr, RMgCl).
Heck reaction	couples an aryl halide or vinylic halide with an alkene.
olefin metathesis (alkene metathesis)	breaks the double bond of an alkene (or the triple bond of an alkyne) and then rejoins the fragments to form a new double (or triple) bond between two carbons that were not previously bonded.
organoboron reagent	an alkyl borane or an alkenyl borane: $R'B(OR)_2$.
organolithium compound	RLi; prepared by adding lithium to an alkyl halide.
organomagnesium compound	RMgBr or RMgCl; prepared by adding an alkyl halide to magnesium shavings.
organometallic compound	a compound with a carbon-metal bond.
organopalladium compound	a compound with palladium inserted between a carbon and a halogen: R-Pd-X.
oxidative addition	insertion of a metal between two atoms.
reductive elimination	elimination of a metal from between two atoms, resulting in a covalent bond between two carbons.
Suzuki reaction	couples an aryl halide or vinylic halide with an organoboron reagent.
transmetallation	metal exchange.

Solutions to Problems

1.   **a.** $CH_3CH_2CH_2CH_2CH_2OH$ **b.** [benzene ring]—$CH_2CH_2CH_2OH$ **c.** [cyclohexane ring]—CH_2CH_2OH

2.   **a.**

   b.

   c.

3.   Solved in the text.

4.   All the reactions occur because, in each case, the reactant acid is a stronger acid than the product acid (methane, $pK_a \sim 60$).

5.   The greater the polarity of the carbon-metal bond, the more reactive is the organometallic compound. Polarity depends on the difference in electronegativity between the atoms forming the bond; the greater the difference in electronegativity, the more polar the bond.

$$C\!-\!Li \qquad\qquad C\!-\!Na$$
$$2.5 - 1.0 = 1.5 \qquad 2.5 - 0.9 = 1.6$$

The electronegativity difference between carbon and sodium is greater than that between carbon and lithium. Therefore, organosodium compounds (with a more polar carbon-metal bond) are more reactive than organolithium compounds.

6.   Because silicon is more electronegative than magnesium, transmetallation will occur.

$$4\ CH_3MgCl\ +\ SiCl_4\ \longrightarrow\ (CH_3)_4Si\ +\ 4\ MgCl_2$$

7.   **a.**   **b.**   **c.**   **d.**

8.   Because the alkyl halide will undergo an elimination reaction instead of reacting with the Gilman reagent in a substitution reaction.

9. $CH_3(CH_2)_3CH_2Br \xrightarrow[\text{hexane}]{\text{Li}} CH_3(CH_2)_3CH_2Li \xrightarrow{\text{CuI}} (CH_3CH_2CH_2CH_2CH_2)_2CuLi$

10.   In a Suzuki reaction, an organoboron reagent reacts with a vinylic halide or an aryl halide in the presence of a palladium catalyst.

a.

organoboron reagent   a vinylic halide

b.

NO₂

organoboron
reagent

+

a vinylic halide

→

11. **a.**

b.

c.

12.

1-pentyne

13. **a.**

b.

14. **a.**

b.

15. $CH_3\overset{O}{C}$—⬡—Br + CH_3O—⬡—$CH=CH_2$

or

CH_3O—⬡—Br + $CH_3\overset{O}{C}$—⬡—$CH=CH_2$

16. **Suzuki reactions**

a.

+

$\xrightarrow[\text{HO}^-]{\text{PdL}_2}$

↑ HB(OR)₂

$CH_3CH_2C{\equiv}CH$

b.

Heck reactions

a.

b.

Notice that only a Suzuki reaction can be used to replace a halogen with an alkyl group. A Heck reaction replaces a halogen wih an alkenyl group. However, the alkenyl group can subsequently be reduced to an alkyl group.

17.   a. $CH_2=CH_2$ +

b. $CH_2=CH_2$ +

c. $CH_2=CH_2$ + $CH_3CH_2CH=CHCH_2CH_3$

E and *Z*

18.   a. $CH_2=CH_2$ +

b. $CH_2{=}CH_2$ +

c. $CH_2{=}CH_2$ +

19. **a.** **b.** **c.**

20. A = a piece containing two carbons and B = a piece containing three carbons. The two pieces can be put together in four different ways.

$$A{-}A \qquad B{-}B \qquad A{-}B \qquad B{-}A$$

Thus, two of the four pieces (A-B and B-A) represent 2-pentene. Therefore, the maximum yield of 2-pentene is 50%.

21.

22. **a.** **d.**

b. **e.**

c. **f.**

23. **C** is the only one that can be used to form a Grignard reagent.

The Grignard reagent formed from **A** will be destroyed by reacting with the proton of the alcohol group.

The Grignard reagent formed from **B** will be destroyed by reacting with the proton of the carboxylic acid group.

The Grignard reagent formed from **D** will be destroyed by reacting with a proton of the amino group.

24. **A** = Mg   **D** = ethylene oxide   **G** = ethylene oxide

B = diethyl ether   **E** = H$^+$ **H** = H$^+$

C = CH$_3$MgBr   **F** = Na

25. **a.**

or

b. CH$_3$CHCH$_2$OH $\xrightarrow[\Delta]{\text{HBr}}$ CH$_3$CHCH$_2$Br $\xrightarrow[\text{Et}_2\text{O}]{\text{Mg}}$ CH$_3$CHCH$_2$MgBr

          | | |
          CH$_3$ CH$_3$ CH$_3$

CH$_3$CHCH$_2$CH$_2$CH$_2$OH

          |
          CH$_3$

c. CH$_3$CH$_2$C≡CH $\xrightarrow{\text{NaNH}_2}$ CH$_3$CH$_2$C≡C$^-$ $\xrightarrow[\text{2. HCl}]{1. \triangle}$ CH$_3$CH$_2$C≡CCH$_2$CH$_2$OH

d.

e.

26.   **a.**   **b.**   **c.**   **d.**

27.   **a.**   **b.**   **c.**

28.   **a.**   **b.**   **c.**

d.

29.   **a.** $CH_3CH_2CH_2CH_2Br \xrightarrow[Et_2O]{Mg} CH_3CH_2CH_2CH_2MgBr \xrightarrow[2.\ HCl]{1.\ \triangle\!\!O} CH_3CH_2CH_2CH_2CH_2CH_2OH$

$$\Big\downarrow \text{PCC} \mid CH_2Cl_2$$

$$CH_3CH_2CH_2CH_2CH_2\overset{\displaystyle O}{\overset{\|}{C}}H$$

b. $CH_3CH{=}CHCH_3 \xrightarrow{\overset{\displaystyle O}{\overset{\|}{R}COOH}} CH_3CH\overset{O}{\overset{}{-}}CHCH_3 \xrightarrow[2.\ HCl]{1.\ CH_3CH_2MgBr}$ $CH_3\underset{}{\overset{OH}{\overset{|}{C}}}HCHCH_2CH_3$

$$\underset{\displaystyle CH_3}{|}$$

$$\Big\downarrow \overset{\Delta}{\underset{}{H_2SO_4}}$$

$$CH_3CH{=}CCH_2CH_3$$
$$\underset{\displaystyle CH_3}{|}$$

c. $CH_3CH{=}CHCH_3$ $\xrightarrow{RCOOH}$ $CH_3CH{-}CHCH_3$ (epoxide) $\xrightarrow[\text{2. HCl}]{\text{1. } CH_3MgBr}$ $CH_3\overset{OH}{\underset{\underset{CH_3}{|}}{CH}}CHCH_3$

$\xrightarrow{H_2CrO_4}$

$CH_3\overset{O}{\overset{||}{C}}\underset{\underset{CH_3}{|}}{C}HCH_3$

30.

$+\ MgBr_2$

31. The student did not get any of the expected product, because the Grignard reagent removed a proton from the alcohol group. Addition of HCl/H_2O protonated the alkoxide ion and opened the epoxide ring.

32. **a.** In the presence of NBS and peroxide, a bromine will substitute for an allylic hydrogen.

b.

c. $CH_3CH_2CH=CH_2$ $\xrightarrow[\text{2. } H_2O_2, HO^-, H_2O]{\text{1. } BH_3/THF}$ $CH_3CH_2CH_2CH_2OH$ $\xrightarrow[\text{pyridine}]{PBr_3}$ $CH_3CH_2CH_2CH_2Br$

CuI | THF

$(CH_3CH_2CH_2CH_2)_2CuLi$

$CH_3CH_2CH_2CH_2Br$
(obtained from the
second step of the
synthesis)

$CH_3CH_2CH_2CH_2CH_2CH_2CH_2CH_3$

d.

$\xrightarrow{PBr_3}$ ... $\xrightarrow[Et_2O]{Mg}$ ... $\xrightarrow{D_2O}$

OH   Br   MgBr   D

33.   **a.** **A** cannot be directly synthesized by a Heck reaction. A Heck reaction couples a vinylic halide and an alkene, so the product will contain at least two double bonds. However, **A** can be made by a Heck reaction followed by a reduction of the double bonds.

+

$\xrightarrow[(CH_3CH_2)_3N]{PdL_2}$

H_2 | Pd/C

C cannot be prepared by a Heck reaction, because the double bond in the non-halogenated reactant always has the *E* configuration in the product. Both double bonds in **C** have the *Z* configuration.

b.   **B**

+

34. Bombykol can be prepared using a Suzuki reaction. First, turn the alkyne starting material around and then make the organoboron reagent. Draw the vinylic bromide so you can see what product will be obtained from the coupling reaction.

Bombykol can also be prepared using a Heck reaction.

35. Only the halogen bonded to the benzylic carbon can undergo a substitution reaction and be replaced by hydroxide ion. Both halogens will react with magnesium to form an organomagnesium group that will react with acid.

36.   **a.**   **b.**   **c.**

Chapter 11 Practice Test

1. Which is more reactive, an organocadmium reagent (electronegativity of Cd = 1.5) or an organolead reagent (electronegativity of lead = 1.6)?

2. Which of the following alkyl halides could be used to form a Grignard reagent?

3. Give the products of each of the following reactions:

    a.

    b.

    c.

4. Give the products of each of the following reactions:

    a.

    b.

    c.

5.   Give the products of each of the following reactions:

a. $\xrightarrow[\text{Catalyst}]{\text{Grubbs}}$

b. $\xrightarrow[\text{Catalyst}]{\text{Grubbs}}$

c. $\xrightarrow[\text{Catalyst}]{\text{Grubbs}}$

6.   Indicate how each of the following compounds could be prepared using the given starting material:

a. $\longrightarrow$

b. $\longrightarrow$

7.   What alcohols would be formed from the reaction of ethylene oxide with the following Grignard reagents?

a. $CH_3CH_2CH_2MgBr$

b.

CHAPTER 12
Radicals • Reactions of Alkanes

Important Terms

alkane	a hydrocarbon that contains only single bonds.
alkoxy radical	an alkoxy group with an unpaired electron on oxygen.
allylic radical	a compound with an unpaired electron on an allylic carbon.
benzylic radical	a compound with an unpaired electron on a benzylic carbon.
combustion	a reaction with oxygen that takes place at high temperatures and converts alkanes (or other organic compounds) to carbon dioxide and water.
free radical	an atom or a molecule with an unpaired electron.
halogenation reaction	the reaction of an alkane (or other organic compounds) with a halogen.
heterolytic bond cleavage (heterolysis)	breaking a bond with the result that both bonding electrons stay with one of the previously bonded atoms.
homolytic bond cleavage (homolysis)	breaking a bond with the result that each of the atoms that formed the bond gets one of the bonding electrons.
initiation step	the step in which radicals are created and/or the step in which the radical needed for the first propagating step is created.
peroxide	a compound with an O—O bond.
peroxide effect	causes the initial step in an addition reaction of HBr to addition of a bromine radical instead of a proton.
primary alkyl radical	an alkyl radical with the unpaired electron on a primary carbon.
propagation step	in the first of a pair of propagation steps, a radical (or an electrophile or a nucleophile) reacts to produce another radical (or an electrophile or a nucleophile) that reacts in the second propagation step to produce the radical (or the electrophile or the nucleophile) that was the reactant in the first propagation step.
radical (often called a free radical)	an atom or a molecule with an unpaired electron.
radical addition reaction	an addition reaction in which the first species that adds is a radical.
radical chain reaction	a reaction in which radicals are formed and react in repeating propagating steps.

radical inhibitor   a compound that traps radicals.

radical initiator   a compound that creates radicals.

radical substitution   a substitution reaction that has a radical intermediate.
reaction

reactivity-selectivity   the greater the reactivity of a species, the less selective it will be.
principle

saturated hydrocarbon   a hydrocarbon that contains only single bonds (it is saturated with hydrogen).

secondary alkyl radical   an alkyl radical with the unpaired electron on a secondary carbon.

termination step   two radicals combine to produce a molecule in which all the electrons are paired.

tertiary alkyl radical   an alkyl radical with the unpaired electron on a tertiary carbon.

Solutions to Problems

1. $Cl_2 \xrightarrow{\ hv\ } 2\,Cl\cdot$ } initiation

2. $Cl_2 \xrightarrow{\ hv\ } 2\,Cl\cdot$

$Cl\cdot + CH_4 \longrightarrow \cdot CH_3 + HCl$

$\cdot CH_3 + Cl_2 \longrightarrow CH_3Cl + Cl\cdot$

$Cl\cdot + CH_3Cl \longrightarrow \cdot CH_2Cl + HCl$

$\cdot CH_2Cl + Cl_2 \longrightarrow CH_2Cl_2 + Cl\cdot$

$Cl\cdot + CH_2Cl_2 \longrightarrow \cdot CHCl_2 + HCl$

$\cdot CHCl_2 + Cl_2 \longrightarrow CHCl_3 + Cl\cdot$

$Cl\cdot + CHCl_3 \longrightarrow \cdot CCl_3 + HCl$

$\cdot CCl_3 + Cl_2 \longrightarrow CCl_4 + Cl\cdot$

$Cl\cdot + \cdot CCl_3 \longrightarrow CCl_4$

3.   **a.**

   b.   6

4.   **a.** 3   **c.** 5   **e.** 5   **g.** 2   **i.** 4

       b. 3   **d.** 1   **f.** 4   **h.** 4

5.   Solved in the text.

6.   Solved in the text.

7.   **a.**   The major product would be the one obtained from removal of a tertiary hydrogen.

$$CH_3\underset{\underset{Br}{|}}{\overset{\overset{CH_3}{|}}{C}}CH_2CH_3$$

   b.   The relative amounts of the four possible products would be:

$$6 \times 1 = 6 \quad 1 \times 1600 = 1600 \quad 2 \times 82 = 164 \quad 3 \times 1 = 3$$

   percentage of major product $= \dfrac{1600}{1773} \times 100 = 90.2\%$

8.   **a.**   Chlorination, because the halogen is substituting for a primary hydrogen.
   b.   Bromination, because the halogen is substituting for a tertiary hydrogen.
   c.   Because the molecule has only one kind of hydrogen, both chlorination and bromination will form only one monohalogenated product.

9.   Solved in the text.

10.   To start peroxide formation, the chain initiating radical removes a hydrogen atom from an α-carbon of the ether.

   a.   **D** is most apt to form a peroxide, because removal of a hydrogen from an α-carbon forms a secondary radical.

   b.   **B** is least apt to form a peroxide, because it does not have any hydrogens bonded to its α-carbons.

11.

12.   **a.**

   b.

c. $CH_3\overset{\overset{\displaystyle CH_3}{|}}{C}=CHCH_3$ + HBr $\xrightarrow{\text{peroxide}}$ $CH_3\overset{\overset{\displaystyle CH_3}{|}}{C}H-\overset{\overset{}{}}{\underset{\underset{\displaystyle Br}{|}}{C}}HCH_3$

d. $CH_3\overset{\overset{\displaystyle CH_3}{|}}{C}=CHCH_3$ + HCl $\xrightarrow{\text{peroxide}}$ $CH_3\overset{\overset{\displaystyle CH_3}{|}}{\underset{\underset{\displaystyle Cl}{|}}{C}}-CH_2CH_3$

13. **a.** $CH_3\overset{\overset{\displaystyle CH_3}{|}}{C}HCH_3$ $\xrightarrow[hv]{Cl_2}$ $CH_3\overset{\overset{\displaystyle CH_3}{|}}{C}HCH_2Cl$ + $CH_3\overset{\overset{\displaystyle CH_3}{|}}{\underset{\underset{\displaystyle Cl}{|}}{C}}CH_3$

achiral   achiral

b. $CH_3CH_2CH_2CH_3$ $\xrightarrow[hv]{Cl_2}$ $CH_3CH_2CH_2CH_2Cl$ +

achiral

chiral   chiral

enantiomers

14. Bromination can occur at either of the two tertiary positions. Because the radical intermediate is sp^2 hybridized and therefore planar, the incoming bromine can add to a tertiary radical from either the top or the bottom of the plane.

15.

16. Solved in the text.

17. The product of the reaction in Problem 16 has one asymmetric center. Therefore, two allylic substituted bromoalkenes are obtained, one with the *R* configuration at the asymmetric center and one with the *S* configuration.

18.

1-methylcyclohexene

a. 1. There are two sets of secondary allylic hydrogens, **a** and **b** (plus a less reactive set of primary allylic hydrogens). Removal of one of the **a** allylic hydrogens forms an intermediate in which the unpaired electron is shared by two secondary carbons. Removal of one of the **b** allylic hydrogens forms an intermediate in which the unpaired electron is shared by a tertiary carbon and a secondary carbon.

Therefore the major products are obtained by removing a **b** allylic hydrogen.

2.   **3.**   **4.**

b. 1. Each of the products has one asymmetric center.
The R and the S isomers will be obtained for each product.

2. The product has two asymmetric centers.
Because addition of Br_2 is anti, only the products with the bromine atoms on opposite sides of the ring are obtained.

3. The product does not have an asymmetric center, so it does not have stereoisomers.

4. The product has two asymmetric centers.
Because radical addition of HBr can be either syn or anti, four stereoisomers are obtained as products.

19. a.

b.

c.

d.

20. Four atoms share the unpaired electron, three carbons and one oxygen.

21. There are several OH groups in a catechin that, upon losing a hydrogen atom as a result of reacting with a reactive radical, form a radical that is stabilized by electron delocalization. This highly stabilized radical is sufficiently unreactive that it does not damage cells. An example of one of the stabilized radicals is shown below.

22. a. CH_2=$CHCHCH_2CH_3$ + $BrCH_2CH$=$CHCH_2CH_3$
                  |
                  Br

b. In the radical intermediate that leads to the major products, the unpaired electron is shared by a primary carbon and a tertiary carbocation. In the radical intermediate that leads to the minor products, the unpaired electron is shared by a primary and a secondary carbocation. It, therefore, is not as stable as the intermediate that leads to the major products.

major products   minor products

In bromination, selectivity is much more important than probability. Therefore, even though twice as many hydrogens are available for removal by a bromine radical that lead to the minor products, they will still be minor products because the easier to remove hydrogens lead to the major products.

c.
major product

d.

e. no reaction
(There is no light or heat, so radicals cannot be formed.)

f.

23. **a.**
dimethylpropane   3-methylhexane

24. **a.**   **b.**   **c.**

25. **a.**   **b.**   **c.**

26.   **a.**

d.

b.

e.

c.   **+**

f.   CH₃

In **c**, because one product is under kinetic control and the other is under thermodynamic control, the major product will depend on the conditions under which the reaction is carried out.

27.

$$\underset{\underset{Cl}{|}}{\overset{\overset{CH_3}{|}}{CH_3CCH_3}}$$

$$\underset{\underset{CH_3CHCH_2Cl}{9 \times 1 = 9}}{\overset{\overset{CH_3}{|}}{}}$$

$1 \times x = x$

$$\begin{array}{l}\text{fraction of the total that is}\\ \text{1-chloro-2-methylpropane}\end{array} = \frac{\text{1-chloro-2-methylpropane}}{\text{1-chloro-2-methylpropane} + \text{2-chloro-2-methylpropane}} = 0.64$$

$$\frac{9}{9 + x} = 0.64$$

$$9 = 0.64\,(9 + x)$$

$$9 = 5.76 + 0.64x$$

$$3.24 = 0.64x$$

$$x = 5$$

It is 5 times easier for a chloride radical to remove a hydrogen atom from a tertiary carbon than from a primary carbon.

28.   Removal of a hydrogen atom from ethane by an iodine radical is a highly endothermic reaction ($\Delta H° = 101 - 71 = 30$ kcal/mol; see Table 3.2 on page 141 of the text), so the iodine radicals will reform I_2 rather than remove a hydrogen atom.

$$CH_3CH_3 + I\cdot \;\not\!\!\longrightarrow\; CH_3\overset{\cdot}{C}H_2 + HI$$

101 kcal/mol   71 kcal/mol

29.

30.   It is easier to break a C—H bond than a C—D bond.
Because a bromine radical is less reactive than a chlorine radical, a bromine radical has a greater preference for the more easily broken C—H bond. Bromination, therefore, would have a greater deuterium kinetic isotope effect than chlorination.

31.   **a.**   Theoretically five monobromination products are possible, but some will be obtained in very small amounts.

b.   A bromine radical is very selective, much preferring to remove a hydrogen atom from a tertiary carbon than from a secondary or primary carbon. 1-Bromo-1-methylcyclohexane would be obtained in greatest yield because it is the only one formed by removing a hydrogen atom from a tertiary carbon.

c.   The number of possible stereoisomers for each compound is indicated below each structure.

A total of 12 stereoisomers can be obtained.

32. **a.** CH₃CHCH₂Cl (CH₃ above)   **c.** CH₃CHCHCH₃ (CH₃ above, Br below)   **e.** [cycloheptene with Br and CH₃]

b. CH₃CCH₃ (CH₃ above, Br below)   **d.** CH₂=CHCH₂CH₂CHCH₂Br (CH₃ above)   **f.** [cyclohexane with CH=CH₂, Br, CH₃]

33. **a.** One difference between this reaction and the monochlorination of ethane is the source of the chlorine radical. This reaction has two sets of propagation steps because two different radicals are generated in the initiation step. This reaction also has several more termination steps because of the two different radicals generated in the initiation step. (Bond dissociation energies can be found in Table 3.2 on page 141 of the text.)

initiation { (CH₃)₃COCl ⟶ (CH₃)₃CO· + Cl·

propagation {
(CH₃)₃CO· + CH₃CH₃ ⟶ (CH₃)₃COH + CH₃ĊH₂
CH₃ĊH₂ + (CH₃)₃COCl ⟶ CH₃CH₂Cl + (CH₃)₃CO·

propagation {
Cl· + CH₃CH₃ ⟶ HCl + CH₃ĊH₂
CH₃ĊH₂ + (CH₃)₃COCl ⟶ CH₃CH₂Cl + (CH₃)₃CO·

$$CH_3\overset{\cdot}{C}H_2 \;+\; Cl\cdot \;\longrightarrow\; CH_3CH_2Cl$$

termination {

$$\underset{\underset{CH_3}{|}}{\overset{\overset{CH_3}{|}}{CH_3CO\cdot}} \;+\; \underset{\underset{CH_3}{|}}{\overset{\overset{CH_3}{|}}{CH_3CO\cdot}} \;\longrightarrow\; \underset{\underset{CH_3}{|}}{\overset{\overset{CH_3}{|}}{CH_3CO}}-\underset{\underset{CH_3}{|}}{\overset{\overset{CH_3}{|}}{OCCH_3}}$$

$$\underset{\underset{CH_3}{|}}{\overset{\overset{CH_3}{|}}{CH_3CO\cdot}} \;+\; Cl\cdot \;\longrightarrow\; \underset{\underset{CH_3}{|}}{\overset{\overset{CH_3}{|}}{CH_3COCl}}$$

$$\underset{\underset{CH_3}{|}}{\overset{\overset{CH_3}{|}}{CH_3CO\cdot}} \;+\; CH_3\overset{\cdot}{C}H_2 \;\longrightarrow\; \underset{\underset{CH_3}{|}}{\overset{\overset{CH_3}{|}}{CH_3COCH_2CH_3}}$$

$$Cl\cdot \;+\; Cl\cdot \;\longrightarrow\; Cl_2$$

$$CH_3\overset{\cdot}{C}H_2 \;+\; CH_3\overset{\cdot}{C}H_2 \;\longrightarrow\; CH_3CH_2CH_2CH_3$$

b.

$$\Delta H° = \text{bonds broken} - \text{bonds formed}$$

Let x = the bond dissociation energy of the O—Cl bond.

$$-42 \text{ kcal/mol} = [101 \text{ kcal/mol} + x \text{ kcal/mol}] - [82 \text{ kcal/mol} + 102 \text{ kcal/mol}]$$

$$-143 \text{ kcal/mol} = x \text{ kcal/mol} - 184 \text{ kcal/mol}$$

$$x = 41 \text{ kcal/mol}$$

c. The $\Delta H°$ for the first propagation step in the first set of propagation steps is $(101 - 105) =$ -4 kcal/mol. The $\Delta H°$ for the first propagation step in the second set of propagation steps is $(101 - 103) = -2$ kcal/mol. (Both sets have the same second propagation step.) Since the first set has the more exothermic first propagation step, it should be the set most likely to occur. However, since the two $\Delta H°$ values are so close and not all O—H bonds have the same bond dissociation energy, the bond dissociation energy of the O—H bond in *tert*-butanol needs to be known to be sure that the first set of propagation steps will be favored.

34. The reaction forms a product with two new asymmetric centers.

$$\underset{\underset{Br}{|}}{CH_3CH_2CH}-\overset{\overset{CH_3\;\;CH_3}{|\;\;\;\;|}}{CCH_2CH_3}$$

Because the reaction involves both syn and anti addition, four stereoisomers are formed.

35. **a.**

b.

c.

d.

36.

$$CH_3CHCH_2CH_3$$
$$\overset{\displaystyle |}{CH_3}$$

2-methylbutane

2-Methylbutane forms two primary alkyl halides, one secondary alkyl halide, and one tertiary alkyl halide. First we need to calculate how much alkyl halide is formed from each primary hydrogen available, from each secondary hydrogen available, and from each tertiary hydrogen available.

a primary alkyl halide

$ClCH_2CHCH_2CH_3$
$\overset{\displaystyle |}{CH_3}$

Substitution of any one of six hydrogens leads to this product.
percentage of this product that is formed = 36%
percentage formed per hydrogen available = 36/6 = 6%

$CH_3CHCH_2CH_2Cl$
$\overset{\displaystyle |}{CH_3}$

Substitution of any one of three hydrogens leads to this product.
percentage of this product that is formed = 18%
percentage formed per hydrogen available = 18/3 = 6%

a secondary alkyl halide

$\overset{\displaystyle Cl}{\overset{\displaystyle |}{CH_3CHCHCH_3}}$
$\overset{\displaystyle |}{CH_3}$

Substitution of any one of two hydrogens leads to this product.
percentage of this product that is formed = 28%
percentage formed per hydrogen available = 28/2 = 14%

a tertiary alkyl halide

$$
\begin{array}{c}
\text{Cl} \\
| \\
\text{CH}_3\text{CCH}_2\text{CH}_3 \\
| \\
\text{CH}_3
\end{array}
$$

Substitution of the one tertiary hydrogen leads to this product.

percentage of this product that is formed = 18%

percentage formed per hydrogen available = 18/1 = 18%

From the above calculations, we see that at 300 °C, the relative rates of removal of a hydrogen atom from a tertiary, secondary, and primary carbocation are:

18:14:6 = 3:2.3:1

In Section 11.4, we saw that at room temperature the relative rates are:

5:3.8:1

Thus we can conclude that at higher temperatures, there is less discrimination by the radical between a tertiary, secondary, and primary hydrogen atom.

37.   The chlorine radical is more reactive at the higher temperature (600 °C), so it is less selective.

38.   **a.**   **b.** $CH_3CH_2CH{=}CHCH_2CH_3$ $\begin{array}{c} \text{CH}_3 \\ | \\ \text{CH}_3\text{C}{=}\text{CCH}_3 \\ | \\ \text{CH}_3 \end{array}$

cis or trans

39.   **a.** $CH_3{-}H$ $Cl{-}Cl$ $\longrightarrow$ $CH_3{-}Cl$ $H{-}Cl$

  105   58   84   103

$\Delta H° = \text{bonds broken} - \text{bonds formed}$

$\Delta H° = [105 \text{ kcal/mol} + 58 \text{ kcal/mol}] - [84 \text{ kcal/mol} + 103 \text{ kcal/mol}]$

$\Delta H° = 163 - 187 = -24 \text{ kcal/mol}$

b. $CH_3{-}H + \dot{C}l$ $\longrightarrow$ $\dot{C}H_3 + H{-}Cl$ $\Delta H° = 105 - 103 = 2 \text{ kcal/mol}$

$\dot{C}H_3 + Cl{-}Cl$ $\longrightarrow$ $CH_3{-}Cl + \dot{C}l$ $\Delta H° = 58 - 84 = -26 \text{ kcal/mol}$

Overall $\Delta H° = 2 + (-26) = -24 \text{ kcal/mol}$

c.   If you cancel the things that are the same on opposite sides of the equations in part **b** and then add the two equations, you are left with the equation in part **a**.

$CH_3{-}H + \dot{C}l$ $\longrightarrow$ $\dot{C}H_3 + H{-}Cl$

$\dot{C}H_3 + Cl{-}Cl$ $\longrightarrow$ $CH_3{-}Cl + \dot{C}l$

$CH_4 + Cl_2$ $\longrightarrow$ $CH_3Cl + HCl$

40.   The overall $\Delta H°$ value is the same for both mechanisms. However, the mechanism that occurs is controlled by the first propagation step. In the case of the alternative mechanism, the first propagation step is very endothermic, so it would not be able to compete with the first propagation step of the mechanism shown in Problem 39b.

$$CH_3—H + ·Cl \longrightarrow CH_3—Cl + ·H \qquad \Delta H° = 105 - 84 = 21 \text{ kcal/mol}$$

$$H· + Cl—Cl \longrightarrow H—Cl + ·Cl \qquad \Delta H° = 58 - 103 = -45 \text{ kcal/mol}$$

$$\text{Overall } \Delta H° = 21 + (-45) = -24 \text{ kcal/mol}$$

41.

42.   a.

b.

43.   The methyl radical that is created in the first propagation step reacts with Br_2, forming bromomethane.

$$\cdot Br + CH_4 \longrightarrow \cdot CH_3 + HBr$$
$$\cdot CH_3 + Br_2 \longrightarrow CH_3Br + \cdot Br$$

If HBr is added to the reaction mixture, the methyl radical that is created in the first propagation step can react with Br_2 or with the added HBr. Because only reaction with Br_2 forms bromomethane (reaction with HBr reforms methane), the overall rate of formation of bromomethane is decreased.

$$\cdot Br + CH_4 \longrightarrow \cdot CH_3 + HBr$$

$$\cdot CH_3 + Br_2 \longrightarrow CH_3Br + \cdot Br$$

$$\cdot CH_3 + HBr \longrightarrow CH_4 + \cdot Br$$

Chapter 12 Practice Test

1. How many monochlorinated products would be obtained from the following reaction? (Ignore stereoisomers.)

    $$\underset{\substack{|\\CH_3}}{CH_3}\underset{\substack{|\\CH_3}}{CHCH_2CH_2}\underset{\substack{|\\CH_3}}{CHCH_3} + Cl_2 \xrightarrow{\Delta}$$

2. Give the first propagation step in the monochlorination of ethane.

3. When (S)-2-bromopentane is brominated, 2,3-dibromopentanes are formed. Which of the following compounds are **not** formed?

CH₃	CH₃	CH₃	CH₃
H——Br	Br——H	Br——H	H——Br
H——Br	H——Br	Br——H	Br——H
CH₂CH₃	CH₂CH₃	CH₂CH₃	CH₂CH₃

4. Determine the $\Delta H°$ of the two propagation steps in the monochlorination of ethane, using Table 3.2 on page 141 of the text.

5. Label the radicals in order of decreasing stability. (Label the most stable #1.)

    $\overset{\bullet}{C}H_2CH_2CH_2CH=CH_2$ $CH_3CH_2\overset{\bullet}{C}HCH=CH_2$

    $CH_3CH_2CH_2CH=\overset{\bullet}{C}H$ $CH_3\overset{\bullet}{C}HCH_2CH=CH_2$

6. Give the product(s) of each of the following reactions, ignoring stereoisomers.

    a. (cyclopentane with =CH₂)   + NBS $\xrightarrow[\text{peroxide}]{\Delta}$

    b. $CH_3CH_2CH=CH_2$ + NBS $\xrightarrow[\text{peroxide}]{\Delta}$

7. a. Give the products that will be obtained from monochlorination of the following alkane at room temperature. (Ignore stereoisomers.)

    $$\underset{\substack{|\\CH_3}}{\overset{\substack{CH_3\\|}}{CH_3C}}CH_2CH_2\underset{\substack{|\\}}{\overset{\substack{CH_3\\|}}{CH}}CH_3 + Cl_2 \xrightarrow{h\nu}$$

    b. What product will be obtained in greatest yield?

    c. What product would be obtained in greatest yield if the alkane were brominated instead of chlorinated?

8.   Calculate the yield of 2-chloro-2-methylbutane (as a percentage of all the monochlorinated products obtained) when 2-methylbutane is chlorinated in the presence of light at room temperature.

9.   What is the major product of each of the following reactions? Ignore stereoisomers.

$$\overset{\overset{\displaystyle CH_3}{\displaystyle |}}{}$$

a. $CH_3CH = CCH_2CH_3$ + HBr $\longrightarrow$ **c.** $CH_3CH = CCH_2CH_3$ + HCl $\longrightarrow$

b. $CH_3CH = CCH_2CH_3$ + HBr $\xrightarrow{\text{peroxide}}$ **d.** $CH_3CH = CCH_2CH_3$ + HCl $\xrightarrow{\text{peroxide}}$

10.   Which of the above reactions forms more than one stereoisomer?

SPECIAL TOPIC VII

Drawing Curved Arrows in Radical Systems

We have seen that chemists use curved arrows:

1. to show how electrons move in going from a reactant to a product.
2. to show how electrons move in going from one resonance contributor to the next.

In Special Topic II (using curved arrows in reactions) and Special Topic V (using curved arrows in drawing resonance contributors), electron movement involves the simultaneous movement of two electrons. Here we will show electron movement that involves the movement of only one electron.

Drawing Curved Arrows in Radical Reactions

When a bond breaks in a way that allows each of the bonded atoms to retain one of the bonding electrons, an arrowhead with one barb is used to represent the movement of the single electron.

$$CH_3\overset{\cdot}{C}HCH_3 \longrightarrow CH_3\overset{\cdot}{C}HCH_3 + \cdot\overset{\cdot\cdot}{\underset{\cdot\cdot}{B}}r\colon$$

Sometimes, the lone pairs are not shown.

When a bond is formed using one electron from one atom and one electron from another, an arrowhead with one barb is used to represent the movement of the single electron.

Sometimes, the lone pairs are not shown.

Problem 1. Draw curved arrows to show the movement of the electrons as the bond breaks.

a. $CH_3CH_2—\overset{\cdot\cdot}{\underset{\cdot\cdot}{C}}l\colon \longrightarrow CH_3\overset{\cdot}{C}H_2 + \cdot\overset{\cdot\cdot}{\underset{\cdot\cdot}{C}}l\colon$

b.

c. $CH_3\overset{\cdot\cdot}{\underset{\cdot\cdot}{O}}—\overset{\cdot\cdot}{\underset{\cdot\cdot}{O}}CH_3 \longrightarrow CH_3\overset{\cdot\cdot}{\underset{\cdot\cdot}{O}}\cdot + \cdot\overset{\cdot\cdot}{\underset{\cdot\cdot}{O}}CH_3$

Problem 2. Draw curved arrows to show the movement of the electrons as the bond forms.

$$\underset{\underset{\displaystyle CH_3}{|}}{CH_3\dot{C}CH_3} + \cdot\ddot{Br}\!: \longrightarrow \underset{\underset{\displaystyle :\ddot{Br}:}{|}}{\underset{\underset{\displaystyle CH_3}{|}}{CH_3\overset{}{C}CH_3}}$$

b. ·Br + ·Br ⟶ Br₂

c.

[cyclohexane ring with $\dot{C}H_2$] + ·Cl ⟶ [cyclohexane ring with CH_2Cl]

Often a bond breaks and a bond forms in the same step.

$$Br\!\cdot + CH_3CHCH_2CH_3 \longrightarrow HBr + CH_3\dot{C}HCH_2CH_3$$
$$\underset{\displaystyle H}{|}$$

$$CH_3CH{=}CH_2 + \cdot\ddot{C}l\!: \longrightarrow \underset{\underset{\displaystyle :\ddot{C}l:}{|}}{CH_3\dot{C}HCH_2}$$

$$RO\!\cdot + H{-}Br \longrightarrow ROH + \cdot Br$$

Problem 3. Draw curved arrows to show the movement of the electrons as one bond breaks and another bond forms.

a.

[cyclopentene ring] + ·$\ddot{Br}$: ⟶ [cyclopentane ring with radical and $\ddot{B}r$:]

b. $CH_3\dot{C}HCH_2CH_3$ + H—Cl ⟶ $CH_3CH_2CH_2CH_3$ + ·Cl

c. $CH_3\dot{C}H_2$ + :$\ddot{Br}$—$\ddot{Br}$: ⟶ CH_3CH_2Br + ·$\ddot{Br}$:

d.

$$\underset{\underset{\displaystyle CH_3}{|}}{\overset{\overset{\displaystyle CH_3}{|}}{CH_3C}}{-}N{=}N{-}\underset{\underset{\displaystyle CH_3}{|}}{\overset{\overset{\displaystyle CH_3}{|}}{CCH_3}} \longrightarrow \underset{\underset{\displaystyle CH_3}{|}}{\overset{\overset{\displaystyle CH_3}{|}}{CH_3\dot{C}}} \quad N{\equiv}N \quad \underset{\underset{\displaystyle CH_3}{|}}{\overset{\overset{\displaystyle CH_3}{|}}{\cdot CCH_3}}$$

Drawing Curved Arrows in Resonance Contributors That Are Radicals

The arrows that represent electron movement in resonance contributors of radicals have only one barb on the arrowhead because the arrow represents the movement of only one electron.

$$CH_3\overset{\frown}{CH}\!\!=\!\!\overset{\frown}{CH}\!\!-\!\!\overset{\frown}{\overset{\displaystyle\cdot}{C}}HCH_3 \quad\longleftrightarrow\quad CH_3\overset{\displaystyle\cdot}{C}H\!\!-\!\!CH\!\!=\!\!CHCH_3$$

Problem 4. Draw curved arrows to show the movement of the electrons as one resonance contributor is converted to the next.

a.   ⟷

b. ⟷   ⟷   ⟷   ⟷

c. ⟷   ⟷   ⟷

d. ⟷   ⟷   ⟷   ⟷

Solutions to Problems

Problem 1

a. $CH_3\dot{C}H_2 \overset{\frown}{\underset{}{}} \ddot{\underset{\cdot\cdot}{C}l} : \longrightarrow CH_3\dot{C}H_2 + \cdot\ddot{\underset{\cdot\cdot}{C}l}:$

b. $\overset{\frown\frown}{\underset{\cdot\cdot}{B}r}: \longrightarrow$ $\cdot + \cdot\ddot{\underset{\cdot\cdot}{B}r}:$

c. $CH_3\ddot{\underset{\cdot\cdot}{O}} \overset{\frown\frown}{\underset{}{}} \ddot{\underset{\cdot\cdot}{O}}CH_3 \longrightarrow CH_3\ddot{\underset{\cdot\cdot}{O}}\cdot + \cdot\ddot{\underset{\cdot\cdot}{O}}CH_3$

Problem 2

a.
$$CH_3\overset{CH_3}{\underset{}{\overset{|}{C}}}CH_3 + \cdot\ddot{B}r: \longrightarrow CH_3\overset{CH_3}{\underset{\underset{:\ddot{B}r:}{|}}{\overset{|}{C}}}CH_3$$

b. $\cdot Br + \cdot Br \longrightarrow Br_2$

c. $\dot{C}H_2 + \cdot Cl \longrightarrow$ CH_2Cl

Problem 3

a. $+ \cdot\ddot{B}r: \longrightarrow$ $\ddot{\underset{\cdot\cdot}{B}}r:$

b. $CH_3\dot{C}HCH_2CH_3 + H \overset{\frown}{\underset{}{}} Cl \longrightarrow CH_3CH_2CH_2CH_3 + \cdot Cl$

c. $CH_3\dot{C}H_2 + :\ddot{B}r \overset{}{\underset{}{}} \ddot{B}r: \longrightarrow CH_3CH_2Br + \cdot\ddot{B}r:$

d.
$$CH_3\overset{CH_3}{\underset{CH_3}{\overset{|}{\underset{|}{C}}}} N=N \overset{CH_3}{\underset{CH_3}{\overset{|}{\underset{|}{C}}}}CH_3 \longrightarrow CH_3\overset{CH_3}{\underset{CH_3}{\overset{|}{\underset{|}{C}}}}\cdot \quad N\equiv N \quad \cdot\overset{CH_3}{\underset{CH_3}{\overset{|}{\underset{|}{C}}}}CH_3$$

Problem 4

a.

b.

c.

d.

CHAPTER 13
Mass Spectrometry, Infrared Spectroscopy, and Ultraviolet/Visible Spectroscopy

Important Terms

absorption band	a peak in a spectrum that occurs as a result of absorption of energy.
auxochrome	a substituent that when attached to a chromophore alters the λ_{max} and intensity of absorption of UV/Vis radiation.
base peak	the peak in a mass spectrum with the greatest intensity.
Beer-Lambert Law	relationship between the absorbance of UV/Vis light, the concentration of the sample, the length of the light path, and the molar absorptivity.
bending vibration	a vibration that does not occur along the line of the bond.
chromophore	the part of a molecule responsible for a UV or visible spectrum.
α-cleavage	homolytic cleavage of an alpha substituent.
electromagnetic radiation	radiant energy that displays wave properties.
electronic transition	promotion of an electron from its HOMO to its LUMO.
fingerprint region	the right-hand third of an IR spectrum (1400–600 cm^{-1}), where the absorption bands are characteristic of the compound as a whole.
fragment ion peak	a positively charged fragment of the molecular ion.
frequency	the velocity of a wave divided by its wavelength. (It has units of cycles/second.)
functional group region	the left-hand two-thirds of an IR spectrum (4000–1400 cm^{-1}), where most functional groups show absorption bands.
highest occupied molecular orbital (HOMO)	the highest energy molecular orbital that contains electrons.
Hooke's law	an equation that describes the motion of a vibrating spring.
λ_{max}	the wavelength at which there is maximum UV/Vis absorbance.
infrared radiation	electromagnetic radiation familiar to us as heat.
infrared spectroscopy	spectroscopy that uses infrared energy to provide a knowledge of the functional groups in a compound.
infrared spectrum (IR spectrum)	a plot of relative absorption versus wavenumber (or wavelength) of absorbed infrared radiation.
lowest unoccupied molecular orbital (LUMO)	the lowest energy molecular orbital that does not contain electrons.

mass spectrometry provides a knowledge of the molecular weight and certain structural features of a compound.

mass spectrum a plot of the relative abundance of the positively charged fragments produced in a mass spectrometer versus their m/z values.

McLafferty rearrangement rearrangement of the molecular ion of a ketone that contains a γ-hydrogen; the bond between the α- and β-carbons breaks, and a γ-hydrogen migrates to the oxygen.

molar absorptivity the absorbance obtained from a 1.00 M solution in a cell with a 1.00 cm light path.

molecular ion (M) the radical cation formed by removing one electron from a molecule.

nominal molecular mass mass to the nearest whole number.

radical cation a species with a positive charge and an unpaired electron.

spectroscopy study of the interaction of matter and electromagnetic radiation.

stretching vibration a vibration occurring along the line of the bond.

ultraviolet light electromagnetic radiation with wavelengths ranging from 180 to 400 nm.

UV/Vis spectroscopy the absorption of electromagnetic radiation that is useful in determining information about conjugated systems.

visible light electromagnetic radiation with wavelengths ranging from 400 to 780 nm.

wavelength distance from any point on one wave to the corresponding point on the next wave.

wavenumber the number of waves in one centimeter.

Solutions to Problems

1.   Only positively charged fragments are accelerated through the analyzer tube.

$$CH_3CH_2\overset{+}{C}H_2 \qquad\qquad [CH_3CH_2CH_3]^{\overset{+}{\bullet}} \qquad\qquad \overset{+}{C}H_2CH{=}CH_2$$

2.   The peak at $m/z = 57$ will be more intense for 2,2-dimethylpropane than for 2-methylbutane or for pentane because the peak at $m/z = 57$ is due to loss of a methyl radical: loss of a methyl radical from 2,2-dimethylpropane results in the formation of a tertiary carbocation, whereas loss of a methyl radical from 2-methylbutane and pentane results in the formation of a secondary and primary carbocation, respectively.

$$
\begin{bmatrix} CH_3 \\ | \\ CH_3\overset{|}{C}CH_3 \\ | \\ CH_3 \end{bmatrix}^{\overset{+}{\bullet}} \longrightarrow
\begin{matrix} CH_3 \\ | \\ CH_3\overset{|}{\underset{+}{C}}CH_3 \end{matrix} \quad + \quad \overset{\bullet}{C}H_3
$$

2,2-dimethylpropane $m/z = 57$
                               a tertiary carbocation

$$
\begin{bmatrix} CH_3 \\ | \\ CH_3\overset{|}{C}HCH_2CH_3 \end{bmatrix}^{\overset{+}{\bullet}} \longrightarrow
CH_3\overset{+}{C}HCH_2CH_3 \quad + \quad \overset{\bullet}{C}H_3
$$

2-methylbutane $m/z = 57$
                           a secondary carbocatio

$$
\begin{bmatrix} CH_3CH_2CH_2CH_2CH_3 \end{bmatrix}^{\overset{+}{\bullet}} \longrightarrow
CH_3CH_2CH_2\overset{+}{C}H_2 \quad + \quad \overset{\bullet}{C}H_3
$$

pentane $m/z = 57$
                     a primary carbocation

Notice that the mass spectrum of 2-methylbutane can be distinguished from those of the other isomers by the peak at $m/z = 43$. The peak at $m/z = 43$ will be most intense for 2-methylbutane, because such a peak is due to loss of an ethyl radical, which results in the formation of a secondary carbocation. Pentane gives a less intense peak at $m/z = 43$, because loss of an ethyl radical from pentane forms a primary carbocation. 2,2-Dimethylpropane will not show a peak at $m/z = 43$, because it does not have an ethyl group.

$$
\begin{bmatrix} CH_3 \\ | \\ CH_3\overset{|}{C}HCH_2CH_3 \end{bmatrix}^{\overset{+}{\bullet}} \longrightarrow
\begin{matrix} CH_3 \\ | \\ CH_3\overset{|}{\underset{+}{C}}H \end{matrix} \quad + \quad CH_3\overset{\bullet}{C}H_2
$$

$m/z = 43$

$$
\begin{bmatrix} CH_3CH_2CH_2CH_2CH_3 \end{bmatrix}^{\overset{+}{\bullet}} \longrightarrow
CH_3CH_2\overset{+}{C}H_2 \quad + \quad CH_3\overset{\bullet}{C}H_2
$$

$m/z = 43$

3. Intense peaks should occur at $m/z = 57$ for loss of an ethyl radical $(86 - 29)$, and at $m/z = 71$ for loss of a methyl radical $(86 - 15)$.

$$\begin{bmatrix} & \text{CH}_3 \\ & | \\ \text{CH}_3\text{CH}_2\text{CHCH}_2\text{CH}_3 \end{bmatrix}^{+\cdot} \longrightarrow \begin{array}{c} \text{CH}_3 \\ | \\ \text{CH}_3\text{CH}_2\overset{+}{\text{CH}} \end{array} + \quad \cdot\text{CH}_2\text{CH}_3$$
$$m/z = 57$$

$$\begin{bmatrix} & \text{CH}_3 \\ & | \\ \text{CH}_3\text{CH}_2\text{CHCH}_2\text{CH}_3 \end{bmatrix}^{+\cdot} \longrightarrow \text{CH}_3\text{CH}_2\overset{+}{\text{C}}\text{HCH}_2\text{CH}_3 + \quad \cdot\text{CH}_3$$
$$m/z = 71$$

A secondary carbocation is formed in both cases. Because an ethyl radical is more stable than a methyl radical, the base peak will most likely be at $m/z = 57$.

4. A hydrocarbon with molecular formula C_9H_{20} has a molecular mass of 128.
Since $C_9H_{20} = C_nH_{2n+2}$, we know that the hydrocarbon has no rings and no π bonds.

$$\begin{array}{c} \text{CH}_3 \qquad\qquad \text{CH}_3 \\ | \qquad\qquad\quad | \\ \text{CH}_3\text{CHCH}_2\text{CH}_2\text{CH}_2\text{CHCH}_3 \end{array} \longrightarrow \begin{array}{c} \text{CH}_3 \\ | \\ \text{CH}_3\overset{+}{\text{CH}} \end{array} + \begin{array}{c} \text{CH}_3 \\ | \\ \cdot\text{CH}_2\text{CH}_2\text{CH}_2\text{CHCH}_3 \end{array}$$
$$m/z = 43$$

$$\begin{array}{c} \text{CH}_3 \\ | \\ \text{CH}_3\overset{\cdot}{\text{CH}} \end{array} + \begin{array}{c} \text{CH}_3 \\ | \\ \overset{+}{\text{CH}}_2\text{CH}_2\text{CH}_2\text{CHCH}_3 \end{array}$$
$$m/z = 85$$

$$\begin{array}{c} \text{CH}_3 \\ | \\ \text{CH}_3\text{CH}\overset{+}{\text{CH}}_2 \end{array} + \begin{array}{c} \text{CH}_3 \\ | \\ \cdot\text{CH}_2\text{CH}_2\text{CHCH}_3 \end{array}$$
$$m/z = 57$$

$$\begin{array}{c} \text{CH}_3 \\ | \\ \text{CH}_3\text{CH}\overset{\cdot}{\text{CH}}_2 \end{array} + \begin{array}{c} \text{CH}_3 \\ | \\ \overset{+}{\text{CH}}_2\text{CH}_2\text{CHCH}_3 \end{array}$$
$$m/z = 71$$

The hydrocarbon below would also be expected to give a base peak of $m/z = 43$, because the piece with $m/z = 43$ is a secondary carbocation and all the others are primary carbocations. However, we would expect fragments with $m/z = 29$ and $m/z = 99$ to be present to the same extent as those with $m/z = 57$, 85, and 71. Since fragments with $m/z = 29$ and $m/z = 99$ are not mentioned, we can conclude that this hydrocarbon is less likely than the one shown above.

$$\begin{array}{c} \text{CH}_3 \\ | \\ \text{CH}_3\text{CHCH}_2\text{CH}_2\text{CH}_2\text{CH}_2\text{CH}_2\text{CH}_3 \end{array} \longrightarrow \begin{array}{c} \text{CH}_3 \\ | \\ \text{CH}_3\text{CHCH}_2\text{CH}_2\text{CH}_2\overset{\cdot}{\text{CH}}_2 \end{array} + \quad \overset{+}{\text{CH}}_2\text{CH}_3$$
$$m/z = 29$$

$$\begin{array}{c} \text{CH}_3 \\ | \\ \text{CH}_3\text{CHCH}_2\text{CH}_2\text{CH}_2\overset{+}{\text{CH}}_2 \end{array} + \quad \cdot\text{CH}_2\text{CH}_3$$
$$m/z = 99$$

5.   a.   1. $15 + (3 \times 14) + 16 = 73$ **2.** $16 + (3 \times 14) + 16 = 74$

b.   An alkane has an even-mass molecular ion. If a CH_2 group (14) of an alkane is replaced by an NH group (15), or if a CH_3 group (15) of an alkane is replaced by an NH_2 group (16), the molecular ion will have an odd mass.
A second nitrogen in the molecular ion will cause it to have an even mass.
Thus, for a molecular ion to have an odd mass, it must have an odd number of nitrogens.

c.   An even-mass molecular ion either has no nitrogens or has an even number of nitrogens.

6. $\text{number of carbon atoms} = \dfrac{\text{relative intensity of M} + 1 \text{ peak}}{0.011 \times \text{relative intensity of M peak}}$

$$= \frac{3.81}{0.011 \times 43.27} = \frac{3.81}{0.467} = 8 \text{ carbon atoms}$$

7.   The ratio is 1:2:1.
To get the M peak, both Br atoms must be ^{79}Br. To get the M + 4 peak, they both must be ^{81}Br. To get the M + 2 peak the first Br atom can be ^{79}Br and the second ^{81}Br, or the first can be ^{81}Br and the second ^{81}Br. So the relative intensity of the M + 2 peak will be twice that of the others.

$\underline{M}$	$\underline{M + 2}$	$\underline{M + 4}$
^{79}Br ^{79}Br	^{79}Br ^{81}Br ^{81}Br ^{79}Br	^{81}Br ^{81}Br

8.   The calculated exact masses show that only C_6H_{14} has an exact mass of 86.10955.

C_6H_{14} $6(12.00000) = 72.00000$ $C_4H_6O_2$ $4(12.00000) = 48.00000$
            $14(1.007825) = \underline{14.10955}$ $6(1.007825) = 6.04695$
                         86.10955 $2(15.9949) = \underline{31.9898}$
                                                                                86.03675

$C_4H_{10}N_2$ $4(12.00000) = 48.00000$
            $10(1.007825) = 10.07825$
             $2(14.0031) = \underline{28.0064}$
                         86.08465

9.   a.   A low resolution spectrometer cannot distinguish between them because they both have the same nominal molecular mass (29).

b.   A high resolution spectrometer can distinguish between these ions, because they have different exact molecular masses; one has an exact molecular mass of 29.039125 and the other an exact molecular mass of 29.002725.

10.   Because the compound contains chlorine, the M + 2 peak is one-third the size of the M peak. Breaking the weak C—Cl bond and therefore losing a chlorine atom from either the M + 2 peak (80 − 37) or the M peak (78 − 35) gives the base peak with $m/z = 43$ ($CH_3CH_2CH_2^+$). It is more difficult to predict the other peaks.

Mass spectrum of 1-chloropropane

11.   The dominant fragmentation pathway in each case is loss of an alkyl radical via α-cleavage to form a cation in which the positive charge is shared by carbon and oxygen.

The base peak at $m/z = 73$, due to loss of a methyl radical (88 − 15), indicates that Figure 13.7a is the mass spectrum of **2-methoxy-2-methylpropane**.

The base peak at $m/z = 59$ (88 − 29), due to loss of an ethyl radical, indicates that Figure 13.7b is the mass spectrum of **2-methoxybutane**.

The base peak at $m/z = 45$ (88 − 43), due to loss of a propyl radical, indicates that Figure 13.7c is the mass spectrum of **1-methoxybutane**.

$$CH_3CCH_3 \atop \overset{\displaystyle CH_3}{\underset{\displaystyle OCH_3}{|}}$$

CH₃CH₂CHCH₃ with OCH₃ below, and CH₃CH₂CH₂CH₂OCH₃

2-methoxy-2-methylpropane   2-methoxybutane   1-methoxybutane

12. $R-CH_2-\overset{..}{\underset{..}{O}}H \xrightarrow{-e^-} R \curvearrowright CH_2 \curvearrowright \overset{+}{\underset{..}{O}}H \longrightarrow R\cdot + CH_2=\overset{+}{\underset{..}{O}}H$
                                                                                        $m/z = 31$

13.   The molecular ions with $m/z = 86$ indicate that both ketones have the molecular formula $C_5H_{10}O$. Figure 13.9a shows a base peak at $m/z = 43$ for loss of a propyl (or isopropyl) radical (86 − 43), indicating that it is the mass spectrum of either 2-pentanone or 3-methyl-2-butanone, since each of these has a propyl or isopropyl group. The fact that the spectrum has a peak at $m/z = 58$, indicating loss of ethene (86 − 28), indicates that the compound has a γ-hydrogen that enables it to undergo a McLafferty rearrangement. Therefore, the ketone must be 2-pentanone, since 3-methyl-2-butanone does not have a γ-hydrogen.

The fact that Figure 13.9b shows a base peak at $m/z = 57$ for loss of an ethyl radical (86 − 29), indicates that it is the mass spectrum of 3-pentanone.

14. All three ketones will have a molecular ion with $m/z = 86$.

3-pentanone   2-pentanone   3-methyl-2-butanone

$86 - 29 = 57$ $86 - 41 = 43$ $86 - 41 = 43$

3-Pentanone will have a base peak at $m/z = 57$, whereas the other two ketones will have base peaks at $m/z = 43$.

2-Pentanone will have a peak at $m/z = 58$ due to a McLafferty rearrangement.

3-Methyl-2-butanone does not have any γ-hydrogens. Therefore, it cannot undergo a McLafferty rearrangement so it will not have a peak at $m/z = 58$.

15.   a. $CH_3CH_2CH_2CH_2CH_2\ddot{O}H$

$\downarrow -e^-$

$CH_3CH_2CH_2CH_2{-}CH_2{-}\overset{+}{\ddot{O}H}$ $\xrightarrow{\alpha\text{-cleavage}}$ $CH_3CH_2CH_2\dot{C}H_2 \ + \ CH_2{=}\overset{+}{\ddot{O}H}$
$m/z = 88$ $m/z = 31$

$CH_3CH_2\dot{C}HCH_2CH_2{-}\overset{+}{\ddot{O}H}$ $\longrightarrow$ $CH_3CH_2\dot{C}HCH_2\overset{+}{C}H_2 \ + \ H_2O$
                                                                          $m/z = 70$

b. $CH_3CH_2CHCH_2CH_2CH_2CH_3$
$\qquad\quad | $
$\qquad\quad :\ddot{O}H$

$\downarrow -e^-$

$CH_3CH_2{-}CHCH_2CH_2CH_2CH_3$ $\xrightarrow{\alpha\text{-cleavage}}$ $CH_3CH_2CH_2CH_2CH{=}\overset{+}{\ddot{O}H} \ + \ CH_3\dot{C}H_2$
$\qquad\quad |$
$\qquad\quad \underset{+\ \cdot\cdot}{\ddot{O}H}$
$\qquad m/z = 116$ $m/z = 87$

$CH_3CH_2CH{-}CH_2CH_2CH_2CH_3$ $\xrightarrow{\alpha\text{-cleavage}}$ $CH_3CH_2CH{=}\overset{+}{\ddot{O}H} \ + \ CH_3CH_2CH_2\dot{C}H_2$
$\qquad\quad |$
$\qquad\quad \underset{+\ \cdot\cdot}{\ddot{O}H}$
                                                                        $m/z = 59$

$CH_2CH_2CHCH_2CH_2CH_2CH_3$ $\longrightarrow$ $\dot{C}H_2CH_2\overset{+}{C}HCH_2CH_2CH_2CH_3 \ + \ H_2O$
$| \qquad\qquad |$
$H \qquad\quad +\ \cdot\cdot\ddot{O}H$ $m/z = 98$

$CH_3CH_2CHCH_2CHCH_2CH_3$ $\longrightarrow$ $CH_3CH_2\overset{+}{C}HCH_2\dot{C}HCH_2CH_3 \ + \ H_2O$
$\qquad |\qquad\quad |$
$\qquad \underset{+}{\ddot{O}H}\ \ H$ $m/z = 98$

c. $CH_3CH_2-\overset{..}{\overset{..}{O}}-\overset{\overset{\displaystyle CH_2CH_3}{|}}{\underset{\underset{\displaystyle CH_3}{|}}{C}}CH_2CH_2CH_3$

$\downarrow -e^-$

$CH_3CH_2-\overset{+}{\overset{.}{\overset{..}{O}}}-\overset{\overset{\displaystyle CH_2CH_3}{|}}{\underset{\underset{\displaystyle CH_3}{|}}{C}}CH_2CH_2CH_3 \longrightarrow CH_3CH_2-\overset{..}{\overset{..}{O}}: \quad + \quad \overset{\overset{\displaystyle CH_2CH_3}{|}}{\underset{\underset{\displaystyle CH_3}{|}}{{}^+C}}CH_2CH_2CH_3$

m/z = 144   *m/z* = 99

$CH_3CH_2-\overset{+}{\overset{.}{\overset{..}{O}}}-\overset{\overset{\displaystyle CH_2CH_3}{|}}{\underset{\underset{\displaystyle CH_3}{|}}{C}}CH_2CH_2CH_3 \xrightarrow{\alpha\text{-cleavage}} CH_3CH_2-\overset{+}{\overset{..}{O}}=\overset{\overset{\displaystyle CH_2CH_3}{|}}{C}CH_2CH_2CH_3 \quad + \quad \overset{.}{C}H_3$

m/z = 129

$CH_3CH_2-\overset{+}{\overset{.}{\overset{..}{O}}}-\overset{\overset{\displaystyle CH_2CH_3}{|}}{\underset{\underset{\displaystyle CH_3}{|}}{C}}CH_2CH_2CH_3 \xrightarrow{\alpha\text{-cleavage}} CH_3CH_2-\overset{+}{\overset{..}{O}}=\underset{\underset{\displaystyle CH_3}{|}}{C}CH_2CH_2CH_3 \quad + \quad CH_3\overset{.}{C}H_2$

m/z = 115

$CH_3CH_2-\overset{+}{\overset{.}{\overset{..}{O}}}-\overset{\overset{\displaystyle CH_2CH_3}{|}}{\underset{\underset{\displaystyle CH_3}{|}}{C}}CH_2CH_2CH_3 \xrightarrow{\alpha\text{-cleavage}} CH_3CH_2-\overset{+}{\overset{..}{O}}=\underset{\underset{\displaystyle CH_3}{|}}{C}CH_2CH_3 \quad + \quad CH_3CH_2\overset{.}{C}H_2$

m/z = 101

$CH_3-CH_2-\overset{+}{\overset{.}{\overset{..}{O}}}-\overset{\overset{\displaystyle CH_2CH_3}{|}}{\underset{\underset{\displaystyle CH_3}{|}}{C}}-CH_2CH_2CH_3 \xrightarrow{\alpha\text{-cleavage}} CH_2=\overset{+}{\overset{..}{O}}-\overset{\overset{\displaystyle CH_2CH_3}{|}}{\underset{\underset{\displaystyle CH_3}{|}}{C}}CH_2CH_2CH_3 \quad + \quad \overset{.}{C}H_3$

m/z = 129

d.

$$\overset{\displaystyle \ddot{O}:}{\underset{\displaystyle \parallel}{CH_3CCH_2CH_2CH_2CH_3}}$$

$\downarrow -e^-$

$$\overset{\displaystyle \overset{+}{\dot{\ddot{O}}}:}{\underset{\displaystyle \parallel}{CH_3-CCH_2CH_2CH_2CH_3}} \xrightarrow{\alpha\text{-cleavage}} CH_3CH_2CH_2CH_2C\equiv\overset{+}{\ddot{O}} \ + \ \dot{C}H_3$$

$m/z = 100$ $m/z = 85$

$$\overset{\displaystyle \overset{+}{\dot{\ddot{O}}}}{\underset{\displaystyle \parallel}{CH_3C-CH_2CH_2CH_2CH_3}} \xrightarrow{\alpha\text{-cleavage}} CH_3C\equiv\overset{+}{\ddot{O}} \ + \ CH_3CH_2CH_2\dot{C}H_2$$

$m/z = 43$

$$\overset{\displaystyle \overset{+}{\dot{\ddot{O}}}}{\underset{\displaystyle \parallel}{CH_3CCH_2-CH_2-CHCH_3}} \overset{\displaystyle H}{} \xrightarrow[\text{rearrangement}]{\text{McLafferty}} \overset{\displaystyle \overset{+}{\ddot{O}H}}{\underset{\displaystyle \parallel}{CH_3C\dot{C}H_2}} \ + \ CH_2{=}CHCH_3$$

$m/z = 58$

e. $CH_3CH_2CH\ddot{C}\ddot{l}: \xrightarrow{-e^-} CH_3CH_2CH-\overset{+}{\underset{\cdot\cdot}{\dot{C}l}}: \longrightarrow CH_3CH_2\overset{+}{CH} \ + \ :\dot{\ddot{C}}l:$

$\underset{CH_3}{}$ $\underset{CH_3}{}$ $\underset{CH_3}{}$

$m/z = 92$ and 94 $m/z = 57$

$CH_3CH_2CH-\overset{+}{\underset{\cdot\cdot}{\dot{C}l}}: \xrightarrow{\alpha\text{-cleavage}} CH_3CH_2CH{=}\overset{+}{\underset{\cdot\cdot}{C}l}: \ + \ \dot{C}H_3$

$\underset{CH_3}{}$

$m/z = 77$ and 79

$CH_3\dot{C}H_2-CH-\overset{+}{\underset{\cdot\cdot}{\dot{C}l}}: \xrightarrow{\alpha\text{-cleavage}} CH_3CH{=}\overset{+}{\underset{\cdot\cdot}{C}l}: \ + \ CH_3\dot{C}H_2$

$\underset{CH_3}{}$

$m/z = 63$ and 65

f. $\overset{\displaystyle CH_3}{\underset{\displaystyle CH_3}{CH_3C-\overset{\cdot\cdot}{\underset{\cdot\cdot}{B}}r}}: \xrightarrow{-e^-} \overset{\displaystyle CH_3}{\underset{\displaystyle CH_3}{CH_3C-\overset{+}{\underset{\cdot\cdot}{\dot{B}}}r}}: \longrightarrow \overset{\displaystyle CH_3}{\underset{\displaystyle CH_3}{CH_3\overset{+}{C}}} \ + \ :\overset{\cdot\cdot}{\underset{\cdot\cdot}{B}}r:$

$m/z = 136$ and 138 $m/z = 57$

16. When (Z)-2-pentene reacts with water and an acid catalyst, 3-pentanol and 2-pentanol are formed. Both alcohols have a molecular weight of 88. (Notice that the first step in solving this problem is to use chemical knowledge to identify the products.) The absence of a molecular ion peak is consistent with the fact that the compounds are alcohols.

$$H_3C \underset{H}{\overset{}{\diagdown}} C = C \underset{H}{\overset{CH_2CH_3}{\diagup}} + H_2O \xrightarrow{H_2SO_4} CH_3CH_2\underset{\underset{OH}{|}}{CH}CH_2CH_3 + CH_3\underset{\underset{OH}{|}}{CH}CH_2CH_2CH_3$$

(Z)-2-pentene   3-pentanol   2-pentanol

Figure 13.10a shows a base peak at $m/z = 59$ due to loss of an ethyl radical (88 − 29), indicating that it is the spectrum of 3-pentanol.

Figure 13.10b shows a base peak at $m/z = 45$ due to loss of a propyl radical (88 − 43), indicating that it is the spectrum of 2-pentanol.

17. The wavelength is the distance from the top of one wave to the top of the next wave. We see that **a** has a longer wavelength than **b**.
Infrared radiation has longer wavelengths than ultraviolet light because infrared radiation is lower in energy. Therefore, **a** depicts infrared radiation and **b** depicts ultraviolet light.

18. **a.** 2000 cm^{-1} (The larger the wavenumber, the higher the energy.)

   b. 8 nm (The shorter the wavelength, the higher the energy.)

   c. 2 nm, because 2 nm = 5000 cm^{-1} (5000 cm^{-1} is a larger wavenumber than 200 cm^{-1}.)

19. **a.** $\tilde{v}$ (cm^{-1}) = $\dfrac{10,000}{4 \text{ (nm)}}$ = 2500 cm^{-1}

   b. $\tilde{v}$ (cm^{-1}) = $\dfrac{10,000}{\lambda \text{ (nm)}}$ = 200 cm^{-1}

   λ = 50 nm

20. **a.** **1.** C≡C stretch   A triple bond is stronger than a double bond, so it takes more energy to stretch a triple bond.

   2. C—H stretch   It requires more energy to stretch a given bond than to bend it.

   3. C=N stretch   A double bond is stronger than a single bond, so it takes more energy to stretch a double bond.

   4. C=O stretch   A double bond is stronger than a single bond, so it takes more energy to stretch a double bond.

   b. **1.** C—O   Vibrations of lighter atoms occur at larger wavenumbers.

   2. C—C   Vibrations of lighter atoms occur at larger wavenumbers.

21.   **a.**   The carbon-oxygen stretch of phenol, because it has partial double-bond character as a result of electron delocalization.

   b.   The carbon-oxygen double-bond stretch of a ketone, because it has more double-bond character. The double bond character of the carbonyl group of the amide is reduced by electron delocalization.

   c.   The C—N stretch of aniline because it has partial double-bond character.

22.   A carbonyl group bonded to an sp^3 carbon will exhibit an absorption band at a larger wavenumber because a carbonyl group bonded to an sp^2 carbon of an alkene will have greater single-bond character as a result of electron delocalization.

23.   The C—O bond of the alcohol is a pure single bond. In contrast, the C—O bond of the carboxylic acid has double bond character, so it is a stronger bond and therefore takes more energy to stretch.

24.   **a.**   Alkyl groups are more electron-donating than a hydrogen because of hyperconjugation. Therefore, the carbonyl group bonded to two relatively electron-withdrawing hydrogens has the largest wavenumber for its C=O absorption band, while the carbonyl group bonded to two relatively electron-donating alkyl groups has the lowest wavenumber.

   b.   The C=O absorption band of an ester occurs at the largest wavenumber because the carbonyl group of an ester has the most double-bond character, since the predominant effect of the ester oxygen atom is inductive electron withdrawal.

The $C=O$ absorption band of an amide occurs at the smallest wavenumber because the carbonyl group of an amide has the least double-bond character, since the predominant effect of the amide nitrogen atom is electron donation by resonance.

25.   Ethanol dissolved in carbon disulfide will show the oxygen-hydrogen stretch at a larger wavenumber, because there is extensive hydrogen bonding in the undiluted alcohol and an oxygen-hydrogen bond is easier to stretch if it is hydrogen bonded.

26.   The absorption band at 1100 cm^{-1} would be less intense if it were due to a C—N bond, because a smaller change in dipole moment is associated with the stretch of a C—N bond compared to the change in dipole moment associated with the stretch of a C—O bond.

27.   **a.**   The absorption band at 1700 cm^{-1} indicates that the compound has a carbonyl group.

The absence of an absorption band at 3300 cm^{-1} indicates that the compound is not a carboxylic acid.

The absence of an absorption band at 2700 cm^{-1} indicates that the compound is not an aldehyde.

The absence of an absorption band at 1100 cm^{-1} indicates that the compound is not an ester or an amide. The compound, therefore, must be a **ketone**.

b.   The absence of an absorption band at 3400 cm^{-1} indicates that the compound does not have an N—H bond.

The absence of a carbonyl absorption band between 1700 cm^{-1} and 1600 cm^{-1} indicates that the compound is not an amide. The compound, therefore, must be **a tertiary amine**.

28.   **a.**   An aldehyde would show absorption bands at 2820 and 2720 cm^{-1}. A ketone would not have these absorption bands.

b.   An open-chain ketone would have a methyl substituent and therefore an absorption band at $1385–1365 \text{ cm}^{-1}$ that a cyclic ketone would not have.

c.   Cyclohexene would show an sp^3 C—H stretch slightly to the right of 3000 cm^{-1}. Benzene would not show an absorption band in this region.

d.   The cis isomer would show a carbon-hydrogen bending vibration at $730–675 \text{ cm}^{-1}$, whereas the trans isomer would show a carbon-hydrogen bending vibration at $960–980 \text{ cm}^{-1}$.

e.   Cyclohexene would show a carbon-carbon double-bond stretching vibration at $1680–1600 \text{ cm}^{-1}$ and an sp^2 carbon-hydrogen stretching vibration at $3100–3020 \text{ cm}^{-1}$. Cyclohexane would not show these absorption bands.

f.   A primary amine would show a nitrogen-hydrogen stretch at $3500–3300 \text{ cm}^{-1}$, and a tertiary amine would not have this absorption band.

29.   **a.**   An absorption band at 1150–1050 cm^{-1} due to a C—O stretching vibration would be present for the ether and absent for the alkane.

b.   An absorption band at 3300–2500 cm^{-1} due to an O—H stretching vibration would be present for the carboxylic acid and absent for the ester.

c.   An absorption band at 1385–1365 cm^{-1} due to C—H bending vibrations of a methyl group would be present for methylcyclohexane and absent for cyclohexane.

d.   Only the terminal alkyne would show an absorption band at 3300 cm^{-1} due to an sp C—H stretching vibration.

e.   An absorption band at 1780–1650 cm^{-1} due to a C=O stretching vibration would be present for the carboxylic acid and absent for the alcohol.

f.   An absorption band at 2960–2850 cm^{-1} due to a sp^3 C—H stretching vibration would be present for the compound with the methyl group and absent for benzene.

30.   2-butyne, H_2, Cl_2, ethene

31.   The absorption bands in the vicinity of 3000 cm^{-1} indicate that the compound has hydrogens attached to both sp^2 and sp^3 carbons. The absence of absorptions at 1600–1800 cm^{-1} and the absence of broad absorptions between 2500–3650 cm^{-1} rules out compounds containing C=O, N—H, and O—H groups. The lack of absorption at 1600 cm^{-1} and 1500 cm^{-1} indicates that the compound does not have a benzene ring. The sp^2 hydrogens, therefore, must be those of an **alkene**.

The lack of significant absorption at 1600 cm^{-1} indicates that the compound must be an alkene with a relatively small (if any) dipole moment change when the vibration occurs. The absorption band at 965 cm^{-1} indicates that the compound is a ***trans*-alkene**.

The molecular ion with $m/z = 84$ suggests that the compound has a molecular formula of C_6H_{12}. The base peak with $m/z = 55$ indicates that the molecular ion most easily loses an ethyl radical ($84 - 29 = 55$), but it can also lose a methyl group ($84 - 15 = 69$) or a propyl radical ($84 - 43 = 41$). Because it most easily loses an ethyl radical, the ethyl group must be attached to an allylic carbon. The compound, therefore, is ***trans*-2-hexene**.

trans-2-hexene

32.   The absorption band at ~1700 cm^{-1} indicates that the compound has a carbonyl group, and the absorption band at ~1600 cm^{-1} indicates that the compound has a carbon-carbon double bond. The absorption bands in the vicinity of 3000 cm^{-1} indicate that the compound has hydrogens attached to both sp^2 and sp^3 carbons. The absorption band at ~1380 cm^{-1} indicates that the compound has a methyl group. Because the compound has only four carbons and one oxygen, it must be **methyl vinyl ketone**. Notice that the carbonyl stretch is at a lower frequency (1700 cm^{-1}) than expected for a ketone (1720 cm^{-1}), because the carbonyl group has partial single-bond character due to electron delocalization.

33. Diethyl ether does not have a π bond, so it does not have a π^* antibonding molecular orbital. Therefore, it cannot have an $n \rightarrow \pi^*$ transition.

34. $A = c\,l\,\varepsilon$

$$c = \frac{A}{l\varepsilon}$$

$$c = \frac{0.52}{12,600} = 4.1 \times 10^{-5}\,M$$

35. $A = c\,l\,\varepsilon$

$$c = \frac{A}{l\varepsilon}$$

$$\varepsilon = \frac{0.40}{4.0 \times 10^{-5}} = 10,000\;M^{-1}cm^{-1}$$

36. **a.**

b.

37. **a.** Blue results from absorption of light that has a longer wavelength than that of the light that when absorbed produces purple. The compound on the right has two $N(CH_3)_2$ auxochromes that will cause it to absorb at a longer wavelength than the compound on the left that has only one $N(CH_3)_2$ auxochrome. Therefore, the compound on the right is the blue compound.

b. They will be the same color at pH = 3 because the $N(CH_3)_2$ groups will be protonated and, therefore, will not possess the lone pair that causes the compound to absorb light of a longer wavelength.

38. NADH is formed as a product; it absorbs light at 340 nm. Therefore, the rate of the oxidation reaction can be determined by monitoring the increase in absorbance at 340 nm as a function of time.

39. The Henderson-Hasselbalch equation (Section 1.25) shows us that when the pH equals the pK_a, the concentration of the species in the acidic form is the same as the concentration of the species in the basic form.

From the data given, we see that the absorbance of the acid is 0. We also see that the absorbance ceases to change with increasing pH after the absorbance reaches 1.60. That means that all of the compound is in the basic form when the absorbance is 1.60. Therefore, when the absorbance is half of 1.60 (or 0.80), half of the compound is in the basic form and half is in the acidic form; in other words, the concentration in the acid form is the same as the concentration in the basic form. We see that the absorbance is 0.80 at pH = 5.0. Therefore, the pK_a of the compound is 5.0.

40.   The molecular ion peak for these compounds is $m/z = 86$; the peak at $m/z = 57$ is due to loss of an ethyl radical ($86 - 29$), and the peak at $m/z = 71$ is due to loss of a methyl radical ($86 - 15$).

    a.   3-Methylpentane will be more apt to lose an ethyl radical (forming a secondary carbocation and a primary radical) than a methyl radical (forming a secondary carbocation and a methyl radical), so the peak at $m/z = 57$ would be more intense than the peak at $m/z = 71$.

3-methylpentane

    b.   2-Methylpentane has two pathways to lose a methyl radical (forming a secondary carbocation and a methyl radical in each pathway), and it cannot form a secondary carbocation by losing an ethyl radical. (Loss of an ethyl radical would form a primary carbocation and a primary radical.) Therefore, it will be more apt to lose a methyl radical than an ethyl radical, so the peak at $m/z = 71$ would be more intense than the peak at $m/z = 57$.

2-methylpentane

41.   **1.**   the change in the dipole moment when the bond stretches or bends
     2.   the number of bonds that give rise to the absorption band
     3.   the concentration of the sample

42.   **a.**   An absorption band at $\sim$1250 cm^{-1} due to a C—O stretching vibration would be present for the ester and absent for the ketone.

    b.   An absorption band at 720 cm^{-1} due to in-phase rocking of the five adjacent methylene groups would be present for heptane and absent for methylcyclohexane.

    c.   An absorption band at 3650–3200 cm^{-1} due to an O—H stretching vibration would be present for the alcohol and absent for the ether.

    d.   An absorption band at 3500–3300 cm^{-1} due to an N—H stretching vibration would be present for the amide and absent for the ester.

    e.   The secondary alcohol would have an absorption band at 1385–1365 cm^{-1} for the methyl group. The primary alcohol does not have a methyl group, so it would not have this absorption band.

    f.   The trans isomer would have a C—H bending absorption band at 980–960 cm^{-1}, whereas the cis isomer would have the absorption band at 730–675 cm^{-1}. In addition, a weak C=C absorption band at 1680–1600 cm^{-1} would be present for the cis isomer and absent for the trans isomer, since only the trans isomer is symmetrical.

    g.   The C=O absorption band will be at a larger wavenumber for the ester (1740 cm^{-1}) than for the ketone (1720 cm^{-1}).

    h.   The C=O absorption band will be at a larger wavenumber for the β,γ-unsaturated ketone (1720 cm^{-1}) than for the α,β-unsaturated ketone (1680 cm^{-1}), since the double bonds in the latter are conjugated.

    i.   The alkene would have absorption bands at 1680–1600 cm^{-1} due to a C=C stretching vibration and at 3100–3020 cm^{-1} due to an sp^2 C—H stretching vibration that the alkyne would not have. The alkyne would have an absorption band at 2260–2100 cm^{-1} that the alkene would not have.

    j.   An absorption band at $\sim$2820 and $\sim$2720 cm^{-1} due to the aldehyde C—H stretching vibration would be present for the aldehyde and absent for the ketone.

k. Absorption bands at 1600 cm^{-1} and 1500 cm^{-1} (aromatic ring stretching vibrations) and at 3100–3020 cm^{-1} (sp^2 C—H stretching vibration) would be present for the compound with the benzene ring and absent for the compound with the cyclohexane ring. An absorption band at 2960–2850 cm^{-1} due to an sp^3 C—H stretching vibration would be present for the compound with the cyclohexane ring and absent for the compound with the benzene ring.

l. Absorption bands at 990 cm^{-1} and 910 cm^{-1} due to an sp^2 C—H bending vibration would be present for the terminal alkene and absent for the internal alkene.

43. **a.** The absorption at 236 nm is due to a $\pi \rightarrow \pi^*$ transition, and the absorption at 314 nm is due to an $n \rightarrow \pi^*$ transition.

b. The one at 236 nm (the $\pi \rightarrow \pi^*$ transition) shows the greater absorbance.

44. **a.** If the reaction had occurred, the intensity of the absorption bands at ~1700 cm^{-1} (due to the carbonyl group) and ~2700 cm^{-1} (due to the aldehyde C—H bond) of the reactant would have decreased. If all the aldehyde had reacted, these absorption bands would have disappeared.

b. If all the NH_2NH_2 had been removed, there would be no N—H absorption at ~3400 cm^{-1}.

45. **a.**   and

The λ_{max} will be at a longer wavelength, because there are 3 conjugated double bonds.

b. CH_2=$CHCH$=$CHCH$=CH_2 and CH_2=$CHCH$=$CHCCH_3$ (with O double-bonded to the final C)

It will show 1 absorption band:
a $\pi \rightarrow \pi^*$ transition

It will show 2 absorption bands:
a $\pi \rightarrow \pi^*$ transition
and
an $n \rightarrow \pi^*$ transition

c.   and

The λ_{max} will be at a longer wavelength because the carbonyl group is conjugated with the benzene ring.

d.   and

The λ_{max} of the phenolate ion is at a longer wavelength than the λ_{max} of phenol. Since the pK_a of phenol is ~10, the λ_{max} will be at a longer wavelength at pH = 11 than at pH = 7.

The λ_{max} is pH independent from pH 0–14.

46. The compound would have an M and an M + 4 peak of equal intensity, and there would be an M + 2 peak that is twice the intensity of the M peak.

47. If the force constants are approximately the same, the lighter atoms absorb at higher frequencies.

$$C-C \quad > \quad C-N \quad > \quad C-O$$

48.

3600 cm^{-1}	3000	1800	1400	1000
OH 3300–3000	sp^3 CH 2950	C=O 1700	C—O 1250–1050	
NH 3600–3200	$\overset{\displaystyle O}{\overset{\|}{RCH}}$ 2700	C=C 1600	C—N 1230–1030	
sp CH 3300	C≡C 2100	⬡ 1500		
sp^2 CH 3050	C≡N 2250			

49. The molecular weight of each of the alcohols is 158. The peak at $m/z = 140$ is due to loss of water ($158 - 18$); each of the alcohols could show such a peak. The peaks at $m/z = 87$, 115, and 143 are due to loss of a group with five carbons (C_5H_{11}), a group with three carbons (C_3H_7), and a methyl group, respectively. Only 2,2,4-trimethyl-4-heptanol would lose all three groups via α-cleavage.

$$\underset{\overset{|}{OH}\;\;\;\overset{|}{CH_3}}{CH_3CH_2CH_2\overset{\overset{\displaystyle CH_3}{|}}{C}CH_2\overset{\overset{\displaystyle CH_3}{|}}{C}CH_3}$$

2,2,4-trimethyl-4-heptanol

$$\underset{}{CH_3\overset{\overset{\displaystyle CH_3}{|}}{C}HCH_2CH_2\overset{\overset{\displaystyle CH_3}{|}}{C}HCH_2CH_2CH_2OH}$$

4,7-dimethyl-1-octanol

$$CH_3CH_2\overset{\overset{\displaystyle CH_3}{|}}{C}HCH_2\underset{\underset{\displaystyle OH}{|}}{C}HCH_2\overset{\overset{\displaystyle CH_3}{|}}{C}HCH_3$$

2,6-dimethyl-4-octanol

50. $CH_2=CHCH_2CH_2CH=CH_2$ $CH_3CH=CHCH=CHCH_3$

1,5-hexadiene   2,4-hexadiene

One way to distinguish the two compounds is by the presence or absence of an absorption band at ~1370 cm^{-1} due to the methyl group that 2,4-hexadiene has but that 1,5-hexadiene does not have. In addition, 2,4-hexadiene has conjugated double bonds; therefore its double bonds have some single-bond character due to electron delocalization. Consequently, they are easier to stretch than the isolated double bonds of 1,5-hexadiene. Thus, the carbon-carbon double bond stretch of 2,4-hexadiene will be at a smaller wavenumber than the carbon-carbon double-bond stretch of 1,5-hexadiene.

51. The fact that the abundance of the M + 2 peak is 30% of the abundance of the M peak indicates that the compound has one chlorine atom. The peak at $m/z = 77$ is due to loss of the chlorine atom ($112 - 35 = 77$). The fact that the peak at $m/z = 77$ does not fragment, indicates it is a phenyl cation. Therefore, the compound is chlorobenzene.

52. A hydrocarbon with eight carbons (C_nH_{2n+2}) and no rings or π bonds has a molecular formula of C_8H_{18} and a molecular weight of 114. Since the molecular ion is 112, the compound must be a hydrocarbon with eight carbons and one double bond or one ring. Possible compounds are cyclooctane, methylcycloheptane, all the dimethylcyclohexanes, ethylcyclohexane, propylcyclopentane, isopropylcyclopentane, all the ethylmethyl cyclopentanes, all the trimethylcyclopentanes, all octenes, all methylheptenes, all ethylhexenes, etc.

53. **a.** The absorption band at $\sim$2100 cm^{-1} indicates a carbon-carbon triple bond, and the absorption band at $\sim$3300 cm^{-1} indicates a hydrogen bonded to an *sp* carbon.

$$CH_3CH_2CH_2CH_2C{\equiv}CH$$

b. The absence of an absorption band at $\sim$2700 cm^{-1} indicates that the compound is not an aldehyde, and the absence of a broad absorption band in the vicinity of 3000 cm^{-1} indicates that the compound is not a carboxylic acid. The ester and the ketone can be distinguished by the absorption band at $\sim$1200 cm^{-1} that indicates the carbon-oxygen single bond of an ester.

c. The absorption band at $\sim$1360 cm^{-1} indicates the presence of a methyl group.

54. The C$=$O absorption band of the three compounds decreases in the following order.

The first compound has the C$=$O absorption band at the largest wavenumber, because a lone pair on the ring oxygen can be delocalized onto two different atoms; thus it is less apt than the lone pair in the other compounds to be delocalized onto the carbonyl oxygen atom.

The third compound has the $C=O$ absorption band at the smallest wavenumber because its carbonyl group has more single bond character due to contributions from two other resonance contributors.

55. 4-Methyl-2-pentanone would show peaks at $m/z = 85$ (loss of a methyl group) and at $m/z = 43$ (loss of an isobutyl group) and a peak at $m/z = 58$ due to a McLafferty rearrangement.

4-methyl-2-pentanone

2-Methyl-3-pentanone would show peaks at $m/z = 71$ (loss of an ethyl group) and at $m/z = 57$ (loss of an isopropyl group). Because it does not have a γ-hydrogen, it cannot undergo a McLafferty rearrangement.

2-methyl-3-pentanone

56. The absorption bands at ~ 2700 cm^{-1} for the aldehydic hydrogen and at ~ 1380 cm^{-1} for the methyl group would distinguish the compounds.

    A would have the band at ~ 2700 cm^{-1} but not the one at ~ 1380 cm^{-1}.
    B would have the band at ~ 1380 cm^{-1} but not the one at ~ 2700 cm^{-1}.
    C would have the both the band at ~ 2700 cm^{-1} and the one at ~ 1380 cm^{-1}.

57. 1-Hexyne will show absorption bands at ~ 3300 cm^{-1} for a hydrogen bonded to an sp carbon and at ~ 2100 cm^{-1} for the triple bond.

$$CH_3CH_2CH_2CH_2C \equiv CH$$
1-hexyne

2-Hexyne will show the absorption band at ~ 2100 cm^{-1} but not the one at ~ 3300 cm^{-1}.

$$CH_3CH_2CH_2C \equiv CCH_3$$
2-hexyne

3-Hexyne will show neither the absorption band at ~ 3300 cm^{-1} nor the one at ~ 2100 cm^{-1} (there is no change in dipole moment when the $C \equiv C$ stretches).

$$CH_2CH_2C \equiv CCH_2CH_3$$
3-hexyne

58. **a.** The broad absorption band at ~3300 cm^{-1} is characteristic of the oxygen-hydrogen stretch of an alcohol, and the absence of absorption bands at ~1600 cm^{-1} and ~3100 cm^{-1} indicates that it is not the alcohol with a carbon-carbon double bond.

b. The absorption band at ~1685 cm^{-1} indicates a carbon-oxygen double bond. The absence of a strong and broad absorption band at ~3000 cm^{-1} rules out the carboxylic acid, and the absence of an absorption band at ~2700 cm^{-1} rules out the aldehyde. Thus, it must be one of the ketones. The conjugated ketone would be expected to show a C=O stretch at ~1685 cm^{-1}, whereas the isolated ketone would show a C=O stretch at ~1720 cm^{-1}. Thus, the compound is the conjugated ketone.

c. The absorption band at ~1700 cm^{-1} indicates a carbon-oxygen double bond. The absence of an absorption band at ~1600 cm^{-1} rules out the ketones with the benzene or cyclohexene rings. The absence of absorption bands at ~2100 cm^{-1} and ~3300 cm^{-1} rules out the ketone with the carbon-carbon triple bond. Thus, it must be 4-ethylcyclohexanone.

59. Since only benzene absorbs light of 260 nm, the concentration of benzene can be determined by measuring the absorbance (A) at 260 nm, using the Beer-Lambert law, since the length of the light path of the cell (1.0 cm) is known and the molar absorptivity of benzene at 260 nm (ε) can be looked up (or can be determined from the absorbance at 260 nm given by a solution of benzene with a known concentration).

the Beer-Lambert law: $A = c\, l\, \varepsilon$

Therefore, the concentration of benzene $= \dfrac{A}{l\varepsilon}$

60. **a.** The absorption bands at 1720 cm^{-1} and ~2700 cm^{-1} indicate that the compound is an aldehyde (carbon-hydrogen stretch of an aldehyde hydrogen). The absence of an absorption band at ~1600 cm^{-1} rules out the aldehyde with the benzene ring. Thus, it must be the other aldehyde.

b. The absorption bands at ~3350 cm^{-1} and ~3200 cm^{-1} indicate that the compound is an amide (nitrogen-hydrogen stretch). The absence of an absorption band at ~3050 cm^{-1} indicates that the compound does not have hydrogens bonded to sp^2 carbons. Therefore, it is not the amide that has a benzene ring. Thus, it must be the other amide.

   c. The absence of absorption bands at ~1600 cm⁻¹ and ~1500 cm⁻¹ indicates that the compound does not have a benzene ring. Thus, it must be the ketone. This is confirmed by the absence of an absorption band at ~1380 cm⁻¹, indicating that the compound does not have a methyl group.

[structure: cyclohexanone]

61. **a.**

[structure: acetaldehyde] 2700 cm⁻¹

$$\underset{C}{\overset{O}{\|}}\quad 1700\ \text{cm}^{-1}$$

C=C   1600 cm⁻¹

c. C—O ~1050 cm⁻¹
     O—H 3600–3200 cm⁻¹ (broad)

e. *sp* CH 3300 cm⁻¹ (narrow)
     C≡C 2100 cm⁻¹

b.

$$\underset{C}{\overset{O}{\|}}\quad 1700\ \text{cm}^{-1}$$

[benzene ring] 1500 cm⁻¹, 1600 cm⁻¹

C—O ~1250 cm⁻¹, ~1050 cm⁻¹

d.

$$\underset{C}{\overset{O}{\|}}\quad 1700\ \text{cm}^{-1}$$

C—N ~1030 cm⁻¹

N—H 3500–3300 cm⁻¹

f.

$$\underset{C}{\overset{O}{\|}}\quad 1700\ \text{cm}^{-1}$$

[carboxylic acid structure, OH] ~3000 cm⁻¹ (broad)

C—O ~1250 cm⁻¹

62. Calculating the term in Hooke's law that depends on the masses of the atoms joined by a bond for a C—H bond and for a C—C bond shows why smaller atoms give rise to greater wavenumbers.

for a carbon-hydrogen bond

$$\frac{m_1 + m_2}{m_1 \times m_2} = \frac{12 + 1}{12 \times 1}$$

$$= \frac{13}{12}$$

$$= 1.08$$

for a carbon-carbon bond

$$\frac{m_1 + m_2}{m_1 \times m_2} = \frac{12 + 12}{12 \times 12}$$

$$= \frac{24}{144}$$

$$= 0.17$$

63. The absorption band at ~1740 cm⁻¹ indicates that the compound has a carbonyl group, and the absence of an absorption band at ~1380 cm⁻¹ indicates that it has no methyl groups. The absence of an absorption band at ~1600 cm⁻¹ indicates the compound does not have a carbon-carbon double bond, and the absence of an absorption band at ~3050 cm⁻¹ indicates the compound does not have hydrogens bonded to *sp²* carbons.

From the molecular formula you can deduce that the compound is **cyclopentanone**.

[structure: cyclopentanone]

64.   Before the addition of acid, the compound is colorless because the benzene rings are not conjugated with each other, since they are separated from each other by two single bonds. In the presence of acid, a carbocation is formed in which the three benzene rings are conjugated with each other. The conjugated carbocation is highly colored.

65.   The broad absorption band at ~3300 cm^{-1} indicates that the compound has an OH group. The absorption bands at ~2900 cm^{-1} indicate that the compound has hydrogens attached to an sp^3 carbon. The compound, therefore, is **benzyl alcohol**.

66.   The broad absorption band at ~3300 cm^{-1} indicates that the compound has an OH group. The absence of absorbance at ~1700 cm^{-1} shows that the compound does not have a carbonyl group. The absence of absorption at ~2950 cm^{-1} indicates the compound does not have any hydrogens bonded to sp^3 carbons. Therefore, the compound is phenol.

67.   The M peak at 100 indicates that the compound has a molecular weight of 100, and the relative intensities of the M and M + 1 peaks indicate that the compound has seven carbons.

$$\text{number of carbon atoms} = \frac{M + 1}{0.011 \times M} = \frac{2.10}{0.011 \times 27.32} = \frac{2.10}{0.301} = 7$$

When $n = 7$, $(C_nH_{2n+2}) = C_7H_{16}$. A compound with that molecular formula has a molecular weight of 100 in agreement with the M peak.

68. $$\tilde{v} = \frac{1}{2 \times 3.1416 \times 3 \times 10^{10} \text{ cm s}^{-1}} \sqrt{\frac{10 \times 10^5 \text{ g s}^{-2}\left(\dfrac{12}{6.02} + \dfrac{12}{6.02}\right) \times 10^{-23} \text{ g}}{\dfrac{12}{6.02} \times 10^{-23} \text{ g} \times \dfrac{12}{6.02} \times 10^{-23} \text{ g}}}$$

$$\tilde{v} = \frac{1}{18.85 \times 10^{10}} \sqrt{10.0 \times 10^{28}}$$

$$\tilde{v} = \frac{1}{18.85 \times 10^{10}} \times 3.16 \times 10^{14}$$

$$\tilde{v} = 1676 \text{ cm}^{-1}$$

69.   **a.** The IR spectrum indicates that the compound is an aliphatic ketone with at least one methyl group. The M peak at $m/z = 100$ indicates that the ketone is a hexanone. The peak at 43 $(100 - 57)$ for loss of a butyl radical and the peak at 85 for loss of a methyl radical $(100 - 15)$ suggest that the compound is **2-hexanone**.

$$
\begin{array}{c}
\overset{\displaystyle O}{\overset{\|}{\underset{}{C}}} \\
CH_3 \qquad CH_2CH_2CH_2CH_3
\end{array}
$$

This is confirmed by the peak at 58 for loss of propene $(100 - 42)$ as a result of a McLafferty rearrangement.

$$
CH_3\overset{+\,\cdot}{\overset{\|}{C}}CH_2{-}CH_2{-}\overset{H}{\underset{}{C}}HCH_3 \xrightarrow[\text{rearrangement}]{\text{McLafferty}} CH_3\overset{+}{\overset{\|}{C}}CH_2 \ + \ CH_2{=}CHCH_3
$$

$$m/z = 58$$

b. The equal heights of the M and M + 2 peaks at 162 and 164 indicate that the compound contains bromine. The peak at $m/z = 83$ $(162 - 79)$ is for the carbocation that is formed when the bromine atom is eliminated. The IR spectrum does not indicate the presence of any functional groups, and it shows that there are no methyl groups present. The m/z peak $= 83$ indicates a carbocation with a formula of C_6H_{11}. The fact that the compound does not contain a methyl group indicates that the compound is **bromocyclohexane**.

bromocyclohexane

c. The absorption bands at $\sim 1700 \ \text{cm}^{-1}$ and $\sim 2700 \ \text{cm}^{-1}$ indicate that the compound is an aldehyde. The molecular ion peak at $m/z = 72$ indicates that the aldehyde contains four carbons (C_4H_8O). The peak at $m/z = 44$ for loss of a group with molecular weight 28 indicates that ethene has been lost as a result of a McLafferty rearrangement.

$$
\begin{array}{c}
\overset{H}{\underset{}{}}\\
CH_2{-}CH_2{-}CH_2\overset{\cdot\,+}{\overset{\|}{C}}H \xrightarrow[\text{rearrangement}]{\text{McLafferty}} CH_2\overset{+}{\overset{\|}{C}}H \ + \ CH_2{=}CH_2
\end{array}
$$

$$m/z = 44$$

A McLafferty rearrangement can occur only if the aldehyde has a γ-hydrogen. The only four-carbon aldehyde that has a γ-hydrogen is **butanal**.

$$
\begin{array}{c}
\overset{\displaystyle O}{\overset{\|}{\underset{}{C}}} \\
CH_3CH_2CH_2 \qquad H
\end{array}
$$

70.   **a.** At pH < 4, the lone pair of the nitrogen of the N(CH$_3$)$_2$ group is protonated, so it cannot interfere with the fully conjugated system, which is responsible for the red color.

pH < 4

At pH > 4, the nitrogen of the N(CH$_3$)$_2$ group is not protonated and the lone pair can be delocalized into the benzene ring. This decreases the extent of conjugation and, therefore, light of shorter wavelengths is absorbed; hence, the change in color from red to yellow.

pH > 4

b. In acidic solutions, the three benzene rings are isolated from one another, so phenolphthalien is colorless.

In basic solutions, loss of the proton from one of the OH groups, causes the five-membered ring to open. As a result, the number of conjugated double bonds increases, which causes the solution to become colored.

Chapter 13 Practice Test

1.   Give one IR absorption band that could be used to distinguish each of the following pairs of compounds. Indicate the compound for which the band would be present.

a.   (aldehyde) CH₃CH₂CH₂CHO   and   (ketone) CH₃CH₂CH₂COCH₃

b.   (amide) CH₃CH₂CONH₂   and   (ester) CH₃CH₂COOCH₃

c. $CH_3CH_2CH_2CH_2OH$ and $CH_3CH_2CH_2OCH_3$

d.   (cyclohexyl)—CHCH₃ with OH   and   (cyclohexyl)—CH₂CH₂OH

e.   CH₃CH₂CH₂COOCH₃   and   CH₃CH₂CH₂COCH₃

f. $CH_3CH_2CH{=}CHCH_3$ and $CH_3CH_2C{\equiv}CCH_3$

g. $CH_3CH_2C{\equiv}CH$ and $CH_3CH_2C{\equiv}CCH_3$

2.   Indicate whether each of the following is true or false:

a.   The O—H stretch of a concentrated solution of an alcohol occurs at a higher frequency than the O—H stretch of a dilute solution.   T   F

b.   Light of 2 μm is of higher energy than light of 3 μm.   T   F

c.   It takes more energy for a bending vibration than for a stretching vibration.   T   F

d.   Propyne will not have an absorption band at 3100 cm⁻¹, because there is no change in the dipole moment.   T   F

e.   Light of 8 μm has the same energy as light of 1250 cm⁻¹.   T   F

f.   The M + 2 peak of an alkyl chloride is half the height of the M peak.   T   F

g.   A chromophore that exhibits both $n \to \pi^*$ and $\pi \to \pi^*$ transitions will have the $n \to \pi^*$ transition at a longer wavelength.   T   F

3.   The major peaks shown in the mass spectrum of a tertiary alcohol are at m/z = 73, 87, 98, and 101. Identify the alcohol.

4.   A 3.8×10^{-4} M solution of cyclohexanone shows an absorbance of 0.75 at 280 nm in a 1.00 cm cell. What is the molar absorptivity of cyclohexanone at 280 nm?

5.   How could you distinguish between the IR spectra of the following compounds?

a.

b.

c.

d.

e.

6.   Which compound has the greater λ_{max}? (300 nm is a greater λ_{max} than 250 nm.)

a.   and

b.   and

c.   and

7.   A solution of a compound with a molar absorptivity of 1200 $M^{-1}cm^{-1}$ at 297 nm gives an absorbance of 0.76 at that wavelength in a 1.0 cm quartz cell. What is the concentration of the solution?

CHAPTER 14
NMR Spectroscopy

Important Terms

applied magnetic field	the externally applied magnetic field.
chemically equivalent protons	protons with the same connectivity relationship to the rest of the molecule.
chemical shift	location of a signal occurring in an NMR spectrum. It is measured downfield from a reference compound (most often TMS).
^{13}C NMR	nuclear magnetic resonance that shows carbon (^{13}C) nuclei.
COSY spectrum	a 2-D NMR spectrum showing ^{1}H-^{1}H correlations.
coupled protons	protons that split each other. Coupled protons have the same coupling constant.
coupling constant	the distance (in hertz) between two adjacent peaks of a split NMR signal.
DEPT ^{13}C NMR spectrum	a group of four ^{13}C NMR spectra that distinguish CH_3, CH_2, and CH groups.
diamagnetic anisotropy	the term used to describe the greater freedom of π electrons to move in response to a magnetic field as a consequence of their greater polarizability compared with σ electrons.
diamagnetic shielding	shielding by the local magnetic field that opposes the applied magnetic field.
diastereotopic hydrogens	two hydrogens bonded to the same carbon that will result in a pair of diastereomers when each of them is replaced in turn with deuterium.
2-D NMR	two-dimensional nuclear magnetic resonance.
doublet	an NMR signal that is split into two peaks.
doublet of doublets	an NMR signal that is split into four peaks of approximately equal height. A doublet of doublets is caused by splitting a signal into a doublet by one hydrogen and into another doublet by another (nonequivalent) hydrogen.
downfield	farther to the left-hand side of the spectrum.
effective magnetic field	the magnetic field that a nucleus "senses" through the surrounding cloud of electrons.
enantiotopic hydrogens	two hydrogens bonded to a carbon that is bonded to two other groups that are nonidentical.
Fourier transform NMR (FT-NMR)	a technique in which all the nuclei are excited simultaneously by an rf pulse, their relaxation monitored, and the data mathematically converted to a frequency spectrum.

geminal coupling	the mutual splitting of two nonidentical protons bonded to the same carbon.
gyromagnetic ratio	the ratio of the magnetic moment of a rotating charged particle to its angular momentum.
HETCOR spectrum	a 2-D NMR spectrum showing ^{13}C-^{1}H correlations.
high-resolution NMR spectroscopy	NMR spectroscopy that uses a spectrometer with a high operating frequency.
^{1}H NMR	nuclear magnetic resonance that shows hydrogen nuclei.
long-range coupling	splitting of a proton by a proton more than 3 σ bonds away.
magnetic resonance imaging (MRI)	NMR used in medicine. The difference in the way water is bound in different tissues produces the signal variation between organs as well as between healthy and diseased states.
methine hydrogen	a tertiary hydrogen.
MRI scanner	an NMR spectrometer used in medicine for whole-body NMR.
multiplet	an NMR signal split by two (or more) nonequivalent sets of protons.
multiplicity	the number of peaks in an NMR signal.
$N + 1$ rule	an ^{1}H NMR signal for a hydrogen with N equivalent hydrogens bonded to an adjacent carbon is split into $N + 1$ peaks; a proton-coupled ^{13}C NMR signal for a carbon bonded to N hydrogens is split into $N + 1$ peaks.
NMR spectroscopy	the absorption of rf radiation in an applied magnetic field to determine the structural features of an organic compound. In the case of ^{1}H NMR spectroscopy, it determines the carbon-hydrogen framework.
operating frequency	the frequency at which an NMR spectrometer operates.
prochiral carbon	a carbon (bonded to two hydrogens) that will become an asymmetric center if one of the hydrogens is replaced by deuterium.
pro-R-hydrogen	replacing this hydrogen with deuterium creates an asymmetric center with the R configuration.
pro-S-hydrogen	replacing this hydrogen with deuterium creates an asymmetric center with the S configuration.
proton exchange	the transfer of a proton from one molecule to another.
quartet	an NMR signal that is split into four equally spaced peaks with an integral ratio of 1:3:3:1.

reference compound	a compound added to the sample whose NMR spectrum is to be taken. The position of the signals in the NMR spectrum are measured from the position of the signal given by the reference compound.
rf radiation	radiation in the radiofrequency region of the electromagnetic spectrum.
shielding	the electrons around a proton shield the proton from the full effect of the applied magnetic field. The more a proton is shielded, the farther to the right its signal appears in an NMR spectrum.
singlet	an unsplit NMR signal.
spin-coupled ^{13}C NMR spectrum	a ^{13}C NMR spectrum in which each signal for a carbon is split by the hydrogens bonded to that carbon.
spin-spin coupling	the splitting of a signal in an NMR spectrum described by the $N + 1$ rule.
α-spin state	nuclei in this spin state have their magnetic moments oriented in the same direction as the applied magnetic field.
β-spin state	nuclei in this spin state have their magnetic moments oriented opposite to the direction of the applied magnetic field.
splitting diagram (splitting tree)	a diagram that describes the splitting of a set of protons.
triplet	an NMR signal that is split into three equally spaced peaks with an integral ratio of 1:2:1.
upfield	farther to the right-hand side of the spectrum.

Solutions to Problems

1. $v = \dfrac{\gamma}{2\pi}B_0$

$= \dfrac{26.75 \times 10^7 \text{ rad T}^{-1} \text{ sec}^{-1} \times 1.0 \text{ T}}{2(3.1416) \text{ rad}}$

$= 43 \times 10^6 \text{ sec}^{-1}$

$= 43 \times 10^6 \text{ Hz} = 43 \text{ MHz}$

2. **a.** $v = \dfrac{\gamma}{2\pi}B_0$

$B_0 = \dfrac{v \times 2\pi}{\gamma}$

$B_0 = \dfrac{360 \times 10^6 \times 2(3.1416)}{26.75 \times 10^7}$

$B_0 = \dfrac{226.2}{26.75}$

$B_0 = 8.46 \text{ T}$

b. $B_0 = \dfrac{v \times 2\pi}{\gamma}$

$B_0 = \dfrac{500 \times 10^6 \times 2(3.1416)}{26.75 \times 10^7}$

$B_0 = \dfrac{314.2}{26.75}$

$B_0 = 11.75 \text{ T}$

From these calculations, you can see that the greater the operating frequency of the instrument (360 MHz versus 500 MHz), the more powerful is the magnet (8.46 T versus 11.75 T) required to operate it.

3.

a. 2	**e.** 1	**i.** 5	**m.** 3
b. 1	**f.** 4	**j.** 4	**n.** 4
c. 1	**g.** 3	**k.** 2	**o.** 3
d. 3	**h.** 3	**l.** 3	

4. **A** would give two signals, **B** would give one signal, and **C** would give three signals.

5.

#1

All the H's are equivalent.

#2

The H's attached to the front of the molecule are equivalent, and the methylene H's are equivalent.

#3

The H's attached to the front of the molecule are equivalent, and the methylene H's are not equivalent.

6. chemical shift in ppm $= \dfrac{\text{downfield from TMS in Hertz}}{\text{operating frequency in Megahertz}}$

a. $1 \text{ ppm} = \dfrac{x \text{ Hz}}{300 \text{ MHz}}$

$x = 300 \text{ Hz}$

b. $1 \text{ ppm} = \dfrac{x \text{ Hz}}{500 \text{ MHz}}$

$x = 500 \text{ Hz}$

7. **a.** $\dfrac{600\,\text{Hz}}{300\,\text{MHz}} = 2.0\ \text{ppm}$

b. The answer would still be 2.0 ppm because the chemical shift is independent of the operating frequency of the spectrometer.

c. $\dfrac{x\,\text{Hz}}{100\,\text{MHz}} = 2.0\ \text{ppm}$

$x = 200$

200 Hz downfield from TMS

8. **a.** The chemical shift is independent of the operating frequency. Therefore, if the two signals differ by **1.5 ppm** in a 300-MHz spectrometer, they will still differ by **1.5 ppm** in a 100-MHz spectrometer.

b. $\dfrac{\text{Hz}}{\text{MHz}} = \text{ppm}$

$\dfrac{90\,\text{Hz}}{300\,\text{MHz}} = 0.3\ \text{ppm}$

$\dfrac{x\,\text{Hz}}{100\,\text{MHz}} = 0.3\ \text{ppm}$

$x = 30\ \text{Hz}$

9. Magnesium is less electronegative than silicon. (See Table 11.1 on page 451 of the text.) Therefore, the peak for $(CH_3)_2Mg$ would be upfield from the TMS peak.

10. **a. and b.**

| CH₃CH₂CH₂Cl | **b.** CH₃CH₂COCH₃ | **c.** CH₃CHCHBr |

| most shielded / least shielded | most shielded / least shielded | most shielded / least shielded |

11. Recall that the highest frequency signal is the one that is farthest to the left on the spectrum. The proton(s) that is underlined in the answer gives the higher frequency signal.

a. CH₃CHCHBr (Br Br)

b. CH₃CHOCH₃ (CH₃)

c. CH₃CH₂CHCH₃ (Cl)

d. CH₃CH—CH₂CH₃ (O, CH₃)

e. CH₃CH₂CH=CH₂

f. CH₃OCH₂CH₂CH₃

12. **a.** CH₃CH₂CH₂Cl

b. CH₃CH₂CHCH₃ (Cl)

c. CH₃CH₂—H (O)

13. a.
$$\begin{array}{c} O \\ \parallel \end{array}$$
$$\overset{a}{CH_3}\overset{b}{CH_2}\overset{C}{\underset{}{}}\overset{c}{\underset{H}{}}$$

d.
$$\overset{a}{CH_3}\overset{b}{CH_2}\overset{d}{CH}\overset{b}{CH_2}\overset{a}{CH_3}$$
$$\underset{\underset{c}{OCH_3}}{|}$$

g.
$$\begin{array}{c} O \\ \parallel \end{array}$$
$$\overset{a}{CH_3}\overset{b}{CH_2}\overset{d}{CH_2}\overset{C}{\underset{}{}}\overset{c}{\underset{CH_3}{}}$$

b.
$$\overset{a}{CH_3}\overset{c}{CH_2}\overset{e}{CH}\overset{b}{CH_3}$$
$$\underset{\underset{d}{OCH_3}}{|}$$

e.
$$\begin{array}{c} O \\ \parallel \end{array}$$
$$\overset{a}{CH_3}\overset{b}{CH_2}\overset{c}{CH_2}\overset{C}{\underset{}{}}\overset{d}{\underset{OCH_3}{}}$$

h.
$$\overset{a}{CH_3}\overset{b}{CH}\overset{d}{CH_2}\overset{c}{OCH_3}$$
$$\underset{\underset{a}{CH_3}}{|}$$

c.
$$\overset{b}{ClCH_2}\overset{a}{CH_2}\overset{b}{CH_2Cl}$$

f.
$$\overset{a}{CH_3}\overset{c}{CH_2}\overset{d}{CH_2}\overset{e}{OCH}\overset{b}{CH_3}$$
$$\underset{\underset{b}{CH_3}}{|}$$

i.
$$\underset{Cl}{}$$
$$\overset{a}{CH_3}\overset{c}{CH}\overset{d}{|}\overset{b}{CHCH_3}$$
$$\underset{\underset{a}{CH_3}}{|}$$

14. From the direction of the electron flow around the benzene ring pictured in Figure 14.6 on page 565 of the text, you can see that the magnetic field induced in the region of the hydrogens that protrude out from the compound shown in this problem is in the same direction as the applied magnetic field, whereas the magnetic field induced in the region of the hydrogens that protrude into the center of the compound is in the opposite direction of the applied magnetic field.

Thus, the signal at 9.25 ppm is for the hydrogens that protrude out because they need a smaller applied magnetic field to come into resonance due to the fact that the induced and applied magnetic fields are in the same direction. The signal at -2.88 ppm is for the hydrogens that protrude inward, because a greater applied field is necessary to make up for the magnetic field that is induced in the opposite direction.

15. Each of the compounds would show two signals, but the ratio of the integrals for the two signals would be different for each of the compounds. The ratio of the integrals for the signals given by the first compound would be 2:9 (or 1:4.5), the ratio of the integrals for the signals given by the second compound would be 1:3, and the ratio of the integrals for the signals given by the third compound would be 1:2.

16. Solved in the text.

17. The heights of the integrals for the signals in the spectrum shown in Figure 14.10 are about 3.5 and 5.2. The ratio of the integrals, therefore, is 5.2/3.5 = 1.5. This matches the ratio of the integrals calculated for **B**. (Later we will see that a signal at ~7 ppm is characteristic of a benzene ring.)

$$HC\equiv C-\!\!\!\!\bigcirc\!\!\!\!-C\equiv CH$$

$$\frac{4}{2} = 2$$

$$ClCH_2-\!\!\!\!\bigcirc\!\!\!\!-CH_2Cl$$

$$CH_3-\!\!\!\!\bigcirc\!\!\!\!-CH_3 \qquad \frac{4}{4} = 1.0$$

$$\frac{6}{4} = 1.5$$

$$Br_2CH-\!\!\!\!\bigcirc\!\!\!\!-CHBr_2$$

$$\frac{4}{2} = 2$$

18.   The signal farthest downfield in both spectra is the signal for the hydrogens bonded to the carbon that is also bonded to the halogen. Because chlorine is more electronegative than iodine, the farthest downfield signal should be farther downfield in the ^{1}H NMR spectrum for 1-chloropropane than in the ^{1}H NMR spectrum for 1-iodopropane. Therefore, the **first spectrum** in Figure 14.5 is the ^{1}H NMR spectrum for **1-iodopropane**, and the **second spectrum** in Figure 14.5 is the ^{1}H NMR spectrum for **1-chloropropane**.

19.   C is easiest to distinguish because it will have **two** signals, whereas **A** and **B** will each have **three** signals.

The signals in the H^1 NMR spectrum of **A** will be (left to right across the spectrum): **triplet, triplet, multiplet**.

The signals in the H^1 NMR spectrum of **B** will be (left to right across the spectrum): **doublet, multiplet, doublet**.

20.   **a.**   A triplet is caused by coupling to two equivalent protons on an adjacent carbon. The two protons can be aligned in three different ways: both with the field, one with the field and one against the field, or both against the field. That is why the signal is a triplet. There is only one way to align two protons that are with the field or two protons that are against the field. However, there are two ways to align two protons if one is with and one is against the field: with and against or against and with. Consequently, the peaks in a triplet have relative intensities of 1:2:1.

$$\uparrow\uparrow \qquad \uparrow\downarrow \qquad \downarrow\downarrow$$
$$\downarrow\uparrow$$

b.   A quintet is caused by coupling to four equivalent protons on adjacent carbons. The four protons can be aligned in five different ways. The following possible arrangements for the alignment of four protons explain why the relative intensities of a quintet are 1:4:6:4:1.

$$\uparrow\uparrow\uparrow\uparrow \quad \uparrow\uparrow\uparrow\downarrow \quad \uparrow\uparrow\downarrow\downarrow \quad \uparrow\downarrow\downarrow\downarrow \quad \downarrow\downarrow\downarrow\downarrow$$

$$\uparrow\uparrow\downarrow\uparrow \quad \downarrow\downarrow\uparrow\uparrow \quad \uparrow\downarrow\uparrow\downarrow$$

$$\uparrow\downarrow\uparrow\uparrow \quad \uparrow\downarrow\uparrow\downarrow \quad \downarrow\downarrow\uparrow\downarrow$$

$$\downarrow\uparrow\uparrow\uparrow \quad \downarrow\uparrow\downarrow\uparrow \quad \downarrow\downarrow\downarrow\uparrow$$

$$\uparrow\downarrow\downarrow\uparrow$$

$$\downarrow\uparrow\uparrow\downarrow$$

21.   The signal at 12.2 ppm indicates that the compound is a carboxylic acid. From the molecular formula and the splitting patterns of the signals, the spectra can be identified as the ^{1}H NMR spectrum of:

a. CH_3CH attached to $\overset{\displaystyle O}{\overset{\displaystyle \|}{C}}$—OH, with Cl on the CH **b.** $ClCH_2CH_2$ attached to $\overset{\displaystyle O}{\overset{\displaystyle \|}{C}}$—OH

22.   **a.** 3 signals   **b.** 2 signals   **c.** 3 signals

$ICH_2CH_2CH_2Br$ $ClCH_2CH_2CH_2Cl$ $ICH_2CH_2CHBr_2$

triplet | triplet   triplet | triplet | triplet

multiplet   quintet   multiplet

23.   The contributing resonance structures show that the nitro group's ability to withdraw electrons by electron delocalization causes the C-2 and C-4 positions of the resonance hybrid (the actual structure of the compound) to have a partial positive charge. The nitro group's ability to withdraw electrons inductively (through the sigma bonds) causes the partial positive charge to be greater at C-2 than at C-4.

Therefore, the H_c proton has the least electron dense (least shielded) environment, so its signal appears at the highest frequency. The C-3 position does not have a partial positive charge. Therefore, the proton attached to it (the H_a proton) has the most electron dense (most shielded) environment, so its signal appears at the lowest frequency.

24.   Each compound will have two doublets. In addition, **A** will have two doublet of doublets, **B** will have one doublet of doublets and a singlet, and **C** will have no other signals.

A	**B**	**C**
4 signals	4 signals	2 signals
(2 doublets and 2 doublets of doublets)	(2 doublets, 1 singlet, and 1 doublet of doublets)	(2 doublets)

25.

1 signal	**2 signals**	**3 signals**	**3 signals**
a singlet	a quintet (a) and a triplet (b)	2 triplets (a and c) and a multiplet (b)	2 doublets (a and b) and a multiplet (c)

26.   **a.**   The signal at ~7.2 ppm indicates the presence of a benzene ring. If the ring has a single substituent, the ring has a formula of C_6H_5. Subtracting C_6H_5 from the molecular formula of the compound $(C_9H_{12} - C_6H_5)$ gives a substituent with a formula of C_3H_7. The triplet at ~0.9 ppm indicates a methyl group adjacent to a methylene group. The identical integration under the signals at 1.6 ppm and 2.6 ppm indicate two methylene groups. Thus, the compound is propylbenzene.

   b.   The triplet and quartet indicate a CH_3CH_2 group bonded to an atom that is not bonded to any hydrogens. The molecular formula of $C_5H_{10}O$ indicates that the compound is diethyl ketone.

   c.   The signals between ~7 and 8 ppm indicate the presence of a benzene ring. The observation that there are 3 benzene ring signals indicates the compound has a mono-substituted benzene ring or a 1,3-disubstituted benzene ring. Since none of the 3 benzene ring signals is a singlet, we know that the compound has a mono-substituted benzene ring. The integration of the aromatic signals (5H) is consistent with mono-substitution. Subtracting C_6H_5 from the molecular formula of the compound $(C_9H_{10}O_2 - C_6H_5)$ shows that the substituent has a formula of $C_3H_5O_2$.

   The triplet and quartet suggest a CH_3CH_2 group. Therefore, we can conclude that the compound is one of the following.

   The substituent attached to the benzene ring in compound **A** withdraws electrons from the ring, whereas the substituent attached to the benzene ring in compound **B** donates electrons into the benzene ring. Because one of the peaks of the benzene ring signal is at a higher frequency (farther downfield) than usual (>8 ppm) suggests that an electron-withdrawing substituent is present on the benzene ring. This is confirmed by the signal for the CH_2 group at 4.3 ppm, indicating that the methylene group is adjacent to the electron-withdrawing oxygen. Thus, the compound is compound **A**.

27.

a. $\overset{t}{\downarrow}\overset{q}{\downarrow}$ $CH_3CH_2CH_2CH_3$

b. $\overset{s}{\downarrow}$ $BrCH_2CH_2Br$

c.

(Remember that equivalent hydrogens do not split each other's signal.)

d. $d\rightarrow$

The H on C-2 will not be split by the H on C-3, since they are equivalent. The H's on C-6 will not be split by the H's on C-5, since the H's on C-5 and C-6 are equivalent.

e.

f. $\overset{t}{\downarrow}\overset{m}{\downarrow}\overset{t}{\downarrow}\overset{O}{\overset{\parallel}{}}\overset{s}{\downarrow}$ $CH_3CH_2CH_2CCH_3$

g. $\overset{quin}{\downarrow}\overset{m}{\downarrow}\overset{t}{\downarrow}$ $CH_3CH_2CHCH_2CH_3$ $\underset{\overset{|}{Cl}}{}$

h. $\overset{d}{\downarrow}\overset{m}{\downarrow}\overset{t}{\downarrow}$ $CH_3CHCH_2CHCH_3$ $\underset{CH_3\quad CH_3}{}$

i. $\overset{d}{\downarrow}\overset{q}{\downarrow}$ CH_3CH- $\underset{\overset{|}{Br}}{}$

j. $\overset{s}{\downarrow}$ CH_3-

$-OCH_3$ $\overset{s}{\downarrow}$

k. $\overset{s}{\downarrow}$ CH_3-

$-CH_3$ $\overset{s}{\downarrow}$

l. $t\rightarrow$

$\leftarrow s$

m. $t\rightarrow$

$-NO_2$

n.

o.

The double bond allows the methyl group to split the H trans to it.

28.

a. $CH_3OCH_2CH_2OCH_3$

b. There are three possibilities:

c.

$$CH_3\overset{O}{\overset{\parallel}{C}}CH_2CH_2\overset{O}{\overset{\parallel}{C}}CH_3$$

$$CH_3OCH_2C\equiv CCH_2OCH_3$$

29.   Each spectrum is described going from left to right.

   a.   BrCH$_2$CH$_2$CH$_2$CH$_2$Br

       triplet   triplet (This triplet will not be split by the adjacent equivalent protons.)

   b.   two triplets (close to each other)   singlet   multiplet
   (Appendix VI on page A-15 of the text indicates that a methylene adjacent to an RO and a
   methylene adjacent to a Br appear at about the same place.)

   c.   singlet (Equivalent H's do not split each other's signals.)

   d.   quartet singlet triplet   **j.**   triplet quintet

   e.   three singlets   **k.**   doublet doublet of septets doublet

   f.   three doublets of doublets   **l.**   triplet singlet quintet

   g.   quartet triplet   **m.**   singlet

   h.   singlet quartet triplet   **n.**   singlet (Equivalent H's do not split each other's signals.)

   i.   doublet multiplet doublet   **o.**   singlet

30.   There is no coupling between H$_a$ and H$_c$, because they are separated by four σ bonds. (We will see in
   Section 14.16 that unless the sample is pure and dry, a proton bonded to an oxygen is not split and does
   not split other protons.)
   There is no coupling between H$_b$ and H$_c$, because they are separated by four σ bonds and one double bond.

31.   The two singlets in the ^{1}H NMR spectrum with the same integration suggest two methyl groups,
   one of which is adjacent to an oxygen. That the benzene ring protons (6.7–7.1 ppm) consist of two
   doublets (each integrating to two hydrogens) indicates a 1,4-disubstituted benzene ring. The IR spectrum
   also indicates a benzene ring (1600 cm^{-1} and 1500 cm^{-1}) with hydrogens bonded to sp^2 carbons and no
   carbonyl group. The absorption bands in the 1250 cm^{-1}–1000 cm^{-1} region suggests there are two C-O
   single bonds, one with no double bond character and one with some double bond character.

CH$_3$O ⟨benzene ring⟩ CH$_3$

32.   **a.**   **b.**

33.   **a. A**   **b.   B and D**

34.   Solved in the text.

35.   The OH proton of a carboxylic acid is more deshielded than the OH proton of an alcohol as a result of
   electron delocalization. In addition, the extent of hydrogen bonding affects the chemical shift of a proton
   bonded to an oxygen. Carboxylic acids exist as tightly hydrogen bonded dimers (see page 714 in the
   text), which strongly deshields the OH proton.

36.   The greater the extent of hydrogen bonding, the greater the chemical shift. Therefore, the ^{1}H NMR spectrum of pure ethanol would have the signal for the OH proton at a greater chemical shift because it would be hydrogen bonded to a greater extent.

37.   RÖ—H + :ÖH ⇌ RÖ:⁻ + H—ÖH

38.   Figure 14.31 is the ^{1}H NMR spectrum of propanamide. Notice that the signals for the N-H protons are unusually broad. Because of the partial double bond character of the C-N bond, there is no free rotation about the C-N bond, so the two N-H protons are not chemically equivalent. The quartet and triplet are characteristic of an ethyl group.

39.   **a.**   **1.** 3   **4.** 4   **7.** 3
                    2. 3   **5.** 3   **8.** 3
                    3. 2   **6.** 4   **9.** 2

  b.   An arrow is drawn to the carbon that gives the signal at the lowest frequency.

1.   CH₃CH₂CH₂Br

2.   (CH₃)₂C=CH₂ structure

3.   CH₃CHCH₃ with Br

4.   CH₃CH₂C(=O)OCH₃

5.   CH₃CH(CH₃)C(=O)H

6.   chlorobenzene structure

7.   (CH₃)₃C—O—CH₃ structure

8.   CH₃C(=O)CH₂C(=O)CH₃ structure

9.   vinyl bromide structure

40.   Each spectrum is described going from left to right:

  1.   triplet triplet quartet   **3.**   doublet quartet   **5.**   doublet doublet quartet

41.

NO_2 (ortho-dinitrobenzene with NO_2) NO_2 (meta-dinitrobenzene with NO_2) $NO_2 \cdot$ (para-dinitrobenzene with NO_2)

a. ^1H NMR   2 signals   3 signals   1 signal

b. ^{13}C NMR   3 signals   4 signals   2 signals

42.   **a.**   The signal at 210 ppm is for the carbonyl carbon of a ketone. There are 10 other carbons in the compound and five other signals. That suggests the compound is a symmetrical ketone with identical five-carbon alkyl groups.

$$CH_3CH_2CH_2CH_2CH_2 \overset{\overset{O}{\|}}{C} CH_2CH_2CH_2CH_2CH_3$$

b.   Because there are only four signals for the six carbons of the benzene ring (the signals between 110–117 ppm), the compound must be a 1,4-disubstituted benzene.

$$Br - \bigcirc - CH_2CH_3$$

c.   The signal at 212 ppm is for the carbonyl carbon of a ketone. The degree of unsaturation in this compound is 2. Since the carbonyl group accounts for one degree of unsaturation and no other sp^2 carbons are evident from the spectrum, the second degree of unsaturation must be due to a ring. There are five other carbons in the compound and three other signals. This suggests the compound is a symmetrical cyclic ketone.

$$\bigcirc = O$$

d.   The molecular formula indicates the compound has one double bond or one ring. The presence of a signal at 130 ppm shows the presence of sp^2 carbons; thus, the compound must have a double bond. Each of the two sp^2 carbons must be bonded to the same groups, because the six carbons exhibit only three signals. Whether the compound is *cis*-3-hexene or *trans*-3-hexene cannot be determined from the spectrum. An ^1H NMR spectrum could distinguish the compounds because the coupling constant for trans protons is greater than the coupling constant for cis protons.

$$CH_3CH_2CH = CHCH_2CH_3$$

43.

If the triangles shown on the previous page are drawn on the spectrum on page 599 of the text, you will be able to easily identify the coupled protons.

Point **A** shows that the **a** protons are split by the **c** protons.
Point **B** shows that the **b** protons are split by the **d** protons.

44.   **a.** **1.** 5   **4.** 2   **b.** **1.** 7   **4.** 2

              2. 5   **5.** 3   **2.** 7   **5.** 2

              3. 4   **6.** 3   **3.** 5   **6.** 4

45.   The H_b proton will be split into a quartet by the H_a protons and into a doublet by the H_c protons.

a.

5 peaks (a quintet)

b.

8 peaks (a double of quartets)

46.   **a.**

          d   septet
          a   b
          CH_3CHNO_2
               CH_3
               a

c.
   d   septet   t   m   t
   b   e   d   c   a
   $CH_3CH \quad CH_2CH_2CH_3$
        CH_3
         b

e.
               s
       s   a
           CH_3
       b
       $ClCH_2CCHCl_2$
             c
          CH_3 s
          a

b.
   t   m   t   s
   a   b   d   c
   $CH_3CH_2CH_2OCH_3$

d.
   t   m   t   s
   a   b   c   d
   $CH_3CH_2CH_2 \quad CH_2Cl$
              O

f.
   t   m   quintet
   c   b   a   b   c
   $ClCH_2CH_2CH_2CH_2CH_2Cl$

47.   **a.**   The spectrum must be that of **2-bromopropane**, because the NMR spectrum has two signals and the lowest frequency (farthest upfield) signal is a doublet and the other signal is given by a single hydrogen.

       b.   The spectrum must be that of **1-nitropropane**, because the NMR spectrum has three signals and both the lowest frequency (farthest upfield) and highest frequency (farthest downfield) signals are triplets.

       c.   The spectrum must be that of **ethyl methyl ketone**, because the NMR spectrum has three signals and the signals are a triplet, a singlet, and a quartet.

48.   By dividing the value of the integration steps by the smallest one, the ratios of the hydrogens are found to be 3:2:1:9.

$$\frac{40.5}{13} = 3.1 \qquad \frac{27}{13} = 2.1 \qquad \frac{13}{13} = 1 \qquad \frac{118}{13} = 9.1$$

Assuming that the ratios are given in the highest frequency to lowest frequency direction, a possible compound could be the following ester:

$$\begin{array}{cc} CH_3 & O \\ | & \| \\ CH_3C\!-\!CHCOCH_3 \\ | & | \\ CH_3 & CH_2Cl \end{array}$$

49.   **a.**   [structure]   and   [structure]

4 signals   2 signals

b.   Br[structure]Br   and   Br[structure]NO$_2$

2 signals   3 signals

c. $\begin{array}{cc} CH_3 & CH_3 \\ | & | \\ CH_3CH\!-\!CHCH_3 \end{array}$ and $\begin{array}{c} CH_3 \\ | \\ CH_3CCH_2CH_3 \\ | \\ CH_3 \end{array}$

2 signals   3 signals

d. $\begin{array}{c} CH_3\ O \\ |\ \ \| \\ CH_3C\!-\!COCH_3 \\ | \\ CH_3 \end{array}$ and $\begin{array}{c} OCH_3 \\ | \\ CH_3CCH_3 \\ | \\ OCH_3 \end{array}$

2 signals with   2 signals with
integration 3:1   integration 1:1

e.   CH$_3$O—[ring]—CH$_2$CH$_3$ and   CH$_3$—[ring]—OCH$_2$CH$_3$

The singlet would be at a higher   The quartet would be at a higher
frequency than the quartet.   frequency than the singlet.

f.   [ring]   and   [ring]

2 signals   3 signals

g. $\begin{array}{c} CH_3CHCl \\ | \\ CH_3 \end{array}$ and $\begin{array}{c} CH_3CDCl \\ | \\ CH_3 \end{array}$,

2 signals   1 signal

h.

Cl
H——CH₃
D——H
Cl

and

Cl
D——CH₃
H——H
Cl

lowest frequency signal = doublet   lowest frequency signal = singlet

i.

H   H
│╱╲│
Cl   Cl

and

H   Cl
│╱╲│
Cl   H

3 signals   2 signals

j.

CH₃

CH₃——⟨benzene ring⟩——CCH₃

CH₃

and

CH₃

⟨benzene ring⟩——CH₂CCH₃

CH₃

the signals for the benzene ring protons   the signals for the benzene ring protons
plus two signals with integration 3:1   plus two signals with integration 4.5:1

50. **a.** Chemical shift in ppm is independent of the operating frequency.
b. Chemical shift in hertz is proportional to the operating frequency.
c. The coupling constant in hertz is independent of the operating frequency.
d. The frequency required for NMR spectroscopy is lower than that required for IR spectroscopy,
which is lower than that required for UV/Vis spectroscopy.

51. **a.** CH₃CH₂CHBr   **b.** CH₃CH₂CH₂CH₂Br   **c.** CH₃CHCH₂Br
                  │   │
                 CH₃   CH₃

52. **a.** CH₃CCH₂Br   **b.**   **c.** CH₃CH₂—C(=O)—OCH₂CH₃
                  │
              CH₃ │ Br (CH₃CCH₂Br with CH₃ above and Br below)

where **b.** is ⟨benzene ring⟩—CHCH₃ with Br above the CH

53. The singlet at 210 ppm indicates the carbonyl group of a ketone. The splitting of the other two signals
indicates an isopropyl group. The molecular formula indicates that it must have two isopropyl groups.

O
‖
CH₃CH—C—CHCH₃
       │   │
      CH₃   CH₃

54.

55.

56. $CH_3CH_2CH{=}CH_2$

1-butene   *cis*-2-butene   2-methylpropene

It would be better to use ^{13}C NMR, because you would only have to look at the number of signals in each spectrum: 1-butene will show four signals, *cis*-2-butene will show two signals, and 2-methylpropene will show three signals. (In the ^{1}H NMR spectrum, 1-butene will show four signals, and *cis*-2-butene and 2-methylpropene will both show two signals.)

57.   **a.** $CH_3CH_2CH_2CH_2OCH_3$

58.

A	**B**	**C**	**D**
3 signals	3 signals	4 signals	3 signals
singlet, quartet, triplet (singlet farthest downfield)	singlet, quartet, triplet (quartet farthest downfield)	singlet, triplet multiplet, triplet	singlet, septet doublet

C can be distinguished from **A**, **B**, and **D**, because **C** has four signals and the others have three signals.

D can be distinguished from **A** and **B**, because the three signals of **D** are a singlet, a septet, and a doublet, whereas the 3 signals of **A** and **B** are a singlet, a quartet, and a triplet.

A and **B** can be distinguished, because the highest frequency signal in **A** is a singlet, whereas the highest frequency signal in **B** is a quartet.

59.   It is the ^{1}H NMR spectrum of *tert*-butyl methyl ether.

$$CH_3Br \ + \ CH_3\overset{\underset{\displaystyle CH_3}{|}}{\overset{\displaystyle CH_3}{|}}C-O^- \longrightarrow CH_3\overset{\underset{\displaystyle CH_3}{|}}{\overset{\displaystyle CH_3}{|}}C-OCH_3$$

tert-butyl methyl ether

60.   a.   **b.**

The numbers indicate the ppm value of the signal given by that carbon.

61.   The broad signal at ~2.9 ppm is for the H_a proton that is bonded to the oxygen. The signal for the H_b protons at ~4.2 ppm is split into a doublet by the H_e proton. The H_c proton and the H_d proton are each split by the H_e proton; the coupling constant is greater for the trans protons. Thus, the H_c proton is at ~5.1 ppm and the H_d proton is at ~5.3 ppm. The H_e proton at ~6.0 ppm is split by the H_b protons and by the H_c proton and the H_d proton.

62.   In the 6.5–8.1 ppm region, each compound will have two doublets. In addition, **A** will have 2 doublet of doublets, **B** will have one doublet of doublets and a second singlet, and **C** will have no other signals.

In the 6.5–8.1 ppm region: 4 signals   4 signals   2 signals
(2 doublets and   (1 singlet,   (2 doublets)
2 doublets of doublets)   2 doublets, and
1 doublet of doublets)

63.   The signals at ~7.2 ppm indicate that the compounds whose spectra are shown in parts **a** and **b** both contain a benzene ring. From the molecular formula, you know that both compounds have five additional carbons.

a. The hydrogens bonded to the five additional carbons in part **a** must all be accounted for by two singlets with integral ratios of ~2:9, indicating that the compound is **2,2-dimethyl-1-phenylpropane**.

2,2-dimethyl-1-phenylpropane

b. The two singlets in the spectrum in part **b** have integral ratios of ~1:3, indicating a methyl substituent and a *tert*-butyl substituent. The signals at ~7.1 and ~7.3 ppm indicate that the benzene ring is 1,4-disubstituted and has two kinds of hydrogens. Thus, the compound is **1-*tert*-butyl-4-methylbenzene**.

1-*tert*-butyl-4-methylbenzene

64.

5 signals

65.   Oxygen is more electronegative than chlorine. (See Table 1.3 on page 10 of the text.)

a.

b.

c.

d.

e.

CH₃ carbons

CH₂ carbons

CH carbons

all carbons

66. If addition of HBr to propene follows the rule that says the electrophile adds to the *sp²* carbon that is bonded to the greater number of hydrogens, the product of the reaction will give an NMR spectrum with two signals (a doublet and a septet). If addition of HBr does not follow the rule, the product will give an NMR spectrum with three signals (two triplets and a multiplet).

CH₃CH=CH₂ + HBr ⟶ CH₃CHCH₃ (Br)   CH₃CH₂CH₂Br
follows the rule   does not follow the rule
2 signals   **3 signals**

67. **a.** **b.** **c.** **d.**

68. Bromomethane gives a signal at 2.7 ppm for its three methyl protons, and 2-bromo-2-methylpropane gives a signal at 1.8 ppm for its nine methyl protons. If equal amounts of each were present in solution, the ratio of the hydrogens (and, therefore, the ratio of the relative integrals) would be 3:1. Because the relative integrals are 1:6, there must be an 18-fold greater concentration of bromomethane in the mixture.

69.   Using the formula given on page 554 of the text:

$$\Delta E = h\nu = \frac{h\gamma}{2\pi} B_0$$

Given: Planck's constant $= h = 6.626 \times 10^{-34}\,\text{Js}$ (page 153 of the text)

$$1\,\text{cal} = 4.184\,\text{J} \qquad\qquad \text{(page 138 of the text)}$$

$$60\,\text{MHz} = 1.4092\,\text{T} \qquad\quad \text{(page 555 of the text)}$$

$$\gamma \text{ for } {}^1\text{H} = 2.675 \times 10^8\,\text{T}^{-1}\text{s}^{-1}$$

$$\Delta E = \frac{h\gamma}{2\pi} B_0$$

$$= \frac{6.626 \times 10^{-34}\,\text{Js} \times 2.675 \times 10^8\,\text{T}^{-1}\text{s}^{-1}}{2(3.1416)} \times 1.4092\,\text{T} \times \frac{1\,\text{cal}}{4.184\,\text{J}}$$

$$= 9.50 \times 10^{-27}\,\text{cal}$$

70.   All four spectra show a singlet at ~2.0 ppm suggesting they are all esters with a methyl group attached to the carbonyl group. So now the problem becomes determining the nature of the group attached to the oxygen.

$$\underset{CH_3 \qquad\quad O-R}{\overset{\displaystyle O \atop \displaystyle \|}{C}}$$

a.   The highest frequency signal in the first spectrum is a 2-hydrogen triplet, indicating that the carbon attached to the oxygen is bonded to a CH_2 group. The lowest frequency signal is also a 2-hydrogen triplet, indicating that it too is attached to a CH_2 group. The presence of two 2-hydrogen multiplets confirms the structure.

$$\underset{CH_3 \qquad\quad O-CH_2CH_2CH_2CH_3}{\overset{\displaystyle O \atop \displaystyle \|}{C}}$$

b.   The highest frequency signal in the second spectrum is a 1-hydrogen multiplet, indicating that the carbon attached to the oxygen is attached to two nonequivalent carbons bonded to hydrogens. The lowest frequency signal is a 3-hydrogen triplet, indicating that it is attached to a CH_2 group. The 3-hydrogen doublet at ~1.2 ppm is due to a methyl group attached to a carbon bonded to one hydrogen.

$$\underset{CH_3 \qquad\quad O-CHCH_2CH_3 \atop {\scriptstyle|} \atop {\scriptstyle CH_3}}{\overset{\displaystyle O \atop \displaystyle \|}{C}}$$

c. The highest frequency signal and the lowest frequency signal in the third spectrum are doublets, indicating that the carbon attached to the oxygen and the terminal carbon of the OR group are both attached to a CH group.

$$\underset{CH_3}{\overset{\displaystyle \overset{O}{\parallel}}{}}\;\;\;\;O{-}CH_2CHCH_3$$

(structure: acetate ester — CH₃–C(=O)–O–CH₂CH(CH₃)CH₃, with CH₃ below the CH)

d. The group attached to the oxygen in the fourth spectrum has only one kind of hydrogen, and the carbon attached to the oxygen is not bonded to any hydrogens, because there is no signal at ~4.0.

(structure: CH₃–C(=O)–O–C(CH₃)₃, tert-butyl acetate)

71.

$$\underset{\underset{OH}{|}}{CH_3CHCHCH_3}\overset{CH_3}{\underset{\quad}{|}}\;\;\xrightarrow[\Delta]{HBr}\;\;\underset{\underset{Br}{|}}{CH_3CH_2CCH_3}\overset{CH_3}{\underset{\quad}{|}}$$

compound A   compound B

72. **a.** The IR spectrum indicates that the compound is a ketone. The 6-hydrogen doublet in the NMR spectrum at ~0.9 ppm suggests an isopropyl group. There is a singlet at ~2.1 ppm on top of a multiplet, which appears to consist of the methine proton of an isopropyl group plus a 3-hydrogen singlet for a methyl adjacent to the carbonyl carbon. The 2-hydrogen doublet at ~2.2 ppm suggests a CH_2 group adjacent to the carbonyl group that couples to an adjacent methine proton. Knowing that the compound contains 6 carbons helps in the identification.

(structure: 4-methyl-2-pentanone)

b. The IR spectrum indicates that this oxygen-containing compound is not a carbonyl compound or an alcohol; the absorption band at ~1000 cm^{-1} suggests that it is an ether. The 12-hydrogen doublet at ~1.1 ppm in the ^1H NMR spectrum suggests an isopropyl group. Since there is only one other signal in the NMR spectrum, the compound must be a symmetrical ether. The compound is diisopropyl ether.

(structure: diisopropyl ether)

c. The IR absorption band at ~3400 cm^{-1} indicates an alcohol. The two ^{1}H NMR signals at 7.3 and 8.1 ppm indicate a 1,4-disubstituted benzene ring with a strongly electron-withdrawing substituent. The two 2-hydrogen triplets indicate that the four-carbon substituent is not branched. The broad signal at 2.1 ppm is due to the OH proton.

d. The IR absorption bands at ~1700 cm^{-1} and 2700 cm^{-1} indicate the compound is an aldehyde. The two doublets at ~7.0 and 7.8 ppm in the ^{1}H NMR spectrum indicate a 1,4-disubstituted benzene ring. That none of the remaining NMR signals is a doublet suggests that the aldehyde group is attached directly to the benzene ring. The two triplets and two multiplets indicate an unbranched substituent. The triplet at ~4.0 ppm indicates that the group giving this signal is next to an electron-withdrawing group.

73.   a. The IR absorption band at ~1730 cm^{-1} and the absence of an aldehyde C-H stretch suggests the compound is a ketone. The signals in the ^{1}H NMR spectrum are consistent with a CH$_3$CH$_2$CH$_2$(C=O)- group. Since there are no other signals in the ^{1}H NMR spectrum suggests it is a symmetrical ketone. The mass spectrum shows a molecular ion with $m/z = 114$; this is consistent with the ketone shown below. Also, the large fragments at $m/z = 71$ (M − propyl) and at $m/z = 43$ (a propyl cation) are consistent with the expected α-cleavage fragmentation. The fact that there are only three signals in the NMR spectrum suggests it is a symmetrical ketone. The splitting pattern confirms the structure.

b. The M + 2 peak at $m/z = 156$ in the mass spectrum tells you the compound contains chlorine; the IR spectrum indicates it is a ketone; the NMR spectrum indicates it has a monosubstituted benzene ring. The singlet at ~4.7 ppm indicates that the CH$_2$ group giving this signal is in a strongly electron-withdrawing environment. The major fragment ions at $m/z = 105$ (M − CH$_2$Cl) and at $m/z = 77$ (C$_6$H$_5^+$) are consistent with the structure below.

74.   The DEPT ^{13}C NMR spectrum indicates that the compound has six carbons. The molecular formula tells us that the compound has six carbons and an oxygen. The signal at 220 ppm suggests a ketone. The following ketones have the single CH_3 group, the single CH group, and the three CH_2 groups indicated by the DEPT spectrum. The position of the CH_3 signal indicates that it is not next to an electron-withdrawing group; the position of the CH signal indicates that it is next to an electron-withdrawing group. Thus, the compound on the right is responsible for the spectrum.

or

75.   There is a singlet for the three methyl groups that are attached to the quaternary carbon. The singlet is on top of the triplet for the other methyl group. The methylene group shows the expected quartet.

Chapter 14 Practice Test

1. How many signals would you expect to see in the ^{1}H NMR spectrum of each of the following compounds?

2. Indicate the multiplicity of each of the indicated sets of protons. (That is, indicate whether it is a singlet, doublet, triplet, quartet, quintet, multiplet, or doublet of doublets.)

3. How could you distinguish the following compounds using ^{1}H NMR spectroscopy?

4. Indicate whether each of the following statements is true or false:

    a. The signals on the right of an NMR spectrum are deshielded compared to the signals on the left.   T   F

    b. Dimethyl ketone has the same number of signals in its ^{1}H NMR spectrum as in its ^{13}C NMR spectrum.   T   F

c. In the ^{1}H NMR spectrum of the compound shown below, the lowest frequency signal (the one farthest upfield) is a singlet and the highest frequency signal (the one farthest downfield) is a doublet.   T   F

d. The greater the frequency of the signal, the greater its chemical shift in ppm.   T   F

5.   For each compound:

    a. Indicate the number of signals you would expect to see in ^{1}H NMR spectrum.

    b. Indicate the hydrogen or set of hydrogens that would give the highest frequency (farthest downfield) signal.

    c. Indicate the multiplicity of that signal.

$$\text{1. } CH_3CH_2CH_2Cl \qquad \text{2. } CH_3CH_2\overset{\displaystyle O}{\overset{\|}{C}}OCH_3 \qquad \text{3. } CH_3\underset{\underset{\displaystyle Br}{|}}{C}HCH_3$$

6.   For each compound of Problem 5:

    a. Indicate the number of signals you would expect to see in its proton-decoupled ^{13}C NMR spectrum.

    b. Indicate the carbon that would give the highest frequency (farthest downfield) signal.

    c. Indicate the multiplicity of that signal in a spin-coupled ^{13}C NMR spectrum.

CHAPTER 15
Aromaticity • Reactions of Benzene

Important Terms

aliphatic compound	a saturated or unsaturated organic compound that does not contain an aromatic ring.
annulene	a monocyclic hydrocarbon with alternating double and single bonds.
antiaromatic compound	a cyclic and planar compound with an uninterrupted cloud of electrons containing an even number of pairs of π electrons.
arene	an aromatic compound.
arene oxide	an aromatic compound that has had one of its double bonds converted to an epoxide.
aromatic compound	a cyclic and planar compound with an uninterrupted cloud of electrons containing an odd number of pairs of π electrons.
benzyl group	
Clemmensen reduction	a reaction that reduces the carbonyl group of a ketone to a methylene group using Zn(Hg)/HCl, Δ.
electrophilic aromatic substitution reaction	a reaction in which an electrophile substitutes for a hydrogen of an aromatic ring.
Friedel-Crafts acylation	an electrophilic aromatic substitution reaction that puts an acyl group on an aromatic ring.
Friedel-Crafts alkylation	an electrophilic aromatic substitution reaction that puts an alkyl group on an aromatic ring.
fused rings	rings that share two adjacent carbons.
Gatterman-Koch reaction	a reaction that uses a high-pressure mixture of carbon monoxide and HCl to form benzaldehyde.
halogenation	reaction with a halogen (Br_2, Cl_2).
heteroatom	an atom other than a carbon atom or a hydrogen atom.
heterocyclic compound (heterocycle)	a cyclic compound in which one or more of the atoms of the ring are heteroatoms.
Hückel's rule or the $4n + 2$ rule	gives the number of π electrons in a cyclic uninterrupted π cloud that a cyclic and planar compound must have in order to be aromatic.

nitration   substitution of a nitro group (NO_2) for a hydrogen of an aromatic ring.

phenyl group C_6H_5 —

principle of microscopic   states that the mechanism for a reaction in the forward direction has the same
reversibility   intermediates and the same transition states as the mechanism for the reaction in
                                     the reverse direction.

sulfonation   substitution of a hydrogen of an aromatic ring with a sulfonic acid group (SO_3H).

Suzuki reaction   a reaction that will replace a bromine substituent on an aromatic ring with an alkyl,
                                     alkenyl, or aryl substituent.

Wolff-Kishner reduction   a reaction that reduces the carbonyl group of a ketone to a methylene group using
                                     NH_2NH_2/HO^-, Δ.

Solutions to Problems

1. **a.** In the case of 9 pairs of π electrons, there are 18 electrons. Therefore, $4n + 2 = 18$ and $n = 4$.

    b. Because it has an odd number of pairs of π electrons, it will be aromatic if it is cyclic, planar, and if every atom in the ring has a p orbital.

2. **a.** Notice that each resonance contributor has a charge of -1.

    b. All five ring atoms share the negative charge.

3. **a.** This is the only one that is aromatic; it is cyclic, planar, every ring atom has a p orbital, and it has one pair of π electrons.

    The first compound is not aromatic, because one of the atoms is sp^3 hybridized and therefore does not have a p orbital.

    The third compound is not aromatic, because it has two pairs of π electrons.

    b. This is the only one that is aromatic; it is cyclic, planar, every ring atom has a p orbital, and it has three pairs of π electrons.

    The first compound is not aromatic, because one of the atoms is sp^3 hybridized and therefore does not have a p orbital.

    The third compound is not aromatic, because it has four pairs of π electrons.

4. **d** and **e** are aromatic.

    d is aromatic, because it is cyclic, planar, every atom in the ring has a p orbital, and it has seven pairs of π electrons.

    e is aromatic, because it is cyclic, planar, every atom in the ring has a p orbital, and it has nine pairs of π electrons.

    c is not aromatic, because it has two pairs of π electrons.

    a and **b** are not aromatic, because each compound has two pairs of π electrons and every atom in the ring does not have a p orbital.

    f is not aromatic, because it is not cyclic.

5. **a.** Solved in the text.

b. There are five monobromophenanthrenes.

6. To be aromatic, a compound must be cyclic and planar, every atom in the ring must have a *p* orbital, and the π cloud must contain an odd number of pairs of π electrons.

[10]-Annulene is cyclic, every atom in the ring has a *p* orbital, and it has the correct number of π electrons to be aromatic (five pairs). Knowing it is not aromatic, we can conclude that it is not planar.

[12]-Annulene has an even number of pairs of π electrons (six pairs) in its π cloud, so it cannot be aromatic.

7. **a.** sp^3 $CH_3CH_2\ddot{N}H_2$ **b.** sp^2 ⬡—CH=NCH_2CH_3   **c.** sp $CH_3CH_2C\equiv N:$

8. **a.** Notice that each resonance contributor has a net charge of 0.

b. Four ring atoms share the negative charge.

9.

10. **A** and **D**: the lone-pair electrons on these compounds are sp^2 electrons; therefore, they are not part of the π cloud. The lone-pair electrons on **B** and **C** are π electrons and are part of the π cloud. Thus, **B** has three pairs of π electrons and **C** has five pairs of π electrons, making them aromatic compounds. If any of

these π electrons are used to pick up a proton, they are no longer part of the π cloud (they cannot be delocalized) and the compounds will not be aromatic.

11. **a.** The nitrogen atom (the atom at the bottom of the epm) in pyrrole has a partial positive charge because it donates electrons by resonance into the ring.
  b. The nitrogen atom (the atom at the bottom of the epm) in pyridine cannot donate electrons by resonance; it withdraws electrons from the ring inductively because it is the most electronegative atom in the molecule. Thus, this nitrogen is electron rich.
  c. The relatively electronegative nitrogen atom in pyridine withdraws electrons from the ring.

12. Solved in the text.

13. **a.** In fulvene, the electrons in the exocyclic double bond (a double bond attached to a ring) move toward the five-membered ring, because the resonance contributor that results has a relatively large predicted stability since it is aromatic.

  fulvene

  b. In calicene, the electrons in the double bond between the two rings move toward the five-membered ring, because that results in a resonance contributor with two aromatic rings.

  calicene

14. **a.** Cyclopentadiene has a lower pK_a. When cyclopentadiene loses a proton, a relatively stable aromatic compound is formed. When cycloheptatriene loses a proton, an unstable antiaromatic compound is formed (Section 14.6). It is, therefore, easier to lose a proton from cyclopentadiene.

b. Cyclopropane has a lower pK_a, because an antiaromatic compound is formed when cyclopropene loses a proton.

antiaromatic

15. 3-Bromocyclopropene is more soluble in water, because it is more apt to ionize since heterolytic cleavage of its carbon-bromine bond forms an aromatic compound.

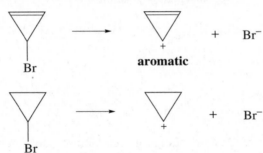

16. **c** is antiaromatic, because it is cyclic, planar, every atom in the ring has a *p* orbital, and it has two pairs of π electrons.

When you did Problem 4, you found that **d** and **e** are aromatic.

a, b, and **f** are neither aromatic nor antiaromatic: **a** and **b** have ring atoms that do not have *p* orbitals, and **f** is not cyclic.

17. Cyclobutadiene has one bonding molecular orbital, two nonbonding molecular orbitals, and one antibonding molecular orbital. It has four π electrons; two π electrons are in the bonding π molecular orbital, and each of the two nonbonding molecular orbitals contains one π electron.

18. It is not possible for an atom that bears a single electron to be part of the π cloud.

Remember that to be aromatic, all the bonding molecular orbitals that make up the π cloud must be filled (each containing two electrons) and there must be no partially filled orbitals. Since a radical has an unpaired electron, there will be an orbital with a single electron (a partially filled molecular orbital), so it cannot be aromatic. (A radical can, however, be attached to an aromatic compound. For example, the benzyl radical has a methylene radical attached to an aromatic benzene ring.)

19. The Frost circles show that compounds with completely filled bonding orbitals, and with electrons in no other orbitals, are aromatic (the cycloheptatrienyl cation and the cyclopropenyl cation).

cycloheptatrienyl cation   cyclopropenyl cation

The compound with unpaired electrons in degenerate orbitals is antiaromatic (the cycloheptatrienyl anion).

cycloheptatrienyl anion

20.   **a.** $CH_3CHCH_2CH_2CH_2CH_3$ **b.** ⟨benzene⟩CH_2OH **c.** $CH_3CH_2CHCH_2CH_3$ **d.** ⟨benzene⟩CH_2Br

with phenyl substituent under a, and CH_2 with phenyl under c.

21. Electrophilic addition to benzene is an endergonic reaction, because benzene is more stable than the substituted benzene that is formed as the product of the reaction. (See Figure 15.5 on page 638 of the text.)

An alkene is much less stable than benzene because it is not aromatic. Electrophilic addition to an alkene is an exergonic reaction because the alkene is less stable than the product of the reaction. (The overall reaction converts a σ bond and a π bond into two σ bonds; a σ bond is stronger and, therefore, of lower energy than a π bond.)

22. Ferric bromide activates Br_2 for nucleophilic attack by accepting a pair of electrons from it. Hydrated ferric bromide cannot do this because it has already accepted a pair of electrons from water.

$$H-\overset{+}{\underset{H}{\ddot{O}}}-\overset{Br}{\underset{Br}{\overset{|}{\underset{|}{Fe^-}}}}-Br$$

23. Solved in the text.

24. In sulfonation, **A** to **B** has a smaller rate constant (a higher energy barrier and, therefore, a slower reaction) than **B** to **C**; **A** to **B** is the rate-determining step.

In desulfonation (the reverse reaction), **C** to **B** has a smaller rate constant than **B** to **A**, but **B** to **A** is the rate-determining step. That is because once **B** is formed, it is easier for it to go back to **C** than to proceed to **A**, so **B** to **A** is the bottleneck (or rate-determining) step.

This example shows that the rate-determining step of a reaction is not necessarily the step with the smallest rate constant; it is the step that has the transition state with the highest energy on the reaction coordinate.

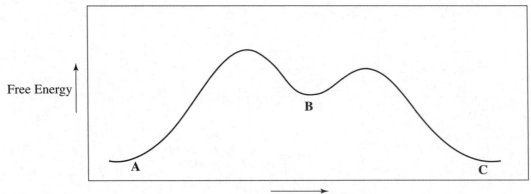

25.

26.

27.   **a.**   **c.**   **e.**

b.   **d.**   **f.**

28.   **a.**

b.

or

or

29.   The carbocation leading to 1-naphthol can be stabilized without destroying the aromaticity of the intact benzene ring. The carbocation leading to 2-naphthol can be stabilized by electron delocalization only by destroying the aromaticity of the intact benzene ring. Therefore, the carbocation leading to 1-naphthol is more stable.

carbocation that leads to 1-naphthol

carbocation that leads to 2-naphthol

30.   <u>without an NIH shift</u>

<u>with an NIH shift</u>

31.   The products obtained from the two reactions would not differ. The epoxide opens in the direction that gives the most stable carbocation. The carbocation undergoes an NIH shift and, as a result of the NIH shift, both reactants form the same ketone intermediate. Because they form the same intermediate, they form the same products. The deuterium-containing product is the major product, because in the last step of the reaction it is easier to break a carbon-hydrogen bond than a carbon-deuterium bond.

major product   minor product

32.   **a.**   Solved in the text.

b.   The compound without the double bond in the second ring is more apt to be carcinogenic. It opens to form a less stable carbocation than the other compound, because it can be stabilized by electron delocalization only if the aromaticity of the benzene ring is destroyed. Because the carbocation is less stable, it is formed more slowly, giving the carcinogenic pathway a better chance to compete with ring-opening.

less stable carbocation

more stable carbocation

33.   Each arene oxide will open in the direction that forms the most stable carbocation. Thus, the methoxy-substituted arene oxide opens so the positive charge can be stabilized by electron donation from the methoxy group.

more stable   **less stable**

The nitro-substituted arene oxide opens in the direction that forms the most stable carbocation intermediate, which is the one where the positive charge is farther away from the electron-withdrawing NO_2 group.

more stable   **less stable**

34.   **a.**   Note that a bond shared by two rings cannot be epoxidized.

I   **II**   **III**

b. The epoxide ring in phenanthrene oxides II and III can open in two different directions to give two different carbocations and, therefore, two different phenols.

c. The two different carbocations formed by phenanthrene oxides II and III differ in stability. One carbocation is more stable than the other because it can be stabilized by electron delocalization without disrupting the aromaticity of the adjacent ring. The more stable carbocation leads to the major product.

major product   **minor product**

major product   **minor product**

d. Phenanthrene oxide I is the most carcinogenic, because it is the only one that opens to form a carbocation that cannot be stabilized without disrupting the aromaticity of the other ring(s).

35. **a.** (phenol, OH)   **d.** (benzaldehyde, O=C–H)   **g.** (toluene, CH₃)

b. (C₆H₅–CH₂O–C₆H₅)   **e.** (anisole, OCH₃)   **h.** (tert-butylbenzene, CH₃/CCH₃/CH₃)

c. (C₆H₅–C≡N)   **f.** (styrene, CH=CH₂)   **i.** (C₆H₅–CH₂Cl)

36. **aromatic**

antiaromatic

neither aromatic or antiaromatic

Each of the following compounds has an sp^3 carbon, so they do not have an uninterrupted π cloud.

Cyclooctatetraene is not planar, so the p orbitals cannot overlap with the p orbitals on both of the adjacent carbons.

37. **a.** (C₆H₅–CCH₃ with CH₃/CH₃)   **b.** (C₆H₅–CCH₂CH₃ with CH₃/CH₃)   **c.** (C₆H₅–CH₂–C₆H₅)

38.   **a.** (cyclopropenyl cation structure)   **b.** (cycloheptatrienyl cation structure)   **c.** (cyclopentadienyl anion structure)   **d.** (cyclobutadiene square structure)

It is aromatic.   It is aromatic.   It is aromatic.   It is not antiaromatic.

39.   The methyl group on benzene can lose a proton easier than the methyl group on cyclohexane, because the electrons left behind in the former case can be delocalized into the benzene ring.

Electrons left behind   Electrons left behind
can be delocalized.   cannot be delocalized.

40.   **a.**

b.

41.   The ^{1}H NMR spectrum is the spectrum of butylbenzene: the benzene ring protons show a signal at ~7.2 ppm. The two triplets and two multiplets indicate a butyl substituent.

Therefore, the acyl chloride has a straight chain propyl group and a carbonyl group that will be reduced to a methylene group.

42.   The first compound is the most stable (weakest) base, because it has a resonance contributor that is aromatic. Therefore, the other base (the one on the right) is the stronger base.

aromatic

43.   The following alkyl halide reacts more rapidly in an S_N1 reaction, because it forms the more stable carbocation; the carbocation is more stable, because it has a resonance contributor that is aromatic.

+ Br$^-$ aromatic

44. The resonance contributor that is shown indicates which nitrogen is most apt to be protonated (the one with the greatest negative charge) and which nitrogen is least apt to be protonated (the one with the greatest positive charge).

45. The compound is 1,3-dibromobenzene. 1,4-Dibromobenzene would form only one product and 1,2-dibromobenzene would form only two.

46. Boron trifluoride is a Lewis acid that can form a complex with oxygen. This allows the alcohol to be converted to an electrophile that can react with the benzene ring.

47.   **a.**   Hydrogen chloride adds to the alkene, forming a secondary carbocation that undergoes a 1,2-hydride shift to form a tertiary carbocation. The tertiary carbocation is an electrophile that can either add to the alkene in a second molecule of the reactant (in an intermolecular reaction) or add to the benzene ring in the same molecule of the reactant (in an intramolecular reaction).

The intramolecular reaction is favored because it forms a stable five-membered ring. (See Section 8.10 in the text). After the electrophile adds to the benzene ring, a base (B:) in the reaction mixture removes a proton and the aromaticity of the benzene ring is restored.

an intramolecular reaction

b.   As in part **a**, an electrophile is formed that can react in either an intermolecular reaction or in an intramolecular reaction. Seeing that the product of the reaction has two benzene rings and that there are twice as many carbons in the product as in the reactant suggests that two reactant molecules react in an intermolecular reaction. In this case, the intermolecular reaction is favored, because the intramolecular reaction cannot occur since it would lead to a highly strained three-membered ring. The electrophile that is formed in the intermolecular reaction can add to the benzene ring in an intramolecular reaction to form a stable five-membered ring.

48.   The signal at ~7 ppm that integrates to 5H indicates a monosubstituted benzene ring. A monosubstituted benzene ring contains six carbons and five hydrogens. Subtracting these atoms from the molecular formula of $C_{13}H_{20}$ gives us a substituent with 7 carbons and 15 hydrogens. The compound shown below has the correct number of carbons and hydrogens that will give two singlets with one (9H) having 1.5 times the area of the second (6H).

49.   **a.**   If an NIH shift occurs, both carbocations will form the same intermediate ketone. Because it is about four times easier to break a C—H bond (k_3) compared with a C—D bond (k_4), about 80% of the deuterium will be retained.

b.   If an NIH shift does not occur, 50% of the deuterium will be retained, because the epoxide can open equally easily in either direction (k_1 is equal to k_2) and subsequent loss of H$^+$ or D$^+$ is fast.

50. The following compound is the strongest acid, because it is the only one that forms a conjugate base that is aromatic. Recall that the more stable (weaker) the base, the stronger is its conjugate acid.

aromatic

51. a.

HB^+

b.

HB^+

52. a. The three resonance contributors marked with an X are the least stable, because in these contributors the two negative charges are on adjacent carbons.

b. Because these contributors are the least stable, they make the smallest contribution to the hybrid.

53.   We have classified cyclobutadiene as antiaromatic. However, the recent observation that cyclobutadiene is rectangular and the observation that there are two different 1,2-dideuterio-1,3-cyclobutadienes both indicate that the π electrons are localized rather than delocalized. Localization of the π electrons prevents the compound from being antiaromatic. Apparently, the extreme instability associated with being antiaromatic causes cyclobutadiene to be rectangular.

54.   From the three resonance structures for the cyclohexadienyl anion, the bond orders in the resonance hybrid can be calculated. (For example, the C1-C2 bond is represented by a single bond in one resonance contributor and by a double bond in two resonance contributors, which gives it a bond order $1 + 2/3$ in the resonance hybrid.)

contributing resonance structures

resonance hybrid

1,4-cyclohexadiene   1,3-cyclohexadiene

Comparing the bond order difference in each bond in the anion and in the corresponding bond in the possible products, there is a $(1/3 + 1/3 + 1/3 + 1/3 = 4/3)$ 4/3 difference in bond order when 1,4-cyclohexadiene is formed and a $(1/3 + 1/3 + 2/3 + 2/3 = 6/3)$ 6/3 difference in bond order when 1,3-cyclohexadiene is formed. Thus, the principle of least motion predicts that 1,4-cyclohexadiene will be the major product, since its formation from the anion involves less of a change in electronic configuration.

Chapter 15 Practice Test

1.   Which are aromatic compounds?

2.   Which are antiaromatic compounds?

3.   Which compound has the greater delocalization energy?

4.   Draw the resonance contributors for the carbocation intermediate that is formed when benzene reacts with an electrophile (Y^+).

5.   Which of the following is a stronger acid?

   a.

   b.

6.   What acid anhydride would you use in a synthesis of propylbenzene?

7.   Which in each of the following pairs is more stable?

a.   **or**

b. $CH_3\overset{+}{C}HCH_2CH_3$ **or** $CH_3\overset{+}{C}HCH=CH_2$

c. $CH_3\overset{-}{C}HCH_2\overset{O}{\overset{||}{C}}CH_3$ **or** $CH_3\overset{-}{C}H\overset{O}{\overset{||}{C}}CH_3$

8.   Give the mechanism for formation of the nitronium ion from nitric acid and sulfuric acid.

CHAPTER 16
Reactions of Substituted Benzenes

Important Terms

activating substituent	a substituent that increases the reactivity of an aromatic ring. Electron-donating substituents activate aromatic rings toward electrophilic attack, and electron-withdrawing substituents activate aromatic rings toward nucleophilic attack.
arenediazonium salt	$Ar \overset{+}{N} \equiv N \ X^-$
azo linkage	a $-N{=}N-$ bond.
benzyne intermediate	a compound with a triple bond in place of one of the double bonds of benzene.
cine substitution	substitution at the carbon adjacent to the carbon that was bonded to the leaving group.
deactivating substituent	a substituent that decreases the reactivity of an aromatic ring. Electron-withdrawing substituents deactivate aromatic rings toward electrophilic attack, and electron-donating substituents deactivate aromatic rings toward nucleophilic attack.
direct substitution	substitution at the carbon that was bonded to the leaving group.
donate electrons by resonance (resonance electron donation)	donation of electrons through p orbital overlap with neighboring π bonds.
inductive electron withdrawal	withdrawal of electrons through a σ bond.
meta director	a substituent that directs an incoming substituent meta to an existing substituent.
nitrosamine (*N*-nitroso compound)	an amine with a nitroso ($N{=}O$) substituent bonded to its nitrogen atom.
nucleophilic aromatic substitution	a reaction in which a nucleophile substitutes for an atom or group bonded to an aromatic ring.
ortho-para director	a substituent that directs an incoming substituent ortho and para to an existing substituent.
Sandmeyer reaction	the reaction of an arenediazonium ion with a cuprous salt.
Schiemann reaction	the reaction of an arenediazonium ion with HBF_4.
S$_N$Ar reaction	a nucleophilic aromatic substitution reaction.
withdraw electrons by resonance (resonance electron withdrawal)	withdrawal of electrons through p orbital overlap with neighboring π bonds.

Solutions to Problems

1. **a.** benzene-COOH

c. benzene-CH$_2$OCH$_3$

b. benzene-COOH, COOH (1,3)

d. benzene-CH$_2$CH$_2$NH$_2$

2. **a.** Solved in the text.

b.

benzene $\xrightarrow[\text{AlCl}_3]{\text{CH}_3\text{CH}_2\text{Cl}}$ CH$_2$CH$_3$ $\xrightarrow[\text{peroxide}]{\text{NBS, }\Delta}$ CHBrCH$_3$ $\xrightarrow{\textit{tert}\text{-BuO}^-}$ CH=CH$_2$

or Br$_2$ | FeBr$_3$

benzene-Br $\xrightarrow[\substack{\text{PdL}_2\\(\text{CH}_3\text{CH}_2)_3\text{N}}]{\text{CH}_2\text{=CH}_2}$ CH=CH$_2$

c. CH=CH$_2$ (product of **b**) $\xrightarrow[\text{peroxide}]{\text{HBr}}$ CH$_2$CH$_2$Br

d. CH=CH$_2$ (product of **b**) $\xrightarrow[\text{2. HO}^-,\text{ H}_2\text{O}_2,\text{ H}_2\text{O}]{\text{1. BH}_3}$ CH$_2$CH$_2$OH

e. benzene $\xrightarrow[\text{H}_2\text{SO}_4]{\text{HNO}_3}$ NO$_2$ $\xrightarrow[\substack{\textbf{or}\\\text{1. Sn/HCl}\\\text{2. HO}^-}]{\substack{\text{H}_2\\\text{Pd/C}}}$ NH$_2$

f. benzene $\xrightarrow[\text{AlCl}_3]{\text{CH}_3\text{Cl}}$ CH$_3$ $\xrightarrow[\Delta]{\text{Na}_2\text{Cr}_2\text{O}_7,\text{ HCl}}$ COOH

3. **a.** *ortho*-ethylphenol or 2-ethylphenol

b. *meta*-bromochlorobenzene or 1-bromo-3-chlorobenzene

c. *meta*-bromobenzaldehyde or 3-bromobenzaldehyde

d. *ortho*-ethyltoluene or 2-ethyltoluene

4. **a.**   **b.**   **c.** CH_3 ... CH_3 **d.** SO_3H ... Cl

5. **a.**   **c.** NH_2 ... NO_2 **e.** OH ... Br ... I   **g.** $HC{=}O$ ... NO_2 ... O_2N

b. OH ... Br   **d.** $C{\equiv}N$ ... Cl   **f.** Cl ... Cl   **h.** NH_2 ... Cl ... Br

6. **a.** 1,3,5-tribromobenzene   **c.** *para*-bromotoluene or 4-bromotoluene

b. *meta*-nitrophenol or 3-nitrophenol   **d.** *ortho*-dichlorobenzene or 1,2-dichlorobenzene

7. **a.** donates electrons by resonance and withdraws electrons inductively

b. donates electrons by hyperconjugation

c. withdraws electrons by resonance and withdraws electrons inductively

d. donates electrons by resonance and withdraws electrons inductively

e. donates electrons by resonance and withdraws electrons inductively

f. withdraws electrons inductively

8. **a.** phenol > toluene > benzene > bromobenzene > nitrobenzene

b. toluene > chloromethylbenzene > dichloromethylbenzene > difluoromethylbenzene

9. Solved in the text.

10.   a.

b.

11.   a.

c.

e.

b.

d.

f.

12.   They are all meta directors:

a.   this group withdraws electrons by resonance from the ring. The relatively electronegative nitrogen atom causes it to also withdraw electrons inductively from the ring.

b. NO_2 withdraws electrons inductively and withdraws electrons by resonance.

c. CH_2OH withdraws electrons inductively from the ring.

d. COOH withdraws electrons inductively and withdraws electrons by resonance.

e. CF_3 withdraws electrons inductively from the ring.

f. N=O withdraws electrons inductively and withdraws electrons by resonance.
You could draw resonance contributors for electron donation into the ring by resonance. However, the most stable resonance contributors are obtained by electron flow out of the benzene ring toward oxygen, the most electronegative atom in the compound.

**resonance electron donation
into the ring**   **resonance electron withdrawal
out of the ring**

13.   For each compound, determine which benzene ring is more highly activated. The more highly activated ring is the one that undergoes electrophilic aromatic substitution.

a. Solved in the text.

b.

c. Because the right hand ring is highly activated and a catalyst is employed, monobrominated, dibrominated, and tribrominated compounds can be obtained. However, since only one equivalent of Br_2 is available, there is not enough bromine to brominate all the reactive positions. Therefore, a mixture of unreacted starting material and the four products shown below will be obtained.

d. Because the left hand ring is highly activated and a catalyst is employed, both ortho positions can be brominated. Because there is not enough Br_2 available to brominate both reactive positions, a mixture of unreacted starting material and the two products shown below will be obtained.

14. For **a, b** and **c**: the closer the electron withdrawing substituent is to the COOH group, the stronger the acid.

a. $ClCH_2COOH$

b. O_2NCH_2COOH

c. $\overset{+}{H_3N}CH_2COOH$

d.

The predominant effect of the CH_3O group is to donate electrons by resonance, so it decreases the acidity of the carboxylic acid.

e. $HOOCCH_2COOH$

It is easier to lose the first proton from a dicarboxylic acid than the second proton. (See Table 17.2 on page 754.)

f. HCOOH

A hydrogen is electron-withdrawing compared to a methyl group, since a methyl group can donate electrons by hyperconjugation.

g. FCH_2COOH

Fluorine is more eletronegative than chlorine.

h.

We know that F withdraws electrons inductively more than Cl does, because F is more electronegative than Cl.
We know that F donates electrons by resonance better than Cl does, because F donation involves 2p-2p overlap, whereas Cl donation involves 3p-2p overlap.

Because Table 16.1 on page 669 of the text shows that F is less deactivating than Cl toward electrophilic aromatic substitution, we know that overall F donates electrons better than Cl. Thus, resonance electron donation is more important than inductive electron withdrawal. (See the answer to Problem 9.)

15. When *para*-nitrophenol loses a proton, the negative charge in the conjugate base can be delocalized onto the nitro substituent. Therefore, the *para*-nitro substituent decreases the pK_a by resonance electron withdrawal and by inductive electron withdrawal.

When *meta*-nitrophenol loses a proton, the negative charge in the conjugate base cannot be delocalized onto the nitro substituent. Therefore, the *meta*-nitro substituent can decrease the pK_a only by inductive electron withdrawal. Therefore, the para isomer has a lower pK_a.

16. **a.** Because fluorine and chlorine both deactivate the benzene ring toward electrophilic aromatic substitution, they both increase acidity compared to unsubstituted benzoic acid. Problem 9 explains that fluorine is the weakest deactivator of the halogens, because electron donation by resonance is more important than inductive electron withdrawal. Therefore, *o*-chlorobenzoic acid is the strongest acid of the three.

b. Benzoic acid is the weakest acid of the three, because it does not have a deactivating (overall electron withdrawing) substituent on the ring.

17. Notice that in all three syntheses, the Friedel-Crafts reaction has to be done first. Both NO_2 and SO_3H are meta directors, and a Friedel-Crafts reaction cannot be carried out if there is a meta director on the ring.

18. No reaction will occur in **a** and **c**, because a Friedel-Crafts reaction cannot be carried out on a ring that possesses a meta director.

a. no reaction

c. no reaction

b. 2,4,6-tribromoaniline (NH_2 ring with Br at 2, 4, 6 positions)

d. p-xylene + o-xylene (CH_3 substituted benzenes)

19. **a.** Unlike *m*-dipropylbenzene that requires a coupling reaction to be used in its synthesis, a coupling reaction does not have to be used in the synthesis of *p*-dipropylbenzene. The propyl group is an ortho/para director, so propylbenzene can undergo a Friedel-Crafts reaction.

benzene $\xrightarrow[\text{2. }H_2O]{\text{1. }CH_3CH_2CCl/AlCl_3}$ propiophenone ($C_6H_5COCH_2CH_3$) $\xrightarrow[\text{Pd/C}]{H_2}$ propylbenzene ($C_6H_5CH_2CH_2CH_3$)

$\xrightarrow[\text{2. }H_2O]{\text{1. }CH_3CH_2CCl/AlCl_3}$ p-(CH_3CH_2CO)($CH_2CH_2CH_3$)benzene $\xrightarrow[\text{Pd/C}]{H_2}$ *p*-dipropylbenzene ($CH_3CH_2CH_2$ and $CH_2CH_2CH_3$ para)

b. A coupling reaction can be used in the synthesis of *p*-dipropylbenzene. Notice that bromination occurs after the reduction of the carbonyl group. In contrast, bromination occurs before the reduction reaction in the synthesis of *m*-dipropylbenzene.

benzene $\xrightarrow[\text{2. }H_2O]{\text{1. }CH_3CH_2CCl/AlCl_3}$ propiophenone ($C_6H_5COCH_2CH_3$) $\xrightarrow[\text{Pd/C}]{H_2}$ propylbenzene ($C_6H_5CH_2CH_2CH_3$)

$\xrightarrow[\text{FeBr}_3]{Br_2}$ 4-bromopropylbenzene (Br and $CH_2CH_2CH_3$ para)

$\xleftarrow[\text{PdL}_2\text{, NaOH}]{CH_3CH_2CH_2B(OR)_2}$ *p*-dipropylbenzene ($CH_3CH_2CH_2$ and $CH_2CH_2CH_3$ para)

or

$\xleftarrow[\text{THF}]{(CH_3CH_2CH_2)_2CuLi}$

20. In Problems **c**, **g**, and **h**, the carbonyl group can be converted to a methylene group by H₂Pd/C (as is used in **c**), or by a Wolff-Kishner reduction (as is used in **g**), or by a Clemmensen reduction (as is used in **h**).

a.

b.

c. Note that benzaldehyde is formed by the Gatterman-Koch reaction (page 645 in the text.)

d.

e.

f.

g.

h.

i. Notice that in the second step of the synthesis, oxymercuration-reduction (Section 4.9) is used to add water to the double bond in order to avoid the carbocation rearrangement that would occur with the acid-catalyzed addition of water.

21. a.

COOH is a meta director and Br is an ortho-para director, so both direct to the same position.

b.

COOH directs to the meta position. The same product will be obtained regardless of which COOH is the director.

Less of this compound will be obtained because of steric hindrance.

c.

COOH directs to its meta
position and Cl directs
to its ortho position,
so they both direct to the
same position on the ring.

d.

A methoxy substituent is strongly activating
whereas a fluorine substituent is deactivating,
so the methoxy substituent will do the directing.

e.

The aldehyde group is a meta director and
the methoxy group is an ortho-para director,
so both direct to the same position.

f.

Less of this product will be obtained,
because of steric hindrance.

22.   **a.**   Two products will be obtained.

b. One product will be obtained.

c. Two products could be obtained, but little, if any, of one of the products will be obtained because of steric hindrance.

23. Solved in the text.

24. Because a diazonium ion is electron withdrawing, it deactivates the benzene ring. A deactivated benzene ring would be too unreactive to undergo an electrophilic substitution reaction at the cold temperature necessary to keep the benzenediazonium ion from decomposing.

25. More of the desired bromo-substituted compound will be obtained if the halide ion in solution is only Br^- than if it is both Br^- and Cl^-.

26. $FeBr_3$ would complex with the amino group, converting it into a meta director. The NH_2 group is a strongly activating substituent, so a Lewis acid is not needed.

27.

28.   **a.**   Solved in the text.

b.   The reaction steps are the same as those in **a** (answered on page 686 of the text) except for the last step. In the last step, the meta-bromo-substituted diazonium ion should undergo reaction with water rather than with CuBr.

c.   The first approach is longer but it will generate a higher yield of the ortho isomer.

or

d. The nitro group cannot be placed on the benzene ring first, because a Friedel-Crafts reaction cannot be carried out on a ring with a meta director. Because formyl chloride is too unstable to be purchased, benzaldehyde is prepared by the Gatterman-Koch reaction (page 645 in the text).

e.

f.

or

29. Because the para position is occupied, the electrophile will add to the ortho position. (Attack at the ortho position is slower than attack at the para position because of steric hindrance, but since the para position is not available, the slow reaction will prevail.)

30. You can see why nucleophilic attack occurs on the neutral nitrogen if you compare the products of nucleophilic attack on the two nitrogens. Nucleophilic attack on the neutral nitrogen forms a stable product, whereas nucleophilic attack on the positively charged nitrogen would form an unstable compound with two charged nitrogen atoms.

compared to

The terminal nitrogen is electrophilic because of electron withdrawal by the positively charged nitrogen. If you draw the resonance contributors, you can see that the "neutral" nitrogen is electron deficient.

31. **a.**

activated ring + diazonium ion   **b.** activated ring + diazonium ion

32. The reaction of an amine with sodium nitrite and HCl converts the amino group into an excellent leaving group. Substitution and elimination reactions can then occur by both $S_N1/E1$ and $S_N2/E2$ pathways. The same products are obtained from both pathways. Because the reaction is carried out in an aqueous solution, the nucleophiles are Cl^- and H_2O.

33. Note: Diazomethane is both explosive and toxic, so it should be synthesized only in small amounts by experienced laboratory workers.

The first step in the reaction is formation of the methyldiazonium ion as a result of removal of a proton from the carboxylic acid by diazomethane.

In the second step of the reaction, the carboxylate ion displaces nitrogen gas (N_2) from the methyldiazonium ion. High yields are obtained, since the only side product is a gas.

34. From the resonance contributors, you can see that the reason that *meta*-chloronitrobenzene does not react with hydroxide ion is because the negative charge that is generated on the benzene ring cannot be delocalized onto the nitro substituent.

Electron delocalization onto the nitro substituent can occur only if the nitro substituent is ortho or para to the site of nucleophilic attack.

35. **a.** 1-chloro-2,4-dinitrobenzene > *p*-chloronitrobenzene > chlorobenzene

b. chlorobenzene > *p*-chloronitrobenzene > 1-chloro-2,4-dinitrobenzene

36. **a.**

b.

c.

The target compound can also be prepared by the method shown in Problem 78a, omitting the last step.

d. This synthesis would be carried out exactly as **c** (above) except for the last step, which would be:

37.

38. **a.**

c.

b.

d.

39. **a.**

c.

e.

g.

b.

d.

f.

h.

40. **a.** *m*-bromobenzoic acid   **d.** *p*-nitrostyrene   **g.** *o*-bromotoluene
   b. 1,2,4-tribromobenzene   **e.** *m*-ethylanisole   **h.** *p*-cyclohexyltoluene
   c. 2,6-dimethylphenol   **f.** 3,5-dichlorobenzenesulfonic acid   **i.** 2-chloro-4-ethylazobenzene

41.

42.   **a.**   The halogens withdraw electrons inductively and donate electrons by resonance. Because they all deactivate the benzene ring toward electrophilic aromatic substitution, we know that their electron withdrawing effect is stronger than their electron donating effect. Therefore, an *ortho*-halo-substituted benzoic acid will be a stronger acid than benzoic acid.

b.   Because fluorine is the weakest deactivator of the halogens (Table 16.1), we know that overall it donates electrons better than the other halogens. Therefore, *ortho*-fluorobenzoic acid is the weakest of the *ortho*-halo-substituted benzoic acids.

c.   The smaller the halogen, the more electronegative it is and therefore the better it is at withdrawing electrons inductively. The smaller the halogen, the better it is at donating electrons by resonance because a $2p$ orbital of carbon overlaps better with a $2p$ orbital of a halogen than with a $3p$ orbital of a halogen and overlaps better with a $3p$ orbital than with a $4p$ orbital. Therefore, Br does not withdraw electrons as well as Cl, and Br does not donate electrons as well as Cl, so their pK_a values are similar.

43.

44. CH_3O—⬡—NH_2 > CH_3—⬡—NH_2 > Br—⬡—NH_2 > $CH_3\overset{O}{\underset{\|}{C}}$—⬡—$NH_2$

45.   Because the $2p$ orbital of oxygen overlaps better than the $3p$ orbital of sulfur with the $2p$ orbital of carbon, oxygen is better at donating electrons by resonance than sulfur. Therefore, the benzene ring of anisole is more activated toward electrophilic aromatic substitution that is the benzene ring of thioanisole.

46. CH_3O—⬡—OCH_3

47.   These are intramolecular Friedel-Crafts acylation reactions.

a.   **b.**

48.   These are intramolecular Friedel-Crafts alkylation reactions. Notice that the major product in part **a** results from a carbocation rearrangement.

a. **major** +   **minor**   **b.**

49.   **a.**

b.

c.

d.

e.

f.

g.

h.

50.   **1.**

most reactive

least reactive
highest % meta product

2.

least reactive
highest % meta product

most reactive

3.

most reactive

least reactive
highest % meta product

51.   **a.**   anisole > ethylbenzene > benzene > chlorobenzene > nitrobenzene

b.   2,4-dinitrophenol > 2,4-dinitrotoluene > 1-chloro-2,4-dinitrobenzene

c.   *p*-cresol > *p*-xylene > toluene > benzene

d.   phenol > propylbenzene > benzene > benzoic acid

e.   *p*-chlorotoluene > *p*-nitrotoluene > 2-chloro-4-nitrotoluene > 2,4-dinitrotoluene

f.   fluorobenzene > chlorobenzene > bromobenzene > iodobenzene

52.   **a.**

b.

c.

d.

e.

f.

53. **a.** CH_2CH_3 donates electrons by hyperconjugation and does not donate or withdraw electrons by resonance.

b. NO_2 withdraws electrons inductively and withdraws electrons by resonance.

c. Br deactivates the ring and directs ortho/para.

d. OH withdraws electrons inductively, donates electrons by resonance, and activates the ring.

e. $^+NH_3$ withdraws electrons inductively and does not donate or withdraw electrons by resonance.

54. Three ways to synthesize anisole from benzene are shown below.

1.

2.

3.

55.

a.

c.

e.

g.

b.

d.

f.

h.

56.

1.

2.

57.

a.

b.

$$\text{benzene} \xrightarrow[\text{AlCl}_3]{\text{CH}_3\text{Cl}} \text{CH}_3 \xrightarrow[\text{peroxide}]{\text{NBS, }\Delta} \text{CH}_2\text{Br} \xrightarrow{\text{CH}_3\text{O}^-} \text{CH}_2\text{OCH}_3$$

c.

$$\text{benzene} \xrightarrow[\text{FeCl}_3]{\text{Cl}_2} \text{Cl} \xrightarrow[\text{AlCl}_3]{\text{CH}_2\text{Cl}_2} \text{Cl}\text{—}\text{CH}_2\text{Cl} \xrightarrow[\text{AlCl}_3]{} \text{Cl}\text{—}\text{CH}_2\text{—}\text{Ph}$$

d.

$$\text{benzene} \xrightarrow[\text{AlCl}_3]{\text{CH}_3\text{Cl}} \text{CH}_3 \xrightarrow[\text{H}_2\text{SO}_4]{\text{HNO}_3} \text{CH}_3, \text{NO}_2 \xrightarrow[\text{or}]{\substack{1.\ \text{Sn/HCl} \\ 2.\ \text{HO}^-}}_{\text{H}_2,\ \text{Pd/C}} \text{CH}_3, \text{NH}_2$$

$$\xrightarrow[\text{HCl}]{\text{NaNO}_2}\ 0°$$

$$\text{O}_2\text{N}\text{—}\text{CH}_3,\text{OH} \xleftarrow[\text{H}_2\text{SO}_4]{\text{HNO}_3} \text{CH}_3,\text{OH} \xleftarrow[\substack{\text{Cu(NO}_3)_2, \\ \text{H}_2\text{O} \\ \textbf{or} \\ \text{H}_3\text{O}^+,\ \Delta}]{\text{Cu}_2\text{O}} \text{CH}_3, \text{N}^+\!\!\equiv\!\text{N}\ \text{Cl}^-$$

e.

$$\text{benzene} \xrightarrow[\text{FeBr}_3]{\text{Br}_2} \text{Br} \xrightarrow[\text{H}_2\text{SO}_4]{\text{HNO}_3} \text{O}_2\text{N}\text{—}\text{Br} \xrightarrow[\Delta]{^-\text{NH}_2} \text{O}_2\text{N}\text{—}\text{NH}_2$$

f.

$$\text{benzene} \xrightarrow[\text{H}_2\text{SO}_4]{\text{HNO}_3} \text{NO}_2 \xrightarrow[\text{FeBr}_3]{\text{Br}_2} \text{NO}_2, \text{Br} \xrightarrow[\substack{\textbf{or} \\ \text{H}_2,\ \text{Pd/C}}]{\substack{1.\ \text{Sn/HCl} \\ 2.\ \text{HO}^-}} \text{NH}_2, \text{Br}$$

$$\xrightarrow[]{0°\ \substack{\text{NaNO}_2 \\ \text{HCl}}}$$

$$\text{I},\text{Br} \xleftarrow{\text{KI}} \text{N}^+\!\!\equiv\!\text{N}\ \text{Cl}^-,\text{Br}$$

g.

h.

i.

58. *para*-Xylene is more reactive than benzene toward electrophilic aromatic substitution. The methyl groups activate the ortho positions, so all four positions on the ring are activated. Attack on any one of the four leads to the same compound, so the product of the reaction will be the one shown below.

para-xylene

59. The compound with the methoxy substituent is the more reactive because it forms the more stable carbocation intermediate. The carbocation intermediate is stabilized by resonance electron donation.

60. *meta*-Xylene will react more rapidly. In *meta*-xylene both methyl groups activate the same position, whereas in *para*-xylene each methyl group activates a different position.

61. **a.** **b.** **c.**

62.   Recall that the highest frequency signal is the one with the most deshielded proton, that is, the one with the least electron density. (The signal farthest to the left on the spectrum.)

    a.   The oxygen withdraws electrons inductively from the carbon bonded to the hydrogens being compared, so the oxygen-containing compound has the signal at the highest frequency.

$$CH_3O - CH_2C\underline{H}_3$$

    b.   The oxygen donates electrons by resonance, so the oxygen-containing compound has the signal at the lowest frequency.

$$CH_3\ddot{O} - CH = CH_2 \longleftrightarrow CH_3\overset{+}{O} = CH - \ddot{C}H_2$$

Therefore, the non-oxygen-containing compound has the signal at the highest frequency.

$$CH_3CH = C\underline{H}_2$$

63.   Yes, the advice is sound. The para isomer will form one product, because the formyl and ethyl groups both direct to the same positions and both positions result in the same product. The ortho isomer will form two products, because the formyl and ethyl groups both direct to the same positions but different products are obtained from each position. The meta isomer will form as many as four products, because the formyl and ethyl groups direct to four different positions and a different product is obtained from each position.

64.

65.

A   B   C   D   E

F   G   H   I

J   K   L

66. The carbocation formed by putting an electrophile at the ortho or para positions can be stabilized by resonance electron donation from the phenyl substituent.

The carbocation formed by putting an electrophile at the meta position cannot be stabilized by resonance electron donation from the phenyl substituent.

67. The chloro substituent primarily withdraws electrons inductively. (It only minimally donates electrons by resonance.) The closer it is to the COOH group, the more it can withdraw electrons from the OH bond and the stronger the acid. Therefore, the ortho isomer is the strongest acid and the para isomer is the weakest acid.

The nitro substituent withdraws electrons inductively. It also withdraws electrons by resonance if it is ortho or para to the COOH group. Therefore, the ortho and para isomers are the strongest acids, and the ortho isomer is a stronger acid than the para isomer because of the greater inductive electron withdrawal from the closer position.

The amino substituent primarily donates electrons by resonance, but it can donate electrons by resonance to the COOH group only if it is ortho or para to it. Therefore, the meta isomer is the strongest acid. Seeing that the ortho isomer is the weakest acid tells us that resonance electron donation to the COOH group is more efficient from the ortho position.

68. The spectrum indicates that the benzene ring has a substituent with two different kinds of hydrogens. The doublet and multiplet indicate that the substituent is an isopropyl group. Therefore, compound A is isopropylbenzene.

isopropylbenzene

69. **1.**

2.

70.   a. The hydroxy-substituted carbocation intermediate is more stable, because the positive charge can be stabilized by resonance electron donation from the OH group.

b. The carbanion with the negative charge meta to the nitro group is more stable, because a negative charge in the meta position can be delocalized onto the nitro group but a negative charge in the ortho position cannot.

71. Phenol will undergo electrophilic aromatic substitution (D^+ is the electrophile) primarily at the ortho and para positions. Notice that the OH proton will also be exchanged for a deuterium.

72.   **a.**

or

b.

c.

d.

73. The unsplit signal at ~7.1 ppm suggests that all the benzene ring hydrogens in the product are chemically equivalent.

CH₃CH₂—⟨benzene⟩ $\xrightarrow[\text{AlCl}_3]{\text{CH}_3\text{CH}_2\text{Cl}}$ CH₃CH₂—⟨benzene⟩—CH₂CH₃

74. **a.** The first three compounds will not show a carbonyl stretch at 1700 cm⁻¹ and the bottom four will show this absorption band. The first three can be distinguished from one another by the presence or absence of the indicated absorption bands.

⟨cyclohexane⟩—CH₂OH   ⟨benzene⟩—CH₂OH   ⟨benzene⟩—CH₂OCH₃

band at 3300 cm⁻¹   band at 3300 cm⁻¹   no band at 3300 cm⁻¹
no band at 1600 cm⁻¹   band at 1600 cm⁻¹   band at 1600 cm⁻¹

The last four compounds all have an absorption band at 1700 cm⁻¹.
They can be distinguished by the presence or absence of the indicated absorption bands.

⟨benzene⟩—C(=O)—OH   ⟨benzene⟩—C(=O)—H   ⟨benzene⟩—C(=O)—OCH₃   ⟨benzene⟩—C(=O)—OCH₃

large broad band at   no band at   no band at   no band at
~2500–3500 cm⁻¹   ~2500–3500 cm⁻¹   ~2500–3500 cm⁻¹   ~2500–3500 cm⁻¹
band at 1250 cm⁻¹   no band at 1250 cm⁻¹ band at 1250 cm⁻¹   no band at 1250 cm⁻¹
no band at 2700 cm⁻¹   band at 2700 cm⁻¹   no band at 2700 cm⁻¹ no band at 2700 cm⁻¹

b. This is the only compound without the characteristic benzene ring hydrogens at ~7–8 ppm.

⟨cyclohexane⟩—CH₂OH

Only two compounds will have two signals other than the signals for the benzene ring hydrogens. They can be distinguished by integration (3:2 versus 2:1), or by the two sharp singlets for the ether, versus the somewhat broader singlet for the hydrogen bonded to oxygen.

⟨benzene⟩—CH₂OCH₃   ⟨benzene⟩—CH₂OH

Each of the following four compounds has only one signal (a singlet) in addition to the benzene ring hydrogens. The four can be distinguished by the positions of the signals.

⟨benzene⟩—C(=O)—CH₃   ⟨benzene⟩—C(=O)—OCH₃   ⟨benzene⟩—C(=O)—H   ⟨benzene⟩—C(=O)—OH

~2 ppm   ~4 ppm   ~9–10 ppm   ~10–12 ppm

75.   The rate-determining step in the S$_N$1 reaction is the formation of the tertiary carbocation. An electron-donating substituent will stabilize the carbocation and cause it to be more easily formed. An electron-withdrawing substituent will destabilize the carbocation and cause it to be less easily formed.

76.   A fluoro substituent is more electronegative than a chloro substituent. Therefore, nucleophilic attack on the carbon bearing the fluoro substituent will be easier than nucleophilic attack on the carbon bearing the chloro substituent. In addition, the smaller fluoro substituent provides less steric hindrance to attack by the nucleophile.

A fluoro substituent is a stronger base than a chloro substituent, so elimination of the halogen in the second step of the reaction will be harder for a fluoro-substituted benzene than for a chloro-substituted benzene.

The fact that the fluoro-substituted compound is more reactive tells you that attack of the nucleophile on the aromatic ring is the rate-determining step of the reaction.

77.

78.   **a.**

b.

79. **a.** The alkyl diazonium ion is very unstable. Loss of N_2 and a 1,2-hydride shift forms a *tert*-butyl carbocation, which can undergo either substitution or elimination.

b. The cation formed from the diazonium ion will undergo a pinacol rearrangement. (See Chapter 10, Problem 62.)

80.

81. The configuration of the asymmetric center in the reactant will be retained only if the asymmetric center undergoes two successive S_N2 reactions. In the first step of the reaction, the NH_2 group is converted to a good leaving group. The first S_N2 reaction involves the carboxylate ion that is closest to the carbon attached to the leaving group. The product of the reaction is a strained three-membered ring that reacts with the other carboxylate ion in a second S_N2 reaction, forming a more stable five-membered ring.

82.   A chloro group is a better leaving group than the ammonium group, so the product is formed without hydroxide ion catalysis.

A methoxy group is a poorer leaving group than the ammonium group, so the ammonium group is eliminated, reforming starting materials. If hydroxide is added to the solution, hydroxide ion converts the ammonium group into an amino group. Since the amino group is a poorer leaving group than the methoxy group, the methoxy group will be eliminated.

83.

Chapter 16 Practice Test

1.   Give one name for each of the following:

a.   b.   c.   d.

2.   Rank the following compounds in order of decreasing reactivity toward $Br_2/FeBr_3$:

3.   For each of the following pairs of compounds, indicate the one that is the stronger acid:

a.   or   c.   or

b.   or   d.   or

4.   a.   Which is more reactive in a nucleophilic substitution reaction, *para*-bromonitrobenzene or *para*-bromoethylbenzene?

b.   Which is more reactive in an electrophilic substitution reaction, *para*-bromonitrobenzene or *para*-bromoethylbenzene?

5.   Give the major product(s) for each of the following reactions:

a.   + H_2SO_4 $\longrightarrow$

b.

OCH$_3$

(benzene ring) + CH$_3$Cl $\xrightarrow{\text{AlCl}_3}$

c.

CH$_2$CH$_3$

(benzene ring) $\xrightarrow[\text{HCl, }\Delta]{\text{Na}_2\text{Cr}_2\text{O}_7}$

d.

Cl

(benzene ring) + CH$_3$O$^-$ $\xrightarrow{\Delta}$

NO$_2$

e.

Cl

(benzene ring) + HNO$_3$ $\xrightarrow{\text{H}_2\text{SO}_4}$

OCH$_3$

f.

$$\underset{\text{(benzene ring)}}{}\overset{\displaystyle O}{\underset{}{C}}\text{CH}_3$$ + Cl$_2$ $\xrightarrow{\text{FeCl}_3}$

6.   Indicate whether each of the following statements is true or false:

a. Benzoic acid is more reactive than benzene toward electrophilic aromatic substitution.   T   F

b. *para*-Chlorobenzoic acid is more acidic than *para*-methoxybenzoic acid.   T   F

c. A CH=CH$_2$ group is a meta director.   T   F

d. *para*-Nitroaniline is more basic than *para*-chloroaniline.   T   F

CHAPTER 17
Carbonyl Compounds I: Reactions of Carboxylic Acids and Carboxylic Acid Derivatives

Important Terms

acid anhydride

$$\underset{R}{\overset{O}{\|}}C\underset{O}{}\underset{R}{\overset{O}{\|}}C$$

acyl adenylate   a carboxylic acid derivative with AMP as the leaving group.

acyl group

acyl halide

         or

acyl phosphate   a carboxylic acid derivative with a phosphate leaving group.

acyl pyrophosphate   a carboxylic acid derivative with a pyrophosphate leaving group.

acyl transfer reaction   a reaction that transfers an acyl substituent from one group to another.

alcoholysis   reaction with an alcohol that converts one compound into two compounds.

amide

amino acid   an α-amino carboxylic acid.

aminolysis   reaction with an amine that converts one compound into two compounds.

biosynthesis   synthesis in a biological system.

carbonyl carbon   the carbon of a C=O group.

carbonyl compound   a compound that contains a C=O group.

carbonyl group   a carbon doubly bonded to an oxygen (C=O).

carbonyl oxygen   the oxygen of a C=O group.

carboxyl group

         or   —COOH   or   —CO$_2$H

carboxylic acid

$$\underset{R}{\overset{\displaystyle \overset{O}{\parallel}}{\underset{}{C}}}\!\!-\!OH$$

carboxylic acid derivative   a compound that is hydrolyzed to a carboxylic acid.

carboxyl oxygen   the single-bonded oxygen of a carboxylic acid or ester.

catalyst   a species that increases the rate of a reaction without being consumed in the reaction.

detergent   a salt of a benzenesulfonic acid.

ester

$$\underset{R}{\overset{\displaystyle \overset{O}{\parallel}}{\underset{}{C}}}\!\!-\!OR'$$

fat   a triester of glycerol that exists as a solid at room temperature.

fatty acid   a long-chain carboxylic acid.

Fischer esterification reaction   reaction of a carboxylic acid with excess alcohol and an acid catalyst.

Gabriel synthesis   a method used to convert an alkyl halide into a primary amine, involving S_N2 attack of phthalimide ion on an alkyl halide followed by hydrolysis.

hydrolysis reaction   reaction with water that converts one compound into two compounds.

hydrophobic interactions   the attractive forces between hydrocarbon groups in water.

imide   a compound with two acyl groups bonded to a nitrogen.

lactam   a cyclic amide.

lactone   a cyclic ester.

micelle   a spherical aggregation of molecules, each with a long hydrophobic tail and a polar head, arranged so the polar head points to the outside of the sphere.

mixed anhydride   an acid anhydride with two different R groups.

nitrile   a compound that contains a carbon-nitrogen triple bond.

$$R—C\equiv N$$

nucleophilic acyl substitution reaction   a reaction in which a group bonded to an acyl group is substituted by another group.

nucleophilic addition-elimination reaction

another name for a nucleophilic acyl substitution reaction, which emphasizes the two-step nature of the reaction: a nucleophile adds to the carbonyl carbon in the first step, and a group is eliminated in the second step.

oil

a triester of glycerol that exists as a liquid at room temperature.

phosphoanhydride bond

the bond holding two phosphate groups together within a diphosphate or triphosphate.

saponification

hydrolysis of a fat under basic conditions.

soap

a sodium or potassium salt of a fatty acid.

symmetrical anhydride

an acid anhydride with identical R groups.

tetrahedral intermediate

the intermediate formed in an adddition-elimination (or a nucleophilic acyl substitution) reaction.

thioester

the sulfur analog of an ester.

$$\underset{R}{\overset{\displaystyle O}{\underset{\displaystyle }{\overset{\displaystyle \|}{C}}}}\!\!-SR'$$

transesterification reaction

the reaction of an ester with an alcohol to form a different ester.

Solutions to Problems

1. The lactone would be a three-membered ring, which is too strained to form.

2. **a.** butanenitrile
   propyl cyanide

   b. ethanoic propanoic anhydride
   acetic propionic anhydride

   c. potassium butanoate
   potassium butyrate

   d. pentanoyl chloride
   valeryl chloride

   e. isobutyl butanoate
   isobutyl butyrate

   f. *N,N*-dimethylhexanamide

   g. γ-butyrolactam or
   2-azacyclopentanone

   h. cyclopentanecarboxylic acid

   i. β-methyl-δ-valerolactone or
   5-methyl-2-oxacyclohexanone

3. **a.**

   b.

   c. CH₃CH₂CH₂C≡N

   d.

   e.

   f.

   g.

   h.

   i.

   j.

4. **B** is a correct statement. The resonance energy (delocalization energy) is greater for the amide than for the ester because the second resonance contributor of the amide has a greater predicted stability and so contributes more to the overall structure of the amide; nitrogen is less electronegative than oxygen and so better holds a positive charge.

5. The carbon-oxygen single bond in an alcohol is longer because, as a result of electron delocalization, the carbon-oxygen single bond in a carboxylic acid has some double-bond character.

6. a. The bond between oxygen and the methyl group is the longest, because it is a pure single bond, whereas the other two carbon-oxygen bonds have some double-bond character.

    The bond between carbon and the carbonyl oxygen is the shortest, because it has the most double-bond character.

    1 = longest
    3 = shortest

    b. Notice that the longer the bond, the lower its IR stretching frequency.

    1 = highest frequency
    3 = lowest frequency

7. The key factor is the extent to which the lone pair on Y can be delocalized onto the carbonyl oxygen. The smaller the contribution from the resonance contributor on the right, the greater will be the double bond character of the C=O and the greater will be its IR stretching frequency.

    When Y = Cl, delocalization of chlorine's lone pair onto the carbonyl oxygen is minimal, because of the poor overlap of the large $3sp^3$ orbital of chlorine with the small $2sp^3$ orbital of carbon. Thus, the C=O group of the acyl chloride has the highest IR stretching frequency.

acyl chloride	~1800 cm^{-1}
acid anhydride	~1800 and 1750 cm^{-1}
ester	~1730 cm^{-1}
amide	~1640 cm^{-1}

In the case of the acid anhydride, the central oxygen can be delocalized onto two oxygens, lowering the degree to which these electrons are donated onto any one C=O group. Therefore, the C=O bond in an acid anhydride is stronger and shorter than that in an ester.

The resonance hybrid shows why the acid anhydride has two carbonyl IR absorption bands.

resonance hybrid

When Y = NH_2, the resonance contributor on the right on the previous page is more stable than in the case of the other compounds, because nitrogen can better accommodate a positive charge than can oxygen. Therefore, the C=O bond of the amide has the least double bond character, so it stretches at the lowest frequency.

8.   It is a true statement.

If the nucleophile is the stronger base, it will be harder to eliminate the nucleophile, from the tetrahedral intermediate (**B**) than to eliminate the group attached to the acyl group in the reactant. In other words, the hill that has to be climbed from the intermediate back to the reactants (**B** to **A**) is higher than the hill that has to be climbed from the intermediate to the products (**B** to **C**). Since the transition state with the highest energy is the transition state of the rate-limiting step, the first step is the rate-limiting step.

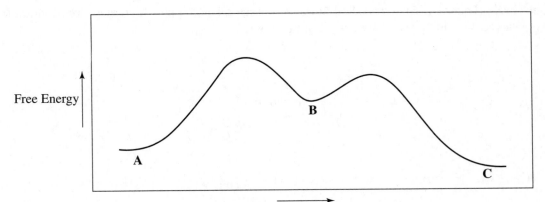

9. **a.** Because HCl is a stronger acid than H_2O, Cl^- is a weaker base than HO^-.
Therefore, Cl^- will be eliminated from the tetrahedral intermediate, so the product of the reaction
will be acetic acid. However, since the solution is basic, acetic acid will be in its basic form as a
result of losing a proton.

acetyl chloride   acetic acid   + Cl^-

b. Because H_2O is a stronger acid than NH_3, HO^- is a weaker base than $^-NH_2$.
Therefore, HO^- will be eliminated from the tetrahedral intermediate, so the reactant will reform.
In other words, no reaction will take place.

acetamide

10. **a.** Acetyl chloride will have the stretching vibration for its carbonyl group at the highest frequency
because it has the most $C=O$ double-bond character, since it has the smallest contribution from the
resonance contributor with a positive charge on Y.

b. Acetamide will have the stretching vibration for its carbonyl group at the lowest frequency
because it has the least $C=O$ double-bond character, since it has the largest contribution from the
resonance contributor with a positive charge on Y.

11. **a.**

b.

c.

d.

12.   **a.**   a new carboxylic acid derivative

b.   no reaction

c.   a mixture of two carboxylic acid derivatives

13.   **a.** $CH_3CH_2CH_2OH$ **c.** $(CH_3)_2NH$ **e.**

b. NH_3 **d.** H_2O **f.** HO—⟨ ⟩—NO_2

14.   Solved in the text.

15.   A protonated amine has a $pK_a \sim 11$. Therefore, the amine will be protonated by the acid that is produced in the reaction, and a protonated amine is not a nucleophile. Excess amine is used in order to have some unprotonated amine available to react as a nucleophile.

A protonated alcohol has a $pK_a \sim -2$. Therefore, the alcohol will not be protonated by the acid that is produced in the reaction.

16.   **a.**

b.

17. a.

CH₃C(=O)—O—C(=O)CH₃ + H₂O: ⇌ tetrahedral intermediate ⇌ products

$$CH_3C(=O)OH + {}^-O-C(=O)CH_3$$

b. The mechanisms are exactly the same.

CH₃C(=O)—O—C(=O)CH₃ + CH₃OH ⇌ tetrahedral intermediate ⇌

$$CH_3C(=O)OCH_3 + {}^-O-C(=O)CH_3$$

18. When an acid anhydride reacts with an amine, the tetrahedral intermediate does not have to lose a proton before it eliminates the carboxylate ion, because the carboxylate ion is a weaker base (pK_a of its conjugate acid is $\sim$5) than the amine (pK_a of its conjugate acid is $\sim$10), so the carboxylate ion is the better leaving group.

CH₃C(=O)—O—C(=O)CH₃ + CH₃NH₂ ⇌
$$CH_3C(O^-)(\!{}^+NH_2CH_3)\!-\!O\!-\!C(=O)CH_3$$

19. a.

$$CH_3CH_2-C(=O)-OCH_3 + H_2O: \rightleftharpoons CH_3CH_2-C(O^-)(\!{}^+OH)-OCH_3$$

B: → H

$$CH_3CH_2-C(O^-)(OH)-OCH_3 + HB^+$$

path a:
$$CH_3CH_2-C(=O)-OH + CH_3O^-$$

path b:
$$CH_3CH_2-C(=O)-OCH_3 + HO^-$$

$$CH_3CH_2-C(=O)-O^- + CH_3OH$$

b.

20.   **a.**   **1.** The carbonyl group of an ester is a weak electrophile.
       2. Water is a weak nucleophile.
       3. $^-OCH_3$ is a strong base and, therefore, a poor leaving group.

      b. Aminolysis is faster, because an amine is a better nucleophile than water.

21.   Solved in the text.

22.   **a.**

      b. $CH_3CH_2CH_2$ $+ CH_3OH$

      c. $HOCH_2CH_2CH_2CH_2$

23.   The mechanism for the acid-catalyzed reaction of acetic acid and methanol is the exact reverse of the mechanism for the acid-catalyzed hydrolysis of methyl acetate.

24. **a.** H_3O^+ $CH_3\overset{+}{O}H_2$ $CH_3\overset{\overset{\displaystyle O}{\|}}{C}OH$ $CH_3\overset{\overset{\displaystyle +OH}{\|}}{C}OH$ $CH_3\underset{\underset{\displaystyle H}{\overset{\displaystyle +}{O}H}}{\overset{\overset{\displaystyle OH}{|}}{C}}OCH_3$ $CH_3\underset{\underset{\displaystyle OH}{|}}{\overset{\overset{\displaystyle OH}{|}}{C}}\overset{+}{\underset{\displaystyle H}{O}}CH_3$

b. H_2O CH_3OH $CH_3\overset{\overset{\displaystyle O}{\|}}{C}OH$ $CH_3\underset{\underset{\displaystyle OH}{|}}{\overset{\overset{\displaystyle OH}{|}}{C}}OCH_3$

c. H_3O^+

d. H_3O^+ if excess water is used, $CH_3\overset{+}{O}H_2$ if excess water is not used.

25. Solved in the text.

26. These are transesterification reactions.

a. PhC(=O)OCH$_2$CH$_3$ + CH$_3$CHCH$_3$ $\overset{HCl}{\rightleftharpoons}$ PhC(=O)OCHCH$_3$ + CH$_3$CH$_2$OH
(with OH on middle carbon of isopropyl; CH$_3$ on product)

b. CH$_3$C(=O)O—Ph + CH$_3$CH$_2$OH $\overset{HCl}{\rightleftharpoons}$ CH$_3$C(=O)OCH$_2$CH$_3$ + Ph—OH

27. (mechanism, transesterification)

28. The mechanism is the same as that on the bottom of page 732 of the text except that the nucleophile is methanol rather than water.

29. **a.** The conjugate base ($CH_3CH_2CH_2O^-$) of the reactant alcohol ($CH_3CH_2CH_2OH$) can be used to increase the rate of the reaction, because it is the nucleophile that we want to become attached to the carbonyl carbon.

b. If H^+ is used as a catalyst, the amine will be protonated in the acidic solution and, therefore, will not be able to react as a nucleophile.

If HO^- is used as a catalyst, HO^- will be the strongest nucleophile present in solution, so it will attack the ester, and the product of the reaction will be a carboxylate ion rather than an amide.

If RO^- is used as a catalyst, RO^- will be the nucleophile, and the product of the reaction will be an ester rather than an amide.

30.

31. **a.** The alcohol (CH_3CH_2OH) contained the ^{18}O label.

b. The carboxylic acid would have contained the ^{18}O label.

32. Solved in the text.

33. Solved in the text.

34. **a.**

b. $CH_3-C(=O)-OH$ $\xrightarrow{SOCl_2}$ $CH_3-C(=O)-Cl$ $\xrightarrow{CH_3(CH_2)_7OH}$ $CH_3-C(=O)-O(CH_2)_7CH_3$

or

$CH_3-C(=O)-O^-$ $\xrightarrow{CH_3(CH_2)_7Br}$ $CH_3-C(=O)-O(CH_2)_7CH_3$

or

$CH_3-C(=O)-OH$ $\xrightarrow[\text{HCl}]{\substack{CH_3(CH_2)_7OH \\ \text{excess}}}$ $CH_3-C(=O)-O(CH_2)_7CH_3$

35. In the first step, protonation occurs on the carbon that results in formation of the most stable carbocation (a tertiary carbocation). In the second step, the pair of electrons of the second π bond provides the nucleophile that adds to the carbocation because, as a result of that reaction, a stable six-membered ring is formed.

36. **a.** $CH_3CH_2CH_2-C(=O)-Cl$ + $2\,CH_3CH_2NH_2$ **b.** $C_6H_5-C(=O)-Cl$ + $2\,CH_3NHCH_3$

37. **a.** #2 and #5 will form amides.

b. For #2: if the nucleophile were $CH_3\bar{N}H$ instead of CH_3NH_2, the better nucleophile will increase the rate of amide formation.

For #5: use a large excess of methylamine to force the equilibrium to the right, increasing the yield.

38. Methicillin is missing the methylene group, placing its more bulky aryl ring closer to the four-membered ring. This may prevent bacterial penicillinase from binding methicillin at its active site; therefore, it would not be able to be used as a substrate, which means its β-lactam ring would not be cleaved.

39. The relative reactivities of the amides depend on the basicities of their leaving groups: the weaker the base, the more reactive the amide. (Recall that the *para*-nitroanilinium ion has a pK_a value of 0.98 and the anilinium ion has a pK_a value of 4.58; page 676 in the text.)

A *para*-nitro-substituted aniline is less basic than a *meta*-substituted aniline because only when the nitro group is in the para (or ortho) position, can electrons be delocalized onto the nitro group.

40. **a.** pentyl bromide   **b.** isohexyl bromide   **c.** benzyl bromide   **d.** cyclohexyl bromide

41. The reaction of an alkyl halide with ammonia leads to primary, secondary, and tertiary amines, and even quaternary ammonium ions. Thus, the yield of primary amine would be relatively low.

In contrast, the Gabriel synthesis forms only primary amines. The reaction of an alkyl halide with azide ion also forms only primary amines, because the compound formed from the initial reaction of the two reagents is not nucleophilic, so polyalkylation does not occur.

42. **a.** $CH_3CH_2CH_2Br$ **b.** CH_3CHCH_2Br **c.**
           $\quad\quad\quad\quad\quad\quad\quad\quad\quad\quad\quad$ CH_3

43. **a.** $BrCH_2CH_2CH_2CHCH_2CH_3$ $\xrightarrow{\text{NaH}}$
           $\quad\quad\quad\quad\quad\quad\quad\quad$ OH

b.

c. In this synthesis and in the synthesis of **d,** a primary alkyl halide cannot be used in the first step because the carbocation will rearrange to a secondary carbocation and will result in the formation of a methyl-substituted five-membered ring. In contrast, a primary allyl halide can be used, since the initially formed allyl cation is stabilized by electron delocalization and is not prone to rearrangement.

d.

e. $BrCH_2CH_2CH_2CHCH_2CH=CH_2$ $\xrightarrow{\text{NaH}}$ $CH_2CH=CH_2$

with OH below the CHCH₂

f.

$CH_2CH=CH_2$ $\xrightarrow[\text{2. HO}^-, \text{H}_2\text{O}_2, \text{H}_2\text{O}]{\text{1. BH}_3}$ $CH_2CH_2CH_2OH$

product of **e**

44.   **a.** $CH_3CH_2\overset{\overset{\displaystyle O}{\|}}{C}OH$ $\xrightarrow{\text{SOCl}_2}$ $CH_3CH_2\overset{\overset{\displaystyle O}{\|}}{C}Cl$ $\xrightarrow{\text{—OH}}$ $CH_3CH_2\overset{\overset{\displaystyle O}{\|}}{C}O$—

b.

45.   **a.**   At neutral pH, both carboxyl groups of succinic acid will be in their basic from.

   b.   Without acetic anhydride, the leaving group would be hydroxide ion. Acetic anhydride causes the reaction to take place via two successive addition-elimination reactions. In both reactions the leaving group is acetate ion, which is much less basic than hydroxide ion and therefore a better leaving group.

46.   PDS lasts longer, because its ester linkage will hydrolyze more slowly since its alkoxide leaving group is a stronger base. You can determine which of the alkoxides is the stronger base by comparing how close the negatively charged oxygen atom is to an electron-withdrawing group. An electron-withdrawing group stabilizes the base, making it a weaker base and therefore a better leaving group. In Dexon, there is a double-bonded oxygen attached to the CH_2 group that is next to the negatively charged oxygen. Thus, the alkoxide leaving group of Dexon is a weaker base and therefore a better leaving group.

47. **a.** $(CH_3CH_2)_2N-\overset{\overset{\displaystyle O}{\|}}{C}-N(CH_2CH_3)_2$

e. $\overset{+}{N}H_4 + HO-\overset{\overset{\displaystyle O}{\|}}{C}-OH \rightleftharpoons$

b. (structure: four carbonyl groups with O linkages)

c. $H_2N-\overset{\overset{\displaystyle O}{\|}}{C}-NHCH_3$

f. CH_3CH_2— (cyclic anhydride structure)

d. no reaction

$CO_2 + H_2O$

48. **a.** $CH_3CH_2CH_2CH_2CH_2-\overset{\overset{\displaystyle O}{\|}}{C}-\underset{\underset{\displaystyle CH_3}{|}}{N}CH_3$

d. $CH_3CH_2C\equiv N$

h. (β-lactone with CH_2CH_3 substituent)

b. $CH_3CH_2CH_2\underset{\underset{\displaystyle CH_3}{|}}{\overset{\overset{\displaystyle CH_3}{|}}{C}}CH_2-\overset{\overset{\displaystyle O}{\|}}{C}-NH_2$

e. $CH_3CH_2-\overset{\overset{\displaystyle O}{\|}}{C}-Br$

i. $CH_3\underset{\underset{\displaystyle CH_3}{|}}{C}HCH_2C\equiv N$

c. (cyclohexane with $-\overset{\overset{\displaystyle O}{\|}}{C}-Cl$)

f. $CH_3-\overset{\overset{\displaystyle O}{\|}}{C}-O^-\ Na^+$

j. (cycloheptane with COOH)

g. (benzoic anhydride structure)

49. **a.** pentanoyl chloride
valeryl chloride

d. *N*-methyl-3-butenamide

g. propanoic anhydride
propionic anhydride

b. *N,N*-dimethylbutanamide
N,N-dimethylbutyramide

e. propyl propanoate
propyl propionate

h. 5-ethylheptanoic acid

c. (*S*)-3-methylpentanoic acid
(*S*)-β-methylvaleric acid

f. acetic benzoic anhydride

i. (*R*)-3-methylhexanenitrile

50.   **a.**

f.

b.

g.

c.

h.

d.

i.

e.

j.

51.   **a.** CH_3CH_2—C(=O)—OH + CH_3OH

c.

—OH + $CH_3CH_2\overset{+}{N}H_3$ Cl^-

b.

—OH + HCl

d. CH_3CH_2—C(=O)—OH + CO_2 + CH_3OH

52.   **a.**   The weaker the base attached to the acyl group, the stronger its electron withdrawing ability; therefore, the easier it is to form the tetrahedral intermediate. (*para*-Chlorophenol is a stronger acid than phenol so the conjugate base of *para*-chlorophenol is a weaker base than the conjugate base of phenol, etc.)

D   >   **A**   >   **C**   >   **B**

b. The tetrahedral intermediate collapses by eliminating the OR group. The lower the basicity of the OR group, the easier it is to eliminate it.

Thus, the rate of both formation of the tetrahedral intermediate and collapse of the tetrahedral intermediate is decreased by increasing the basicity of the OR group.

53. Because acetate ion is a weak base, the S_N2 reaction will form only a substitution product (see page 397 of the text). The product of the S_N2 reaction is an ester which, when hydrolyzed, forms cyclohexanol (the target molecule) and acetic acid.

54. **a.** Methyl acetate has a resonance contributor that butanone does not have, and this resonance contributor causes methyl acetate to be more polar than butanone. Because butanone is less polar, it has the lower dipole moment.

b. Because it is more polar, the intermolecular forces holding methyl acetate molecules together are stronger, so we would expect methyl acetate to have a higher boiling point.

55. Propyl formate is easy to distinguish because it is the only ester that will show four signals. The other three esters each show three signals. Isopropyl formate can be distinguished by its unique splitting pattern: a singlet, a doublet, and a septet. The splitting patterns of the other two esters are the same: a singlet, a triplet, and a quartet. They can be distinguished because the highest frequency signal in ethyl acetate is a quartet, whereas the highest frequency signal in methyl propionate is a singlet.

4 signals

3 signals	**3 signals**	**3 signals**
singlet, doublet, septet	singlet, triplet, quartet	singlet, triplet, quartet
	The signal farthest downfield is a quartet.	The signal farthest downfield is a singlet.

56. The reaction of methylamine with propionyl chloride generates a proton that will protonate unreacted amine, thereby destroying the amine's nucleophilicity. If two equivalents of CH_3NH_2 are used, one equivalent will remain unprotonated and be able to react as a nucleophile with propionyl chloride to form *N*-methylpropanamide.

57. **a.**

b. The carboxyl oxygen will be labeled. Only one isotopically labeled oxygen can be incorporated into the ester because the bond between the methyl group and the labeled oxygen does not break, so there is no way for the carbonyl oxygen to become labeled.

c. In the presence of an acid catalyst, the ester will be hydrolyzed to a carboxylic acid and an alcohol. Both oxygen atoms of the carboxylic acid will be labeled for the same reason that both oxygen atoms of the carboxylic acid are labeled in **a**. The alcohol will not contain any label, because the bond between the methyl group and the oxygen does not break.

58.

a.

b.

c.

d.

59.

a. isopropyl alcohol and HCl

b. aqueous sodium hydroxide

c. ethylamine

d. water and HCl

60.   The offset in the NMR spectrum shows that there is a broad signal at ~10 ppm, which is characteristic of a COOH group. The two triplets and the multiplet are characteristic of a propyl group. The compound is butanoic acid.

$$CH_3CH_2CH_2 \overset{\overset{\displaystyle O}{\|}}{C} OCH_3 \qquad\qquad CH_3CH_2CH_2 \overset{\overset{\displaystyle O}{\|}}{C} OH$$

reactant   hydrolysis product

The molecular formula shows that the unknown reactant has one more carbon than butanoic acid; the IR spectrum shows it is an ester. Since butanoic acid is formed from acid hydrolysis of the ester, the compound must be the methyl ester of butanoic acid.

61.   Aspartame has an amide group and an ester group that will be hydrolyzed in an aqueous solution of HCl. Because the hydrolysis is carried out in an acidic solution, the carboxylic acid groups and the amino groups in the products will be in their acidic forms.

62.   **a.**   1, 3, 4, 6, 7, 9 will not form the indicated products under the given conditions.

b.   9 will form the product shown in the presence of an acid catalyst.

63.   The tertiary amine is a better nucleophile than the alcohol, so formation of the amide with a positively charged nitrogen will be faster than formation of the new ester would have been. The positively charged amide is more reactive than an ester, so formation of the new ester by reaction of the alcohol with the charged amide will be faster than formation of the ester by reaction of the alcohol with the starting ester would have been. In other words, both reactions that occur in the presence of the tertiary amine are faster than the single reaction that occurs in the absence of the tertiary amine.

64.   1-Bromobutane undergoes an S_N2 reaction with NH_3 to form butylamine (**A**). Butylamine can then react with 1-bromobutane to form dibutylamine (**B**). The amines each form an amide upon reaction with acetyl chloride. The IR spectrum of **C** exhibits an NH stretch at about ~3300 cm^{-1}, whereas the IR spectrum of **D** does not exhibit this absorption band, because the nitrogen in **D** is not bonded to a hydrogen.

$$CH_3CH_2CH_2CH_2Br \xrightarrow{NH_3} CH_3CH_2CH_2CH_2NH_2 \xrightarrow{CH_3CH_2CH_2CH_2Br} (CH_3CH_2CH_2CH_2)_2NH$$

                                A   **B**

$$CH_3-\overset{\overset{\displaystyle O}{\|}}{C}-Cl \;+\; CH_3CH_2CH_2CH_2NH_2 \;\longrightarrow\; CH_3-\overset{\overset{\displaystyle O}{\|}}{C}-NHCH_2CH_2CH_2CH_3$$

A · C

$$CH_3-\overset{\overset{\displaystyle O}{\|}}{C}-Cl \;+\; (CH_3CH_2CH_2CH_2)_2NH \;\longrightarrow\; CH_3-\overset{\overset{\displaystyle O}{\|}}{C}-NH(CH_2CH_2CH_2CH_3)_2$$

B · D

65.

1. $Cl-\overset{\overset{\displaystyle O}{\|}}{C}-OCH_3$

3. $CH_3CH_2CH_2NH-\overset{\overset{\displaystyle O}{\|}}{C}-NHCH_2CH_3$

2. $CH_3O-\overset{\overset{\displaystyle O}{\|}}{C}-OCH_3$

4. $CH_3CH_2O-\overset{\overset{\displaystyle O}{\|}}{C}-NHCH_3$

66.

67.

68.

a. $CH_3-\overset{\overset{\displaystyle O}{\|}}{C}-F$

c. (benzamide) $\overset{\overset{\displaystyle O}{\|}}{C}-NHCH_3$

e. (benzodioxole cyclic carbonate)

b. $HO-\overset{\overset{\displaystyle O}{\|}}{C}-CH_2CH_2CH_2\overset{+}{N}H_3 \;\; Cl^-$

d. $HO-\overset{\overset{\displaystyle O}{\|}}{C}-CH_2CH_2-\overset{\overset{\displaystyle O}{\|}}{C}-OH$

f. $HO-\overset{\overset{\displaystyle O}{\|}}{C}-CH_2CH_2CH_2OH$

g.

i.

h.

j.

69.

$$ClCH_2CH_2CH_2 \overset{\overset{\displaystyle O}{\|}}{C} Cl \ + \ CH_3OH \ \longrightarrow \ ClCH_2CH_2CH_2 \overset{\overset{\displaystyle O}{\|}}{C} OCH_3 \ + \ HCl$$

Compound A

70. a. The amino alcohol has two groups that can be acetylated by acetic anhydride, the NH_2 group and the OH group. If excess amino alcohol used, the unreacted amino alcohol will remove a proton from acetic acid that is a product of the reaction.

b. Because the NH_2 group is a better nucleophile than the OH group, it will be the first group that is acetylated. Therefore, if the reaction is stopped before it is half over, the product will have only its NH_2 group acetylated.

71. If the amine is tertiary, the nitrogen in the amide product cannot get rid of its positive charge by losing a proton. An amide with a positively charged nitrogen is very reactive toward addition-elimination, because the positively charged nitrogen-containing group is an excellent leaving group. Thus, if water is added, it will immediately react with the amide, and because the $^+NR_3$ group is a better leaving group than the

OH group, the $^+NR_3$ group will be eliminated. The product will be a carboxylic acid, which will lose its acidic proton to the amine.

72. The amine is a better nucleophile than the alcohol, but since the acyl chloride is very reactive, it can react easily with both nucleophiles. Therefore, steric hindrance will be the most important factor in determining the rate of formation of the products. The amino group is less sterically hindered than the alcohol group, so that will be the group most easily acetylated.

Pyridine (a tertiary amine) is used for the second equivalent of amine (page 724 in the text).

73. **a.** The steric hindrance provided by the methyl groups prevents methyl alcohol from attacking the carbonyl carbon.

b. No, because there would be no steric hindrance to nucleophilic attack.

c.

74. The spectrum shows that the compound has four different kinds of hydrogens with relative ratios 2:1:1:6. The doublet at 0.9 ppm with the ratio of 6 hydrogens suggests that the compound has an isopropyl group. The signal at 3.4 ppm indicates hydrogens on a carbon that is attached to an oxygen. We can conclude that the spectrum is that of isobutyl alcohol. The molecular formula indicates the compound that undergoes hydrolysis is an ester. Subtracting the atoms due to the isobutyl group lets us identify the ester as isobutyl benzoate.

75. The carbonyl IR absorption band decreases in the order:

When Y = Cl, delocalization of chlorine's lone pair onto the carbonyl oxygen is minimal, because of the poor overlap of the large $3sp^3$ orbital of chlorine with the small $2sp^3$ orbital of carbon. Thus, the C=O group of the acyl chloride has the highest IR stretching frequency.

The predominant effect of the oxygen of an ester is inductive electron withdrawal, so its carbonyl group stretches at a higher frequency than the carbonyl group of the aldehyde.

The carbonyl group of the amide stretches at the lowest frequency, because the nitrogen is less electronegative than oxygen so it is better able to accommodate a positive charge. Therefore, the resonance contributor on the right makes a larger contribution to the overall structure. The larger the contribution from the resonance contributor on the right, the greater the single bond character of the C=O group.

76. **a.** The electron withdrawing ester group makes the α-carbon more susceptible to nucleophilic attack than the β-carbon.

b.

The *intramolecular* formation of **B** will occur more rapidly than the *intermolecular* formation of **A** because, in the reaction that forms **B**, the reagents do not have to wander through the solution to find one another.

77.   **a.** $CH_3COOH + CH_3CH_2OH \rightleftharpoons CH_3CO_2CH_3 + H_2O$

$$K_{eq} = \frac{[CH_3CO_2CH_3]\,[H_2O]}{[CH_3COOH]\,[CH_3CH_2OH]}$$

$$4.02 = \frac{x^2}{(1-x)^2}$$

take the square of both sides:

$$2.00 = \frac{x}{(1-x)}$$

$2 - 2x = x$

$2 = 3x$

$x = 0.667$

[ethyl acetate] = 0.667 times the concentration of acetic acid used

b. $$4.02 = \frac{x^2}{(10-x)\,(1-x)}$$

$$4.02 = \frac{x^2}{x^2 - 11x + 10}$$

$$4.02\,(x^2 - 11x + 10) = x^2$$

$$3.02x^2 - 44.22x - 4.02 = 0$$

solving for x using the quadratic formula:

$x = 0.974$

[ethyl acetate] = 0.974 times the concentration of acetic acid used

c. $$4.02 = \frac{x^2}{(100-x)\,(1-x)}$$

$x = 0.997$

[ethyl acetate] = 0.997 times the concentration of acetic acid used

78.   Each of the NMR spectra has signals between about 7–8 ppm (integrating to 5H), indicating that the compound has a benzene ring with one substituent. There is one additional signal that is a singlet. From the molecular formulas it can be determined that the esters have the following structures. The singlet in each spectrum is due to a methyl group. Because the methyl group is at a higher frequency in the spectrum on the right, it is the spectrum of the compound on the right, since its methyl group is adjacent to an electron-withdrawing oxygen.

The equilibrium constant for hydrolysis of the compound on the left is greater than that for the compound on the right. The phenolate ion is a much better leaving group than hydroxide ion, so the tetrahedral intermediate of the compound on the left will form products rather than revert to starting materials. Methoxide ion and hydroxide ion have the same leaving tendency, so the tetrahedral intermediate of the compound on the right will be as likely to revert to starting materials as to form products. However at pH = 10, the carboxylic acid product in both reactions will be deprotonated, which will drive both equilibria to the right.

79.   We know that the electrophile is thionyl chloride and that its electrophilic site is the sulfur atom. We know that the amide is the nucleophile and its nucleophilic site is the oxygen atom, because the nitrogen cannot be a nucleophile since its lone-pair electrons are delocalized.

Therefore, we should let the nucleophile attack the electrophile and then see if that product helps us figure out how we can arrive at the given product (the nitrile).

80.   **a.**

b.

c.

d. toluene $\xrightarrow[\text{HCl, }\Delta]{\text{NaCr}_2\text{O}_7}$ benzoic acid $\xrightarrow{\text{SOCl}_2}$ benzoyl chloride $\xrightarrow[\text{CH}_3\text{NH}_2]{\text{excess}}$ N-methylbenzamide

Top scheme:

$\text{CH}_3(\text{CH}_2)_{10}\text{CH}_2$–benzene $\xleftarrow[\Delta]{\text{H}_2\text{SO}_4}$ ... $\text{CH}_3(\text{CH}_2)_{10}\text{CH}_2$–C$_6H_4$–SO$_3$H

$\xrightarrow{\text{NaOH}}$ $\text{CH}_3(\text{CH}_2)_{10}\text{CH}_2$–C$_6H_4$–SO$_3^-$ Na$^+$

e. aniline $\xrightarrow[\text{pyridine}]{\text{CH}_3\text{CCl (O)}}$ acetanilide $\xrightarrow[\text{2. H}_2\text{O}]{\text{1. CH}_3\text{CCl, AlCl}_3}$ 4′-acetylacetanilide $\xrightarrow[\text{1. HCl, H}_2\text{O, }\Delta \quad\text{2. HO}^-]{}$ 4-aminoacetophenone

f. toluene $\xrightarrow[\text{peroxide}]{\text{NBS, }\Delta}$ benzyl bromide (CH$_2$Br) $\xrightarrow{^-\text{C}\equiv\text{N}}$ phenylacetonitrile (CH$_2$C≡N) $\xrightarrow[\text{HCl, }\Delta]{\text{H}_2\text{O}}$ phenylacetic acid (CH$_2$COOH)

81.   The acid-catalyzed hydrolysis of acetamide forms acetic acid and ammonium ion. It is an irreversible reaction, because the pK_a of an acetic acid is less than the pK_a of the ammonium ion. Therefore, it is impossible to have the carboxylic acid in its reactive acidic form and ammonia in its reactive basic form.

If the solution is sufficiently acidic to have the carboxylic acid in its acidic form, ammonia will also be in its acidic form so it will not be a nucleophile. If the pH of the solution is sufficiently basic to have ammonia in its nucleophilic basic form, the carboxylic acid will also be its basic form; a negatively charged carboxylate ion cannot be attacked by nucleophiles.

$$\text{CH}_3\text{C(O)OH} + \overset{+}{\text{N}}\text{H}_4 \longrightarrow \text{no reaction}$$
The ammonium ion is not nucleophilic.

$$\text{CH}_3\text{C(O)O}^- + \text{NH}_3 \longrightarrow \text{no reaction}$$
The carboxylate ion is not attacked by nucleophiles.

82.   **a.**   **b.**

+ CH₃CH₂OH

c.

83.   Notice that in the sixth step of the reaction sequence, PCl_3 converts the OH group of the sulfonic acid to a Cl just as it converts the OH group of an alcohol and the OH group of a carboxylic acid to a Cl.

84.   **a.**

b. The Ritter reaction does not work with primary alcohols, because primary alcohols do not form carbocations.

c. The only difference in the two reactions is the electrophile that attaches to the nitrogen of the nitrile: it is a carbocation in the Ritter reaction and a proton in the acid-catalyzed hydrolysis of a nitrile.

85.

86. **a.**

b.

c.

87.   An enzyme called penicillinase provides resistance to penicillin by using its CH_2OH group to open the four-membered ring (see page 744 of the text).

The inhibitor of penicillinase has an excellent leaving group, so it readily reacts with nucleophiles. When the CH_2OH group of penicillinase attacks the inhibitor, a relatively stable phosphonyl-enzyme is formed, so its OH group is no longer available to react with penicillin. (The double arrows indicate formation and collapse of the tetrahedral intermediate.)

a phosphonyl-enzyme

Hydroxylamine (H_2NOH) reactivates penicillinase by liberating the enzyme from the phosphonyl-enzyme. (Unlike alkylation reactions with hydroxylamine where nitrogen is the nucleophile, it is known that oxygen is the nucleophile when hydroxylamine reacts with phosphate esters.)

88.   **a.**

b.

89.

90. a.

transition state if formation of the
tetrahedral intermediate is rate-limiting

transition state if collapse of the
tetrahedral intermediate is rate-limiting

b.

c.

91.   Notice that the Friedel-Crafts reaction can be done on a ring with a sulfonic acid group, because its deactivation is offset by the activating amide group. The sulfonic acid group is put on the ring to block the para position and thereby force the methyl groups to go to the ortho positions.

92.

93.   Because electron-withdrawing substituents have positive substituent constants and electron-donating substituents have negative substituent constants, a reaction with a positive ρ value is one in which compounds with electron-withdrawing substituents react more rapidly than compounds with electron-donating substituents, and a reaction with a negative ρ value is one in which compounds with electron-donating substituents react more rapidly than compounds with electron-withdrawing substituents.

**electron withdrawal increases
the rate of the reaction**

log rate

$\rho = +$

−2   0   2

σ

**electron donation increases
the rate of the reaction**

log rate

$\rho = -$

−2   0   2

σ

a. In the hydroxide-ion-promoted hydrolysis of a series of ethyl benzoates, electron-withdrawing substituents increase the rate of the reaction by increasing the amount of positive charge on the carbonyl carbon, thereby making it more readily attacked by hydroxide ion. The ρ value for this reaction is, therefore, positive.

In amide formation with a series of anilines, electron donation increases the rate of the reaction by increasing the nucleophilicity of the aniline. The ρ value for this reaction is, therefore, negative.

b. Because ortho substituents are close to the site of the reaction, they introduce steric factors into the rate constant for the reaction. In other words, the presence of an ortho substituent can slow a reaction down, not because it can donate or withdraw electrons but because it can get in the way of the reactants. Therefore, any change in the rate is due to a combination of steric effects and the electron-donating or electron-withdrawing ability of the substituent.
Because the change in rate cannot be attributed solely to the electron-donating or electron-withdrawing ability of the substituent, ortho-substituted compounds were not included in the study.

c. An electron-withdrawing substituent will make it easier for benzoic acid to lose a proton, so ionization of a series of substituted benzoic acids will show a positive ρ value.

Chapter 17 Practice Test

1.   Circle the compound in each pair that is more reactive toward nucleophilic addition-elimination.

a.   CH₃—C(=O)—OCH₃   or   CH₃—C(=O)—NHCH₃

b.   CH₃—C(=O)—OCH₃   or   CH₃—C(=O)—O—C₆H₅

c.   CH₃—C(=O)—O—C₆H₄—OCH₃   or   CH₃—C(=O)—O—C₆H₄—NO₂

d.   CH₃—C(=O)—O—C(=O)—CH₃   or   CH₃—C(=O)—Cl

2.   Give the systematic name for each of the following:

a.

b.

c.

d.

3.   Give an example of each of the following:

a.   a symmetrical anhydride

b.   a hydrolysis reaction

c.   a transesterification reaction

4. What carbonyl compound would be obtained from collapse of each of the following tetrahedral intermediates?

a.
$$CH_3-\overset{\overset{\displaystyle OH}{|}}{\underset{\underset{\displaystyle OH}{|}}{C}}-\overset{+}{N}H_3$$

c.
$$CH_3-\overset{\overset{\displaystyle O^-}{|}}{\underset{\underset{\displaystyle OH}{|}}{C}}-NH_2$$

b.
$$CH_3-\overset{\overset{\displaystyle OH}{|}}{\underset{\underset{\displaystyle +\overset{\displaystyle OH}{\underset{\displaystyle H}{}}}{|}}{C}}-OCH_3$$

d.
$$CH_3-\overset{\overset{\displaystyle OH}{|}}{\underset{\underset{\displaystyle OH}{|}}{C}}-\overset{+}{O}CH_3 \atop H$$

5. Give the product of each of the following reactions:

a. $CH_3CH_2C\equiv N + H_2O \xrightarrow[\Delta]{HCl}$

b. $CH_3CH_2\overset{\overset{\displaystyle O}{\|}}{C}OCH_2CH_3 + H_2O \xrightarrow[\Delta]{HO^-}$

c. $CH_3CH_2CH_2\overset{\overset{\displaystyle O}{\|}}{C}O\overset{\overset{\displaystyle O}{\|}}{C}CH_2CH_3 + H_2O \longrightarrow$

d. $CH_3CH_2\overset{\overset{\displaystyle O}{\|}}{C}Cl + 2\ CH_3CH_2NH_2 \longrightarrow$

e. ⬡ $CH\overset{\overset{\displaystyle O}{\|}}{C}OCH_3 + \underset{excess}{CH_3CH_2OH} \xrightarrow{HCl}$

f. $CH_3CH_2CH_2\overset{\overset{\displaystyle O}{\|}}{C}OH \xrightarrow[\text{2. } CH_3CH_2CH_2OH]{\text{1. } SOCl_2}$

g. $CH_3CH_2\overset{\overset{\displaystyle O}{\|}}{C}NHCH_3 + H_2O \xrightarrow[\Delta]{HCl}$

h. $CH_3CH_2\overset{\overset{\displaystyle O}{\|}}{C}O$⬡ $+ \underset{excess}{H_2O} \xrightarrow{HCl}$

CHAPTER 18
Carbonyl Compounds II: Reactions of Aldehydes and Ketones • More Reactions of Carboxylic Acid Derivatives • Reactions of α,β-Unsaturated Carbonyl Compounds

Important Terms

acetal	$$\begin{array}{c} OR \\ | \\ R-C-H \\ | \\ OR \end{array}$$
aldehyde	$$\begin{array}{c} O \\ \| \\ C \\ R\quad\quad H \end{array}$$
Borch reaction	reductive animation using $NaBH_3CN$.
conjugate addition (1,4-addition)	nucleophilic addition to the β-carbon of an α,β-unsaturated carbonyl compound.
cyanohydrin	$$\begin{array}{c} OH \\ | \\ R-C-C\equiv N \\ | \\ R'(H) \end{array}$$
deoxygenation	removal of an oxygen from a reactant.
direct addition (1,2-addition)	nucleophilic addition to the carbonyl carbon of an α,β-unsaturated carbonyl compound.
disconnection	breaking a bond to carbon, in a retrosynthetic analysis, to give a simpler molecule.
enamine	an α,β-unsaturated tertiary amine.
***gem*-diol (hydrate)**	a molecule with two OH groups on the same carbon.
hemiacetal	$$\begin{array}{c} OH \\ | \\ R-C-H \\ | \\ OR \end{array}$$
hemiketal	$$\begin{array}{c} OH \\ | \\ R-C-R' \\ | \\ OR \end{array}$$
hydrate	a molecule with two OH groups on the same carbon.

hydrazone

$$\underset{(H)R}{\overset{R}{\diagdown}}C=N-NH_2$$

imine
(Schiff base)

$$\underset{(H)R}{\overset{R}{\diagdown}}C=N-R$$

ketal

$$R-\underset{\underset{OR}{|}}{\overset{\overset{OR}{|}}{C}}-R$$

ketone

$$\underset{R\diagup\ \diagdown R}{\overset{\overset{O}{\|}}{C}}$$

nucleophilic addition reaction   a reaction that involves the addition of a nucleophile to the carbonyl carbon of an aldehyde or a ketone.

nucleophilic addition-elimination reaction   a reaction involving nucleophilic addition to a carbonyl group to form a tetrahedral intermediate, which then undergoes elimination of a leaving group. Imine formation is an example: an amine adds to the carbonyl carbon, and water is eliminated.

oxime

$$\underset{(H)R}{\overset{R}{\diagdown}}C=N-OH$$

pH-rate profile   a plot of the rate constant of a reaction versus the pH of the reaction mixture.

phenylhydrazone

$$\underset{(H)R}{\overset{R}{\diagdown}}C=N-NHC_6H_5$$

prochiral carbonyl carbon   a carbonyl carbon that will become an asymmetric center if a group adds to it that is unlike any of the groups already bonded to it.

protecting group   a group that protects a functional group from a synthetic operation that it would otherwise not survive.

Re and Si faces   the different faces of an unsymmetrically substituted carbonyl compound.

reduction reaction   in the case of an organic molecule, a reaction in which the number of C—H bonds is increased or the number of C—O, C—N, or C—X (X = halogen) bonds is decreased.

reductive amination   the reaction of an aldehyde or a ketone with ammonia or with a primary amine in the presence of a reducing agent.

semicarbazone

$$\underset{\text{(H)R}}{\overset{R}{\diagdown}}C\!=\!NNHC\overset{\overset{\textstyle O}{\|}}{N}H_2$$

synthetic equivalent   the reagent actually used as the source of a synthon.

synthon   a fragment of a disconnection.

Wittig reaction   the reaction of an aldehyde or a ketone with a phosphonium ylide, resulting in the formation of an alkene.

ylide   a compound with opposite charges on adjacent, covalently bonded atoms with complete octets.

Solutions to Problems

1.   If the carbonyl group were anywhere else in these compounds, they would not be ketones and, therefore, would not have the "one" suffix.

2.   **a.** 3-methylpentanal, β-methylvaleraldehyde
      b. 4-heptanone, dipropyl ketone
      c. 2-methyl-4-heptanone, isobutyl propyl ketone
      d. 4-phenylbutanal, γ-phenylbutyraldehyde
      e. 4-ethylhexanal, γ-ethylcaproaldehyde
      f. 1-hepten-3-one, butyl vinyl ketone

3.   **a.** 6-hydroxy-3-heptanone   **b.** 2-oxocyclohexylmethanenitrile   **c.** 4-formylhexanamide

4.   **a.** 2-Heptanone is more reactive, because it has less steric hindrance. There is little difference in the amount of steric hindrance provided at the carbonyl carbon (the site of nucleophilic addition) by a pentyl and a propyl group because they differ at a point somewhat removed from the site of nucleophilic addition. The difference in size between a methyl group and a propyl group is significant at the site of nucleophilic addition.

      2-heptanone   4-heptanone

      b. *para*-Nitroacetophenone is more reactive, because the electron-withdrawing nitro group makes the carbonyl carbon more susceptible to nucleophilic addition compared with the electron-donating methoxy group, which makes the carbonyl carbon less susceptible to nucleophilic attack.

5.   **a.**   **b.**   **c.**

6.

7.   a. Two stereoisomers are obtained, because the reaction creates an asymmetric center in the product.

(S)-3-methyl-3-hexanol   (R)-3-methyl-3-hexanol

b. Only one compound is obtained, because the product does not have an asymmetric center.

8.   a. Solved in the text.

b. 1. CH_3 OR   +   2 CH_3MgBr **4.** CH_3 OR   +   2 CH_3CH_2MgBr

2. Solved in the text.   **6.** CH_3 OR   +   2 ⟨benzene⟩—MgBr

9.   If a secondary alcohol is formed from the reaction of a formate ester with excess Grignard reagent, the two alkyl substituents of the alcohol will be identical because they both come from the Grignard reagent. Therefore, only the following two alcohols (B and D) can be prepared that way.

$$CH_3CHCH_3 \quad CH_3CH_2CHCH_2CH_3$$
$$\;\;\;\;\;|\qquad\qquad\;\;\;\;\;|$$
$$\;\;\;OH\qquad\qquad\;\;\;OH$$

10.   **A** and **C** will not undergo nucleophilic addition with a Grignard reagent. **A** has an H bonded to a nitrogen and **C** has an H bonded to an oxygen; these acidic hydrogens will react with the Grignard reagent, converting it into an alkane, before the Grignard reagent has a chance to react as a nucleophile. **B** will undergo a nucleophilic addition-elimination reaction with the Grignard reagent to form a ketone. This will be followed by a nucleophilic addition reaction of the ketone with the Grignard reagent. The product will be a tertiary alcohol.

11.   a. $HC{\equiv}CH$ $\xrightarrow{NaNH_2}$ $HC{\equiv}C^-$ $\xrightarrow{CH_3CH_2CH, \; O}$ $HC{\equiv}CCHCH_2CH_3$ (O^-) $\xrightarrow{HCl}$ $HC{\equiv}CCHCH_2CH_3$ (OH)

b. $HC\equiv CH$ $\xrightarrow{NaNH_2}$ $HC\equiv C^-$ $\xrightarrow{CH_3Br}$ $HC\equiv CCH_3$ $\xrightarrow{NaNH_2}$ $^-C\equiv CCH_3$

c. $HC\equiv CH$ $\xrightarrow{NaNH_2}$ $HC\equiv C^-$ $\xrightarrow{CH_3CH_2Br}$ $HC\equiv CCH_2CH_3$ $\xrightarrow{NaNH_2}$ $^-C\equiv CCH_2CH_3$

Notice that ethyne must be alkylated before it reacts as a nucleophile with the carbonyl compound. If it is alkylated after nucleophilic addition to the ketone, alkylation on oxygen will compete with alkylation on carbon.

12.   Similar to the way a Grignard reagent reacts with an ester, an acetylide ion first undergoes a nucleophilic addition-elimination reaction with the ester to form a ketone; this is followed by a nucleophilic addition reaction to form a tertiary alcohol.

13. **a.** CH_3CHCH_2OH (with CH_3 substituent)   **b.** cyclohexane—OH   **c.** benzene—CH_2OH **d.** benzene—$CHCH_3$ with OH

14. **a.** $CH_3CH_2CH_2CH_2OH$ + CH_3CH_2OH **c.** benzene—CH_2OH + CH_3OH

b. benzene—CH_2OH **d.** $CH_3CH_2CH_2CH_2CH_2OH$

15. **a.** H—C(=O)—NHCH₂—C₆H₅ **or** C₆H₅—C(=O)—NHCH₃   **c.** CH₃—C(=O)—NHCH₂CH₃

b. CH₃—C(=O)—NH₂   **d.** CH₃—C(=O)—NCH₂CH₃ with CH₂CH₃

16. **a.** 1. LiAlH₄ 2. H₂O

b. HCl, H₂O, Δ

c. 1. HCl, H₂O, Δ 2. LiAlH₄ 3. HCl

17. No, an acid must be present in the reaction mixture in order to protonate the oxygen of the cyanohydrin. Otherwise the cyano group will be eliminated and the reactants will be reformed.

18. Strong acids like HCl and H₂SO₄ have very weak conjugate bases (Cl^- and HSO_4^-), which are excellent leaving groups. When these bases add to the carbonyl group, they are readily eliminated, reforming the starting materials. Cyanide ion is a strong enough base, so it is not eliminated unless the oxygen in the product is negatively charged.

19. Solved in the text.

20. The OH group withdraws electrons inductively and this makes the compound a stronger acid.

$$HO—\overset{+}{N}H_3 \qquad CH_3—\overset{+}{N}H_3$$

21. Figure 18.2 for imine formation when the amine is hydroxylamine (pK_a of its conjugate acid = 6.0) shows that the maximum rate is obtained when the pH is about 1.5 units lower than the pK_a of the amine's conjugate acid. Thus if the amine's conjugate acid has a pK_a of 10, imine formation should be carried out at about pH = 8.5.

22. For the derivations of the equations used to calculate the amount of a compound that is present in either its acidic or basic form, refer to Special Topic I in this study guide.

a. fraction present in the acidic form $= \dfrac{[H^+]}{K_a + [H^+]}$

$$\dfrac{[H^+]}{K_a + [H^+]} = \dfrac{3.2 \times 10^{-5}}{3.2 \times 10^7 + 3.2 \times 10^{-5}}$$

$$= \dfrac{3.2 \times 10^{-5}}{3.2 \times 10^7}$$

$$= 1 \times 10^{-12}$$

b.   fraction present in the acidic form = $\dfrac{[H^+]}{K_a + [H^+]}$

$$\dfrac{[H^+]}{K_a + [H^+]} = \dfrac{3.2 \times 10^{-2}}{3.2 \times 10^7 + 3.2 \times 10^{-2}}$$

$$= \dfrac{3.2 \times 10^{-2}}{3.2 \times 10^7}$$

$$= 1 \times 10^{-9}$$

c.   fraction present in the basic form = $\dfrac{K_a}{K_a + [H^+]}$

$$\dfrac{K_a}{K_a + [H^+]} = \dfrac{1.0 \times 10^{-6}}{1.0 \times 10^{-6} + 3.2 \times 10^{-2}}$$

$$= \dfrac{1.0 \times 10^{-6}}{3.2 \times 10^{-2}}$$

$$= 3.1 \times 10^{-3}$$

23.   The nitrile reacts with the Grignard reagent to form an imine which can then be hydrolyzed to a ketone.

$$R-C\equiv N \;+\; R'-MgBr \longrightarrow \underset{R'}{\overset{R}{\diagup}}C=\bar{N} \xrightarrow{HCl} \left[\underset{R'}{\overset{R}{\diagup}}C=NH\right] \xrightarrow[H_2O]{HCl} \underset{R'}{\overset{R}{\diagup}}C=O \;+\; {}^+NH_4$$

24.   **a.**   **b.**   **c.**

25.   The lone-pair electrons that are on each of the nitrogens attached to the carbonyl group are delocalized onto the carbonyl group. These nitrogens, therefore, cannot act as nucleophiles since their lone-pair electrons are not available. Therefore, only one of the nitrogens in semicarbazide can act as a nucleophile.

26.   **a.**   Notice that both of these mechanisms have a feature that we have seen in many mechanisms: formation of three tetrahedral intermediates—a protonated tetrahedral intermediate, a neutral tetrahedral intermediate, and a second protonated tetrahedral intermediate.

1.

2.

b. The only difference is in the first step of the mechanism: in imine hydrolysis the acid protonates the nitrogen; in enamine hydrolysis, the acid protonates the β-carbon.

27.   **a.**

b.

c.

d.

28. A tertiary amine will be obtained, because the primary amine synthesized in the first part of the reaction by reductive animation will react with the excess carbonyl compound, forming an imine that will be reduced to a secondary amine. The secondary amine will then react with the carbonyl compound, forming an enamine that will be reduced to a tertiary amine.

$$R_2C{=}O \; + \; NH_3 \xrightarrow[Pd/C]{H_2} R_2CHNH_2 \xrightarrow[H_2,\,Pd/C]{R_2C{=}O} R_2CHNHCHR_2 \xrightarrow[H_2,\,Pd/C]{R_2C{=}O} (R_2CH)_3N$$

29. a.

b.

30.

31. Electron-withdrawing groups decrease the stability of the aldehyde (and therefore decrease its concentration), increasing the equilibrium constant for hydrate formation.

$$\underset{\substack{\text{the concentration} \\ \text{is decreased}}}{} K_{eq} = \frac{[\text{hydrate}]}{[\text{aldehyde}][H_2O]}$$

32. Because an electron-withdrawing substituent decreases the stability of a ketone and increases the stability of a hydrate, the compound with the electron-withdrawing *meta*-nitro substituents has the largest equilibrium constant for addition of water.

33. **a.** hemiacetals: 7   **d.** ketals: 5

   b. acetals: 2, 3   **e.** hydrates: 4, 6

   c. hemiketals: 1, 8

34. **a.** Hemiacetals are unstable in basic solution, because the base can remove a proton from an OH group, thereby providing an oxyanion that can expel the OR group (since the charge that resides on the oxygen in the transition state is negative) and form a stable aldehyde.

   b. In order for an acetal to form, the CH_3O group in the hemiacetal must eliminate an OH group. Hydroxide ion is too basic to be eliminated by a CH_3O group (since the charge that develops on the oxygen in the transition state is positive) but water, a much weaker base, can be eliminated by a CH_3O group. In other words, the OH group must be protonated before it can be eliminated by a CH_3O group. Therefore, acetal formation must be carried out in an acidic solution.

   c. Hydrate formation can be catalyzed by hydroxide ion because a group does not have to be eliminated after hydroxide ion attacks the aldehyde or ketone.

35. When a tetrahedral intermediate collapses, the intermediate that is formed is very unstable because of the positive charge on the sp^2 oxygen atom. In the case of an acetal or ketal, the only way to form a neutral species is to reform the acetal or ketal. In the case of a hydrate, a neutral species can be formed by loss of a proton.

36. **a.**

b. NaBH$_4$

37. An acetal has a very poor leaving group (CH$_3$O$^-$).

38. If the OH group of the carboxylic acid is not protected, both OH groups will be converted to COOH groups.

39. If the yield of each step is 80%, the yield of **B** is 80%; the yield of **C** from **B** is 80% times the yield of **B**; the yield of **D** from **C** is 80% times the yield of **C**, etc.

a.

A	→	B	→	C	→	D	→	E	→	F	→	G
		80%		0.80 × 0.80		0.80 × 0.64		0.80 × 0.51		0.80 × 0.41		0.80 × 0.33
				64%		51%		41%		33%		26%

b.

G	→	H	→	I
26%		0.80 × 0.26		0.80 × 0.21
		21%		17%

40. **a.** The OH group has to be protected to prevent its acidic proton from reacting with the Grignard reagent.

$$HOCH_2CH_2CH_2Br \ + \ (CH_3)_3SiCl \xrightarrow{(CH_3CH_2)_3N} (CH_3)_3SiOCH_2CH_2CH_2Br$$

$$\downarrow Mg \ | \ Et_2O$$

$$(CH_3)_3SiOCH_2CH_2CH_2\underset{\underset{O^- \ Mg^+Br}{|}}{C}HCH_3 \xleftarrow{CH_3CH=O} (CH_3)_3SiOCH_2CH_2CH_2MgBr$$

$$\downarrow H_3O^+$$

$$HOCH_2CH_2CH_2\underset{\underset{OH}{|}}{C}HCH_3$$

b. The carbonyl group has to be protected to prevent it from reacting with the Grignard reagent. This synthesis is easier to understand if the CO_2 needed in the third step is written as $O=C=O$.

41. The reason is that there are four ways to carry out the reaction in part **a** but only three ways to carry it out in part **b** is because only a carbonyl group adjacent to a benzene ring can be reduced to a methylene group by H_2, Pd/C.

a.

b.

$$\xrightarrow[\text{HCl, }\Delta]{\text{Zn(Hg)}}$$

$$\xrightarrow[\text{HO}^-, \Delta]{\text{NH}_2\text{NH}_2}$$

$$\xrightarrow[\text{2. H}_2\text{, Raney Ni}]{\text{1. HSCH}_2\text{CH}_2\text{CH}_2\text{SH/HCl}}$$

42.   a.

1.   Solved in the text.

3. $(C_6H_5)_2C{=}O$ + $CH_3CH{=}P(C_6H_5)_3$

or $(C_6H_5)_2C{=}P(C_6H_5)_3$ + $CH_3CH{=}O$

2. $=O$ + $CH_3CH_2CH{=}P(C_6H_5)_3$

4. $-CH{=}O$ + $CH_2{=}P(C_6H_5)_3$

or $=P(C_6H_5)_3$ + $CH_3CH_2CH{=}O$

or $-CH{=}P(C_6H_5)_3$ + $CH_2{=}O$

b.

1.   Solved in the text.

2. $CH_3CH_2CH_2Br$

or

$-Br$

3. CH_3CH_2Br

or

$(C_6H_5)_2CHBr$

4. CH_3Br

or

$-CH_2Br$

43.   To review how to determine whether an asymmetric center has the *R* or *S* configuration, see Section 5.7.

a.

$$\xrightarrow[\text{2. HCl}]{\text{1. CH}_3\text{MgBr}}$$

S

b.

$$\xrightarrow[\text{2. HCl}]{\text{1. CH}_3\text{MgBr}}$$

R

c.

No other stereoisomer exists, because the
compound does not have an asymmetric center.

d.

S

44.   a.

The above synthesis of cyclohexanol will also form cyclohexene, the elimination product.
The following procedure will give a greater yield cyclohexanol, because the substitution product is
favored over the elimination product by a weak base (see page 397 of the text). The substitution
product can be hydrolyzed to the target molecule.

b.

c.

or

d.

e.

$(C_6H_5)_3P{=}O$ +

f.

45.   **a.**

Weak base favors
conjugate addition.

b.

Steric hindrance
at the β-carbon
favors direct addition.

c. $CH_3C{=}CHCCH_3$
with OH and CH_3, CH_3 substituents

d. $CH_3CH{=}CHCH_2OH$

46.   If the initial reaction of the α,β-unsaturated ketone with sodium borohydride is conjugate addition, a ketone will be formed, which is then reduced to a saturated alcohol.

a saturated alcohol

If the initial reaction of the α,β-unsaturated ketone with sodium borohydride is direct addition, the final product will be an α,β-unsaturated alcohol, which will not be further reduced.

unsaturated alcohol

The answer to this problem depends on the location of the steric hindrance. Steric hindrance around the carbonyl carbon will make direct addition less favorable and therefore will favor conjugate addition, which forms the saturated alcohol product. In contrast, steric hindrance around the β-carbon will favor direct addition and therefore will give a higher yield of the unsaturated alcohol product.

47.   Conjugate addition occurs in all four reactions because, in each case, the carbonyl compound reacts with a weak base. Because excess base is used in part **d**, both conjugate addition and direct addition occur.

48.

e. CH_3 $\overset{O}{\overset{\|}{C}}$ CH_2 $\overset{O}{\overset{\|}{C}}$ CH_3

h. cyclopentane ring with $\overset{O}{\overset{\|}{C}}$H aldehyde and CH_2CH_3

f. CH_3CH_2 $\overset{O}{\overset{\|}{C}}$ $\overset{}{CHCH_2CH_2CH_3}$; Br

i. CH_3 $\overset{O}{\overset{\|}{C}}$ $CHCH_2CH_2$ $\overset{O}{\overset{\|}{C}}$ H ; CH_3

g. $CH_3CH_2CHCH_2CH_2$ $\overset{O}{\overset{\|}{C}}$ H ; Br

j. benzene ring (1,3) with two $\overset{O}{\overset{\|}{C}}$H aldehyde groups

49. **a.** CH_3CH_2O , OCH_2CH_3 , $\overset{}{C}$, CH_3CH_2 , H

f. $CH_3CH_2CH_2CH_2OH$ + CH_3CH_2OH

b. benzene with C($=NNH_2$) and CH_2CH_3

g. $CH_3CH_2CH_2$ $\overset{}{C}$ CH_3 with dioxolane (O–O) ring

c. benzene with propyl chain

h. cyclohexanone ring with CH_3 and $C{\equiv}N$

d. $CH_3CH_2CHCH_3$; OH

If sufficient HCN is available, direct addition can also occur.

e. $CH_3CH_2\overset{OH}{\overset{|}{C}}CH_2CH_3$; $C{\equiv}N$

cyclohexane with HO, C≡N, CH_3, $C{\equiv}N$

50.

51.   a.

b. The reaction of a Grignard reagent with an epoxide and the hydroboration/oxidation of an alkene both lead to $R'CH_2CH_2OH$. Since this alcohol does not have the needed methyl group on the β-carbon, these methods cannot be used to synthesize isobutyl alcohol.

c. The reaction of a Grignard reagent with an epoxide requires a Grignard reagent with an R' that has one less carbon than the R of RCH_2OH has. The R' of the alkene also must have one less carbon than R has, so these two syntheses lead to a product with a different carbon skeleton than the starting material.

52.   a.

b.

c.

d.

cyclohexanone + NH₃ (excess) →[H₂, Pd/C] cyclohexylamine (NH₂)

e.

cyclohexyl bromide (product of **c**) →[⁻C≡N] cyclohexyl-C≡N →[H₂, Raney Nickel] cyclohexyl-CH₂NH₂

f.

cyclohexanone + CH₃NH–CH₃ →[trace acid] enamine–N(CH₃)₂ →[NaBH₃CN] cyclohexyl–N(CH₃)₂

g.

cyclohexyl bromide (product of **c**) + (CH₂=CH)₂CuLi →[THF] cyclohexyl–CH=CH₂

or

cyclohexyl bromide (product of **c**) →[⁻C≡CH] cyclohexyl–C≡CH →[H₂, Lindlar catalyst] cyclohexyl–CH=CH₂

h.

cyclohexanone →[NH₂NH₂, HO⁻, Δ or Zn(Hg), HCl, Δ or 1. HSCH₂CH₂CH₂SH/HCl 2. H₂, Pd/C] cyclohexane

or

cyclohexanone →[1. NaBH₄ 2. H₃O⁺] cyclohexanol (OH) →[H₂SO₄, Δ] cyclohexene →[H₂, Pd/C] cyclohexane

i.

cyclohexanone →[1. CH₃CH₂MgBr 2. H₃O⁺] HO–C(CH₂CH₃) →[H₂SO₄, Δ] (CH₂CH₃ alkene) + (CH₂ exocyclic alkene) →[H₂, Pd/C] ethylcyclohexane (CH₂CH₃)

or

cyclohexanone + CH₃CH=P(C₆H₅)₃ → (=CHCH₃) →[H₂, Pd/C] ethylcyclohexane (CH₂CH₃)

53.

54. The greater the electron-withdrawing ability of the para substituent, the greater the K_{eq} for hydrate formation.

55.   a. $CH_3OH \xrightarrow{PBr_3} CH_3Br \xrightarrow[Et_2O]{Mg} CH_3MgBr \xrightarrow[\text{2. HCl}]{\text{1. } H_2C=O} CH_3CH_2OH$

   b. $CH_4 \xrightarrow[hv]{Br_2} CH_3Br \xrightarrow[Et_2O]{Mg} CH_3MgBr \xrightarrow[\text{2. HCl}]{\text{1. ethylene oxide}} CH_3CH_2CH_2OH$

   excess

56.   a.

$+$ $CH_3CH_2\overset{+}{N}H_3$

   c.

   e.

   b. $CH_3CH_2\overset{\overset{\displaystyle OH}{|}}{\underset{\underset{\displaystyle CH_2CH_3}{|}}{C}}CH_3$

   d. $CH_3CH_2\overset{\overset{\displaystyle OH}{|}}{\underset{\underset{\displaystyle CH_2CH_3}{|}}{C}}CH_2CH_3$

   f.

g.

i. CH_3NH ... OCH_3

h.

j. CH_3O ... N ... OCH_3
CH_3

57.   a.

b. $CH_3CH_2CH_2CH_2CH_2SH$

58.   The offset shows there is a signal at ~10.0 ppm, indicating an aldehyde. The ^1H NMR spectrum is that of benzaldehyde. Phenylmagnesium bromide, therefore, must react with a compound with one carbon to form an alcohol that can be oxidized by a mild oxidizing agent (MnO_2) to benzaldehyde. Therefore, Compound Z must be formaldehyde.

$H_2C=O$ $\xrightarrow[\text{2. HCl}]{\text{1. phenylmagnesium bromide}}$ $\bigcirc$—CH_2OH $\xrightarrow[\Delta]{MnO_2}$ $\bigcirc$—CH

formaldehyde
Compound Z

59.   a.

b.

c.

60.   **a.**   three signals in its ^{1}H NMR spectrum   **b.**   three signals in its ^{13}C NMR spectrum

$$\text{CH}_3\overset{\displaystyle O}{\underset{\displaystyle \|}{C}}\text{CH}_2\text{CH}_2\overset{\displaystyle O}{\underset{\displaystyle \|}{C}}\text{OCH}_3 \xrightarrow[\text{2. HCl}]{\text{1. excess CH}_3\text{MgBr}} \text{CH}_3\underset{\underset{\displaystyle \text{CH}_3}{|}}{\overset{\overset{\displaystyle \text{OH}}{|}}{C}}\text{CH}_2\text{CH}_2\underset{\underset{\displaystyle \text{CH}_3}{|}}{\overset{\overset{\displaystyle \text{OH}}{|}}{C}}\text{CH}_3$$

61. $\text{CH}_3\text{CH}_2\overset{\displaystyle O}{\underset{\displaystyle \|}{C}}\text{H} \xrightarrow[\text{2. HCl}]{\text{1. CH}_3\text{MgBr}} \text{CH}_3\text{CH}_2\overset{\overset{\displaystyle \text{OH}}{|}}{C}\text{HCH}_3 \xrightarrow{\text{H}_2\text{CrO}_4} \text{CH}_3\text{CH}_2\overset{\displaystyle O}{\underset{\displaystyle \|}{C}}\text{CH}_3 \xrightarrow[\substack{\text{CH}_3\text{OH}\\ \text{excess}}]{\text{HCl}} \text{CH}_3\text{CH}_2\underset{\underset{\displaystyle \text{OCH}_3}{|}}{\overset{\overset{\displaystyle \text{OCH}_3}{|}}{C}}\text{CH}_3$

62. **a.**

$$\xrightarrow[\text{2. H}_2\text{O}]{\text{1. LiAlH}_4}$$

b.

$$\xrightarrow[\text{H}_2\text{O, }\Delta]{\text{HCl}}$$

c.

$$\xrightarrow[\text{H}_2\text{O, }\Delta]{\text{HCl}}$$

$$\xrightarrow{\text{SOCl}_2}$$

$$\downarrow \text{CH}_3\text{OH}$$

d.

$$\xrightarrow[\text{H}_2\text{O, }\Delta]{\text{HCl}}$$

$$\xrightarrow[\text{2. H}_3\text{O}^+]{\text{1. LiAlH}_4}$$

63. In parts **a**, **b**, and **d**, a new asymmetric center is formed in the product, so a racemic mixture is formed. In part **c**, both cis and trans stereoisomers are formed.

a.

c.

b.

d.

64. **a.**

CH_3CH_2—$\overset{\overset{\displaystyle O}{\|}}{C}$—$CH_2CH_2CH_2CH_3$ + —$MgBr$

—$\overset{\overset{\displaystyle O}{\|}}{C}$—$CH_2CH_3$ + $CH_3CH_2CH_2CH_2MgBr$

—$\overset{\overset{\displaystyle O}{\|}}{C}$—$CH_2CH_2CH_2CH_3$ + CH_3CH_2MgBr

b. CH_3CH_2—$\overset{\overset{\displaystyle O}{\|}}{C}$—$CH_2CH_2CH_3$ + CH_3CH_2MgBr

CH_3CH_2—$\overset{\overset{\displaystyle O}{\|}}{C}$—$CH_2CH_3$ + $CH_3CH_2CH_2MgBr$

$CH_3CH_2CH_2$—$\overset{\overset{\displaystyle O}{\|}}{C}$—$OCH_2CH_3$ + $2\ CH_3CH_2MgBr$

65. **a.** **b.** **c.**

Because there is excess cyanide ion.

d. **e.** **f.**

66. Enovid would have its carbonyl stretch at a higher frequency. The carbonyl group in Norlutin has some single-bond character, because of electron delocalization as a result of having conjugated double bonds.

This causes the carbon-oxygen bond to be easier to stretch than the carbon-oxygen bond in Enovid, which has isolated double bonds and therefore no electron delocalization.

Norlutin

67. Notice that ketal formation can be used to protect either ketones or 1,2-diols. Thus in part **a**, the 1,2-diol is protected using acetone, whereas in part **b**, 1,2-ethanediol is used to protect a ketone.

a.

b.

68. **a.**

c.

b.

d.

69. **a.**

b.

70. The absorption bands at 1600 cm^{-1}, 1500 cm^{-1}, and >3000 cm^{-1} in the IR spectrum indicate that the compound has a benzene ring. The absorption band at 1720 cm^{-1} suggests that it is an aldehyde or a ketone in which the carbonyl group is not conjugated with the benzene ring. The absorption bands at 2720 cm^{-1} and 2820 cm^{-1} suggests that the compound is an aldehyde.

The broad signal at 1.8 ppm in the ^{1}H NMR spectrum of the product indicates an OH group. We also see there are three different kinds of hydrogens (other than the OH group and the benzene ring hydrogens). The doublet at ~1.2 ppm indicates a methyl group adjacent to a carbon bonded to one hydrogen. The doublet at 3.6 ppm is due a methylene group bonded to a CH group. The chemical shift of this signal suggests that the electron-withdrawing OH group is also bonded to this methylene group.

The IR spectrum is the spectrum of 2-phenylpropanal and the ^{1}H NMR spectrum is the spectrum of 2-phenyl-1-propanol.

71.   In the first step of each reaction, cyanide is eliminated, because it is a weak base. In the second step, cyanide ion adds to the β-carbon, again, because it is a weak base.

a.

$$CH_2=CHCH-C\equiv N \longrightarrow CH_2=CHCH \longrightarrow N\equiv CCH_2CH_2CH$$

b.

$$CH_3CCH=CH_2 \longrightarrow CH_3CCH=CH_2 \longrightarrow CH_3CCH_2CH_2C\equiv N$$

72.   The difference in the two reactions is a result of the difference in the leaving ability of a sulfonium group and a phosphonium group. Because the sulfonium group is a weaker base, it is a better leaving group. Therefore, it is eliminated by the oxyanion. In contrast, the phosphonium group is too strongly basic to be eliminated, so the oxyanion forms a four-membered ring by sharing electrons with the positively charged phosphorus. Thus, the two reactions proceed by different mechanisms and form different products.

$$CH_3CH_2CH \longrightarrow CH_3CH_2CH-CH_2-\overset{+}{S}(CH_3)_2 \longrightarrow CH_3CH_2CH-CH_2 + (CH_3)_2S$$

$$(CH_3)_2\overset{+}{S}-\overset{-}{\overset{..}{C}}H_2$$

$$CH_3CH_2CH \longrightarrow CH_3CH_2CH-CH_2-\overset{+}{P}(C_6H_5)_3 \longrightarrow CH_3CH_2CH-CH_2 \longrightarrow CH_3CH_2CH=CH_2 +$$

$$(C_6H_5)_3\overset{+}{P}-\overset{-}{\overset{..}{C}}H_2 \qquad\qquad O=P(C_6H_5)_3$$

73.   a.

$$\xrightarrow[\text{2. HCl}]{\text{1. excess } CH_3MgBr}$$

b.

$$\xrightarrow[\text{2. HCl}]{\text{1. LiAlH}_4} \qquad\text{CH}_2\text{OH} \xrightarrow{\text{PCC}}$$

The synthesis can also be carried out in one step.

$$\xrightarrow[\text{2. H}_2\text{O}]{\text{1. } [(CH_3)_2CHCH_2]_2AlH, -78\text{ °C}}$$

c.

or

d.

e.

74.   **a.**

b.

75.   **a.**   The OH group on C-5 of glucose reacts with the aldehyde group in an intramolecular reaction, forming a cyclic hemiacetal. Because the reaction creates a new asymmetric center, two cyclic hemiacetals can form, one with the *R* configuration at the new asymmetric center and one with the *S* configuration.

b. The two products can be drawn in their chair conformations by putting the largest group (CH_2OH) in the equatorial position and then putting the other groups in axial or equatorial positions depending on whether they are cis or trans to one another. The hemiacetal on the left has all but one of its OH groups in the more stable equatorial position, whereas the hemiacetal on the right has all its OH groups in the equatorial position. Therefore, the hemiacetal on the right is more stable.

less stable   more stable

76. The alkyl bromide is 1-bromo-2-phenylethane. The molecular formula of the product of the Wittig reaction indicates that the ketone that reacts with the phosphonium ylide has 3 carbons (that is, the ketone is acetone).

77.

78. Converting the Fischer projection of (*R*)-mandelonitrile into a wedge-and-dash structure allows you to see that attack of cyanide ion occurred on the *Si* face (decreasing priorities on the face closest to the observer are in a counterclockwise direction).

(*R*)-mandelonitrile

79.   **a.**

b.

c.

d.

80. The first step is conjugate addition of a Gilman reagent and the second step is direct addition of a Grignard reagent.

81.   a.

b.

82.   The compound that gives the ^{1}H NMR spectrum is 2-phenyl-2-butanol.

Therefore, the compound that reacts with methylmagnesium bromide is 1-phenyl-1-propanone.

83.   a.

b.

c.

84.

85.

86. **a.** This carboxyl group is more acidic, because its conjugate base is more stable since it is closest to the electron-withdrawing keto group.

b. The data show that the amount of hydrate decreases with increasing pH until about pH = 6, and that increasing the pH beyond 6 has no effect on the amount of hydrate.

A hydrate is stabilized by electron-withdrawing groups. A COOH group is electron withdrawing, but a COO⁻ group is less so. In acidic solutions, where both carboxylic acid groups are in their acidic (COOH) forms, the compound exists as essentially all hydrate. As the pH of the solution increases and the COOH groups become COO⁻ groups, the amount of hydrate decreases. Above pH = 6, where both carboxyl groups are in their basic (COO⁻) forms, there is only a small amount of hydrate.

oxaloacetic acid   hydrate

87.   **a.** $(EtO)_3P\colon$ + CH$_2$CCH$_3$ ⟶ CH$_3$CCH$_2$—P—O—CH$_2$CH$_3$

Br

:Br:

CH$_3$CCH—P=O ⟵ CH$_3$CCH—P=O + CH$_3$CH$_2$Br

B: H

b.

R$_2$C=O + :CH(CCH$_3$)(P(OEt)$_2$=O) ⟶ intermediate

R   H
 \   /
  C=C   +   (EtO)$_2$PO$_2^-$ ⟵
 /   \
R   CCH$_3$
      ‖
      O

c.   In each synthesis, we start by synthesizing the phosphonium-stabilized carbanion that will add to the aldehyde starting material.

1. $(EtO)_3P\colon$ + BrCH$_2$CCH$_3$ ⟶ $(EtO)_2P$—CH$_2$CCH$_3$ + CH$_3$CH$_2$Br

B:⁻

CH$_3$CH$_2$ H
       \   /
        C=C   ⟵ CH$_3$CH$_2$CH $(EtO)_2P$—CHCCH$_3$
       /   \
      H   CCH$_3$
            ‖
            O
   major product

2.

88. **a.** The negative ρ value obtained when hydrolysis is carried out in a basic solution indicates that electron-donating substituents increase the rate of the reaction. This means that the rate-determining step must be protonation of the carbon (the first step), because the more electron-donating the substituent, the greater the partial negative charge on the carbon and the easier it will be to protonate.

b. The positive ρ value obtained when hydrolysis is carried out in an acidic solution indicates that electron-withdrawing substituents increase the rate of the reaction. This means that the rate-determining step must be nucleophilic addition of water to the iminium carbon to form the tetrahedral intermediate (the second step), because electron withdrawal increases the electrophilicity of the iminium carbon, making it more susceptible to nucleophilic addition.

89.

90.   Notice that in the mechanisms in parts **a** and **b,** there is an equilibrium between a protonated intermediate, a neutral intermediate, and a second protonated intermediate, just as we saw in many of the mechanisms in Chapters 17 and 18.

a.

b.

c.

91.   **a.**

b.

c.

d.

e.

Chapter 18 Practice Test

1.   Give the product of each of the following reactions:

a. $+$ NH_2OH $\xrightarrow{\text{trace acid}}$

b. $+$ $\xrightarrow{\text{trace acid}}$

c. CO_2 $\xrightarrow[\text{2. HCl, H}_2\text{O}]{\text{1. CH}_3\text{CH}_2\text{CH}_2\text{MgBr}}$

d. $+$ CH_3CH_2OH excess $\xrightarrow{\text{HCl}}$

e. $CH_3CH_2\overset{\overset{\displaystyle O}{\|}}{C}CH_2CH_3$ $\xrightarrow[\text{2. HCl}]{\text{1. CH}_3\text{MgBr}}$

f. $+$ NH_3 excess $\xrightarrow[\text{Pd/C}]{\text{H}_2}$

g. $CH_3CH_2\overset{\overset{\displaystyle O}{\|}}{C}CH_2CH_3$ $\xrightarrow[\text{HCl}]{^-C{\equiv}N}$

h. $CH_3CH{=}CH\overset{\overset{\displaystyle O}{\|}}{C}CH_3$ $+$ CH_3SH $\longrightarrow$

i. $\xrightarrow[\text{2. HCl}]{\text{1. 2 CH}_3\text{CH}_2\text{CH}_2\text{MgBr}}$

2.   Which of the following alcohols cannot be prepared by the reaction of an ester with excess Grignard reagent?

$$\underset{\underset{CH_3}{|}}{CH_3CH_2\overset{\overset{OH}{|}}{C}CH_2CH_2CH_3} \qquad \underset{\underset{CH_3}{|}}{CH_3\overset{\overset{OH}{|}}{C}CH_2CH_3} \qquad \underset{\underset{CH_3}{|}}{CH_3\overset{\overset{OH}{|}}{C}CH_3}$$

3.   Which of the following ketones would form the greatest amount of hydrate in an aqueous solution?

4.   Give an example of each the following:

a.   an enamine

b.   an acetal

c.   an imine

d.   a hemiacetal

e.   a phenylhydrazone

5.   Which is more reactive toward nucleophilic addition?

a.   butanal or methyl propyl ketone

b.   2-heptanone or 3-pentanone

6.   Indicate how the following compound could be prepared using the given starting material:

$$CH_3CH_2CH_2Br \longrightarrow \underset{CH_3CH_2CH_2}{\overset{\overset{\displaystyle O}{\|}}{C}}OCH_2CH_3$$

CHAPTER 19
Carbonyl Compounds III: Reactions at the α-Carbon

Important Terms

acetoacetic ester synthesis	synthesis of a methyl ketone using ethyl acetoacetate as the starting material.
aldol addition	a reaction between two molecules of an aldehyde (or two molecules of a ketone) that connects the α-carbon of one with the carbonyl carbon of the other.
aldol condensation	an aldol addition followed by elimination of water.
ambident nucleophile	a nucleophile with two nucleophilic sites.
annulation reaction	a ring-forming reaction.
α-carbon	a carbon adjacent to a carbonyl carbon.
carbon acid	a compound that contains a carbon bonded to a relatively acidic hydrogen.
Claisen condensation	a reaction between two molecules of an ester that connects the α-carbon of one with the carbonyl carbon of the other and eliminates an alkoxide ion.
condensation reaction	a reaction combining two molecules while removing a small molecule (usually water or an alcohol).
crossed aldol addition	an aldol addition using two different aldehydes or ketones.
crossed Claisen condensation	a Claisen condensation using two different esters.
decarboxylation	loss of carbon dioxide.
Dieckmann condensation	an intramolecular Claisen condensation.
β-diketone	a ketone with a second ketone carbonyl group at the β-position.
E1cB	a two-step elimination reaction that proceeds via a carbanion intermediate.
enolization	keto-enol interconversion.
gluconeogenesis	the synthesis of D-glucose from pyruvate.
glycolysis	the breakdown of D-glucose into two molecules of pyruvate.
haloform reaction	the conversion of a methyl ketone to a carboxylic acid and haloform using Cl_2 (or Br_2 or I_2) and HO^-.
Hell-Volhard-Zelinski (HVZ) reaction	conversion of a carboxylic acid into an α-bromocarboxylic acid using $PBr_3 + Br_2$.
α-hydrogen	a hydrogen bonded to the carbon adjacent to a carbonyl carbon.

keto-enol tautomerism
(keto-enol interconversion)   interconversion of keto and enol tautomers.

β-keto ester   an ester with a ketone carbonyl group at the β-position.

Kolbe-Schmidt
carboxylation reaction   reaction of a phenolate ion with carbon dioxide under pressure.

malonic ester synthesis   synthesis of a carboxylic acid using diethyl malonate as the starting material.

Michael reaction   the addition of an α-carbanion to the β-carbon of an α,β-unsaturated carbonyl compound.

mixed aldol addition   an aldol addition using two different aldehydes or ketones (see crossed aldol addition).

mixed Claisen
condensation   a Claisen condensation in which two different esters are used (see crossed Claisen condensation).

Robinson annulation   a Michael reaction followed by an intramolecular aldol condensation.

Stork enamine reaction   uses an enamine as a nucleophile in a Michael reaction.

α-substitution reaction   a reaction that puts a substituent on an α-carbon in place of an α-hydrogen.

tautomers   constitutional isomers that are in rapid equilibrium; for example, keto and enol tautomers. The keto and enol tautomers differ only in the location of a double bond and a hydrogen.

Solutions to Problems

1. The electrons left behind when a proton is removed from propene are delocalized over three carbons. In contrast, the electrons left behind when a proton is removed from an alkane are localized—they belong to a single carbon. Because electron delocalization allows the charge to be distributed over more than one atom, it stabilizes the base. Thus, a base with delocalized electrons is more stable, so it has the stronger conjugate acid. Thus, propene is a stronger acid than an alkane.

$$B\text{:}^- + CH_2{=}CHCH_3 \;\rightleftharpoons\; CH_2{=}CH\ddot{C}H_2 \;\longleftrightarrow\; \bar{C}H_2CH{=}CH_2$$
$$+ \; HB$$

$$B\text{:}^- + CH_3CH_2CH_3 \;\rightleftharpoons\; CH_3\ddot{C}HCH_3 + HB$$

Propene, however, is not as acidic as the carbon acids in Table 19.1, because the electrons left behind when a proton is removed from these carbon acids are delocalized onto an oxygen or a nitrogen, which are more electronegative atoms than carbon and, therefore, are better able to accommodate the electrons.

$$B\text{:}^- + RCH_2-\overset{\overset{\displaystyle :\ddot{O}}{\|}}{C}-R \;\longrightarrow\; RCH{-}\overset{\overset{\displaystyle :\ddot{O}}{\|}}{C}-R \;\longleftrightarrow\; RCH{=}\overset{\overset{\displaystyle :\ddot{O}:^-}{|}}{C}-R$$
$$+ \; HB$$

2. CH₃—C(=O)—CH₂C≡N   a β-keto nitrile

    CH₃O—C(=O)—CH₂—C(=O)—OCH₃   a β-diester

    CH₃—C(=O)—CH₂—C(=O)—H   a β-keto aldehyde

3. A proton cannot be removed from the α-carbon of N-methylethanamide or ethanamide, because these compounds have a hydrogen bonded to the nitrogen and this hydrogen is more acidic than the one attached to the α-carbon. Therefore, the hydrogen attached to the nitrogen will be the one removed by a base. In the case of N,N-dimethylethanamide, there is no N–H proton, so a proton can be removed from the α-carbon.

    CH₃—C(=O)—NHCH₃   *N*-methylethanamide

    CH₃—C(=O)—NH₂   ethanamide

    CH₃—C(=O)—N(CH₃)CH₃   *N,N*-dimethylethanamide

    The following resonance contributors show why the hydrogen attached to the nitrogen is more acidic (the nitrogen has a partial positive charge) than the hydrogen attached to the α-carbon.

    CH₃—C(=O)—ṄH₂ ⟷ CH₃—C(—O⁻)=NH₂⁺

4. Electron delocalization of the lone pair on nitrogen or oxygen competes with electron delocalization of the electrons left behind on the α-carbon when it loses a proton. The lone pair on nitrogen is more delocalized than the lone pair on oxygen, because nitrogen is better able to accommodate a positive charge. Therefore, the amide competes better with the carbanion for electron delocalization, so the α-carbon is less acidic.

5. a.

b.

c.

Electrons can be delocalized onto two oxygens.

The H attached to the N is the most acidic H, so it will come off rather than the α-hydrogen.

Problem 4 explains why the ester is more acidic than the amide.

The ketone is a stronger acid than the ester or the *N*-alkylated amide, because there is no competition for delocalization of the electrons that are left behind when the α-hydrogen is removed.

6. Both the keto and enol tautomers of 2,4-pentanedione can form hydrogen bonds with water. Neither the keto nor the enol tautomer can form hydrogen bonds with hexane, but the enol tautomer can form an intramolecular hydrogen bond. This intramolecular hydrogen bonding stabilizes the enol tautomer. Thus, the enol tautomer is more stable relative to the keto tautomer in hexane than in water and therefore more of the enol tautomer will be present in hexane than in water.

keto tautomer

enol tautomer

2,4-pentanedione

7. $pK_a = pH + \log\dfrac{[enol]}{[enolate]}$

a. $20 = 8 + \log\dfrac{[enol]}{[enolate]}$

$12 = \log\dfrac{[enol]}{[enolate]}$

$10^{12} = \dfrac{[enol]}{[enolate]}$

b. $16 = 8 + \log\dfrac{[enol]}{[enolate]}$

$8 = \log\dfrac{[enol]}{[enolate]}$

$10^{8} = \dfrac{[enol]}{[enolate]}$

c. $10 = 8 + \log\dfrac{[enol]}{[enolate]}$

$2 = \log\dfrac{[enol]}{[enolate]}$

$10^{2} = \dfrac{[enol]}{[enolate]}$

8. **a.**

b.

c.

d.

and

more stable,
because the double bonds
are conjugated

e.

more stable,
because the double bonds
are conjugated

f.

more stable,
because the double bond
is conjugated with the
benzene ring

9. The aldehyde hydrogen cannot be removed by a base. The aldehyde hydrogen is not acidic, because the electrons left behind if it were to be removed cannot be delocalized.

The aldehyde hydrogen is not acidic.

10. The haloform reaction requires that a group be created that is a weaker base than hydroxide ion so that hydroxide ion is not the group eliminated from the tetrahedral intermediate. For an alkyl group to be the weaker base, the α-carbon must be bonded to three halogen atoms. The only alkyl group that can become bonded to three halogen atoms at the α-carbon is a methyl group (because it is bonded to three hydrogens).

11. The hydroxide ion is acting as a nucleophile (not a base) in the haloform reaction, so the term "base" should not be used. The hydroxide ion is consumed by the reaction, so it is not a catalyst. Thus, the reaction is a hydroxide-ion promoted reaction.

12. A Br—Br bond is weaker and easier to break than a Cl—Cl bond, which in turn is weaker and easier to break than a D—O bond. Because the experimentally determined rates of bromination, chlorination, and deuterium exchange are about the same, you know that breaking the Br—Br, Cl—Cl, or D—O bond, which occur at different rates, takes place after the rate-determining step. Therefore, the rate-determining step must be removal of the proton from the α-carbon of the ketone.

13.   **a.**

b.

c.

d.

14.   a.

b.

15.   a.

b.

c.

d.

16.

17.   Alkylation of an alpha carbon is an S$_N$2 reaction. S$_N$2 reactions work best with primary alkyl halides, because there is less steric hindrance in a primary alkyl halide than in a secondary alkyl halide. S$_N$2 reactions do not work at all with tertiary alkyl halides because they are the most sterically hindered of the alkyl halides. Therefore, in the case of tertiary alkyl halides, the S$_N$2 reaction cannot compete with the E2 elimination reaction.

18.   a.

b.

19. a.

b.

20. a.

b.

21. **a.**

HO^-

b. CH_3 ... $CH=CH_2$ CH_3CH_2O ... CH_2 ... OCH_2CH_3 $CH_3CH_2O^-$

22. **a.** $CH_3CH_2CH_2CH_2CHCH$ (OH, CHO, $CH_2CH_2CH_3$)

b. $CH_3CH_2-C-CH-C-CH_2CH_3$ (OH, O, CH_3CH_2 CH_3)

c.

23. **a.** $CH_3CH_2CH_2$ ... H (C=O)   **b.** CH_3 ... CH_3 (C=O)   **c.**

d. CH_3CH_2 ... CH_2CH_3 (C=O)

24.

25.

aldol addition   aldol condensation $+ \ H_2O$

A bond has formed between the α-carbon of one
molecule and the carbonyl carbon of another.

26. **a.** Solved in the text.

b.

$$CH_3 \overset{O}{\underset{}{C}} CH_3 \xrightarrow{\text{LDA}} \overset{O}{\underset{}{C}} CH_3 \xrightarrow{CH_3CH_2Br} CH_3CH_2CH_2 \overset{O}{\underset{}{C}} CH_3$$

c. $CH_3CH_2CH_2Br \xrightarrow{^-C\equiv N} CH_3CH_2CH_2C\equiv N \xrightarrow[\Delta]{HCl, H_2O} CH_3CH_2CH_2 \overset{O}{\underset{}{C}} OH$

or

$CH_3CH_2CH_2Br \xrightarrow[Et_2O]{Mg} CH_3CH_2CH_2MgBr \xrightarrow[2.\ H_3O^+]{1.\ CO_2} CH_3CH_2CH_2 \overset{O}{\underset{}{C}} OH$

27.

There is only one site from which a proton can be removed; all the other hydrogens in the two reactants are attached to sp^2 carbons.

The carbonyl carbon in benzophenone is the only electrophile available for the nucleophile to attack.

The first step in an E1cB reaction is formation of a carbanion. Again there is only one C—H hydrogen that can be removed.

The negatively charged oxygen is the atom with the greatest electron density, so that is the atom that will be protonated.

A group is eliminated in the second step of an E1cB reaction.

28.   **a.**   CH₃CH₂CH₂CHCH $\overset{\overset{\displaystyle O}{\|}}{\underset{\displaystyle}{C}}$ H   +   CH₃CH₂C—CH $\overset{\overset{\displaystyle O}{\|}}{\underset{\displaystyle}{C}}$ H   +

The structures (part a, first row):

CH₃CH₂CH₂CHCH—(C=O)H with OH on the CHCH carbon and CH₂CH₃ below
+
CH₃CH₂C—CH—(C=O)H with OH, and CH₂CH₃ / CH₂CH₃ substituents
+

(part a, second row):

CH₃CH₂CH₂CHCH—(C=O)—CH₂CH₃ with OH, and CH₃ below
+
CH₃CH₂C—CH—(C=O)CH₂CH₃ with OH, and CH₃CH₂ / CH₃ substituents

b.   HOCH₂CHCH(C=O) with CH₃ below
— major product
+
CH₃CH₂CHCHCH(C=O) with OH and CH₃
— minor product

29.   Solved in the text.

30.   **a.**   H—(C=O)—H   +   CH₃CH₂—(C=O)—CH₂CH₃ (add slowly) $\xrightarrow[\text{H}_2\text{O}]{\text{HO}^-}$ HOCH₂CH—(C=O)—CH₂CH₃ with CH₃ below

b.   [benzaldehyde] + [cyclohexanone] (add slowly) $\xrightarrow[\text{H}_2\text{O}]{\text{HO}^-}$ [2-benzylidenecyclohexanone]   +   H₂O

31.

32. **a.** $CH_3CH_2CH_2$ — C(=O) — CH(CH_2CH_3) — C(=O) — OCH_3

b. CH_3CHCH_2 — C(=O) — CH(CH(CH_3)CH_3) — C(=O) — OCH_3

33.   **A**, **B**, and **D** cannot undergo a Claisen condensation. **A** cannot, because a proton cannot be removed from an sp^2 carbon. **B** cannot, because it does not have an α-carbon. **D** cannot, because it does not have an α-hydrogen.

34. **a.**

b.

35.

36. Solved in the text.

37.

CH₃O⁻

a 1,7-diester

+ CH₃OH

+ CH₃O⁻

38. **a.**

b.

39. No, because an intramolecular reaction would lead to a strained four-membered ring. Therefore, an intermolecular reaction will be preferred; the reaction will involve removal of a proton from the more acidic central carbon.

40. Solved in the text.

41.   **a.**   **b.**   **c.**   **d.**

42.   **a.**

$$CH_2=CH-\overset{\overset{\displaystyle O}{\|}}{C}-CH_3 \;+\; CH_3CH_2CH_2-\overset{\overset{\displaystyle O}{\|}}{C}-H \;\xrightarrow[\Delta]{HO^-}$$

b.

$$CH_2=CH-\overset{\overset{\displaystyle O}{\|}}{C}-CH_2CH_3 \;+\; CH_3-\overset{\overset{\displaystyle O}{\|}}{C}-H \;\xrightarrow[\Delta]{HO^-}$$

c.

$$CH_2=\overset{\overset{\displaystyle}{\underset{\underset{\displaystyle CH_3}{|}}{C}}}{}-\overset{\overset{\displaystyle O}{\|}}{C}-CH_3 \;+\; CH_3CH_2-\overset{\overset{\displaystyle O}{\|}}{C}-C_6H_5 \;\xrightarrow[\Delta]{HO^-}$$

d.

43.   **A** and **D** are β-keto acids so they can be decarboxylated on heating.
B cannot be decarboxylated, because it does not have a carboxyl group.
The electrons left behind if **C** were decarboxylated cannot be delocalized onto an oxygen.

44.   **a.**   methyl bromide   **c.**   benzyl bromide
b.   methyl bromide (twice)   **d.**   isobutyl bromide

45.   **a.**   An S$_N$2 reaction cannot be done on bromobenzene. (See Section 8.7 of the text.)

b.   An S$_N$2 reaction cannot be done on vinyl bromide. (See Section 8.7 of the text.)

c.   An S$_N$2 reaction cannot be done on a tertiary alkyl halide. (Only elimination occurs; see Section 9.8 of the text.)

46.   Solved in the text.

47.   **a.**   ethyl bromide   **b.**   pentyl bromide   **c.**   benzyl bromide

48.   **a.**

b.

c.

d.

49. Because the catalyst is a hydroxide ion rather than an enzyme, four stereoisomers will be formed since two asymmetric centers are created in the product.

50. **Seven** moles. The first two carbons in the fatty acid come from acetyl CoA. Each subsequent two-carbon piece comes from malonyl CoA. Because this amounts to fourteen carbons for the synthesis of the 16-carbon fatty acid, seven moles of malonyl CoA are required.

51. **a.** **Three** deuteriums would be incorporated into palmitic acid, because only one CD_3COSR is used in the synthesis.

  b. **Seven** deuteriums would be incorporated into palmitic acid, because seven $^-OOCCD_2COSR$ are used in the synthesis (for a total of 14 D's), and each $^-OOCCD_2COSR$ loses one deuterium in the dehydration step (14 D's – 7 D's = 7 D's).

52. It tells you that an imine is formed as an intermediate. Because an imine is formed, the only source of oxygen for acetone formation is $H_2^{18}O$.

If decarboxylation occurred without imine formation, most of the acetone would contain ^{16}O. There would be some ^{18}O incorporated into acetone because acetone would react with $H_2^{18}O$ and form a hydrate. The ^{18}O would become incorporated into acetone when the hydrate reforms acetone.

53.

a. CH_3–C(=O)–CH_2–C(=O)–OCH_2CH_3

b. HO–C(=O)–$CH(CH_3)$–C(=O)–OH

c. R–C(=O)–CH_2–C(=O)–OR'

d. 1-cyclopentenol (cyclopentene with OH)

e. $CH_3CH_2CH_2CH_2$–C(=O)–OH

54.

a. 2-(ethoxycarbonyl)cyclohexanone

b. $CH_3CH_2CH_2CH(Br)$–C(=O)–OH

c. CH_3–C(=O)–CH_2–C(=O)–CH_3

d. ethyl 3-ethyl-2-oxocyclopentanecarboxylate

e. $CH_3CH(CH_3)CH_2CH_2$–C(=O)–OH

f. C_6H_5–C(=O)–CH_2–C(=O)–OCH_2CH_3

55.

56. The electron withdrawing nitro group will cause the signal to occur at a higher frequency (will have a larger chemical shift). The more acidic the hydrogen, the greater the chemical shift.

$$CH_3NO_2 \qquad CH_2(NO_2)_2 \qquad CH(NO_2)_3$$
$$\delta 4.33 \qquad\quad \delta 6.10 \qquad\qquad \delta 7.52$$

57. The compound on the far right loses CO_2 at the lowest temperature, because it is a β-keto acid. Thus, the electrons left behind when CO_2 is removed can be delocalized.

58.

a.

b.

c.

d.

e.

f.

59. **a.** When the ketone enolizes, the asymmetric center is lost. When the enol reforms the ketone, an asymmetric center is created; it can form the *R* and *S* enantiomers equally easily, so a racemic mixture is obtained.

(*R*)-2-methyl-1-phenyl-1-butanone

b. You need a ketone that has an α-carbon that is an asymmetric center. Racemization will occur when an α-hydrogen is removed from the asymmetric center.

60. The initially formed addition product loses water immediately, because the new double bonds are particularly stable (and therefore easy to form) since each is conjugated with a benzene ring.

$+$ $2 H_2O$

61.

62. a.   **b.**   **c.**

63.

Notice that the β-substituted compounds are prepared via an α,β-unsaturated ketone, which can be prepared in two different ways—either by dehydrohalogenation of α-bromocyclohexanone or by dehydration of α-hydroxycyclohexanone.

64. Remember that there are no positively charged organic reactants, intermediates, or products in a basic solution, and no negatively charged organic reactants, intermediates, or products in an acidic solution.

a.

b.

65.   a.

b.

c.

d.

66. The Reformatsky reaction requires an organozinc reagent obtained by treating an α-bromo ester with zinc.

a.

$$CH_3CH_2CH_2 \overset{O}{\underset{}{C}} H \ + \ BrCH_2 \overset{O}{\underset{}{C}} OCH_3 \xrightarrow[\text{2. H}_2\text{O}]{\text{1. Zn}} CH_3CH_2CH_2\overset{OH}{\underset{}{C}}HCH_2 \overset{O}{\underset{}{C}} OCH_3$$

b.

$$CH_3CH_2 \overset{O}{\underset{}{C}} H \ + \ CH_3CH_2\overset{}{\underset{Br}{C}}H \overset{O}{\underset{}{C}} OCH_3 \xrightarrow[\text{2. H}_2\text{O}]{\text{1. Zn}} CH_3CH_2\overset{OH}{\underset{}{C}}H\overset{}{\underset{CH_2CH_3}{C}}H \overset{O}{\underset{}{C}} OCH_3$$

c.

$$CH_3CH_2 \overset{O}{\underset{}{C}} H \ + \ CH_3\overset{}{\underset{Br}{C}}H \overset{O}{\underset{}{C}} OCH_3 \xrightarrow[\text{2. H}_2\text{O}]{\text{1. Zn}} CH_3CH_2\overset{OH}{\underset{}{C}}H\overset{}{\underset{CH_3}{C}}H \overset{O}{\underset{}{C}} OCH_3$$

$$\Big\downarrow \text{HCl} \quad \overset{\text{H}_2\text{O}}{\text{excess}}$$

$$CH_3CH_2CH=\overset{}{\underset{CH_3}{C}} \overset{O}{\underset{}{C}} OH \xleftarrow[\Delta]{\text{H}_2\text{SO}_4} CH_3CH_2\overset{OH}{\underset{}{C}}H\overset{}{\underset{CH_3}{C}}H \overset{O}{\underset{}{C}} OH$$

d.

$$CH_3CH_2 \overset{O}{\underset{}{C}} CH_2CH_3 \ + \ BrCH_2 \overset{O}{\underset{}{C}} OCH_3 \xrightarrow[\text{2. H}_2\text{O}]{\text{1. Zn}} CH_3CH_2\overset{OH}{\underset{CH_2CH_3}{C}}CH_2 \overset{O}{\underset{}{C}} OCH_3$$

67. The ketone is 2-hexanone. Therefore, the alkyl halide is a propyl halide (propyl bromide, propyl chloride, or propyl iodide).

$$\underbrace{CH_3 \overset{O}{\underset{}{C}} CH_2CH_2CH_2CH_3}$$

This part comes from acetoacetic ester.

68. a.

b. 1.

2.

c.

d.

e.

f.

69. The positive iodoform test indicates that the compound with two singlets with a ratio of 3 : 1 is a methyl ketone. 3,3-Dimethyl-1-butyne will show two singlets with a ratio of 9 : 1, and when it adds water, it will form a methyl ketone with two singlets with a ratio of 3 : 1.

two singlets (9 : 1)   **two singlets (3 : 1)**

70.   a.

b.

c.

d.   CH$_3$—C(=O)—(CH$_2$)$_3$—C(=O)—OCH$_3$ $\xrightarrow[\text{condensation}]{\text{CH}_3\text{O}^- \;\; \text{Claisen}}$ (1,3-cyclohexanedione) $\xrightarrow[\substack{\text{CH}_3\text{I} \\ (2\ \text{equiv.})}]{\text{CH}_3\text{O}^-}$ (2,2-dimethyl-1,3-cyclohexanedione)

e.   (cyclopentanone) $\xrightarrow[\text{2. } \textit{tert}\text{-BuO}^-]{\text{1. Br}_2,\ \text{H}_3\text{O}^+}$ (cyclopentenone) $\xrightarrow[\text{CH}_3\text{O}^-]{\text{CH}_3\text{CCH}_2\text{COCH}_3}$ (addition product with OCH$_3$ ester)

$\xrightarrow[\Delta]{\text{HCl, H}_2\text{O}}$ (product) $+\ \text{CO}_2\ +\ \text{CH}_3\text{OH}$

f.   CH$_3$CH$_2$O—C(=O)—(CH$_2$)$_4$—C(=O)—OCH$_2$CH$_3$ $\xrightarrow[\substack{\text{Claisen} \\ \text{condensation}}]{\text{CH}_3\text{CH}_2\text{O}^-}$ (cyclized β-keto ester, OCH$_2$CH$_3$)

$\xrightarrow[]{\Delta\ \big|\ \text{HCl, H}_2\text{O}}$ (cyclopentanone) $+\ \text{CO}_2\ +\ \text{CH}_3\text{OH}$

(2-propylcyclopentanone)—CH$_2$CH$_2$CH$_3$ $\xleftarrow[\text{2. CH}_3\text{CH}_2\text{CH}_2\text{Br}]{\text{1. LDA}}$

71.   Synthesizing benzaldehyde from benzene would be easy if a one-carbon acyl chloride could be used, but there is no such compound. The synthesis of benzaldehyde is shown below being done via a Friedel-Crafts alkylation reaction. A Gatterman-Koch reaction (page 645 in the text) could also be used.

CH$_3$CCH$_3$
LDA

(benzene) $\xrightarrow[\substack{\text{2. NBS/peroxide} \\ \Delta \\ \text{3. HO}^- \\ \text{4. PCC}}]{\text{1. CH}_3\text{Cl/AlCl}_3}$ (benzaldehyde) —H $\xrightarrow{\bar{\text{C}}\text{H}_2\text{CCH}_3}$ (benzalacetone) $\xrightarrow[\substack{\text{I}_2 \\ \text{excess}}]{\substack{\text{HO}^- \\ \text{excess}}}$ (cinnamate)—O$^-$ $+\ \text{H}_2\text{O}$

$\xrightarrow[]{\text{1. SO}_2\text{Cl}\ \big|\ \text{2. CH}_3\text{CH}_2\text{OH}}$ (ethyl cinnamate)

72.   **a.**   The first reaction is an intermolecular Claisen condensation. An intramolecular condensation reaction forms the final product.

b.   The reaction involves two successive aldol condensations.

c.   The base removes a proton from the carbon that is flanked by two carbonyl groups; this is followed by an intramolecular S_N2 reaction.

d. A proton is removed from the alpha carbon that is flanked by a carbonyl group and a nitrile; then an intramolecular S_N2 reaction occurs.

73.

how the reagents become connected

74. **a.**

b.

75. **A** can be prepared by removing an α-hydrogen from acetone with LDA and then adding formaldehyde, followed by dehydration.

B can be prepared by removing an α-hydrogen from propionaldehyde with LDA and then slowly adding acetaldehyde, followed by dehydration.

C can be prepared by removing an α-hydrogen from 3-hexanone with LDA and then adding formaldehyde, followed by dehydration. The yield is poor because 3-hexanone is an asymmetrical ketone; therefore, two different α-carbanions are formed that lead to two different α,β-unsaturated ketones.

D can be prepared by removing an α-hydrogen from acetaldehyde with LDA and then slowly adding acetone, followed by dehydration.

E cannot be prepared by a mixed aldol condensation. It can be prepared via an intramolecular aldol condensation using 5-oxooctanal, followed by dehydration.

F cannot be prepared by a mixed aldol condensation. It can be prepared via an intramolecular aldol condensation using 1,7-heptanedial, followed by dehydration.

G cannot be prepared by a mixed aldol condensation. A small amount of **G** can be formed via an intramolecular aldol condensation using 7-oxooctanal, but this compound can form two different α-carbanions that can react with a carbonyl group to form a six-membered ring. Because an aldehyde is more reactive than a ketone, the desired compound will be formed only as a minor product.

minor product   **major product**

Theoretically **H** can be prepared by a mixed aldol condensation. However, because of steric hindrance, the reaction of propanal with dimethylpropanal to form the desired product would be very slow.

I can be prepared using propanal and 3,3-dimethylbutanal.

76.   **a.**

b.

c.

77.   At low temperatures (−78 °C), the proton will be more apt to be removed from the methyl group, because its hydrogens are the most accessible and are slightly more acidic (that is, the kinetically controlled product will be formed).

At higher temperatures (25 °C), the proton will be more apt to be removed from the more substituted α-carbon, because in that way the more stable enolate is formed (the one with the more stable double bond). Thus, the thermodynamically controlled product is formed.

78.

79. The compound that gives the ^{1}H NMR spectrum is 4-phenyl-3-buten-2-one. The singlet at 2.3 ppm is the methyl group, the doublet at ~6.7 ppm and the doublet buried under the benzene ring protons are the hydrogens bonded to the sp^2 carbons. The compounds that would form this compound (via an aldol condensation) are benzaldehyde and acetone.

4-phenyl-3-buten-2-one

80. The middle carbonyl group is hydrated, because it is stabilized by the electron-withdrawing carbonyl groups on either side of it.

81.

82.

83. **a.**

b. C_2H_5O ... CH_2 ... OC_2H_5 $\xrightarrow[\text{2. Br(CH}_2)_5\text{Br}]{\text{1. CH}_3\text{CH}_2\text{O}^-}$ C_2H_5O ... CH ... OC_2H_5

$CH_2CH_2CH_2CH_2CH_2Br$

malonic ester
$CH_3CH_2O^-$

C_2H_5O ... CH ... OC_2H_5

$CH_2CH_2CH_2CH_2CH_2CH$ ... $C-OC_2H_5$, $C-OC_2H_5$

$\xrightarrow[\Delta]{\text{HCl, H}_2\text{O}}$

HO ... C ... $(CH_2)_7$ ... C ... OH

$+ \ 2\ CO_2 \ + \ 4\ CH_3CH_2OH$

84.

85.   The mechanism starts with an isomerization that converts isolated double bonds to conjugated double bonds.

b. The carbonyl group on the left is protonated because electron delocalization causes it to be more basic than the other carbonyl group.

86. The benzoin condensation requires the aldehydic hydrogen to be removed. This hydrogen cannot be removed unless the electrons left behind can be delocalized onto an electronegative atom. The nitrogen of the cyano group serves that purpose. Therefore, the reaction will not occur if hydroxide ion is used instead of cyanide ion, because the electrons left behind if the hydrogen were to be removed cannot be delocalized.

87. To arrive at the final product, three equivalents of malonyl thioester are needed. (See page 872 in the text.)

88.

89.

90.

$$CH_3CH_2O-\overset{O}{\underset{}{C}}-CH_2-\overset{O}{\underset{}{C}}-OCH_2CH_3 \xrightarrow[\text{2. } CH_3CH_2Br]{\text{1. } CH_3CH_2O^-} CH_3CH_2O-\overset{O}{\underset{}{C}}-\underset{CH_2CH_3}{CH}-\overset{O}{\underset{}{C}}-OCH_2CH_3$$

1. $CH_3CH_2O^-$
2. $CH_3CHCH_2CH_2Br$
     |
     CH_3

$$\xleftarrow[\underset{\Delta}{CH_3CH_2O^-}]{H_2N\overset{O}{\underset{}{C}}NH_2} \quad CH_3CH_2O-\overset{O}{\underset{}{C}}-\underset{\underset{}{\underset{CH_3CH_2}{\,}}\,\underset{CH_2CH_2CHCH_3}{\,}}{C}-\overset{O}{\underset{}{C}}-OCH_2CH_3$$

Notice that in the last step of the reaction, a diester is converted into a diamide.

91.

$$CH_3-\overset{O}{\underset{}{C}}-CH_2-\overset{O}{\underset{}{C}}-OCH_2CH_3 \xrightarrow{CH_3CH_2O^-} CH_3-\overset{O}{\underset{}{C}}-\overset{\cdot\cdot}{CH}-\overset{O}{\underset{}{C}}-OCH_2CH_3 \longrightarrow$$

$$CH_3CH_2O^- \quad + \quad$$

92.   **a.**

b. The substituent in ibuprofen is placed on the ring by a Friedel-Crafts acylation reaction. A Friedel-Crafts reaction cannot be done in the synthesis of ketoprofen, because the benzene rings are deactivated and deactivated rings cannot undergo Friedel-Crafts reactions.

Chapter 19 Practice Test

1.   Rank the following compounds in order of decreasing acidity (label the most acidic #1):

2.   Give a structure for each of the following:

   a.   the most stable enol tautomer of 2,4-pentanedione

   b.   a β-keto ester

3.   Give the product of each of the following reactions:

   a.   CH₃CH₂—C(=O)—CH₂—C(=O)—OH →^Δ

   b.   CH₃CH₂CH₂—C(=O)—H   →^(Br₂, HO⁻)

   c.   CH₃CH₂CH₂—C(=O)—OCH₃   →^(1. CH₃O⁻ ; 2. HCl)

   d.   CH₂=CH—C(=O)—CH₃ + CH₃—C(=O)—CH₂—C(=O)—CH₃   →^(HO⁻ / H₂O)

   e.   cyclopentanone   →^(1. Br₂, H₃O⁺ ; 2. *tert*-BuO⁻)

   f.   cyclohexanone + Cl₂   →^(H₃O⁺)

g. CH_3CH_2O

$\xrightarrow[\text{2. } CH_3CH_2CH_2Br]{\text{1. } CH_3CH_2O^-}$
$\text{3. HCl, H}_2O, \Delta$

4. Give an example of each of the following:

   a. an aldol addition

   b. an aldol condensation

   c. a Claisen condensation

   d. a Dieckmann condensation

   e. a malonic ester synthesis

   f. an acetoacetic ester synthesis

5. Give the products of the following crossed aldol addition:

6. What ester would be required to prepare each β-ketoester?

   a.

   b.

CHAPTER 20
More About Oxidation-Reduction Reactions

Important Terms

Baeyer-Villiger oxidation	oxidation of aldehydes or ketones with H_2O_2 to form carboxylic acids or esters, respectively.
catalytic hydrogenation	addition of hydrogen to a double or triple bond in the presence of a metal catalyst.
chemoselective reaction	a reaction in which a reagent reacts with one functional group in preference to another.
dissolving metal reduction	a reduction using sodium or lithium metal dissolved in liquid ammonia.
enantioselective reaction	a reaction that forms an excess of one enantiomer.
functional group interconversion	a reaction that converts one functional group into another.
glycol	a compound containing two or more OH groups.
metal-hydride reduction	a reduction using a metal-containing reagent that delivers a hydride ion.
molozonide	an unstable intermediate containing a five-membered ring with three oxygens in a row that is formed from the reaction of an alkene with ozone.
oxidation	in the case of an organic molecule, a reaction in which the number of C—H bonds is decreased or the number of C—O, C—N, or C—X (X = halogen) bonds is increased.
oxidative cleavage	an oxidation reaction that cleaves the reactant into two or more compounds.
oxidizing agent	the reagent that is reduced in a redox reaction as it oxidizes the other compound.
ozonide	a five-membered ring compound formed as a result of rearrangement of a molozonide.
ozonolysis	reaction of a carbon-carbon double or triple bond with ozone.
peroxyacid	a carboxylic acid derivative with an OOH group instead of an OH group.
redox reaction	an oxidation-reduction reaction.
reducing agent	the reagent that is oxdized in a redox reaction as it reduces the other compound.
reduction	in the case of an organic molecule, a reaction in which the number of C—H bonds is increased or the number of C—O, C—N, or C—X (X = halogen) bonds is decreased.

Rosenmund reduction the reduction of an acyl chloride to an aldehyde using H_2/Lindlar catalyst.

Swern oxidation uses dimethyl sulfoxide, oxalyl chloride, and triethylamime to oxidize primary alcohols to aldehydes and secondary alcohols to ketones.

Tollens test a test to determine the presence of an aldehyde. A positive result is indicated by the formation of a silver mirror.

vicinal diol (vicinal glycol) a 1,2-diol.

Solutions to Problems

1. **a.** reduction   **c.** oxidation   **e.** reduction
    b. neither   **d.** oxidation   **f.** neither

2. **a.** [benzene ring]—CH$_2$OH

    d. [cyclohexane ring]—OH

    g. CH$_3$—C(=O)—H

    b. CH$_3$CH$_2$CH$_2$CH$_2$NH$_2$

    e. no reaction

    h. [cyclohexane ring]—NHCH$_3$

    c.
    CH$_3$CH$_2$CH$_2$\C=C/CH$_3$ with H and H below

    f. CH$_3$CH$_2$OH

3. **a.** CH$_3$CH$_2$NHCH$_3$

    b. CH$_3$CH$_2$NCH$_3$ with CH$_3$ below (on N)

4. **a.** [benzene ring]—CH$_2$NH$_2$

    c. CH$_3$CH$_2$CHCH$_2$CH$_3$ with OH above (on middle carbon)

    e. no reaction

    b. no reaction

    d. [cyclohexane ring]—CH$_2$OH + CH$_3$CH$_2$OH

    f. CH$_3$CH$_2$CH$_2$CH$_2$OH

5. Carbon-nitrogen double and triple bonds can be reduced by LiAlH$_4$ because the bonds are polar. The hydride ion will be attracted to the partially positively charged carbon.

    $\overset{\delta_+}{C}=\overset{\delta_-}{N}$—   with :H$^-$

    —$\overset{\delta_+}{C}\equiv\overset{\delta_-}{N}$—   with :H$^-$

6.
    CH$_3$CH=CH—C(=O)—H

    H$_2$/Pd/C → CH$_3$CH$_2$CH$_2$—C(=O)—H

    H$_2$ | Raney Nickel → CH$_3$CH$_2$CH$_2$CH$_2$OH

    1. NaBH$_4$ 2. H$_2$O → CH$_3$CH=CHCH$_2$OH

7. A terminal alkyne cannot be reduced by Na, because Na reacts with a terminal alkyne to form an acetylide ion rather than a radical anion (Section 6.9).

$$2\,CH_3C{\equiv}CH + 2\,Na \longrightarrow 2\,CH_3C{\equiv}C^- + 2\,Na^+ + H_2$$

8.

a. (cyclohexene with CHCH₃ and OH substituent)

c. (cyclohexane with OH and CH₂COCH₃ substituents)

b. (cyclohexane with COCH₃ substituent)

d. (cyclohexane with OH and CH₂CH₂OH substituents) + CH₃OH

9.   **a.** Solved in the text.

b.

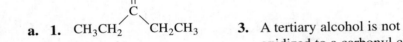

10.   **a.   1.** CH₃CH₂—CO—CH₂CH₃

2. CH₃CH₂CH₂CH₂—CO—OH

3. A tertiary alcohol is not oxidized to a carbonyl compound.

4. CH₃—CO—CH₂—CO—CH₂CH₃

5. (cyclohexanone) =O

6. HO—CO—CH₂CH₂—CO—OH

b.   1. CH₃CH₂—CO—CH₂CH₃

2. CH₃CH₂CH₂CH₂—CO—H

3. A tertiary alcohol is not oxidized to a carbonyl compound.

4. CH₃—CO—CH₂—CO—CH₂CH₃

5. (cyclohexanone) =O

6. H—CO—CH₂CH₂—CO—H

11.   Chromic acid needs to be protonated to convert its poor leaving group (HO^-) into a good leaving group (H_2O). The alcohol displaces water to form a protonated chromate ester, which loses a proton. The aldehyde is then formed in an elimination reaction. (Propanal would undergo further oxidation with chromic acid to form a carboxylic acid.)

12.   a.

c.

e.

b.

d.

f.   no reaction

13.   The major product results from attack of the Grignard reagent on the less sterically hindered carbon of the epoxide (the one bonded to a hydrogen and to a methyl group). It will be easier to answer this question if you use molecular models.

14. Methoxide ion attacks a carbon of the three-membered ring from the backside. (It is an S_N2 reaction.) Either of the two ring carbons can be attacked. Therefore, the products are the trans stereoisomers.

15. The addition of Br_2 to an alkene is a stereoselective reaction, because not all possible stereoisomers are formed: the trans alkene forms the erythro pair of enantiomers, whereas the cis alkene forms the threo pair of enantiomers.

It is a stereospecific reaction, because a cis alkene leads to a different set of products than does a trans alkene.

It is not enantioselective, because if an asymmetric center is created, both enantiomers are obtained in equal amounts, and if two asymmetric centers are created, two pairs of enantiomers (or a meso form and a pair of enantiomers) are obtained.

16. **a.**

  b.

17. **a.** Diol formation is a syn addition reaction. Syn addition to the trans isomer forms the threo pair of enantiomers. (See Section 5.19.)

  b. Syn addition to the cis isomer forms the erythro pair of enantiomers. In this case, the product is a meso compound because each asymmetric center is bonded to the same four substituents, so only one stereoisomer is formed.

  c. Syn addition to the cis isomer forms the erythro pair of enantiomers.

d. Syn addition to the trans isomer forms the threo pair of enantiomers.

18. The fact that only one ketone is obtained means that the alkene must be symmetrical. The cyclic ketone that is obtained is cyclohexanone. Therefore, the alkene is:

19. **a.** The compound on the left is more reactive because, when the bulky *tert*-butyl group is in the more stable equatorial position, the two OH groups are in equatorial positions and can form the required cyclic intermediate. In the other compound, when the *tert*-butyl group is in the equatorial position, the two OH groups are in axial positions and, therefore, are too far away one another to form the required cyclic intermediate.

more reactive   **less reactive**

b. The compound on the right is more reactive because when the bulky *tert*-butyl group is in the more stable equatorial position, the two OH groups are in equatorial positions and can form the required cyclic intermediate. In the other compound, when the *tert*-butyl group is in the equatorial position, the two OH groups are in axial positions and, therefore, cannot form the required cyclic intermediate.

less reactive   **more reactive**

20. Any alkene that forms two ketones when it is cleaved will form the same products when the ozonide is worked up under reducing conditions as it forms when it is worked up under oxidizing conditions. In other words, each sp^2 carbon must be bonded to two alkyl groups.

21.

1. a.

$$\underset{\substack{\text{CH}_3\text{CH}_2\text{CH}_2 \quad\quad \text{CH}_3}}{\overset{\overset{\displaystyle O}{\|}}{\text{C}}} \; + \; \underset{\substack{\text{CH}_3 \quad\quad \text{H}}}{\overset{\overset{\displaystyle O}{\|}}{\text{C}}}$$

b.

$$\underset{\substack{\text{CH}_3\text{CH}_2\text{CH}_2 \quad\quad \text{CH}_3}}{\overset{\overset{\displaystyle O}{\|}}{\text{C}}} \; + \; \underset{\substack{\text{CH}_3 \quad\quad \text{OH}}}{\overset{\overset{\displaystyle O}{\|}}{\text{C}}}$$

2. a.

$$\underset{\substack{\text{H} \quad\quad \text{H}}}{\overset{\overset{\displaystyle O}{\|}}{\text{C}}} \; + \; \underset{\substack{\text{CH}_3\text{CH}_2\text{CH}_2\text{CH}_2 \quad\quad \text{H}}}{\overset{\overset{\displaystyle O}{\|}}{\text{C}}}$$

b.

$$\underset{\substack{\text{H} \quad\quad \text{OH}}}{\overset{\overset{\displaystyle O}{\|}}{\text{C}}} \; + \; \underset{\substack{\text{CH}_3\text{CH}_2\text{CH}_2\text{CH}_2 \quad\quad \text{OH}}}{\overset{\overset{\displaystyle O}{\|}}{\text{C}}}$$

3. a.

$$\underset{\substack{\text{CH}_3 \quad\quad \text{CH}_2\text{CH}_2\text{CH}_2 \quad\quad \text{H}}}{\overset{\overset{\displaystyle O}{\|}}{\text{C}} \quad\quad \overset{\overset{\displaystyle O}{\|}}{\text{C}}}$$

b.

$$\underset{\substack{\text{CH}_3 \quad\quad \text{CH}_2\text{CH}_2\text{CH}_2 \quad\quad \text{OH}}}{\overset{\overset{\displaystyle O}{\|}}{\text{C}} \quad\quad \overset{\overset{\displaystyle O}{\|}}{\text{C}}}$$

4. a.

$$\underset{\substack{\text{H} \quad\quad \text{H}}}{\overset{\overset{\displaystyle O}{\|}}{\text{C}}} \; + \; \text{(cyclopentanone)}$$

b.

$$\underset{\substack{\text{H} \quad\quad \text{OH}}}{\overset{\overset{\displaystyle O}{\|}}{\text{C}}} \; + \; \text{(cyclopentanone)}$$

5. a.

$$\underset{\substack{\text{CH}_3\text{CH}_2\text{CH}_2 \quad\quad \text{H}}}{\overset{\overset{\displaystyle O}{\|}}{\text{C}}}$$
(2 equivalents)

b.

$$\underset{\substack{\text{CH}_3\text{CH}_2\text{CH}_2 \quad\quad \text{OH}}}{\overset{\overset{\displaystyle O}{\|}}{\text{C}}}$$
(2 equivalents)

6. a.

$$\underset{\substack{\text{CH}_3 \quad\quad \text{C} \quad\quad \text{H}}}{\overset{\overset{\displaystyle O}{\|}}{\text{C}}} \underset{\displaystyle \|}{\underset{\displaystyle O}{}} \; + \; \underset{\substack{\text{H} \quad\quad \text{CH}_2\text{CH}_2 \quad\quad \text{H}}}{\overset{\overset{\displaystyle O}{\|}}{\text{C}} \quad\quad \overset{\overset{\displaystyle O}{\|}}{\text{C}}}$$

b.

$$\underset{\substack{\text{CH}_3 \quad\quad \text{C} \quad\quad \text{OH}}}{\overset{\overset{\displaystyle O}{\|}}{\text{C}}} \underset{\displaystyle \|}{\underset{\displaystyle O}{}} \; + \; \underset{\substack{\text{HO} \quad\quad \text{CH}_2\text{CH}_2 \quad\quad \text{OH}}}{\overset{\overset{\displaystyle O}{\|}}{\text{C}} \quad\quad \overset{\overset{\displaystyle O}{\|}}{\text{C}}}$$

22.

a.

$$\underset{\substack{\text{H}_3\text{C} \\ \text{H}_3\text{C}}}{} \text{C} = \text{C} \underset{\substack{\text{CH}_3 \\ \text{CH}_3}}{}$$

2,3-dimethyl-2-butene

b.

$$\underset{\substack{\text{CH}_3\text{CH}_2\text{CH}_2 \\ \text{H}}}{} \text{C} = \text{C} \underset{\substack{\text{CH}_2\text{CH}_2\text{CH}_3 \\ \text{H}}}{}$$

cis-4-octene

or

$$\underset{\substack{\text{CH}_3\text{CH}_2\text{CH}_2 \\ \text{H}}}{} \text{C} = \text{C} \underset{\substack{\text{H} \\ \text{CH}_2\text{CH}_2\text{CH}_3}}{}$$

trans-4-octene

23.   Solved in the text.

24.   It does not tell you whether the double bond has the cis or the trans configuration.

25.   Solved in the text.

26.   **a.**

    b.

27.   **a.**

    b.

28.   **a.   1.** $CH_3CH_2Br \xrightarrow[Et_2O]{Mg} CH_3CH_2MgBr \xrightarrow{H_2C=O} CH_3CH_2CH_2O^- \xrightarrow{HCl} CH_3CH_2CH_2OH$

        2. $CH_3CH_2Br \xrightarrow[Et_2O]{Mg} CH_3CH_2MgBr \xrightarrow{CO_2} CH_3CH_2\overset{\overset{\displaystyle O}{\|}}{C}O^- \xrightarrow[2.\ H_3O^+]{1.\ LiAlH_4} CH_3CH_2CH_2OH$

        3. $CH_3CH_2Br \xrightarrow[HCl]{^-C\equiv N} CH_3CH_2C\equiv N \xrightarrow[\Delta]{HCl,\ H_2O} CH_3CH_2\overset{\overset{\displaystyle O}{\|}}{C}OH \xrightarrow[2.\ H_3O^+]{1.\ LiAlH_4} CH_3CH_2CH_2OH$

b. $CH_3CH_2Br \xrightarrow{\ ^-C\equiv N\ } CH_3CH_2C\equiv N \xrightarrow[\text{Raney Nickel}]{H_2} CH_3CH_2CH_2NH_2$

c. $CH_3CH_2CH_2Br \xrightarrow{\text{tert-BuO}^-} CH_3CH=CH_2 \xrightarrow[\text{2. Zn, H}_2O]{\text{1. O}_3, -78\ ^\circ C} CH_3CH=O \xrightarrow[\substack{\text{2. H}_2 \\ \text{H}_2, \text{Pd/C}}]{\text{1. NH}_3} CH_3CH_2NH_2$

29.   **a.**

or

b.

c.

A bulky base is used to encourage elimination.

d.

30.

aldehyde	acyl halide	alkyl halide
carboxylic acid	acid anhydride	ether
ester	alkene	epoxide

31.   **a.** oxidized   **c.** reduced   **e.** oxidized   **g.** reduced
         b. reduced   **d.** oxidized   **f.** oxidized

32.   a.   CH$_3$CH$_2$CH$_2$CH$_2$—C(=O)—OH
oxidation

b.   CH$_3$CH$_2$CH$_2$ / H
C=C
H / CH$_3$
reduction

c.   CH$_3$CH$_2$—C(=O)—H   **reduction**

d.   CH$_3$CH$_2$CH$_2$NHCH$_3$ **reduction**

e.   CH$_3$CH$_2$CH$_2$CH$_2$OH

Formation of the aldehyde after steps 1 and 2
is neither oxidation or reduction.
Formation of the alcohol after steps 3 and 4
is a **reduction**.

f.   CH$_3$CH$_2$—C(=O)—H   **oxidation**

g.   [benzene ring]—CH$_2$CH$_2$CH$_3$
reduction

h.   [cyclopentane ring]=O   +   HCH(=O)
oxidation

33.   a.   CH$_3$CH$_2$CH$_2$CH(=O) $\xrightarrow{H_2CrO_4}$ CH$_3$CH$_2$CH$_2$COH(=O)

b.   CH$_3$CH$_2$CH$_2$CH$_2$OH $\xrightarrow{H_2CrO_4}$ CH$_3$CH$_2$CH$_2$COH(=O)

c.   CH$_3$CH$_2$CH$_2$CH$_2$Br $\xrightarrow{HO^-}$ CH$_3$CH$_2$CH$_2$CH$_2$OH $\xrightarrow{H_2CrO_4}$ CH$_3$CH$_2$CH$_2$COH(=O)

d.   CH$_3$CH$_2$CH=CH$_2$ $\xrightarrow[\text{2. HO}^-,\ H_2O_2,\ H_2O]{\text{1. BH}_3}$ CH$_3$CH$_2$CH$_2$CH$_2$OH $\xrightarrow{H_2CrO_4}$ CH$_3$CH$_2$CH$_2$COH(=O)

34.

35.

a.

oxidation

e.

oxidation

b.

reduction

f.

oxidation

c.

oxidation

g.

oxidation

d.

+ CH_3OH

reduction

36.   **a.** $CH_3CH_2CH_2CH{=}CCH_3$
$\quad\quad\quad\quad\quad\quad\quad\quad\quad\quad\quad$ |
$\quad\quad\quad\quad\quad\quad\quad\quad\quad\quad$ CH_3

c. $={=}CHCH_2CH_3$

e.

b.

or

d.

f.

37.

the primary amine

butanal   2-methylpropanal

38.   **a.** $CH_3CH_2CH{=}CH_2 \xrightarrow[\text{2. }H_2O_2,\ HO^-,\ H_2O]{\text{1. }BH_3} CH_3CH_2CH_2CH_2OH$

$$\Big\updownarrow \begin{array}{l} H_2SO_4 \\ \Delta \end{array}$$

$$2\ CH_3\overset{\displaystyle O}{\overset{\|}{C}}OH \xleftarrow[\text{2. }H_2O_2]{\text{1. }O_3,\ -78\ °C} CH_3CH{=}CHCH_3 \quad + \quad H_2O$$

b. $CH_3CH_2Br \xrightarrow[\text{Et}_2O]{Mg} CH_3CH_2MgBr \xrightarrow[\text{2. }H_3O^+]{\text{1. } \triangle^O} CH_3CH_2CH_2CH_2OH$

$$\Big\downarrow PCC$$

$$CH_3CH_2CH_2\overset{\displaystyle O}{\overset{\|}{C}}H$$

c.

$$\text{cyclohexane} \xrightarrow[hv]{Br_2} \text{bromocyclohexane} \xrightarrow{HO^-} \underset{\text{substitution product}}{\text{cyclohexanol}} \quad + \quad \underset{\text{elimination product}}{\text{cyclohexene}}$$

substitution product $\xrightarrow{H_2CrO_4}$ cyclohexanone

elimination product $\xrightarrow[\text{2. } H_2O_2]{\text{1. } O_3,\ -78\ ^\circ C}$ HO$_2$C(CH$_2$)$_3$CO$_2$H

cyclohexanone + glutaric acid

39. a. 1.

b.

or

2.

3.

4.

5.

40. a.

$$\text{cyclohexene} \xrightarrow[H_2O]{H_2SO_4} \text{cyclohexanol} \xrightarrow{H_2CrO_4} \text{cyclohexanone} \xrightarrow{F_3CCOOH} \text{lactone}$$

b.

$$\text{cyclohexene} \xrightarrow[\text{2. Zn, } H_2O]{\text{1. } O_3} \underset{}{\text{H}}\overset{O}{\underset{}{C}}\text{CH}_2\text{CH}_2\text{CH}_2\text{CH}_2\overset{O}{\underset{}{C}}\text{H} \xrightarrow[\text{2. } H_3O^+]{\text{1. NaBH}_4} HOCH_2CH_2CH_2CH_2CH_2CH_2OH$$

c.

41. Because the compound is produced as a result of ozonolysis of an alkene under oxidizing conditions, it must be a ketone or a carboxylic acid. Since there is no signal for an OH group, the compound must be a ketone. The NMR spectrum shows two signals with splitting that is characteristic of an ethyl group. Therefore, the compound produced as a result of ozonolysis must be 3-pentanone. Since the ozonolysis reaction forms only one product, the alkene must be 3,4-diethyl-3-hexene.

3,4-diethyl-3-hexene

42.

43. The rate-determining step in the chromic acid oxidation of an alcohol is the E2 elimination reaction of the chromate ester. 2-Propanol is oxidized more rapidly than 2-deutero-2-propanol because it is easier to break the carbon-hydrogen bond than the stronger carbon-deuterium bond.

44.

45. a.

b.

c.

d.

e.

46. Two structures are possible candidates. The integration of the signals will allow you to determine which alkene was used in the ozonolysis reaction.

singlet, a doublet,
and a quartet

formic acid

47. A cyclic intermediate is formed when HIO_4 cleaves a 1,2-diol. It is easier to form a cyclic intermediate if the two OH groups are on the same side of the molecule.
Therefore, **A** is cleaved more easily because it is a cis diol, whereas **B** is a trans diol.

48.

49. Because oxidative cleavage formed only one four-carbon carboxylic acid, you know the compound is either a symmetrical eight-carbon alkene or a symmetrical eight-carbon alkyne, since both alkenes and alkynes undergo oxidative cleavage to form carboxylic acids. Therefore, the compound has a molecular formula of C_8H_{14} or C_8H_{16} and, consequently, a molecular weight of 110 or 112.

0.50 g of the hydrocarbon was hydrogenated; this corresponds to 0.0045 mol (0.5/112 = 0.0045, 0.5/110 = 0.0045). The key here is to remember from general chemistry that 1 mol of an ideal gas has a volume of 24 liters at 25 °C and 1 atmosphere pressure. Knowing this, it can be calculated that 0.0045 mol of H_2 = 108 mL H_2. Since about 200 mL of H_2 were consumed, the compound must have two π bonds. Thus, the unknown compound is an alkyne; it is either 4-octyne or 2,5-dimethyl-3-hexyne.

$$CH_3CH_2CH_2C \equiv CCH_2CH_2CH_3$$

4-octyne

$$\underset{\underset{CH_3}{|}}{CH_3CHC} \equiv \underset{\underset{CH_3}{|}}{CCHCH_3}$$

2,5-dimethyl-3-hexyne

50.   **a.**   Only #4 was successful.

  b.   **1.** no reaction   **2.** no reaction

3.

c.   **1.**   1. OsO_4
2. H_2O_2
3. HIO_4
   2.   1. $LiAlH_4$
2. H_3O^+
   3.   1. OsO_4
2. H_2O_2

51.   The absorption bands at 1600 cm⁻¹, 1500 cm⁻¹, and >3000 cm⁻¹ in the IR spectrum indicate the compound has a benzene ring. The absorption band at 2250 cm⁻¹ indicates that the compound has a triple bond, and the absence of an absorption band 3200 cm⁻¹ tells you that it is not a terminal alkyne.

The two triplets and the 2H multiplet in the ¹H NMR spectrum indicate a propyl group. The 5H multiplet suggests a phenyl group.

Compound **A** is 1-phenyl-1-propyne and compound **B** is propylbenzene.

52.   **a.**   The student has enough information to identify diols **A**, **C**, **E**, and **F**, but not enough information to distinguish between diols **B** and **D**.

Only the fourth compound will form two products upon cleavage with periodic acid. So it must be **F**.

The first and fifth compounds will not react with periodic acid. Because the first compound is optically inactive and the fifth compound is optically active, the first compound must be **C** and the fifth compound must be **E**.

The third compound is the only optically active compound that forms one compound with periodic acid. Therefore, it must be **A**.

The student could distinguish between **B** and **D** if she analyzed the products obtained from the reaction of **B** and **D** with periodic acid. Compound **#2** will form acetaldehyde (a 2-carbon aldehyde), whereas compound **#6** will form propanal (a 3-carbon aldehyde).

b.   A = (2S,4R)-2,4-pentanediol   D = (2S,3S)-2,3-pentanediol

   B = (2S,3R)-2,3-butanediol   E = (2S,4S)-2,4-pentanediol

   C = (2S,3S)-2,3-butanediol   F = (3S,4R)-2,3-hexanediol

53.

54.

55.

CH₃   CH₃
two singlets

⟶

CH₃   CH₃
cis-1,2-dimethylcyclopropane

56.   **a.** (2*R*,3*S*)-Tartaric acid is a meso compound. It will be formed by syn addition to the cis isomer (maleic acid). Syn addition of two OH groups can be carried out using osmium tetroxide followed by H_2O_2.

(2*R*,3*S*)-tartaric acid

b. (2*R*,3*S*)-Tartaric acid can also be formed by anti addition to the trans isomer (fumaric acid). Anti addition of two OH groups can be carried out by first forming an epoxide and then treating the epoxide with an aqueous solution of hydroxide.

(2*R*,3*S*)-tartaric acid

c. (2*R*,3*R*)-Tartaric acid and (2*S*,3*S*)-tartaric acid are the threo isomers. They will be formed by anti addition to the cis isomer.

(2*R*,3*R*)-tartaric acid   (2*S*,3*S*)-tartaric acid

or

d. The threo isomers can also be formed by syn addition to the trans isomer.

(2R,3R)-tartaric acid   (2S,3S)-tartaric acid

or

57.

58. a.

b. Note that 2-bromopropane (obtained from the first step) is divided into two portions; one portion is used to form propene, and the other is used to form the Grignard reagent.

$$CH_3CH_2CH_3 \xrightarrow[hv]{Br_2} \underset{\underset{Br}{|}}{CH_3CHCH_3} \xrightarrow{tert\text{-}BuO^-} CH_3CH{=}CH_2$$

$$\downarrow Mg \mid Et_2O \qquad\qquad\qquad \downarrow \begin{array}{l}1.\ O_3\\2.\ Zn,\ H_2O\end{array}$$

$$\underset{\underset{MgBr}{|}}{CH_3CHCH_3} \quad + \quad \overset{O}{\overset{\|}{CH_3CH}} \longrightarrow \underset{\underset{CH_3}{|}}{CH_3\overset{O^-}{\overset{|}{CH}}CHCH_3}$$

$$\downarrow H_2SO_4 \mid \Delta$$

$$\underset{\underset{CH_3}{|}}{CH_3CH{=}CCH_3}$$

c. $$CH_3CH_2CH_3 \xrightarrow[hv]{Br_2} \underset{\underset{Br}{|}}{CH_3CHCH_3} \xrightarrow[Et_2O]{Mg} \underset{\underset{MgBr}{|}}{CH_3CHCH_3} \xrightarrow[2.\ HCl]{1.\ \overset{O}{\overset{\|}{CH_3CH}}} \underset{\underset{CH_3}{|}}{CH_3\overset{OH}{\overset{|}{CH}}CHCH_3}$$

$$\downarrow H_2CrO_4$$

$$\underset{\underset{CH_3}{|}}{CH_3\overset{O}{\overset{\|}{CH}}CCH_3}$$

d. $$CH_3CH_3 \xrightarrow[hv]{Br_2} CH_3CH_2Br \xrightarrow[Et_2O]{Mg} CH_3CH_2MgBr \xrightarrow[2.\ HCl]{1.\ \triangle O} CH_3CH_2CH_2CH_2OH$$

$$\downarrow PBr_3$$

$$\underset{CH_3CH_2CH}{\overset{O}{\overset{\|}{}}} + \underset{HCH}{\overset{O}{\overset{\|}{}}} \xleftarrow[2.\ Zn,\ H_2O]{1.\ O_3,\ -78\ °C} CH_3CH_2CH{=}CH_2 \xleftarrow{tert\text{-}BuO^-} CH_3CH_2CH_2CH_2Br$$

The ethylene oxide needed for the reaction can be obtained from some of the bromoethane prepared in the first step.

$$CH_3CH_2Br \xrightarrow{tert\text{-}BuO^-} CH_2{=}CH_2 \xrightarrow{RCOOH} \triangle O$$

59.   **a.**   **b.**

60.

dimethylchlorosulfonium ion

61.   **a.**

   b.

   c.

   d.

e.

f.

62. a.

b.

c.

Chapter 20 Practice Test

1.   Indicate whether each of the following reactions is an oxidation, a reduction, or neither:

a. CH_3CH_2—C(=O)—CH_3 $\longrightarrow$ $CH_3CH_2CHCH_3$ (with OH on the CH)

b. CH_3CH_2—C(=O)—H $\longrightarrow$ CH_3CH_2—C(=O)—OH

2.   Give the product(s) of each of the following reactions, or if no reaction will occur, so state:

a. $CH_3CH_2CH_2CH_2OH$ $\xrightarrow{H_2CrO_4}$

b. $CH_3CH_2CH_2$—C(=O)—OCH_3 $\xrightarrow[Pd/C]{H_2}$

c. CH_3CH_2—C(=O)—$NHCH_2CH_3$ $\xrightarrow[2.\ H_2O]{1.\ LiAlH_4}$

d.   (phenyl)$CH{=}CHCH_3$ $\xrightarrow[2.\ Zn,\ H_2O]{1.\ O_3,\ -78\ °C}$

e. $CH_3CH_2CH_2CH_2$—C(=O)—H $+$ F_3C—C(=O)—OOH $\longrightarrow$

f.   (phenyl)—C(=O)—OCH_2CH_3 $\xrightarrow[2.\ H_3O^+]{1.\ LiAlH_4}$

g. $CH_3CH_2CH_2$—C(=O)—CH_2CH_3 $\xrightarrow[2.\ H_3O^+]{1.\ NaBH_4}$

h. $CH_3CH_2CH_2\overset{\displaystyle CH_3}{C}{=}CHCH_2CH_3$ $\xrightarrow[2.\ H_2O_2]{1.\ O_3,\ -78\ °C}$

3.   Which of the following sets of reagents **cannot** be used to convert an aldehyde into an alcohol?

1. $NaBH_4$	1. $LiAlH_4$	1. Ag_2O, NH_3	H_2, Raney Nickel
2. H_3O^+	2. H_3O^+	2. H_3O^+	

4.   Two alkenes, when treated with ozone and then with $(CH_3)_2S$, both form only the ketone shown below. Identify the alkenes.

$$\underset{CH_3CH_2 \qquad CH_3}{\overset{\displaystyle O}{\underset{\displaystyle \|}{C}}}$$

5.   Indicate whether each of the following statements is true or false:

a. $NaBH_4$ is a weaker reducing agent than $LiAlH_4$.   T   F

b. Esters are easier to reduce than ketones.   T   F

c. In an oxidation-reduction reaction the oxidizing agent is oxidized.   T   F

d. Ketones are reduced to primary alcohols.   T   F

e. Aldehydes are oxidized to carboxylic acids.   T   F

f. Acyl halides are oxidized to aldehydes.   T   F

g. Alkenes cannot be reduced with $NaBH_4$.   T   F

6.   Describe how the following compound can be prepared from the given starting material:

$$CH_3CH_2CH_2CH_3 \longrightarrow \underset{CH_3CH_2 \qquad CH_3}{\overset{\displaystyle O}{\underset{\displaystyle \|}{C}}}$$

7.   Which reagents could be used to carry out the following reaction?

$$\underset{CH_3C=CHCH_3}{\overset{\displaystyle CH_3}{\underset{\displaystyle |}{}}} \longrightarrow \underset{CH_3 \quad CH_3}{\overset{\displaystyle O}{\underset{\displaystyle \|}{C}}} + \underset{CH_3 \quad H}{\overset{\displaystyle O}{\underset{\displaystyle \|}{C}}}$$

a. 1. O_3 2. Zn, H_2O and   1. O_3 2. H_2O_2

b. 1. O_3 2. Zn, H_2O and   1. OsO_4 2. H_2O_2

c. 1. O_3 2. Zn, H_2O and   1. O_3 2. $(CH_3)_2S$

CHAPTER 21
More About Amines • Heterocyclic Compounds

Important Terms

furan	a five-membered ring aromatic compound containing an oxygen heteroatom.
heteroatom	an atom other than a carbon atom or a hydrogen atom.
heterocyclic compound (heterocycle)	a cyclic compound in which one or more of the atoms of the ring are heteroatoms.
imidazole	a five-membered ring aromatic compound containing two nitrogen heteroatoms.
ligation	sharing of nonbonded electrons with a metal.
porphyrin ring system	consists of four pyrrole rings joined by one carbon bridges.
purine	a pyrimidine ring fused to an imidazole ring.
pyrimidine	a benzene ring with nitrogens at the 1- and 3-positions.
pyrrole	a five-membered ring aromatic compound containing a nitrogen heteroatom.
thiophene	a five-membered ring aromatic compound containing a sulfur heteroatom.

Solutions to Problems

1. **a.** 2,2-dimethylaziridine
    b. 4-ethylpiperidine
    c. 3-methylazacyclobutane
    d. 2-methylthiacyclopropane
    e. 2,3-dimethyltetrahydrofuran
    f. 2-ethyloxacyclobutane

2. The electron-withdrawing oxygen in morpholine stabilizes the conjugate base, making protonated morpholine the stronger acid. Recall that the more stable the base, the stronger is its conjugate acid.

$pK_a = 9.28$ morpholine $pK_a = 11.12$ piperidine

3. **a.**

    b. The conjugate acid of 3-quinuclidinone has a lower pK_a than the conjugate acid of morpholine because the electron-withdrawing oxygen in 3-quinuclidinone is closer to the nitrogen and therefore can better stabilize its conjugate base.

conjugate acid
of 3-quinuclidinone
$pK_a = 7.46$

conjugate acid
of morpholine
$pK_a = 9.28$

    c. The conjugate acid of 3-chloroquinuclidine has a lower pK_a than the conjugate acid of 3-bromoquinuclidine, because chlorine is more electronegative than bromine and therefore can better stabilize its conjugate base. Because stable bases are weak bases, 3-bromoquinuclidine is a stronger base than 3-chloroquinuclidine.

conjugate acid of
3-chloroquinuclidine

conjugate acid of
3-bromoquinuclidine

4.   Solved in the text.

5.   **a.**

c.

b.

d.

6.

7.

2-deuteriopyrrole

8.   Protonation on C-2 leads to a cation with three resonance contributors. Protonation on nitrogen leads to a cation with no resonance contributors. (The positively charged nitrogen cannot accept electrons, because that would put ten electrons around the nitrogen.)

protonation on C-2

protonation on nitrogen

9. Pyrrole and its conjugate base are both aromatic. Cyclopentadiene does not become aromatic until it loses a proton. It is the drive for the nonaromatic compound to become a stable aromatic compound that causes cyclopentadiene to be more acidic than pyrrole.

10. Solved in the text.

11. Pyridine will act as an amine with the alkyl bromide, forming a quaternary ammonium salt.

12. The first reaction is a two-step nucleophilic aromatic substitution reaction (S_NAr). The second reaction is a one-step nucleophilic substitution reaction (S_N2).

13. **a.**

2-pyridinone

b. 4-Pyridinone is also formed because nucleophilic addition of hydroxide ion can take place at the 4-position as well as at the 2-position. It proceeds by the same mechanism as the one shown in part **a** for formation of 2-pyridinone.

4-pyridinone

14. It is easiest to remove a proton from the methyl group of the *N*-alkylated pyridine because the electrons left behind when the proton is removed can be delocalized onto the positively charged nitrogen.

It is easier to remove a proton from 4-methylpyridine than from 3-methylpyridine because in the former, the electrons left behind when the proton is removed can be delocalized onto the electronegative nitrogen. In contrast, in 3-methylpyridine the electrons can be delocalized onto only carbons.

15. The second nitrogen in imidazole, onto which the electrons left behind when the proton is removed can be delocalized, causes it to be a stronger acid than pyrrole.

imidazole
pK_a = 14.4

pyrrole
pK_a = 17

16. There are three possible sites for electrophilic substitution: C-2, C-4, and C-5. To determine the major product, compare the relative stabilities of the carbocations formed in the first step of the reaction.

Substitution at C-2 leads to a carbocation with three contributing resonance structures.
One of them is particularly unstable, because the positive charge is on a nitrogen atom that does not have a complete octet.

relatively unstable

Substitution at C-4 leads to a carbocation with only two contributing resonance structures.

Substitution at C-5 leads to a carbocation with three contributing resonance structures. Because none of them are as unstable as one of the contributing resonance structures obtained from substitution at C-2, it is the most stable of the three possible carbocation intermediates.

Therefore, the major product of the reaction is 5-bromo-*N*-methylimidazole.

5-bromo-*N*-methylimidazole

17.   Pyrrole and imidazole are both more reactive than benzene, because each reacts with an electrophile to form a carbocation intermediate that can be stabilized by resonance electron donation into the ring by a nitrogen atom. Pyrrole is more reactive than imidazole, because the second nitrogen atom of imidazole cannot donate electrons into the ring by resonance but can withdraw electrons from the ring inductively.

18.   Imidazole forms intermolecular hydrogen bonds, whereas *N*-methylimidazole cannot form hydrogen bonds. Because the hydrogen bonds have to be broken in order for the compound to boil, imidazole has a higher boiling point.

N-methylimidazole cannot form hydrogen bonds, because it does not have a hydrogen bonded to a nitrogen.

19.   fraction of imidazole in the acidic form $= \dfrac{[H^+]}{K_a + [H^+]}$

$$pH = 7.3; [H^+] = 5.0 \times 10^{-8}$$

$$pK_a = 6.8; K_a = 1.58 \times 10^{-7} \text{ (See Table 21.1.)}$$

fraction of imidazole in the acidic form $= \dfrac{5.0 \times 10^{-8}}{1.58 \times 10^{-7} + 5.0 \times 10^{-8}}$

$$= \dfrac{5.0 \times 10^{-8}}{2.08 \times 10^{-7}}$$

$$= 0.24$$

percent of imidazole in the acidic form $= 24\%$

20.   The second nitrogen in protonated pyrimidine (that withdraws electrons inductively) causes it to be a stronger acid than protonated pyridine.

protonated pyrimidine   protonated pyridine

$pK_a = 1.0$ $pK_a = 5.2$

21.   Yes, porphyrin is aromatic, because it fulfills the two requirements for aromaticity. It has an uninterrupted ring of *p* orbital-bearing atoms (it is cyclic and planar and each ring atom has a *p* orbital), and the π cloud contains an odd number of pairs (thirteen pairs) of π electrons.

22.   **a.** 2-methylazacyclobutane   **c.** 3-chloropyrrole
        b. 2,3-dimethylpiperidine   **d.** 2-ethyl-5-methylpiperidine

23.   **a.**

e.

b.

f. $\overset{+}{N}H_2CH_2CH_2CH_3 \quad Br^-$

c. $CH_3CH_2CH_2CH_2CH_2NH_2$

$\overset{+}{N}H(CH_2CH_2CH_3)_2 \quad Br^-$

d.

$\overset{+}{N}(CH_2CH_2CH_3)_3 \quad Br^-$

g.

h.

i.

24.

A > B > C > D > E > F > G

A is the most acidic, because it becomes aromatic when it loses a proton.

B, C, and **D** are the next most acidic, because in all three, the proton is bonded to a positively charged nitrogen.

B and **C** are more acidic than **D**, because in **B** and **C** the proton to be lost is bonded to an sp^2 nitrogen, which is more electronegative than the sp^3 nitrogen in **D**.

B is more acidic than **C**, because the uncharged nitrogen in **C** can donate electrons by resonance to the positively charged nitrogen.

Neutral compounds **E, F,** and **G** are the least acidic. **E** and **F** are more acidic than **G**, because **E** and **F** lose a proton from an sp^2 nitrogen, whereas a proton is lost from a less electronegative sp^3 nitrogen in **G**.

E is more acidic than **F**, because the negative charge on **E**'s conjugate base can be delocalized onto the second nitrogen.

25.   The compound on the right (the one shown below) is easier to decarboxylate, because the electrons left behind when CO_2 is removed can be delocalized onto nitrogen, forming a stable neutral species as a result of proton donation to nitrogen. The electrons left behind when the other compound loses CO_2 cannot be delocalized.

26.   The *N*-, *O*-, and *S*-substituted benzenes have the same relative reactivity toward electrophilic aromatic substitution as the *N*-, *O*-, and *S*-containing five-membered heterocyclic rings and for the same reason.

The *N*-substituted benzene is more reactive than the *O*-substituted benzene, because nitrogen is more effective than oxygen at donating electrons to the benzene ring, since it is less electronegative than oxygen. (Recall that electrophilic substitution is aided by electron donation to the ring.)

The *S*-substituted benzene is the least reactive because the lone-pair electrons of sulfur are in a $3p$ orbital, whereas the lone-pair electrons of nitrogen and oxygen are in a $2p$ orbital. Electron delocalization by overlap of a $3p$ orbital with the $2p$ orbital of carbon is less effective than electron delocalization by overlap of a $2p$ orbital with the $2p$ orbital of carbon.

27.   By comparing the carbocation intermediates formed in the rate-determining step when the amino-substituted compound undergoes electrophilic substitution at C-3 and C-4, you can see that the carbocation formed in the case of substitution at C-3 is more stable; one of its resonance contributors is particularly stable because it is the only one that does not have an atom with an incomplete octet. As a result, the amino-substituted compound undergoes electrophilic substitution predominantly at C-3. Therefore, the keto-substituted compound is the one that undergoes electrophilic substitution predominantly at C-4.

relatively stable

28.   **a.** The Lewis acid, AlCl₃, complexes with nitrogen, causing the aziridine ring to open when it is attacked by the nucleophilic benzene ring. The ring will open in the direction that puts the partial positive charge on the more substituted carbon, because that will form the more stable carbocation.

$$CH_3$$
$$\delta+ \, \overset{|}{N} \cdots \overset{\delta-}{AlCl_3}$$
$$\underset{\delta+}{\triangle} \; CH_3$$

Therefore, the major and minor products are those shown.

$$CH_3$$
$$\underset{}{\overset{|}{CHCH_2NHCH_3}}$$

major

$$CH_3$$
$$\underset{}{\overset{|}{CH_2CHNHCH_3}}$$

minor

b. Yes, epoxides can undergo similar reactions.

29.   Oxygen is the negative end of the dipole in both compounds. Tetrahydrofuran has the greater dipole moment because in furan the effect of the electron-withdrawing oxygen is mitigated by the ability of oxygen to donate electrons by resonance into the ring.

tetrahydrofuran
1.73 D

furan
0.70 D

30.   The products of the reaction show that the hydrogen is removed from the β-carbon that is bonded to the most hydrogens. Therefore, we can conclude that the transition state has a carbanion-like transition state. (See page 381 in the text.)

31.   **a.** $CH_3\overset{\overset{\displaystyle CH_3}{|}}{\underset{\underset{\displaystyle OH}{|}}{N}}$ + $CH_2{=}CHCH_3$

c. $CH_2{=}CH_2$ + $\overset{\overset{\displaystyle CH_3}{|}}{\underset{\underset{\displaystyle OH \;\; CH_3}{|\;\;\;\;|}}{NCH_2CHCH_3}}$

b. $CH_3\overset{\overset{\displaystyle OH}{|}}{\underset{\underset{\displaystyle \bigcirc}{|}}{N}}$ + $CH_2{=}CHCH_3$

d. $CH_2{=}CHCH_2CH_2CH_2CH_2\overset{\overset{\displaystyle}{}}{\underset{\underset{\displaystyle OH}{|}}{NCH_3}}$

The carbon of the methyl group is the β-carbon that is bonded to the greatest number of hydrogens.

32.   Pyrrolidine is a saturated nonaromatic compound, whereas pyrrole and pyridine are unsaturated aromatic compounds. The C-2 hydrogens of pyrrolidine are at δ 2.82 ppm, about where one would expect the signal for hydrogens bonded to an sp^3 carbon adjacent to an electron-withdrawing amino group.

The C-2 hydrogens of pyrrole and pyridine are expected to be at a higher frequency because of magnetic anisotropy. Because the nitrogen of pyrrole donates electrons into the ring and the nitrogen of pyridine withdraws electrons from the ring, the C-2 hydrogens of pyrrole are in an environment with a greater electron density, so they should show a signal at a lower frequency relative to the C-2 hydrogens of pyridine. Thus, the C-2 hydrogens of pyrrole are at δ 6.42 ppm, and the C-2 hydrogens of pyridine are at δ 8.50 ppm.

33.   A UV spectrum results from the π electron system. The lone-pair electrons on the nitrogen in aniline are delocalized into the benzene ring and, thus, are part of the π system. Protonation of aniline removes two electrons from the π system. This has a significant effect on its UV spectrum.

The lone-pair electrons on the nitrogen atom in pyridine are sp^2 electrons and thus are not part of the π system. Protonation of pyridine, therefore, does not remove any electrons from the π system and has only a minor effect on the UV spectrum.

34.   When ammonia loses a proton, the electrons left behind remain on nitrogen.

$$\ddot{N}H_3 \; + \; :B \longrightarrow \; \bar{:}\ddot{N}H_2 \; + \; HB^+$$

When pyrrole loses a proton, the electrons left behind on the nitrogen can be delocalized onto the four ring carbons. Electron delocalization stabilizes the anion and makes it easier to form. Recall that stabilizing the base increases the acidity of its conjugate acid.

35.

36.   Before we can answer the questions, we must figure out the mechanism of the reaction. Once the mechanism is known, it will be easy to determine how a change in a reactant will affect the product. The mechanism is shown below.

Propenal, an α,β-unsaturated aldehyde, undergoes a conjugate addition reaction with aniline. This is followed by an intramolecular electrophilic aromatic substitution reaction. Dehydration of the alcohol results in 1,2-dihydroquinoline, which is oxidized to quinoline by nitrobenzene.

a.

b.

c.

meta-ethylaniline   2- methyl-2-pentenal   2,7-diethyl-3-methylquinoline

37.   **a.**

b.

38.   a.

You can see why the nitro substituent goes to this position by examining the relative stabilities of the possible carbocation intermediates.

The carbocation has three resonance contributors.

The carbocation has two resonance contributors.

relatively unstable   The carbocation has three resonance contributors, but the first one is relatively unstable, because the positive charge is on the carbon attached to the electron-withdrawing substituent.

b.

You can understand why the bromo substituent goes to this position by examining the relative stabilities of the possible carbocation intermediates.

relatively unstable

relatively unstable

c.

The other nitrogen is not alkylated, because its lone pair is delocalized into the pyridine ring, so it is not available to react with the alkyl halide.

d.

PCl$_3$ substitutes a Cl for an OH, as it does in alcohols and carboxylic acids.

e.   In the first step, hydroxide ion removes the most acidic hydrogen from the compound. A hydrogen bonded to the C-4 methyl group is the most acidic hydrogen, because the electrons left behind when the proton is removed can be delocalized onto the positively-charged nitrogen of the pyridine ring.

f.

39.   The increase in the electron density of the ring as a result of resonance donation of electrons by oxygen causes pyridine-*N*-oxide to be more reactive toward electrophilic aromatic substitution than pyridine, since the extra electron-density will stabilize the carbocation intermediate.

In Section 16.4, we saw that substituents that are able to donate electrons by resonance into the ring are ortho/para directors. Therefore, pyridine-*N*-oxide will undergo electrophilic aromatic substitution at the 2- and 4-positions. Because the 2-position is somewhat sterically hindered, pyridine-*N*-oxide undergoes electrophilic substitution primarily at the 4-position.

40.

41.

pyrrole   *para*-(*N*,*N*-dimethylamino)-
benzaldehyde

Repeat the previous series of reactions at the 5-position of the pyrrole ring.

colored compound

42.

43.   **a.**   C₆H₅—NHNH₂ + CH₃CCH₂CH₃

c.   C₆H₅—NHNH₂ + (cyclohexanone)

b.   C₆H₅—NHNH₂ + CH₃CH₂CH₂CH

44.

Repeat two more times.

1. Repeat with benzaldehyde.
2. Instead of using another molecule of pyrrole, do an intramolecular reaction with the pyrrole at the end of the chain.

Chapter 21 Practice Test

1.   Give two names for each of the following compounds:

a.   b.   c.   d.

2.   Give the product of each of the following reactions:

a.   +   2 $CH_3CH_2NH_2$ $\longrightarrow$

b.   + $\xrightarrow{\text{trace acid}}$

c.   + Br_2 $\xrightarrow[\text{300 °C}]{\text{FeBr}_3}$

d.   + CH_3I $\longrightarrow$

e.   + Cl_2 $\longrightarrow$

3.   Which of the following is a stronger acid?

a.   or

b.

 or

c.

 or

d.

 or

4.   Indicate whether each of the following statements is true or false:

a.   4-Chloropyridine is more reactive toward nucleophilic aromatic substitution
than is 3-chloropyrrole.   T   F

b.   Pyrrole is more reactive toward electrophilic aromatic substitution than
is furan.   T   F

c.   Pyrrole is more reactive toward electrophilic aromatic substitution
than is benzene.   T   F

d.   Pyridine is more reactive toward electrophilic aromatic substitution
than is benzene.   T   F

CHAPTER 22
The Organic Chemistry of Carbohydrates

Important Terms

aldaric acid	a dicarboxylic acid with an OH group bonded to each carbon. Obtained by oxidizing the aldehyde and primary alcohol groups of an aldose.
alditol	a compound with an OH group bonded to each carbon. Obtained by reducing an aldose or a ketose.
aldonic acid	a carboxylic acid with an OH group bonded to each carbon. Obtained by oxidizing the aldehyde group of an aldose.
aldose	a polyhydroxy aldehyde.
amino sugar	a sugar in which one of the OH groups is replaced by an NH_2 group.
anomeric carbon	the carbon in a cyclic sugar that is the carbonyl carbon in the straight-chain form.
anomeric effect	preference for the axial position by certain substituents bonded to the anomeric carbon.
anomers	two cyclic sugars that differ in configuration only at the carbon that is the carbonyl carbon in the straight-chain form.
bioorganic compound	an organic compound that is found in a biological system.
carbohydrate	a sugar, a saccharide. Naturally occurring carbohydrates have the D-configuration.
complex carbohydrate	contains two or more sugar molecules linked together; it can be hydrolyzed to simple sugars.
deoxy sugar	a sugar in which one of the OH groups has been replaced by a hydrogen.
disaccharide	a compound containing two sugar molecules linked together.
enediol rearrangement	a base-catalyzed reaction that interconverts monosaccharides.
epimerization	changing the configuration of a carbon by removing a proton and then reprotonating it.
epimers	monosaccharides that differ in configuration at only one carbon.
furanose	a five-membered ring sugar.
furanoside	a five-membered ring glycoside.
glycoprotein	a protein that is covalently bonded to an oligosaccharide.
glycoside	the acetal of a sugar.

N-glycoside	a glycoside with a nitrogen instead of an oxygen at the glycosidic linkage.
glycosidic bond	the bond between the anomeric carbon of one sugar and an alcohol residue of a second sugar in a glycoside.
α-1,4'-glycosidic linkage	a glycosidic linkage between the C-1 of one sugar and the C-4 of a second sugar with the oxygen atom at C-1 in the axial position.
α-1,6'-glycosidic linkage	a glycosidic linkage between the C-1 of one sugar and the C-6 of a second sugar with the oxygen atom at C-1 in the axial position.
β-1,4'-glycosidic linkage	a glycosidic linkage between the C-1 of one sugar and the C-4 of a second sugar with the oxygen atom at C-1 in the equatorial position.
Haworth projection	a way to show the structure of a sugar in which the five- and six-membered rings are represented as being flat.
heptose	a monosaccharide with seven carbons.
hexose	a monosaccharide with six carbons.
ketose	a polyhydroxy ketone.
Kiliani-Fischer synthesis	a method used to increase the number of carbons in an aldose by one, resulting in the formation of a pair of C-2 epimers.
molecular recognition	the ability of molecules to recognize one another.
monosaccharide	a single sugar molecule.
mutarotation	a slow change in optical rotation to an equilibrium value.
nonreducing sugar	a sugar that cannot be oxidized by reagents such as Ag^+ and Cu^+. Nonreducing sugars are not in equilibrium with the open-chain aldose or ketose.
oligosaccharide	three to ten sugar molecules linked by glycosidic bonds.
osazone	the product obtained by reacting an aldose or a ketose with excess phenylhydrazine. An osazone contains two imine bonds.
oxocarbenium ion	an ion in which the positive charge is shared by a carbon and an oxygen.
pentose	a monosaccharide with five carbons.
polysaccharide	a compound containing ten or more sugar molecules linked together.
pyranose	a six-membered ring sugar.
pyranoside	a six-membered ring glycoside.

reducing sugar   a sugar that can be oxidized by reagents such as Ag^+ and Cu^+.
                          Reducing sugars are in equilibrium with the open-chain aldose or ketose forms.

simple carbohydrate   a single sugar molecule.

tetrose   a monosaccharide with four carbons.

triose   a monosaccharide with three carbons.

Wohl degradation   a method used to shorten an aldose by one carbon.

Solutions to Problems

1.   D-Ribose is an aldopentose.
      D-Sedoheptulose is a ketoheptose.
      D-Mannose is an aldohexose.

2.   Notice that an L-sugar is the mirror image of a D-sugar.

L-glucose   L-fructose

3.   The easiest way to answer this question is to determine first whether the structure represents
      R-glyceraldehyde or S-glyceraldehyde. Then, since R-glyceraldehyde = D-glyceraldehyde and
      S-glyceraldehyde = L-glyceraldehyde, we can answer the question.

L-glyceraldehyde   L-glyceraldehyde   D-glyceraldehyde

4.   **a.**   enantiomers   **b.**   diastereomers

5.   **a.**   D-ribose   **b.**   L-talose   **c.**   L-allose

6.   **a.**   D-glucose = $(2R,3S,4R,5R)$-2,3,4,5,6-pentahydroxyhexanal

      b.   Because D-mannose is a C-2 epimer of D-glucose, the systematic name of D-mannose can be obtained
             just by changing the configuration of the C-2 carbon in the systematic name of D-glucose.

             D-mannose = $(2S,3S,4R,5R)$-2,3,4,5,6-pentahydroxyhexanal

      c.   D-Galactose is the C-4 epimer of D-glucose. Therefore, each of its carbon atoms, except C-4, has the
             same configuration as it has in D-glucose.

             D-galactose = $(2R,3S,4S,5R)$-2,3,4,5,6-pentahydroxyhexanal

      d.   L-Glucose is the mirror image of D-glucose, so each carbon in L-glucose has the opposite configuration
             of that in D-glucose.

             L-glucose = $(2S,3R,4S,5S)$-2,3,4,5,6-pentahydroxyhexanal

7.   D-psicose

8.   **a.**   A ketoheptose has four asymmetric centers ($2^4 = 16$ stereoisomers).

   b.   An aldoheptose has five asymmetric centers ($2^5 = 32$ stereoisomers).

   c.   A ketotriose does not have an asymmetric center; therefore, it has no stereoisomers.

9.

D-fructose   an enolate ion   an enediol

10.   Removal of an α-hydrogen creates an enol that can enolize back to the ketone (using the OH at C-2) or can enolize to an aldehyde (using the OH at C-1). The aldehyde has a new asymmetric center (indicated by an *); one of the epimers is D-glucose, and the other is D-mannose.

D-fructose

D-glucose
D-mannose

11.   D-galactose and D-talose in addition to D-tagatose

D-tagatose

D-galactose   D-talose

12.   a.   When D-idose is reduced, D-iditol is formed.

D-idose   D-iditol

b.   When D-sorbose is reduced, C-2 becomes an asymmetric center, so both D-iditol and the C-2 epimer of D-iditol (D-gulitol) are formed.

D-sorbose   D-iditol   D-gulitol

13.   a.   1.   D-Altrose is reduced to the same alditol as D-talose is reduced to.
The easiest way to answer this question is to draw D-talose and its alditol. Then draw the monosaccharide with the same configuration at C-2, C-3, C-4, and C-5 as D-talose, reversing the functional groups at C-1 and C-6. (Put the primary alcohol group at the top and the aldehyde group at the bottom.) The resulting Fischer projection can be rotated 180° in the plane of the paper, and the monosaccharide can then be identified.

D-talose   the alditol of D-talose   D-altrose

2. L-Gulose is reduced to the same alditol as D-glucose is reduced to.

D-glucose →(1. NaBH₄, 2. H₃O⁺)→ alditol of D-glucose ←(1. NaBH₄, 2. H₃O⁺)— [structure] ←rotate 180°→ L-gulose

3. L-Galactose is reduced to the same alditol as D-galactose is reduced to.

D-galactose →(1. NaBH₄, 2. H₃O⁺)→ the alditol of D-galactose ←(1. NaBH₄, 2. H₃O⁺)— [structure] ←rotate 180°→ L-galactose

b. 1. The ketohexose (D-tagatose) with the same configuration at C-3, C-4, and C-5 as D-talose will give the same alditol as D-talose. The other alditol (D-galactitol) is the one with the opposite configuration at C-2.

D-tagatose →(1. NaBH₄, 2. H₃O⁺)→ D-talitol + D-galactitol

2. Similarly, the ketohexose (D-psicose) with the same configuration at C-3, C-4, and C-5 as D-allose will give the same alditol as D-allose. The other alditol (D-altritol) is the one with the opposite configuration at C-2.

14.   a. L-gulose (Note the similarity of this problem to Problem **13. a. 2.**)

b. L-Gularic acid, because it is also the oxidation product of L-gulose.

c. D-allose and L-allose, D-altrose and D-talose, L-altrose and L-talose, D-galactose and L-galactose

15. The other aldose is the C-2 epimer. The ketose is the one whose asymmetric centers have the same configuration as those in the aldoses.

a. D-arabinose and D-ribulose   **c.** L-gulose and L-sorbose

b. D-allose and D-psicose   **d.** D-talose and D-tagatose

16. The monosaccharides are the aldoses that are C-2 epimers whose asymmetric centers have the same configuration as the ketose.

D-gulose and D-idose

17. The monosaccharides are the aldoses with one additional carbon that are C-2 epimers and whose other asymmetric centers have the same configuration as the given aldose.

a. D-gulose and D-idose   **b.** L-xylose and L-lyxose

18. The monosaccharides are the aldoses whose C-1 has been removed, whose C-2 carbon has been converted to an aldehyde, and whose other asymmetric centers have the same configuration as the given aldose.

a. D-allose and D-altrose   **b.** D-glucose and D-mannose   **c.** L-allose and L-altrose

19. Solved in the text.

20.

D-glucose

A

D-mannose

B

D-arabinose

C

D-erythrose

D

21. The hemiacetals in **a** and **b** have one asymmetric center; therefore, each has two stereoisomers. The hemiacetals in **c** and **b** have two asymmetric centers; therefore, each has four stereoisomers.

a. Solved in the text.

b.

c.

d.

22. **a.**

c.

D-glucose

L-glucose
the mirror image of D-glucose

b.

23.

24. First recall that, in the more stable chair conformation, an α-anomer has the anomeric OH group in the axial position, whereas a β-anomer has the anomeric OH group in the equatorial position. Then recall that glucose has all its OH groups in equatorial positions. Now this question can be answered easily.

    a. Solved in the text.

    b. D-Idose differs in configuration from D-glucose at C-2, C-3, and C-4. Therefore, the OH groups at C-2, C-3, and C-4 in β-D-idose are in the axial position.

    c. D-Allose is a C-3 epimer of D-glucose. Therefore, the OH group at C-3 is in the axial position and, since it is the α-anomer, the OH group at C-1 (the anomeric carbon) is also in the axial position.

25. If more than a trace amount of acid is used, the amine that acts as a nucleophile when it forms the *N*-glycoside becomes protonated, and a protonated amine is not a nucleophile.

26.   **a.** Solved in the text.

    b. α-D-talose (or α-D-talopyranose)   (reducing; it is a hemiacetal)

    c. methyl α-D-galactoside (or methyl α-D-galactopyranoside)   (nonreducing; it is an acetal)

    d. ethyl β-D-psicoside (or ethyl β-D-psicofuranoside)   (nonreducing; it is an acetal)

27.   specific rotation of glucose + specific rotation of fructose = -22.0

$$+ 52.7 + \text{specific rotation of fructose} = -22.0$$

$$\text{specific rotation of fructose} = -22.0 + (-52.7)$$

$$\text{specific rotation of fructose} = -74.7$$

28.   **a.** Amylose has α-1,4'-glycosidic linkages, whereas cellulose has β-1,4'-glycosidic linkages.

    b. Amylose has α-1,4'-glycosidic linkages, whereas amylopectin has both α-1,4'-glycosidic linkages and α-1,6'-glycosidic linkages.

    c. Glycogen and amylopectin have the same kind of linkages, but glycogen has a higher frequency of α-1,6'-glycosidic linkages.

    d. Cellulose has a hydroxyl at C-2, whereas chitin has an *N*-acetyl amino group at that position.

29. A proton is more easily lost from the C-3 OH group, because the anion that is formed when the proton is removed is more stable than the anion that is formed when a proton is removed from the C-2 OH group. When a proton is removed from the C-3 OH group, the electrons that are left behind can be delocalized onto another oxygen. When a proton is removed from the C-2 OH group, the electrons that are left behind are delocalized onto a carbon. Because oxygen is more electronegative than carbon, a negatively charged oxygen is more stable than a negatively charged carbon.

30.   **a.** People with Type O blood can receive blood only from other people with Type O blood, because Types A, B, and AB blood have sugar components that Type O blood does not have.

   b. People with Type AB blood can give blood only to other people with Type AB blood, because Type AB blood has sugar components that Types A, B, and O blood do not have.

31.   **a.**

$$
\begin{array}{c}
\text{COOH}\\
\text{H}\!-\!\text{OH}\\
\text{HO}\!-\!\text{H}\\
\text{HO}\!-\!\text{H}\\
\text{H}\!-\!\text{OH}\\
\text{COOH}
\end{array}
$$

   b.

$$
\begin{array}{c}
\text{COO}^-\\
\text{H}\!-\!\text{OH}\\
\text{HO}\!-\!\text{H}\\
\text{HO}\!-\!\text{H}\\
\text{H}\!-\!\text{OH}\\
\text{CH}_2\text{OH}
\end{array}
$$

   c.

$$
\begin{array}{c}
\text{CH}_2\text{OH}\\
\text{H}\!-\!\text{OH}\\
\text{HO}\!-\!\text{H}\\
\text{HO}\!-\!\text{H}\\
\text{H}\!-\!\text{OH}\\
\text{CH}_2\text{OH}
\end{array}
$$

   d.

$$
\begin{array}{c}
\text{CH}\!=\!\text{NNHC}_6\text{H}_5\\
\text{C}\!=\!\text{NNHC}_6\text{H}_5\\
\text{HO}\!-\!\text{H}\\
\text{HO}\!-\!\text{H}\\
\text{H}\!-\!\text{OH}\\
\text{CH}_2\text{OH}
\end{array}
$$

   e.

$$
\begin{array}{c}
\text{COO}^-\\
\text{H}\!-\!\text{OH}\\
\text{HO}\!-\!\text{H}\\
\text{HO}\!-\!\text{H}\\
\text{H}\!-\!\text{OH}\\
\text{CH}_2\text{OH}
\end{array}
$$

   f. (two pyranose ring structures with CH_2OH, OH, HO, OH and OCH_2CH_3 substituents) +

   g.

$$
\begin{array}{c}
\text{HC}\!=\!\text{O}\\
\text{HO}\!-\!\text{H}\\
\text{HO}\!-\!\text{H}\\
\text{H}\!-\!\text{OH}\\
\text{CH}_2\text{OH}
\end{array}
$$

32.   **a.** D-lyxose   **b.** D-tagatose   **c.** D-talose   **d.** D-psicose

33.

$$
\begin{array}{c}
\text{H}\diagup\text{C}\diagdown\text{O} \\
\text{HO}\!-\!\!-\!\text{H} \\
\text{H}\!-\!\!-\!\text{OH} \\
\text{CH}_2\text{OH} \\
\text{D-threose}
\end{array}
\xrightarrow[\text{synthesis}]{\text{Kiliani-Fischer}}
\begin{array}{c}
\text{H}\diagup\text{C}\diagdown\text{O} \\
\text{H}\!-\!\!-\!\text{OH} \\
\text{HO}\!-\!\!-\!\text{H} \\
\text{H}\!-\!\!-\!\text{OH} \\
\text{CH}_2\text{OH} \\
\text{sugar A} \\
\text{D-xylose}
\end{array}
+
\begin{array}{c}
\text{H}\diagup\text{C}\diagdown\text{O} \\
\text{HO}\!-\!\!-\!\text{H} \\
\text{HO}\!-\!\!-\!\text{H} \\
\text{H}\!-\!\!-\!\text{OH} \\
\text{CH}_2\text{OH} \\
\text{sugar B} \\
\text{D-lyxose}
\end{array}
\xrightarrow{\text{HNO}_2}
\begin{array}{c}
\text{COOH} \\
\text{H}\!-\!\!-\!\text{OH} \\
\text{HO}\!-\!\!-\!\text{H} \\
\text{H}\!-\!\!-\!\text{OH} \\
\text{COOH} \\
\text{optically inactive}
\end{array}
+
\begin{array}{c}
\text{COOH} \\
\text{HO}\!-\!\!-\!\text{H} \\
\text{HO}\!-\!\!-\!\text{H} \\
\text{H}\!-\!\!-\!\text{OH} \\
\text{COOH} \\
\text{optically active}
\end{array}
$$

34.   **a.** D-ribose and L-ribose, D-arabinose and L-arabinose, D-xylose and L-xylose, D-lyxose and L-lyxose

   b. D-ribose and D-arabinose, L-ribose and L-arabinose, D-xylose and D-lyxose, L-xylose and L-lyxose

   c. D-arabinose, L-arabinose, D-lyxose, and L-lyxose

35.

methyl α-D-ribofuranose   methyl β-D-ribofuranose

methyl α-D-ribopyranose   methyl β-D-ribopyranose

36.   D-Altrose and D-talose are reduced to the same alditol and would form different osazones.

$$
\begin{array}{c}
\text{H}\diagup\text{C}\diagdown\text{O} \\
\text{HO}\!-\!\!-\!\text{H} \\
\text{H}\!-\!\!-\!\text{OH} \\
\text{H}\!-\!\!-\!\text{OH} \\
\text{H}\!-\!\!-\!\text{OH} \\
\text{CH}_2\text{OH} \\
\text{D-altrose}
\end{array}
\xrightarrow[\text{2. H}_3\text{O}^+]{\text{1. NaBH}_4}
\begin{array}{c}
\text{CH}_2\text{OH} \\
\text{HO}\!-\!\!-\!\text{H} \\
\text{HO}\!-\!\!-\!\text{H} \\
\text{HO}\!-\!\!-\!\text{H} \\
\text{H}\!-\!\!-\!\text{OH} \\
\text{CH}_2\text{OH} \\
\text{alditol}
\end{array}
\xleftarrow[\text{2. H}_3\text{O}^+]{\text{1. NaBH}_4}
\begin{array}{c}
\text{H}\diagup\text{C}\diagdown\text{O} \\
\text{HO}\!-\!\!-\!\text{H} \\
\text{HO}\!-\!\!-\!\text{H} \\
\text{HO}\!-\!\!-\!\text{H} \\
\text{H}\!-\!\!-\!\text{OH} \\
\text{CH}_2\text{OH} \\
\text{D-talose}
\end{array}
$$

If D-altrose is **A** and D-talose is **B**, then **C** must be D-galactose, the C-2 epimer of **B**.

If D-altrose is **B** and D-talose is **A**, then **C** must be D-allose, the C-2 epimer of **B**.

37. **1.** As Fischer did, we can narrow our search to eight 6-carbon sugars, since there are eight pairs of enantiomers. First we need to find an aldopentose that forms (+)-galactose as a product of a Kiliani-Fischer synthesis. That sugar is the one known as (−)-lyxose. The Kiliani-Fischer synthesis on (−)-lyxose yields two sugars with melting points that show them to be the sugars known as (+)-galactose and (+)-talose. Now we know that (+)-galactose and (+)-talose are C-2 epimers. The eight 6-carbon sugars are sugars 1 and 2, 3 and 4, 5 and 6, or 7 and 8. (See page 952 of the text.)

2. When (+)-galactose and (+)-talose react with HNO_3, (+)-galactose forms an optically inactive product and (+)-talose forms an optically active product. Thus, (+)-galactose and (+)-talose are sugars 1 and 2 or 7 and 8. Since (+)-galactose is the one that forms the optically inactive oxidation product, it is either sugar 1 or 7.

3. To determine the structure of (+)-galactose, we can go back to (−)-lyxose, the sugar that forms sugars 7 and 8 by a Kiliani-Fischer synthesis, and oxidize it with HNO_3. Finding that the aldaric acid is optically active allows us to conclude that (+)-galactose is sugar 7, because the aldopentose that leads to sugars 1 and 2 would give an optically inactive aldaric acid.

38. A monosaccharide with a molecular weight of 150 must have five carbons (five C's = 60, five O's = 80, and 10 H's = 10 for a total of 150). All aldopentoses are optically active. Therefore, the compound must be a ketopentose. The following is the only ketopentose that would not be optically active.

39. The hydrogen that is bonded to the anomeric carbon will be the hydrogen that has its signal at the highest frequency, because it is the only hydrogen that is bonded to a carbon that is bonded to two electronegative oxygen atoms. So the two anomeric hydrogens, one on the α-anomer and one on the β-anomer, are responsible for the two high-frequency doublets.

40.

41. The hexose is D-altrose.

1. Knowing that (+)-glyceraldehyde is D-glyceraldehyde gives the configuration at C-5.

2. Knowing that the second Wohl degradation gives D-erythrose gives the configuration at C-4.

3. Knowing that the first Wohl degradation followed by oxidation gives an optically inactive aldaric acid, gives the configuration at C-3.

4. Knowing that the original hexose forms an optically active aldaric acid gives the configuration at C-2.

D-altrose

42.

the β-D-glucuronide   the α-D-glucuronide

43. **D-arabinose**. The only D-aldopentoses that are oxidized to optically active aldaric acids are D-arabinose and D-lyxose. A Wohl degradation of D-arabinose forms D-erythrose, whereas a Wohl degradation of D-lyxose forms D-threose. Since D-erythrose forms an optically inactive aldaric acid but D-threose does not, the D-aldopentose is D-arabinose.

44.

hyaluronic acid

45. She can take a sample of one of the sugars and oxidize it with nitric acid to an aldaric acid or reduce it with sodium borohydride to an alditol. If the product is optically active, the sugar was D-lyxose; if the product is not optically active, the sugar was D-xylose.

D-lyxose $\xrightarrow[\Delta]{HNO_3}$ **optically active**   D-xylose $\xrightarrow[\Delta]{HNO_3}$ **optically inactive**

D-lyxose → optically active   D-xylose → optically inactive

1. NaBH₄
2. H₃O⁺

46.

aldonic acid of
D-glucose

47.

aldaric acid of
D-glucose

rotate 180°

48.   **10** aldaric acids

Each of the following pairs forms the same aldaric acid:

D-allose and L-allose   L-altrose and L-talose

D-galactose and L-galactose   D-gulose and L-gulose

D-altrose and D-talose   L-gulose and D-glucose

Thus twelve aldohexoses form six aldaric acids. The other four aldohexoses each form a distinctive aldaric acid, and (6 + 4 = 10).

49.   We know the hexose is a ketohexose since it does not react with Br_2; we know it is a 2-ketohexose because if it were a 3-ketohexose it would not be able to form a hemiacetal (because the hemiacetal would have an unstable four-membered ring), and therefore would not under mutarotation. Knowing that it is oxidized by Tollens reagent to the aldonic acids D-talonic acid and D-galactonic acid tells us that the aldonic acids and the ketohexose have the same configuration at C-3, C-4, and C-5. Therefore, the hexose is D-tagatose.

D-talonic acid   D-galactonic acid   D-tagatose

50.

D-fructose with one deuterium

D-fructose with two deuteriums

51. **a.**

CH$_2$OH

HO   O   OH

OHHO

differs from D-glucose
at C-2 and C-4

b.

CH$_2$OH

HO   O   OH

HO

HO

differs from D-glucose
at C-2, C-3, and C-4

52.   Let A = the fraction of D-glucose in the α-form and B = the fraction of D-glucose in the β-form.

$$A + B = 1$$
$$B = 1 - A$$

specific rotation of A = 112.2
specific rotation of B = 18.7
specific rotation of the equilibrium mixture = 52.7
specific rotation of the mixture = specific rotation of A × fraction of D-glucose in the α-form +
specific rotation of B × fraction of D-glucose in the β-form

$$52.7 = 112.2\,A + (1 - A)\,18.7$$
$$52.7 = 112.2\,A + 18.7 - 18.7\,A$$
$$34.0 = 93.5\,A$$
$$A = 0.36$$
$$B = 0.64$$

This calculation shows that 36% is in the α-form and 64% is in the β-form.

53.   Silver oxide increases the leaving tendency of the iodide ion from methyl iodide, thereby allowing the nucleophilic substitution reaction to take place with the poorly nucleophilic alcohol groups. Because mannose is missing the methyl substituent on the oxygen at C-6, the disaccharide must be formed using the C-6 OH group of mannose and the anomeric carbon of galactose.

HO   CH$_2$OH

O

HO

OH   HO   O   CH$_2$ HO   O

D-galactose   HO

HO   D-mannose   OH

54.   D-Altrose will most likely exist as a furanose because:
(1) the furanose is particularly stable since all the large substituents are trans to each other, and
(2) the pyranose has two of its OH groups in the unstable axial position.

D-altrofuranose   D-altropyranose

55. **a.** From its molecular formula and the fact that only gulose is formed when trehalose is hydrolyzed, we know that it is a disaccharide. Trehalose can be a nonreducing sugar only if the anomeric carbon of one glucose is connected to the anomeric carbon of the other glucose. Because it can be hydrolyzed by maltase, its glycosidic linkage must have the same geometry as an α-1,4'-glycosidic linkage. Therefore, the OH group attached to the anomeric carbon of the glucose on the left must be axial, and the OH group attached to the anomeric carbon of the glucose on the right must be equatorial. (Notice that the glucose on the right is drawn upside down and is reversed horizontally.)

b. Silver oxide increases the leaving tendency of the iodide ion from methyl iodide, thereby allowing the nucleophilic substitution reaction to take place with the poorly nucleophilic alcohol groups.

56.

57. Because all the glucose units have six-membered rings, the 5-position is never methylated.

2,3,4,6-tetra-*O*-methyl-D-glucose has only its 1-position in an acetal linkage.

2,4,6-tri-*O*-methyl-D-glucose has its 1-position and its 3-position in an acetal linkage.

2,3,4-tri-*O*-methyl-D-glucose has its 1-position and its 6-position in an acetal linkage.

2,4-di-*O*-methyl-D-glucose has its 1-position, its 3-position, and its 6-position in an acetal linkage.

58. In the case of D-idose, the chair conformer with the OH substituent at C-1 and the CH₂OH substituent both in axial positions (which is necessary for formation of the anhydro form) has the OH substituents at C-2, C-3, and C-4 all in equatorial positions. Thus, this is a relatively stable conformer, because three of the five large substituents are in the more stable equatorial position.

In contrast, the case of D-glucose, the chair conformer with the OH substituent at C-1 and the CH₂OH substituent both in axial positions has the OH substituents at C-2, C-3, and C-4 all in axial positions. This is a relatively unstable conformer, because all the large substituents are in less stable axial positions. Therefore, a large percentage of D-idose but only a small percentage of D-glucose exists in the anhydro form at 100 °C.

anhydro form of D-idose   anhydro form of D-glucose

59. An abbreviated mechanism is shown so you can easily see how the one sugar can be converted into the other.

D-glucose

D-allose   D-altrose

Chapter 22 Practice Test

1.   Give the product(s) of each of the following reactions:

a.

$$
\begin{array}{c}
\text{H}\diagdown\quad\diagup\text{O} \\
\text{C} \\
\text{H} \quad\text{---OH} \\
\text{HO} \quad\text{---H} \\
\text{H} \quad\text{---OH} \\
\text{H} \quad\text{---OH} \\
\text{CH}_2\text{OH}
\end{array}
\xrightarrow[\Delta]{\text{HNO}_3}
$$

b.

$$
\xrightarrow[\text{CH}_3\text{OH}]{\text{HCl}}
$$

c.

$$
\begin{array}{c}
\text{H}\diagdown\quad\diagup\text{O} \\
\text{C} \\
\text{HO} \quad\text{---H} \\
\text{HO} \quad\text{---H} \\
\text{H} \quad\text{---OH} \\
\text{CH}_2\text{OH}
\end{array}
\xrightarrow{\begin{array}{l}1.\ \text{HC}\equiv\text{N}\\ 2.\ \text{H}_2/\text{Pd}/\text{BaSO}_4\\ 3.\ \text{HCl, H}_2\text{O}\end{array}}
$$

d.

$$
\begin{array}{c}
\text{H}\diagdown\quad\diagup\text{O} \\
\text{C} \\
\text{H} \quad\text{---OH} \\
\text{H} \quad\text{---OH} \\
\text{H} \quad\text{---OH} \\
\text{CH}_2\text{OH}
\end{array}
\quad+\quad \text{Br}_2 \xrightarrow{\text{H}_2\text{O}}
$$

2.   Indicate whether each of the following statements is true or false:

a.   Glycogen contains α-1,4'- and β-1,6'-glycosidic linkages.   T   F

b.   D-Mannose is a C-1 epimer of D-glucose.   T   F

c.   D-Glucose and L-glucose are anomers.   T   F

d.   D-Erythrose and D-threose are diastereomers.   T   F

e.   Wohl degradations of D-glucose and D-gulose form the same aldopentose.   T   F

3.   Which of the following sugars will form an optically active aldaric acid?

4.   When crystals of D-fructose are dissolved in a basic aqueous solution, two aldohexoses are obtained. Identify the aldohexoses.

5.   A sugar forms the same osazone as D-galactose but is not oxidized by an aqueous solution of Br_2. Identify the sugar.

6.   D-Talose and _____ are reduced to the same alditol.

7.   What is the main structural difference between amylose and cellulose?

8.   What aldohexoses are formed from a Kiliani-Fischer synthesis starting with D-xylose?

9.   What aldohexose is the C-3 epimer of D-glucose?

10.   Draw the most stable chair conformer of β-D-allose, a C-3 epimer of β-D-glucose.

CHAPTER 23
The Organic Chemistry of Amino Acids, Peptides, and Proteins

Important Terms

amino acid	an α-amino carboxylic acid. Naturally occurring amino acids have the L-configuration.
D-amino acid	the configuration of an amino acid drawn in a Fischer projection such that the carboxyl group is on top, the hydrogen is on the left, and the amino group is on the right.
L-amino acid	the configuration of an amino acid drawn in a Fischer projection such that the carboxyl group is on top, the hydrogen is on the right, the amino group is on the left.
amino acid analyzer	an instrument that automates the ion-exchange separation of amino acids.
anion-exchange resin	a resin that binds anions.
antiparallel β-pleated sheet	a type of secondary structure in which the adjacent hydrogen-bonded peptide chains in a β-pleated sheet run in opposite directions.
automated solid-phase peptide synthesis	an automated technique that synthesizes a peptide (in the C-terminal to N-terminal direction) while its C-terminal amino acid is attached to a solid support.
cation-exchange resin	a resin that binds cations.
coil conformation (loop conformation)	that part of a protein that is highly ordered but not in an α-helix or a β-pleated sheet.
C-terminal amino acid	the terminal amino acid of a peptide (or protein) that has a free carboxyl group.
denaturation	destruction of the highly organized tertiary structure of a protein.
dipeptide	two amino acids linked together by an amide bond.
disulfide	a compound with an S—S bond.
disulfide bridge	a disulfide (S—S) bond formed by two cysteine residues in a peptide or protein.
Edman's reagent	phenyl isothiocyanate. A reagent used to determine the N-terminal amino acid of a polypeptide.
electrophoresis	a technique that separates amino acids on the basis of their pI values.
endopeptidase	an enzyme that hydrolyzes a peptide bond that is not at the end of a peptide chain.

essential amino acid   an amino acid that humans must obtain from their diet because they either cannot synthesize it at all or they cannot synthesize it in adequate amounts.

exopeptidase   an enzyme that hydrolyzes a peptide bond at the end of a peptide chain.

fibrous protein   a water insoluble protein that has its polypeptide chains arranged in bundles.

globular protein   a water soluble protein that tends to have a roughly spherical shape.

α-helix   the backbone of a polypeptide coiled in a right-handed spiral with hydrogen bonding occurring within the helix.

hydrophobic interactions   interactions between nonpolar groups. They increase stability by decreasing the amount of structured water (increasing entropy).

interchain disulfide bridge   a disulfide bridge between two cysteine residues in different peptide chains.

intrachain disulfide bridge   a disulfide bridge between two cysteine residues in the same peptide chain.

ion-exchange chromatography   a technique that uses a column packed with an insoluble resin to separate compounds on the basis of their charge and polarity.

isoelectric point (pI)   the pH at which there is no net charge on an amino acid.

kinetic resolution   separating enantiomers based on the difference in their rate of reaction with an enzyme.

loop conformation   see coil conformation.

N-terminal amino acid   the terminal amino acid of a peptide (or protein) that has a free amino group.

oligomer   a protein with more than one peptide chain.

oligopeptide   three to ten amino acids linked by amide bonds.

paper chromatography   a technique that separates amino acids based on polarity.

parallel β-pleated sheet a type of secondary structure in which the adjacent hydrogen-bonded peptide chains in a β-pleated sheet run in the same direction.

partial hydrolysis   a technique that hydrolyzes only some of the peptide bonds in a polypeptide.

peptidase   an enzyme that catalyzes the hydrolysis of a peptide bond.

peptide   polymer of amino acids linked together by amide bonds.

peptide bond   the amide bond that links the amino acids in a peptide or protein.

β-pleated sheet   a type of secondary structure in which the backbone of a polypeptide extends in a zigzag structure with hydrogen bonding between neighboring chains.

polypeptide   many amino acids linked by amide bonds.

primary structure   the sequence of amino acids and the location of the disulfide bridges in a protein.

protein   a naturally occurring polymer of 40 to 4000 amino acids linked together by amide bonds.

quaternary structure   a description of the way in which the individual polypeptide chains of an oligomeric protein are arranged with respect to one other.

random coil   the conformation of a totally denatured protein.

secondary structure   a description of the conformation of the backbone of a protein.

side chain   the substituent attached to the α-carbon of an amino acid.

structural protein   a protein that gives strength to a biological structure.

subunit   an individual chain of an oligomeric protein.

tertiary structure   a description of the three-dimensional arrangement of all the atoms in a protein.

thin-layer chromatography   a technique that separates compounds on the basis of their polarity.

tripeptide   three amino acids linked by amide bonds.

zwitterion   a compound with a negative charge and a positive charge on nonadjacent atoms.

Solutions to Problems

1.   **a.**   When the imidazole ring is protonated, the double-bonded nitrogen is the one that accepts the proton. The lone-pair electrons on the singly-bonded nitrogen are delocalized and therefore are not available to be protonated.

In contrast, the lone-pair electrons on the double-bonded nitrogen are not delocalized, because the constraints of the ring will not allow two adjacent double bonds.

Furthermore, if the lone pair on the single-bonded nitrogen were protonated, the protonated compound would not be aromatic.

In contrast, when the double-bonded nitrogen is protonated, the protonated compound is still aromatic.

   b.   The lone-pair electrons on the double-bonded nitrogen are protonated, because the lone-pair electrons on the other nitrogens are delocalized and therefore are not available for protonation. In addition, protonation of the double-bonded nitrogen leads to a highly resonance-stabilized conjugate acid.

2.   **a.**   D-alanine is (R)-alanine

   b.   D-asparate is (R)-aspartate

D-alanine
(R)-alanine

D-aspartate
(R)-aspartate

c. The α-carbons of all the D-amino acids except cysteine have the *R*-configuration. Similarly, the α-carbons of all the L-amino acids except cysteine have the *S*-configuration. In all the amino acids except cysteine, the amino group has the highest priority and the carboxyl group has the second-highest priority. In cysteine, the thiomethyl group has a higher priority than the carboxyl group, because sulfur has a greater atomic number than oxygen, causing the counterclockwise arrow to be a clockwise arrow.

$$
\begin{array}{c}
\overset{-}{O}\diagdown\underset{C}{}\diagup^{O} \\
H-\overset{|}{\underset{|}{C}}-\overset{+}{N}H_3 \\
CH_2SH
\end{array}
$$

D-cysteine
(*S*)-cysteine

3. Solved in the text.

4. Isoleucine is the only other amino acid that has more than one asymmetric center. Like threonine, it has two asymmetric centers.

5. The electron-withdrawing $^+NH_3$ substituent on the α-carbon increases the acidity of the carboxyl group.

6. **a.** Solved in the text.

c. $H_2N-\overset{O}{\underset{}{C}}-CH_2CH_2\overset{}{\underset{\underset{^+NH_3}{|}}{CH}}-\overset{O}{\underset{}{C}}-O^-$

e. $H_2N-\overset{^+NH_2}{\underset{}{C}}-NHCH_2CH_2CH_2\overset{}{\underset{\underset{^+NH_3}{|}}{CH}}-\overset{O}{\underset{}{C}}-O^-$

b. imidazole$-CH_2\overset{}{\underset{\underset{^+NH_3}{|}}{CH}}-\overset{O}{\underset{}{C}}-O^-$

d. $\overset{+}{H_3}NCH_2CH_2CH_2CH_2\overset{}{\underset{\underset{^+NH_3}{|}}{CH}}-\overset{O}{\underset{}{C}}-O^-$

f. $HO-\text{(benzene ring)}-CH_2\overset{}{\underset{\underset{^+NH_3}{|}}{CH}}-\overset{O}{\underset{}{C}}-O^-$

7. **a.** $HO-\overset{O}{\underset{}{C}}-CH_2CH_2\overset{}{\underset{\underset{^+NH_3}{|}}{CH}}-\overset{O}{\underset{}{C}}-OH$

c. $^-O-\overset{O}{\underset{}{C}}-CH_2CH_2\overset{}{\underset{\underset{^+NH_3}{|}}{CH}}-\overset{O}{\underset{}{C}}-O^-$

b. $HO-\overset{O}{\underset{}{C}}-CH_2CH_2\overset{}{\underset{\underset{^+NH_3}{|}}{CH}}-\overset{O}{\underset{}{C}}-O^-$

d. $^-O-\overset{O}{\underset{}{C}}-CH_2CH_2\overset{}{\underset{\underset{NH_2}{|}}{CH}}-\overset{O}{\underset{}{C}}-O^-$

8. **a.** The carboxyl group of the aspartic acid side chain is a stronger acid than the carboxyl group of the glutamic acid side chain, because the aspartic acid side chain is closer to the electron-withdrawing protonated amino group.

b. The protonated lysine side chain is a stronger acid than the protonated arginine side chain. The protonated arginine side chain has less of a tendency to lose a proton, because its positive charge is delocalized over three nitrogens.

9. In order for the amino acid to have no net charge, the two amino groups must have a $+1$ charge between them in order to cancel out the -1 charge of the carboxylate group. Because they are positively charged in their acidic forms and neutral in their basic forms, the sum of their charges will be $+1$ at the midpoint of their pK_a values.

10. **a.** asparagine $pI = \dfrac{2.02 + 8.84}{2} = \dfrac{10.86}{2} = 5.43$

b. arginine $pI = \dfrac{9.04 + 12.48}{2} = \dfrac{21.52}{2} = 10.76$

c. serine $pI = \dfrac{2.21 + 9.15}{2} = \dfrac{11.36}{2} = 5.68$

d. aspartate $pI = \dfrac{2.09 + 3.86}{2} = \dfrac{5.95}{2} = 2.98$

11. **a.** aspartate ($pI = 2.98$)

b. arginine ($pI = 10.76$)

c. Aspartate, because its pI is lower than that of glutamate. Therefore, pH 6.20 is farther away from the pH at which aspartate has no net charge. The farther an amino acid with no net charge has to "move" to get to the given pH value, the more charged it will become.

d. Methionine, because at pH $= 6.20$ methionine is farther away from the pH at which it has no net charge (pI of methine $= 5.75$, pI of glycine $= 5.97$).

12. Tyrosine and cysteine each have two groups that are neutral in their acidic forms and negatively charged in their basic forms. Unlike other amino acids that have similarly ionizing groups, the pK_a value for one of the two similarly ionizing groups in tyrosine and in cysteine are close to the pK_a value of the group that ionizes differently. Therefore, the group that ionizes differently cannot be ignored in calculating the pI.

13.

$$
\begin{array}{c}
\text{O} \\
\| \\
\text{C} \\
\end{array}
$$

CH$_3$CH   H
 |
CH$_3$

14. Leucine and isoleucine both have C_4H_9 side chains and, therefore, have the same polarity. Consequently, the spots for both amino acids appear at the same place on the chromatographic plate. Therefore, the chromatographic plate has one less spot than the number of amino acids.

15. Because the amino acid analyzer contains a cation-exchange resin (it binds cations), the less positively charged the amino acid, the less tightly it is bound to the column. Using buffer solutions of increasingly higher pH to elute the column causes the amino acids bound to the column to become increasingly less positively charged, so they can be released from the column.

16. Cation-exchange chromatography releases amino acids in order of their pI values. The amino acid with the lowest pI is released first, because at a given pH it will be the amino acid with the highest concentration of negative charge, and negatively charged molecules are not bound by the negatively charged resin. The relatively nonpolar resin will release polar amino acids before nonpolar amino acids.

    a. Asp (pI = 2.98) is more negative at pH 4 than is Ser (pI = 5.68).

    b. Gly is more polar than Ala.

    c. Val is more polar than Leu.

    d. Tyr is more polar than Phe.

17. A column containing an anion-exchange resin releases amino acids in reverse order of their pI values. The amino acid with the highest pI is released first, because at a given pH it will be the amino acid with the highest concentration of positive charge.

$$\text{His} > \text{Val} > \text{Ser} > \text{Asp}$$

18. The first equivalent of ammonia will react with the acidic proton of the carboxylic acid to form ammonium ion, which is not nucleophilic and, therefore, cannot substitute for Br. Thus, a second equivalent of ammonia is needed for the desired nucleophilic substitution reaction.

19.   **a.** The reactions below show that pyruvic acid forms alanine, oxaloacetic acid forms aspartate, and α-ketoglutarate forms glutamate. Remember that imine formation is best carried at a pH about 1.5 units lower than the pK_a of the protonated ammonium ion (Section 18.8), i.e., at about pH = 8. Therefore, the carboxyl groups will be in their basic forms.

If reductive amination (page 795) is carried out in the cell, only the L-isomer of each amino acid will be formed.

    b. If reductive amination is carried out in the laboratory, both the D- and L-isomer (a racemic mixture) of each amino acid will be formed.

20.   Notice that the group attached to the Br is the same as the group attached to the α-carbon of the amino acid.

    a. leucine

    b. methionine

21.   Notice that the group attached to the Br is the same as the group attached to the α-carbon of the amino acid.

    a. 4-bromo-1-butanamine

    b. benzyl bromide

22.   Notice that the R group attached to the carbonyl group of the aldehyde is the same as the group attached to the α-carbon of the amino acid.

    a. alanine   **b.** isoleucine   **c.** leucine

23.   Convert the amino acids into esters using $SOCl_2$ followed by ethanol. Then treat the esters with pig liver esterase. Since the enzyme hydrolyzes only esters of L-amino acids, the products will be the L-amino acid, ethanol and the ester of the D-amino acid. These compounds can be readily separated. After they are separated, the D-amino acid can be obtained by acid-catalyzed hydrolysis of the ester of the D-amino acid. This separation technique is called a kinetic resolution, because the enantiomers are separated (resolved) as a result of reacting at different rates in the enzyme-catalyzed reaction.

24.

25.   **A-G-M   A-M-G   M-G-A   M-A-G   G-A-M   G-M-A**

26.

27.   The bonds on either side of the α-carbon can freely rotate. In other words, the bond between the α-carbon and the carbonyl carbon and the bond between the α-carbon and the nitrogen (the bonds indicated by arrows) can freely rotate.

28.

29.   **a.**   glutamate, cysteine, and glycine

   b.   In forming the amide bond between glutamate and cysteine, the amino group of cysteine reacts with the γ-carboxyl group of glutamate rather than with the α-carboxyl group.

glutamate

amide bond   glutathione

30.   **Leu-Val** and **Val-Val** will be formed because the amino group of leucine is not reactive (so leucine could not be the C-terminal amino acid) but the amino group of valine would react equally easily with the carboxyl group of leucine and the carboxyl group of valine.

N-protected leucine   valine   Leu-Val

valine   valine   Val-Val

31.   If valine's carboxyl group is activated with thionyl chloride, the OH group of serine, as well as the NH$_2$ group of serine, would react readily with the very reactive acyl halide, forming both an ester (with its OH group) and an amide (with its NH$_2$ group).

If valine's carboxyl group is activated with DCC, an imidate will be formed. Because an imidate is less reactive than an acyl chloride, the imidate will react with the more reactive NH$_2$ group in preference to the OH group.

32.

$(CH_3)_3CO$ ... $OC(CH_3)_3$ + H_2NCH ... O^-
CH_2
$CH(CH_3)_2$
Leu

$\longrightarrow$

$(CH_3)_3CO$ ... NHCH ... O^-
CH_2
$CH(CH_3)_2$
N-protected Leu

1. DCC
2. H_2NCH ... O^-
CH_2
Phe

$(CH_3)_3CO$ ... NHCH ... NHCH ... O^-
CH_2 CH_2
$CH(CH_3)_2$

1. DCC
2. H_2NCH ... O^-
CH_3 Ala

$(CH_3)_3CO$ ... NHCH ... NHCH ... NHCH ... O^-
CH_2 CH_2 CH_3
$CH(CH_3)_2$

1. DCC
2. H_2NCH ... O^-
$CH(CH_3)_2$ Val

$(CH_3)_3CO$ ... NHCH ... NHCH ... NHCH ... NHCH ... O^-
CH_2 CH_2 CH_3 $CH(CH_3)_2$
$CH(CH_3)_2$

CF_3CO_2H | CH_2Cl_2

Leu   Phe   Ala   Val

33.   **a.** 5.8%

2	**3**	**4**	**5**	**6**	**7**	**8**	**9**
70%	49%	34%	24%	17%	12%	8.2%	5.8%

b. 4.4%

2	**3**	**4**	**5**	**6**	**7**	**8**	**9**	**10**	**11**	**12**	**13**	**14**	**15**
80%	64%	51%	41%	33%	26%	21%	17%	13%	11%	8.6%	6.9%	5.5%	4.4%

34.

35. It is an S_N2 reaction followed by dissociation of a proton.

36. Since insulin has two peptide chains, treatment with Edman's reagent would release two PTH-amino acids in approximately equal amounts.

37. Knowing that the N-terminal amino acid is Gly, look for a peptide fragment that contains Gly.

"Fragment 6" tells you that the 2nd amino acid is Arg.

"Fragment 5" tells you the next two are Ala-Trp or Trp-Ala.

"Fragment 4" tells you that Glu is next to Ala, so the 3rd and 4th amino acids must be Trp-Ala and the 5th is Glu.

"Fragment 7" tells you the 6th amino acid is Leu.

"Fragment 8" tells you the next two are Met-Pro or Pro-Met.

"Fragment 3" tells you that Pro is next to Val, so the 7th and 8th amino acids must be Met-Pro and the 9th is Val.

"Fragment 2" tells you the last amino acid is Asp.

$$\text{Gly-Arg-Trp-Ala-Glu-Leu-Met-Pro-Val-Asp}$$

38.   Cysteine can react with cyanogen bromide, but the sulfur would not be positively charged, so it would be a poor leaving group. In addition, the lactone will not be formed, because it would have a strained four-membered-ring. Without lactone formation, the imine would not be formed, so cleavage cannot occur.

39.   **a.**   His-Lys   Leu-Val-Glu-Pro-Arg   Ala-Gly-Ala

      b.   Leu-Gly-Ser-Met-Phe-Pro-Tyr   Gly-Val

40.   Solved in the text.

41.   The data from treatment with Edman's reagent and carboxypeptidase A identify the first and last amino acids.

Leu __ __ __ __ __ __ Ser

The data from cleavage with cyanogen bromide identify the position of Met and identify the other amino acids in the pentapeptide and tripeptide but not their order.

cleavage with cyanogen bromide

Arg, Lys, Tyr   Arg, Phe

Leu __ __ __ Met | __ __ Ser

The data from treatment with trypsin put the remaining amino acids in the correct positions.

Leu   Tyr   Lys | Arg | Met   Phe   Arg | Ser

42.   **a.**   Trypsin cleaves at Arg and Lys. There are two possible primary structures:

$$\text{Val-Gly-Asp-Lys-Leu-Glu-Pro-Ala-Arg-Ala-Leu-Gly-Asp}$$

or

$$\text{Leu-Glu-Pro-Ala-Arg-Val-Gly-Asp-Lys-Ala-Leu-Gly-Asp}$$

The two possible primary structures can be distinguished by Edman's reagent. Edman's reagent would release Val in one case and Leu in the other.

b. Trypsin cleaves at Arg and Lys. There are two possible primary structures:

Ala-Glu-Pro-Arg-Ala-Met-Gly-Lys-Val-Leu-Gly-Glu

or

Ala-Met-Gly-Lys-Ala-Glu-Pro-Arg-Val-Leu-Gly-Glu

The two possible primary structures can be distinguished by treatment with cyanogen bromide. Cyanogen bromide would cleave the polypeptide into two hexamers in one case and into a dimer and a decamer in the other.

43. 74 amino acids/3.6 amino acids per turn of the helix = 20.6 turns of the helix

20.6 × 5.4 Å = **110 Å** in an α-helix

74 amino acids × 3.5 Å = **260 Å** in fully extended polypeptide chain

44. It would fold so that its nonpolar residues are on the outside of the protein in contact with the nonpolar membrane and its polar residues are on the inside of the protein.

45. **a.** A cigar-shaped protein has the greatest surface area to volume ratio, so it has the highest percentage of polar amino acids.

b. A subunit of a hexamer would have the smallest percentage of polar amino acids, because part of the surface of the subunit can be on the inside of the hexamer and therefore have nonpolar amino acids on its surface.

46. An amino acid is insoluble in diethyl ether (a relatively nonpolar solvent), because an amino acid exists as a highly polar zwitterion at neutral pH. In contrast, carboxylic acids and amines are either neutral or have a single charge depending on the extent of dissociation in diethyl ether. They are therefore less polar.

47. **a.** pH = 9.69   **b.** pH = 6.02 (its pI)   **c.** pH = 2.34

48. **a.** Val-Arg-Gly-Met-Arg-Ala   Ser

b. Ser-Phe-Lys-Met   Pro-Ser-Ala-Asp

c. Arg   Ser-Pro-Lys   Lys   Ser-Glu-Gly

49. As an amino acid moves from a solution with a pH equal to its pI to a more basic solution, the amino acid becomes more and more negatively charged. Because asparagine has a lower pI than leucine, in a solution of pH = 7.3, asparagine has moved farther from its pI than has leucine. Asparagine therefore will have a higher percentage of negative charge at pH 7.3.

50.

51.   a.   **c.**

b.   **d.**

52.   The student is correct. At the pI, the total of the positive charges on the tripeptide's amino groups must be 1 to balance the one negative charge of the carboxyl group. When the pH of the solution is equal to the pK_a of a lysine residue, the three lysine groups each have one-half a positive charge for a total of one and one-half positive charges. Thus, the solution must be more basic than this in order to have just one positive charge.

53.   Since the mixture of amino acids is in a solution of pH = 5, **His** will have an overall positive charge and **Glu** will have an overall negative charge. **His**, therefore, will migrate to the cathode, and **Glu** will migrate to the anode. **Ser** is more polar than **Thr** (both have OH groups, but **Thr** has an additional carbon); **Thr** is more polar than **Met** (**Met** has SCH_2CH_2 instead of CHOH and an additional carbon); and **Met** is more polar than **Leu** (**Leu** has $CHCH_3$ instead of S).

54.   We know that "Fragment 3" (Leu-Pro-Phe) is at the C-terminal end of the polypeptide. Thus, the only question we need to answer is: Which of the two fragments obtained by cleavage with cyanogen bromide, that begin with Gly and end with Met, is nearest the C-terminal end of the polypeptide? The answer is obtained from "Fragment 4" obtained from trypsin cleavage. The Met nearest to the C-terminal end must be preceded by Arg or Lys. Therefore, the polypeptide has the following sequence:

Gly-Leu-Try-Phe-Lys-Ser-Met-Gly-Leu-Tyr-Lys-Val-Ile-Arg-Met-Leu-Pro-Phe

55.   You would (correctly) expect serine and cysteine to have lower pK_a values than alanine, since a hydroxymethyl and a thiomethyl group are more electron-withdrawing than a methyl group. Because oxygen is more electronegative than sulfur, one would probably expect serine to have a lower pK_a than cysteine. The fact that cysteine has a lower pK_a than serine can be explained by stabilization of serine's carboxyl proton by hydrogen bonding to the β-OH group of serine, which causes it to have less of a tendency to be removed by a base.

56.   Each compound has two groups that can act as a buffer, one amino group and one carboxyl group. Thus, the compound in higher concentration (0.2 M glycine) will be a more effective buffer.

57.   Groups that are not fully charged at the given pH are shown in the form that predominates. For example, tyrosine has a $pK_a = 9.11$, so at pH = 9 it is shown in its acidic (neutral) form.

a.

b.

c.

d.

58.

lysine

maleic
anhydride

59.   **a.**   When the polypeptide is treated with maleic anhydride, lysine reacts with maleic anhydride (see Problem 58), but the amino group of arginine is not sufficiently nucleophilic to react with maleic anhydride. Therefore, trypsin will cleave only at arginine residues, since it will no longer recognize lysine residues.

b.   Four fragments will be obtained from the polypeptide. Remember that trypsin will not cleave the Arg-Pro bond.

c.   The N-terminal end of each fragment will be positively charged because of the $^+NH_3$ group; the C-terminal end will be negatively charged because of the COO^- group; arginine residues will be positively charged; aspartate and glutamate residues will be negatively charged; and lysine residues will be negatively charged because they are attached to the maleic acid group.

Elution order: **A > D > C > B**

A Gly-Ser-Asp-Ala-Leu-Pro-Gly-Ile-Thr-Ser-Arg   overall charge = 0
    \+   −   \+ −

B Asp-Val-Ser-Lys-Val-Glu-Tyr-Phe-Glu-Ala-Gly-Arg   overall charge = −3
    \+ −   −   −   −   \+ −

C Ser-Glu-Phe-Lys-Glu-Pro-Arg   overall charge = −2
    \+ − −   −   \+ −

D Leu-Tyr-Met-Lys-Val-Glu-Gly-Arg-Pro-Val-Ser-Ala-Gly-Leu-Trp   overall charge = −1
    \+   −   −   \+   −

60.   First mark off where the chains would have been cleaved by chymotrypsin (C-side of Phe, Trp, Tyr).

Val-Met-Tyr ├-Ala-Cys-Ser-Phe ├-Ala-Glu-Ser

Ser-Cys-Phe│-Lys-Cys-Trp │-Lys-Tyr ├Cys-Phe ├Arg-Cys-Ser

Then, from the fragments given, you can determine where the disulfide bridges are in the original intact peptide.

Val-Met-Tyr-Ala-Cys-Ser-Phe-Ala-Glu-Ser
                              |
                              S
                              |
                              S
                              |
Ser-Cys-Phe-Lys-Cys-Trp-Lys-Tyr-Cys-Phe-Arg-Cys-Ser
            └————————S—S————————┘

61.   The methyl ester of phenylalanine rather than phenylalanine itself should be added in the peptide bond-forming step, because if esterification of phenylalanine is done after amide bond formation, both the carboxyl group of phenylalanine and the γ-carboxyl group of aspartate could be esterified. Both the carboxyl group of phenylalanine and the γ-carboxyl group of aspartate can be activated by DCC. Therefore, some product will be obtained in which the amide bond is formed with the γ-carboxyl group of aspartate rather than with the α-carboxyl group.

aspartame

62. **a.**

$$(CH_3)_2CHCH_2 - \overset{\overset{\displaystyle O}{\|}}{C} - OH \xrightarrow[\text{2. H}_2\text{O}]{\text{1. Br}_2,\ \text{PBr}_3} (CH_3)_2CHCH - \overset{\overset{\displaystyle O}{\|}}{C} - OH \xrightarrow[\text{excess}]{\text{NH}_3} (CH_3)_2CHCH - \overset{\overset{\displaystyle O}{\|}}{C} - O^- + \overset{+}{N}H_4$$

with Br below the second structure, and $\overset{+}{N}H_3$ below the third structure, labeled **valine**

b.

$$\overset{(CH_3)_2CH}{\underset{H}{\diagdown}}C=O \xrightarrow[\text{trace}\atop\text{acid}]{\text{NH}_3} \overset{(CH_3)_2CH}{\underset{H}{\diagdown}}C=NH \xrightarrow[\text{HCl}]{^-C\equiv N} (CH_3)_2CHCH-C\equiv N \xrightarrow[\Delta]{\text{HCl, H}_2\text{O}} (CH_3)_2CHCH - \overset{\overset{\displaystyle O}{\|}}{C} - OH$$

an aldehyde   an imine   with $\overset{+}{N}H_3$ below,   with $\overset{+}{N}H_3$ below, labeled **valine**

c.

$$(CH_3)_2CHC\overset{\overset{\displaystyle O}{\|}}{\underset{\underset{\displaystyle O}{\|}}{}}OH \xrightarrow[\text{trace}\atop\text{acid}]{\text{NH}_3\atop\text{excess}} (CH_3)_2CHC - \overset{\overset{\displaystyle O}{\|}}{C} - O^- \xrightarrow[\text{Pd/C}]{\text{H}_2} (CH_3)_2CHCH - \overset{\overset{\displaystyle O}{\|}}{C} - O^-$$

with NH below second structure, NH_2 below third structure labeled **valine**

d. $C_2H_5O\overset{\overset{\displaystyle O}{\|}}{C} - CH - \overset{\overset{\displaystyle O}{\|}}{C}OC_2H_5$ + potassium phthalimide ($\overset{\bigcirc}{N}{}^-K^+$) $\longrightarrow$ (N-substituted phthalimide with $-C(H)(COC_2H_5)(COC_2H_5)$)

with Br below, labeled **α-bromomalonic ester**

$$\xrightarrow{\text{CH}_3\text{CH}_2\text{O}^-} \text{(phthalimide N–C carbanion with two } COC_2H_5) + CH_3CH_2OH$$

$$\xrightarrow[\text{CH}_3]{\text{CH}_3\text{CHBr}} \text{(phthalimide N–C with } CH_3CH,\ CH_3,\ \text{two } COC_2H_5)\ Br^-$$

$$\xrightarrow[\Delta]{\text{HCl, H}_2\text{O}}$$

phthalic acid (benzene ring with two COOH) + CO₂ + $H_3\overset{+}{N} - CH - \overset{\overset{\displaystyle O}{\|}}{C} - OH$ with $CH(CH_3)_2$ below, labeled **valine**

e.

acetamidomalonic ester   CH₃CH₂OH

$CH_3CH_2O^-$ →

CH_3CHBr / CH_3 →

Br⁻

HCl, H₂O, Δ ↓

acetic acid + CO₂ + valine

---

63.   Ser-Glu-Leu-Trp-Lys-Ser-Val-Glu-His-Gly-Ala-Met

From the experiment with carboxypeptidase A, we know the C-terminal amino acid is Met.

"Fragment 12" tells us the amino acid adjacent to Met is Ala.

"Fragment 5" tells us the next amino acid is Gly.

"Fragment 2" tells us the next amino acid is His.

"Fragment 7" tells us the next amino acid is Glu.

"Fragment 10" tells us the next amino acid is Val.

"Fragment 3" tells us the next amino acid is Ser.

"Fragment 9" tells us the next amino acid is Lys.

"Fragment 1" tells us the next amino acid is Trp.

"Fragment 8" tells us the next amino acid is Leu.

"Fragment 11" tells us the next amino acid is Glu.

"Fragment 4" tells us the next (first) amino acid is Ser.

---

64.   **a.** $20^8 = 25,600,000,000$ **b.** 20^{100}

---

65.   The pK_a of the carboxylic acid group of the amino acid is lower than the pK_a of the carboxylic acid group of the dipeptide, because the positively charged ammonium group of the amino acid is more strongly electron withdrawing than the amide group of the peptide. This causes the amino acid to be a stronger acid and therefore have a lower pK_a.

The pK_a of the ammonium group of the dipeptide is lower than the pK_a of the ammonium group of the amino acid, because the amide group of the dipeptide is more strongly electron-withdrawing than the carboxylate group of the amino acid.

66. Finding that there is one less spot than the number of amino acids tells you that the spots for two of the amino acids superimpose. Since leucine and isoleucine have identical polarities, they are good candidates for being the amino acids that migrate to the same location.

Val (pI = 5.97), Trp (pI = 5.89), and Met (pI = 5.75) can be ordered based on their pI values, because the one with the greatest pI will be the one with the greatest amount of positive charge at pH = 5. (See Problem 49.)

67.

68. Oxidation of dithiothreitol is an intramolecular reaction, so it occurs with a larger rate constant than the oxidation of 2-mercaptoethanol, which is an intermolecular reaction. The reverse reduction reaction should occur with about the same rate constant in both cases. Increasing the rate of the oxidation reaction while keeping the rate of the reduction reaction constant is responsible for the greater equilibrium constant, because $K_{eq} = k_1/k_{-1}$, where k_1 is the rate constant for oxidation and k_{-1} is the rate constant for reduction.

$$2 \ HOCH_2CH_2SH \ \underset{k_{-1}}{\overset{k_1}{\rightleftharpoons}} \ HOCH_2CH_2S-SCH_2CH_2OH$$

69. **a.**

intermediate I   intermediate II

b.

3-methylbutanal   leucine

c.

2-methylpropanal   valine

70.

amino acid	absorbance at 280 nm	molar absorptivity = absorbance/concentration
tryptophan	0.61	610
tyrosine	0.15	150
phenylalanine	0.01	10

71.   Because the native enzyme has four disulfide bridges, the denatured enzyme has eight cysteine residues. The first cysteine has a 1 in 7 chance of forming a disulfide bridge with the correct cysteine. The first cysteine of the next pair has a 1 in 5 chance, and the first cysteine of the third pair has a 1 in 3 chance.

$$\frac{1}{7} \times \frac{1}{5} \times \frac{1}{3} = 0.0095$$

If disulfide bridge formation were entirely random, the recovered enzyme should have 0.95% of its original activity. The fact that the enzyme the chemist recovered had 80% of its original activity supports his hypothesis that disulfide bridges form after the minimum energy conformation of the protein has been achieved. In other words, disulfide bridge formation is not random, but is determined by the tertiary structure of the protein.

72. Formation of the tetrahedral intermediate and collapse of the tetrahedral intermediate are shown together in the second step.

a thiazoline

a PTH-amino acid

73. The spot marked with an **X** is the peptide that is different in the normal and mutant polypeptide. The spot is closer to the cathode and farther to the right, indicating that the substituted amino acid in the mutant has a greater pI and is less polar.

The fingerprints are those of hemoglobin (normal) and sickle-cell hemoglobin (mutant). In sickle-cell hemoglobin, a glutamate in the normal polypeptide is substituted with a valine. This agrees with our observation that the substituted amino acid is less negative and more nonpolar. (See the discussion of sickle cell anemia on page 1164 of the text.)

normal

mutant

74. **a.** Acid-catalyzed hydrolysis indicates the peptide contains 12 amino acids.

— — — — — — — — — — — —

b. Treatment with Edman's reagent indicates that Val is the N-terminal amino acid.

<u>Val</u> — — — — — — — — — — —

c. Treatment with carboxypeptidase A indicates that Ala is the C-terminal amino acid.

<u>Val</u> — — — — — — — — — <u>Ala</u>

d. Treatment with cyanogen bromide indicates that Met is the 5th amino acid with Arg, Gly, Ser in an unknown order in positions 2, 3, and 4.

<u>Val</u> __ __ __ <u>Met</u> — — — — — — <u>Ala</u>
　　　　Arg, Gly, Ser

e. Treatment with trypsin indicates that Arg is the 3rd amino acid, and Ser is 2nd, Gly is 4th, Tyr is 6th, and Lys is 7th. Since Lys is in the terminal fragment, cleavage did not occur at Lys, so Pro must be at lysine's cleavage site.

<u>Val</u> <u>Ser</u> <u>Arg</u> <u>Gly</u> <u>Met</u> <u>Tyr</u> <u>Lys</u> __ __ __ __ __ <u>Ala</u>
　　　　　　　　　　　　　　　　　Lys-Pro,　　Phe, Ser

f. Treatment with chymotrypsin indicates that Phe is the 10th amino acid and Ser is the 11th.

<u>Val</u> <u>Ser</u> <u>Arg</u> <u>Gly</u> <u>Met</u> <u>Tyr</u> <u>Lys</u> <u>Lys</u> <u>Pro</u> <u>Phe</u> <u>Ser</u> <u>Ala</u>

Chapter 23 Practice Test

1. Give the structure of the following amino acids at pH = 7:

    a. glutamic acid   **b.** lysine   **c.** isoleucine   **d.** arginine   **e.** asparagine

2. Draw the form of histidine that predominates at:

    a. pH = 1   **b.** pH = 4   **c.** pH = 8   **d.** pH = 11

3. Answer the following:

    a. Alanine has a pI = 6.02 and serine has a pI = 5.68. Which would have the highest concentration of positive charge at pH = 5.50?

    b. Which amino acid is the only one that does not have an asymmetric center?

    c. Which are the two most nonpolar amino acids?

    d. Which amino acid has the lowest pI?

4. Why does the carboxyl group of alanine have a lower pK_a than the carboxyl group of propanoic acid?

$$CH_3CH \quad \overset{\overset{\displaystyle O}{\|}}{C} \quad OH \quad pK_a = 2.2 \qquad CH_3CH_2 \quad \overset{\overset{\displaystyle O}{\|}}{C} \quad OH \quad pK_a = 4.7$$

    $^+NH_3$

    alanine   propanoic acid

5. Indicate whether each of the following statements is true or false:

    a. A cigar-shaped protein has a greater percentage of polar residues than a spherical protein.   T   F

    b. Naturally occurring amino acids have the L-configuration.   T   F

    c. There is free rotation about a peptide bond.   T   F

6. Give the compound obtained from mild oxidation of cysteine.

7. Define each of the following:

    a. the primary structure of a protein

    b. the tertiary structure of a protein

    c. the quaternary structure of a protein

706   Chapter 23

8.   Identify the spots.

Asp
Ile
Lys
Phe
Ser
Tyr

9.   Calculate the pI of each of the following amino acids:

a. phenylalanine (pK_a's = 2.16, 9.18)   **b.** arginine (pK_a's = 2.17, 9.04, 12.48)

10.   From the following information, determine the primary sequence of the decapeptide:

a. Acid hydrolysis gives: Ala, 2 Arg, Gly, His, Ile, Lys, Met, Phe, Ser

b. Reaction with Edman's reagent liberated Ala

c. Reaction with carboxypeptidase A liberated Ile

d. Reaction with cyanogen bromide (cleaves on the C-side of Met)

    1. Gly, 2 Arg, Ala, Met, Ser

    2. Lys, Phe, Ile, His

e. Reaction with trypsin (cleaves on the C-side of Arg and Lys)

    1. Arg, Gly

    2. Ile

    3. Phe, Lys, Met, His

    4. Arg, Ser, Ala

f. Reaction with thermolysin (cleaves on the N-side of Leu, Ile, Phe, Trp, Tyr)

    1. Lys, Phe

    2. 2 Arg, Ser, His, Gly, Ala, Met

    3. Ile

CHAPTER 24
Catalysis

Important Terms

acid catalyst	a catalyst that increases the rate of a reaction by donating a proton.
active site	a pocket or cleft in an enzyme where the substrate is bound.
acyl-enzyme intermediate	an amino acid residue of an enzyme that has been acylated while catalyzing a reaction.
amino acid side chain	the substituent on the α-carbon of an amino acid.
base catalyst	a catalyst that increases the rate of a reaction by removing a proton.
catalyst	a substance that increases the rate of a reaction without itself being consumed in the overall reaction.
covalent catalysis (nucleophilic catalysis)	catalysis that occurs as a result of a nucleophile forming a covalent bond with one of the reactants.
effective molarity	the concentration of the reagent that would be required in an intermolecular reaction for it to have the same rate as an intramolecular reaction.
electrophilic catalyst	an electrophile that facilitates a reaction.
electrostatic catalysis	stabilization of a charge by an opposite charge.
enzyme	a protein that is a catalyst.
general-acid catalysis	catalysis in which a proton is transferred to the reactant during the slow step of the reaction.
general-base catalysis	catalysis in which a proton is removed from the reactant during the slow step of the reaction.
induced fit model	a model that describes the specificity of an enzyme for its substrate: the shape of the active site does not become completely complementary to the shape of the substrate until after the enzyme has bound the substrate.
intramolecular catalysis	catalysis in which the catalyst that facilitates the reaction is part of the molecule undergoing reaction.
lock-and-key model	a model that describes the specificity of an enzyme for its substrate: the substrate fits the enzyme like a key fits into a lock.
metal-ion catalysis	catalysis in which the species that facilitates the reaction is a metal ion.
molecular recognition	the recognition of one molecule by another as a result of specific interactions; for example, the specificity of an enzyme for its substrate.

**nucleophilic catalysis
(covalent catalysis)**
catalysis that occurs as a result of a nucleophile forming a covalent bond with one of the reactants.

nucleophilic catalyst
a catalyst that increases the rate of a reaction by acting as a nucleophile.

**pH-activity profile
       or
pH-rate profile**
a plot of the activity of an enzyme as a function of the pH of the reaction mixture.

relative rate
obtained by dividing the actual rate constant by the rate constant of the slowest reaction in the group being compared.

site-specific mutagenesis
a technique that substitutes one amino acid of a protein for another.

specific-acid catalysis
catalysis in which the proton is fully transferred to the reactant before the slow step of the reaction.

specific-base catalysis
catalysis in which the proton is completely removed from the reactant before the slow step of the reaction.

substrate
the reactant of an enzyme-catalyzed reaction.

Solutions to Problems

1. $\Delta H^{\ddagger}$, E_a, $\Delta S^{\ddagger}$, $\Delta G^{\ddagger}$, k_{rate}. A catalyst increases the rate of a reaction by decreasing the energy between the reactant and the transition state of the rate-limiting step. That is, it decreases the height of the energy hill of the rate-limiting step. The following parameters would be different for a reaction carried out in the presence of a catalyst: $\Delta H^{\ddagger}$, E_a, $\Delta S^{\ddagger}$, $\Delta G^{\ddagger}$, k_{rate}. These are the parameters that reflect the difference in energy between the reactant and the transition state.

2.   Notice that (1) and (2) have only the first part (because the final product of the reaction is a tetrahedral intermediate), (3) has only the second part (because the initial reactant is a tetrahedral intermediate), and (4) has both the first and second parts.

Mechanism for acid-catalyzed ester hydrolysis: first part

a.   **Similarities:** the first step is protonation of the carbonyl compound, the second step is attack of a nucleophile on the protonated carbonyl compound, and the third step is loss of a proton.

b.   **Differences:** the carbonyl compound that is used as the starting material; the nucleophile used in (2) is an alcohol rather than water.

Mechanism for acid-catalyzed ester hydrolysis: second part

acid-catalyzed ester hydrolysis

(3)

(4)

a. **Similarities:** the first step is protonation of the tetrahedral intermediate, and the second step is elimination of a group from the tetrahedral intermediate. In two of the three reactions, the third step is loss of a proton.

b. **Differences:** the nature of the group that is eliminated from the tetrahedral intermediate. The third step in acetal formation is not loss of a proton, because the intermediate does not have a proton to lose. Instead, the third step is attack of a nucleophile, and the fourth step is loss of a proton.

3.   Both slow steps are specific-acid catalyzed; the compound is protonated before the slow step in each case.

4. a. In general-acid catalysis, the proton is added to the reactant and the O—C bond forms in the same step.

b. In specific-acid catalysis, the proton is added to the reactant before the O—C bond forms.

5. Solved in the text.

6.   **a.** In general-base catalysis, the proton is removed from the reactant and the O—C bond forms in the same step.

$$CH_3OH \ + \ :B$$

   b. In specific-base catalysis, the proton is removed from the reactant before the O—C bond forms.

$$CH_3OH \ + \ :B$$

7. The metal ion catalyzes the decarboxylation reaction by complexing with the negatively charged oxygen of the carboxyl group and carbonyl oxygen of the β-keto group, thereby making it easier for the carbonyl oxygen to accept the electrons that are left behind when CO_2 is eliminated.

Because acetoacetate and the monoethyl ester of dimethyloxaloacetate do not have a negatively charged oxygen on one carbon and a carbonyl group on an adjacent carbon with which to form a complex, a metal ion does not catalyze decarboxylation of these compounds.

8. Co²⁺ can catalyze the reaction in three different ways. It can complex with the reactant, increasing the susceptibility of the carbonyl group to nucleophilic attack. It can also complex with water, increasing the tendency of water to lose a proton, resulting in a better nucleophile for hydrolysis. And it can complex with the leaving group, decreasing its basicity and thereby making it a better leaving group.

$$Co^+ \!\!-\!\! \overset{\delta+}{O}H_2 \rightleftharpoons Co^+ \cdots \overset{\delta-}{O}H + H^+$$

9. Because the reacting groups in the trans isomer are pointed in opposite directions, they cannot react in an intramolecular reaction. Because they can react only via an intermolecular pathway, they will have approximately the same rate of reaction as they would have if the reacting groups were in separate molecules. Consequently, the relative rate would be expected to be close to one.

10. a. The nucleophile can attack the back side of either of the two ring carbons to which the sulfur is bonded in the intermediate, thereby forming two trans products. There are two nucleophiles (water and ethanol), so a total of four products will be formed.

b.

11. The C_6H_5S- group withdraws electrons inductively. If the cis-substituted compound undergoes an S_N2 reaction, it should be slightly more reactive than the unsubstituted compound, because the electron-withdrawing C_6H_5S- group makes the carbon attached to the Cl more susceptible to nucleophilic attack and the substituent would not block back-side attack by the nucleophile. If, however, the cis-substituted compound undergoes an S_N1 reaction, it should be slightly less reactive than the unsubstituted compound, because the electron-withdrawing C_6H_5S- group makes the carbocation intermediate less stable and therefore it forms more slowly. Thus, the cis-substituted compound undergoes an S_N1 reaction, which we would have predicted since the nucleophiles are poor (H_2O and CH_3CH_2OH) and the reaction is carried out in a polar solvent (H_2O, CH_3CH_2OH).

12. Solved in the text.

13. The tetrahedral intermediate has two leaving groups, a carboxylate ion and a phenolate ion. When there are no nitro substituents on the aromatic ring, the carboxylate ion is a weaker base (a better leaving group) than the phenolate ion, so the tetrahedral intermediate reforms **A**. Thus, hydrolysis proceeds through general-base catalyzed hydrolysis of **A**. In contrast, 2,4-dinitrophenoxide ion is a weaker base (better leaving group) than the carboxylate ion, so the tetrahedral intermediate forms **B**. Thus, hydrolysis proceeds through a hydrolysis of an anhydride form by nucleophilic catalysis.

14. If the *ortho*-carboxyl substituent acts as an intramolecular base catalyst, ^{18}O would not be incorporated into salicylic acid.

salicylate

If the *ortho*-carboxyl substituent acts as an intramolecular nucleophilic catalyst, ^{18}O would be incorporated into salicylic acid.

Not all the salicylic acid would contain ^{18}O, because the anhydride intermediate has two different carbonyl groups. If water attacks the benzyl carbonyl group (above), salicylic acid will contain ^{18}O. If, however, water attacks the acetyl carbonyl group (below), acetic acid will contain ^{18}O.

15. Solved in the text.

16. **2**, **3**, and **4** are bases, so they can help remove a proton.

17. Ser-Ala-Phe would be more readily cleaved by carboxypeptidase A, because the phenyl substituent of phenylalanine would be more attracted to the hydrophobic pocket of the enzyme than would the negatively charged substituent of aspartate.

18. Glu 270 attacks the carbonyl group of the ester, forming a tetrahedral intermediate. Collapse of the tetrahedral intermediate is most likely catalyzed by an acid group of the enzyme in order to increase the leaving ability of the RO group. (Perhaps the HO substituent of tyrosine is close enough in the esterase to act as the catalyst.) The group that donates the proton can then act as a base catalyst to remove a proton from water as it hydrolyzes the anhydride.

19. Because arginine extends farther into the binding pocket, it must be the one that forms direct hydrogen bonds. Lysine, which is shorter, needs the mediation of a water molecule in order to engage in bond formation with aspartate.

20. The side chains of D-Arg and D-Lys are not positioned to bind correctly at the active site; they would be able to bind at a mirror image of the active site.

21. NAM would contain ^{18}O because it is the ring that would undergo nucleophilic attack by $H_2^{18}O$.

22. Lemon juice contains citric acid. Some of the side chains of an enzyme will become protonated in an acidic solution. This will change the charge of the group (for example a negatively charged aspartate, when protonated, becomes neutral; a neutral lysine, when protonated, becomes positively charged), and because the shape of an enzyme is determined by the interaction of the side chains, changing the charges of the side chains will cause the enzyme to undergo a conformational change that leads to denaturation. When the enzyme is denatured, it loses its ability to catalyze the reaction that causes apples to turn brown.

23.   In the presence of base in the absence of an enzyme, D-fructose is in equilibrium with D-glucose and D-mannose as a result of an enediol rearrangement (Section 22.5). Both C-2 epimers are formed, because a new asymmetric center is formed at C-2 and it can have either the *R* or the *S* configuration. Enzyme catalyzed reactions are typically highly stereoselective—the enzyme catalyzes the formation of a single stereoisomer. Thus, D-fructose is in equilibrium with only D-glucose in the presence of the enzyme that catalyzes the enediol rearrangement.

24.   The fact that the pK_a is given by a descending leg means that the side chain of the amino acid is functioning as an acid catalyst. From the pK_a value it could be lysine ($pK_a = 10.79$), tyrosine ($pK_a = 10.07$), or cysteine ($pK_a = 8.33$).

From what we are told about the mechanism on page 1049 of the text, the amino acids functioning as catalysts are histidine, lysine, and glutamate. Thus, the amino acid is lysine.

25.   Because **2** is a primary amine, it can form an imine. Notice that **1** cannot form an imine because the lone pair on the NH_2 group is delocalized onto the oxygen, so this NH_2 group is not a nucleophile. Although two of the nitrogens in **3** are not nucleophilic because their lone pairs are delocalized, the resonance contributors show that the third nitrogen is nucleophilic, so it should be able to form an imine-like structure.

26.   The positively charged nitrogen atom of the protonated imine can accept the electrons that are left behind when the C3—C4 bond breaks.

In the absence of the protonated imine, the electrons would be delocalized onto a neutral oxygen. The neutral oxygen is not as electron withdrawing as the positively charged nitrogen of the protonated imine. In other words, imine formation makes it easier to break the C—C bond.

27.   In order to break the C3—C4 bond, the carbonyl group has to be at the 2-position as it is in fructose, so it can accept the electrons that result from C3—C4 cleavage; the carbonyl group at the 1-position in glucose cannot accept these electrons. Therefore, glucose must isomerize to a ketose in order for the carbonyl group to be at the 2-position.

glucose-6-phosphate   fructose-6-phosphate

28.   Cysteine residues react with iodoacetic acid. If a cysteine residue is at the active site of an enzyme, adding an acetate group to the sulfur atom could interfere with the enzyme's being able to bind the substrate or it could interfere with positioning the tyrosine residue that is involved in catalyzing the reaction. Adding a substituent to cysteine might also cause a conformational change in the enzyme that could destroy its activity.

$$Cys-CH_2\ddot{S}H \ + \ I-CH_2 \quad O^- \ \longrightarrow \ Cys-CH_2SCH_2 \quad O^- \ + \ I^- \ + \ H^+$$

29.   The following compound will lose HBr more rapidly, because the negatively charged oxygen is in position to act as an intramolecular general-base catalyst.

30.   The compounds shown below will be the more reactive, because the methyl substituent causes the reacting groups (COOH and OH) to stay in a more favorable conformation for reaction.

a.   b.

31. The compound shown below will form a lactone more rapidly, because it forms a five-membered-ring lactone, which is less strained and than the seven-membered-ring lactone formed by the other compound. The greater stability of the five-membered-ring product causes the transition state leading to its formation to be more stable than the transition state leading to formation of the seven-membered-ring product.

32. In order to hydrolyze an amide, the NH_2 group in the tetrahedral intermediate has to leave in preference to the less basic OH group. This can happen if the NH_2 group is protonated because $^+NH_3$ is a weaker base and, therefore, easier to eliminate than OH. Of the four compounds, two have substituents that can protonate the NH_2 by acting as acid catalysts, *ortho*-carboxy-benzamide and *ortho*-hydroxybenzamide.

ortho-carboxybenzamide

ortho-hydroxybenzamide

Because the carboxy group withdraws electrons from the ring and the OH group donates electrons to the ring, formation of the tetrahedral intermediate will be faster for carboxy-substituted compound. Because the carboxy group is a stronger acid than the OH group, the tetrahedral intermediate of the carboxy-substituted compound will collapse to products faster. Therefore, the *o*-carboxybenzamide has the faster rate of hydrolysis.

33.   a. $CH_3CH_2\ddot{S}CH_2CH_2\!-\!Cl$ intramolecular nucleophilic catalysis

b.

intramolecular general-acid catalysis

The OH substituent is protonating the leaving group as it departs, causing it to be a weaker base and, thus, a better leaving group.

34. If the *ortho*-carboxy substituent is acting as a general base catalyst, the kinetic isotope effect will be greater than 1.0, because an O—H (or O—D) bond is broken in the slow step of the reaction, and an O—D bond is stronger than an O—H bond and therefore is harder to break.

If the *ortho*-carboxyl substituent is acting as a nucleophilic catalyst, the kinetic isotope effect will be about 1.0, because an O—H (or O—D) bond is not broken in the slow step of the reaction.

35. $\dfrac{1.5 \times 10^6 \ \text{s}^{-1}\text{M}^{-1}}{0.6 \ \text{s}^{-1}\text{M}^{-1}} = 2.5 \times 10^6$

36. Because the catalytic group is an acid catalyst, it will be active in its acidic form and inactive in its basic form. The pH at the midpoint of the curve corresponds to the pK_a of the group responsible for the catalysis, since $pH = pK_a$ when $[HA] = [A^-]$ (see page 57 of the text).

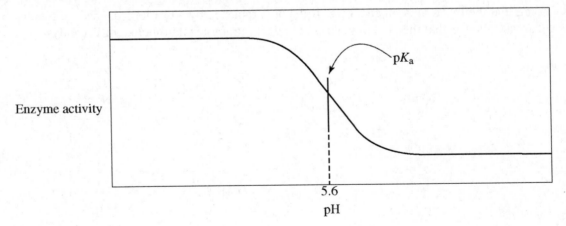

37.   Co^{2+} can catalyze the hydrolysis reaction by complexing with three nitrogen atoms in the substrate as
well as with water. Complexation increases the acidity of water, thereby providing a better nucleophile
(metal-bound hydroxide) for the hydrolysis reaction. Complexation with the substrate locks the
nucleophile into the correct position for addition to the carbonyl carbon.

38.   Class I aldolases use a protonated imine as a site to which electrons can be delocalized. Class II aldolases
use a metal ion complexed with the carbonyl group as a site to which electrons can be delocalized.

39.   Knowing that the reaction is much slower if the N is replaced by a CH tells us that the N must be acting
as a catalyst. We see that the N is in position to be an intramolecular nucleophilic catalyst.

40.   There are two possible mechanisms that can account for the observed catalysis by morpholine. In the absence of morpholine, the first step of the reaction is addition of water to the **ester**, because addition of water to the aldehyde would form a hydrate that would be in equilibrium with the aldehyde. In the presence of morpholine, the first step of the reaction (in both mechanisms) is addition of morpholine to the **aldehyde**. The reaction of morpholine with the aldehyde is a faster reaction that the reaction of water with the ester, because morpholine is a better nucleophile than water and an aldehyde is more susceptible to nucleophilic addition than an ester.

In one mechanism, the negatively charged aldehyde oxygen (which is a much better nucleophile than water) is the nucleophile that adds to the ester. The tetrahedral intermediate collapses to form a lactone. Iminium ion formation followed by hydrolysis gives the final product.

In the other mechanism, the reaction of morpholine with the aldehyde forms an iminium ion. The positively charged iminium ion makes it easier for water to add to the ester by stabilizing the negative charge that develops on the oxygen. Collapse of the tetrahedral intermediate followed by imine hydrolysis forms the final product.

41.

42.   At pH = 12, the nucleophile is hydroxide ion. Attack of hydroxide ion on the carbonyl group is faster in **A**, because the negative charge that is created in the tetrahedral intermediate is stabilized by the positively charged nitrogen.

At pH = 8, the nucleophile is water. Attack of water on the carbonyl group is faster in **B**, because the amino group can act as a general base catalyst to make water a better nucleophile.

43.   **a.**   The cis reactants each undergo an S_N2 reaction. Because the acetate displaces the tosylate group by backside attack, each cis reactant forms a trans product.

b.   The acetate group in a trans reactant is positioned to be able to displace the tosylate leaving group by an intramolecular S_N2 reaction. Acetate ion then attacks in a second S_N2 reaction from the back side of the group it displaces, so trans products are formed. Because both trans reactants form the same intermediate, they both form the same products. Because the acetate ion can attack either of the carbons in the intermediate equally easily, a racemic mixture will be formed.

c.   The trans reactants are more reactive, because the tosylate leaving group is displaced in an intramolecular reaction, and the resulting positively charged cis intermediate is considerably more reactive than the neutral cis reactants. Thus, a trans isomer undergoes two successive S_N2 reactions, each of which is faster than the one step S_N2 reaction that a cis isomer undergoes.

44.   Reduction of the imine linkage with sodium borohydride causes fructose to become permanently attached to the enzyme, because the hydrolyzable imine bond has been lost. Acid-catalyzed hydrolysis removes the two phosphate groups and hydrolyzes the peptide bonds, so the radioactive fragment that is isolated after hydrolysis is the lysine residue (covalently attached to fructose) of the enzyme that originally formed the imine.

45. **a.** 3-Amino-2-oxindole catalyzes the decarboxylation of an α-keto acid by first forming an imine and then losing a proton from the α-carbon. Now the compound can be decarboxylated, because the electrons left behind can be delocalized onto an oxygen.

b. 3-Aminoindole would not be as effective a catalyst, because the electrons left behind when CO_2 is eliminated cannot be delocalized onto an electronegative atom.

3-aminoindole   α-keto acid   imine formation

46.   **a.** Intramolecular nucleophilic attack on an alkyl halide occurs more rapidly than intermolecular attack on an alkyl halide, because the reacting groups are tethered together in the former. The intramolecular reaction is followed by another relatively rapid S_N2 reaction, because the strain in the three-membered ring causes it to break easily.

b.

47.

dihydroxyacetone
phosphate

glyceraldehyde-
3-phosphate

Chapter 24 Practice Test

1. Indicate whether each of the following statements is true or false:

    a. A catalyst increases the equilibrium constant of a reaction.   T   F

    b. An acid catalyst donates a proton to the substrate and a base catalyst
        removes a proton from the substrate.   T   F

    c. The reactant of an enzyme-catalyzed reaction is called a substrate.   T   F

    d. Complexing with a metal ion increases the pK_a of water.   T   F

2. In lysozyme, glutamate 35 is a catalyst that is active in its acid form. Explain its catalytic function.

3. a. Draw a pH-rate profile for an enzyme that has one catalytic group at the active site with a
        $pK_a = 5.7$ that functions as a general base catalyst.

    b. Draw a pH-rate profile for an enzyme that has one catalytic group at the active site with a
        $pK_a = 5.7$ that functions as a general acid catalyst.

4. Using arrows, show the mechanism arrows for the first step in the mechanism for chymotrypsin.

    a. What kind of catalyst is histidine in this step?

    b. What kind of catalyst is serine in this step?

    c. How does aspartate catalyze the reaction?

CHAPTER 25
The Organic Chemistry of the Coenzymes, Compounds Derived from Vitamins

Important Terms

apoenzyme	an enzyme (that needs a cofactor to catalyze a reaction) without its cofactor.
biotin	the coenzyme required by enzymes that catalyze carboxylation of a carbon adjacent to an ester or a keto group.
coenzyme	a cofactor that is an organic molecule.
coenzyme A	a thiol used by biological organisms to form thioesters.
coenzyme B$_{12}$	the coenzyme required by enzymes that catalyze certain rearrangement reactions.
cofactor	an organic molecule or a metal ion that an enzyme needs in order to catalyze a reaction.
competitive inhibitor	a compound that inhibits an enzyme by competing with the substrate for binding at the active site.
dehydrogenase	an enzyme that carries out an oxidation reaction by removing hydrogen from the substrate.
electron sink	a site to which electrons can be delocalized.
flavin adenine dinucleotide (FAD)	a coenzyme required in certain oxidation reactions. It is reduced to FADH$_2$, which is a coenzyme required in certain reduction reactions.
flavin mononucleotide (FMN)	a coenzyme required in certain oxidation reactions. It is reduced to FMNH$_2$, which is a coenzyme required in certain reduction reactions.
heterocyclic compound	a cyclic compound in which one or more of the ring atoms is an atom other than carbon.
holoenzyme	an enzyme plus its cofactor.
lipoate	a coenzyme required in certain oxidation reactions.
mechanism-based inhibitor	an inhibitor that inactivates an enzyme by undergoing part of the normal catalytic mechanism.
molecular recognition	the recognition of one molecule by another as a result of specific interactions.
nicotinamide adenine dinucleotide (NAD$^+$)	a coenzyme required in certain oxidation reactions. It is reduced to NADH, which is a coenzyme required in certain reduction reactions.
nicotinamide adenine dinucleotide phosphate (NADP$^+$)	a coenzyme required in certain oxidation reactions. It is reduced to NADPH, which is a coenzyme required in certain reduction reactions.

nucleotide　　　　a heterocycle attached in the β-position to the anomeric carbon of a phosphorylated ribose.

pyridoxal phosphate　　　　the coenzyme required by enzymes that catalyze certain transformations of amino acids.

suicide inhibitor (mechanism-based inhibitor)　　　　a compound that inactivates an enzyme by undergoing part of its normal catalytic mechanism.

tetrahydrofolate (THF)　　　　the coenzyme required by enzymes that catalyze a reaction that donates a group containing a single carbon to its substrate.

thiamine pyrophosphate (TPP)　　　　the coenzyme required by enzymes that catalyze a reaction that transfers an acyl group from one species to another.

transamination　　　　a reaction in which an amino group is transferred from one compound to another.

transimination　　　　the reaction of a primary amine with an imine to form a new imine and a primary amine derived from the original imine.

vitamin　　　　a substance needed in small amounts for normal body function that the body cannot synthesize or cannot synthesize in adequate amounts.

vitamin KH$_2$　　　　the coenzyme required by the enzyme that catalyzes the carboxylation of glutamate side chains.

Solutions to Problems

1. The metal ion (Zn^{2+}) makes the carbonyl carbon more susceptible to nucleophilic attack, increases the nucleophilicity of water by making it more like a hydroxide ion, and stabilizes the negative charge on the transition state.

2. The alcohol is oxidized to a ketone.

3. The ketone is reduced to an alcohol.

4. FAD contains a diphosphate linkage. One of the phosphate groups comes from FMN and the other comes from ATP. The other product of the reaction must be pyrophosphate to account for loss of the two phosphate groups from ATP. In other words, the phosphate group of FMN must attack the α-phosphorus of ATP, eliminating pyrophosphate. (See pages 752 and 1094 of the text.)

5.   **a.**   FAD has seven conjugated double bonds. (The conjugated double bonds are indicated by *.)

   b.   FADH$_2$ has three conjugated double bonds. It also has two conjugated double bonds that are isolated from the other three.

6.   The nitrogen that is the stronger base is the one most apt to be protonated. N-1 is a weaker base than N-5, because the π electrons on N-1 can be delocalized onto an oxygen. The π electrons on N-5 can be delocalized only into the benzene ring. Nucleophilic attack therefore occurs at the position (4a) that results in the stronger base (N-5) being protonated.

7.   The mechanism for reduction of lipoate by FADH$_2$ is the reverse of the mechanism for oxidation of dihydrolipoate by FAD. Because the pK_a of the N-1 hydrogen is 7, FADH$_2$ is present in both the acidic and basic forms at physiological pH (7.3).

8.   Solved in the text.

9.   When a proton is removed from the methyl group at C-8, the electrons that are left behind can be delocalized onto the oxygen at the 2-position or onto the oxygen at the 4-position.

When a proton is removed from the methyl group at C-7, the electrons that are left behind can be delocalized only onto carbons that, being less electronegative than oxygen, are less able to accommodate the electrons.

10.

pyruvate-TPP
intermediate

pyruvate

11.   Notice that the only difference in the mechanisms for pyruvate decarboxylase and acetolactate synthase is the species the two-carbon fragment is transferred to: a proton in the case of pyruvate carboxylase and pyruvate in the case of acetolactate synthase.

acetolactate

12.   Notice that the only difference in this reaction and that in Problem 11 is that the species to which the two-carbon fragment is transferred has an ethyl group in place of the methyl group.

13. Solved in the text.

14. Acetaldehyde is responsible for the physiological effects known as a hangover. A way to "cure" a hangover, therefore, is to get rid of the acetaldehyde that is formed from the oxidation of ethanol. Thiamine pyrophosphate (the coenzyme form of vitamin B_1) can convert acetaldehyde into acetyl-CoA. It does this by attacking the carbonyl carbon of acetaldehyde to form an intermediate. Loss of a proton from the intermediate forms the same reactive intermediate that is formed in the reaction catalyzed by the pyruvate dehydrogenase system. The final product is acetyl-CoA.

Notice that, except for the second step of the reaction, the reaction is the same as the reaction catalyzed by the pyruvate dehydrogenase system. In the second step of the reaction catalyzed by the pyruvate dehydrogenase system, CO_2 is removed from the carbon that formerly was the carbonyl carbon; in this reaction it is a proton rather than CO_2 that is removed in that step.

15. The α-keto group that accepts the amino group from pyridoxamine is converted into an amino group.

a.

pyruvate   alanine

b.

oxaloacetate   aspartate

16.   For clarity, the substituents are not put on the three bonds attached to the pyridinium ring in Problems 16–19.

transimination

glycine

17.   In the first step, a proton is removed from the α-carbon; in the next step, the leaving group is eliminated from the β-carbon. Transimination produces an enamine. The enamine tautomerizes to an imine that is hydrolyzed to the final product.

transimination

18. The compound on the right is more easily decarboxylated, because the electrons left behind when CO_2 is eliminated are delocalized onto the positively charged nitrogen of the pyridine ring. The electrons left behind if the other compound is decarboxylated cannot be delocalized.

19. The first step (after transimination) in all amino acid transformations catalyzed by PLP is removal of a substituent from the α-carbon of the amino acid. The electrons left behind when the substituent is removed are delocalized onto the positively charged nitrogen of the pyridine ring. If the ring nitrogen were not protonated, it would be less attractive to the electrons. In other words, it would be a less effective electron sink.

20.   The hydrogen of the OH substituent forms a hydrogen bond with the nitrogen of the imine linkage (see Problem 19). This puts a partial positive charge on the nitrogen, which makes it easier for the amino acid to attack the imine carbon in the transimination reaction that attaches the amino acid to the coenzyme. It also makes it easier to remove a substituent from the α-carbon of the amino acid. If the OH substituent is replaced by an OCH_3 substituent, there is no longer a proton available to form the hydrogen bond.

21.   The mechanism is the same as that shown in the text for dioldehydrase. The tetrahedral intermediate that is formed as a result of the coenzyme B_{12}-catalyzed isomerization is unstable and loses ammonia to give acetaldehyde, the final product of the reaction.

22.

23. The methyl group in thymidine comes from the methylene group of N^5,N^{10}-methylene-THF. The methylene group of N^5,N^{10}-methylene-THF comes from the CH_2OH group of serine by means of a PLP-catalyzed C_α-C_β cleavage.

24. Two thiol groups are oxidized in each of the two reactions that overall reduces vitamin K epoxide to vitamin KH_2. Since dihydrolipoate has two thiol groups, each thiol oxidation involves an intramolecular reaction. When thiols such as ethanethiol or propanethiol are used, each thiol oxidation involves an intermolecular reaction. These thiols react more slowly than dihydrolipoate, because the two thiol groups are not in the same molecule. Therefore, they have to find each other in order to react.

25.   **a.** NAD$^+$, NADP$^+$, FAD, FMN, lipoate, vitamin K epoxide

      b. N^5-methyl-THF; N^5,N^{10}-methylene-THF; N^5,N^{10}-methenyl-THF; N^5-formyl-THF; N^{10}-formyl-THF; N^5-formimino-THF; S-adenosylmethionine

      One could include biotin; it accepts a one carbon group (from HCO_3^-) and transfers it to the substrate.

      c. thiamine pyrophosphate

      d. FAD oxidizes dihydrolipoate back to lipoate.

      e. NAD$^+$ oxidizes FADH$_2$ back to FAD.

      f.

      g. thiamine pyrophosphate and pyridoxal phosphate

      h. Thiamine pyrophosphate is used for the decarboxylation of α-keto acids. Pyridoxal phosphate is used for the decarboxylation of amino acids.

      i. biotin and vitamin KH_2

      j. Biotin carboxylates a carbon adjacent to a carbonyl group (that is, an α-carbon). Vitamin KH_2 carboxylates the γ-carbon of a glutamate.

26. **a.** NAD$^+$, NADP$^+$, FAD, FMN, thiamine pyrophosphate, carboxybiotin, pyridoxal phosphate

     b. FAD, FMN, FADH$_2$, FMNH$_2$, thiamine pyrophosphate, biotin, pyridoxal phosphate, coenzyme B$_{12}$, some tetrahydrofolate coenzymes (the ones that do not deliver a one-carbon fragment in one step), and vitamin KH$_2$

     c. thiamine pyrophosphate

     d. vitamin KH$_2$

27. **a.** acetyl-CoA carboxylase; biotin

     b. dihydrolipoyl dehydrogenase; FAD

     c. methylmalonyl-CoA mutase; coenzyme B$_{12}$

     d. lactate dehydrogenase; NADH (This enzyme is named according to the reaction it catalyzes in the reverse direction.)

     e. aspartate transaminase; pyridoxal phosphate

     f. propionyl-CoA carboxylase; biotin

28.

29.   **a.**   thiamine pyrophosphate, lipoate, coenzyme A, FAD, NAD$^+$, indicated by numbers 1–5 in the mechanism below

b.

30.   The products of the reaction are shown below.

The following scheme explains the products:

CH₃
HO—C—T
T—C—T
OH

AdCH₂
Co(III)

→

AdCH₂
Co(II)

CH₃
HO—C—T
T—C—T
OH

→

AdCH₂T
Co(II)

CH₃
HO—C—T
·C—T
OH

↓

AdCHT ← AdCHT   H—C—T ← AdCHT   ·C—T
Co(III)   Co(II) HO—C—T   Co(II) HO—C—T
                  OH   OH
                  CH₃   CH₃
                  H   H

‖ dehydration

CH₃
H—C—T
C
O   T

+   H₂O

If there is only a limited amount of coenzyme, the second H of the coenzyme will be replaced by T in a subsequent round of catalysis, which means that the final product will be:

O
‖
C
Ad—CT₂   +   CH₃CHT   T   +   H₂O
|
Co(III)

31.

O   SCoA
‖   /
C
⁻OOC—CH
H—CH₂

AdCH₂
Co(III)

→

O   SCoA
‖   /
C
⁻OOC—CH
H—CH₂

AdCH₂
Co(II)

→

O   SCoA
‖   /
C
⁻OOC—CH
·CH₂

AdCH₃
Co(II)

↓

AdCH₂ ← AdCH₂   O   SCoA ← AdCH₂   O   SCoA
Co(III) Co(II) ‖ /   Co(II)   ‖ /
                 C   C
                 CH₂   ·CH
⁻OOC—CH₂   ⁻OOC—CH₂   H

32. **a.**

derived from Val   derived from Leu   derived from Ile

b.

c. thiamine pyrophosphate, lipoate, coenzyme A, FAD, NAD⁺

d. The disease can be treated by a diet low in branched-chain amino acids.

33. **a.** —CH₃   —CH₂—   —C—H

methyl   methylene   formyl

b.

34.

ribose-5′-phosphate   ribose-5′-phosphate   ribose-5′-phosphate

35. For clarity, the substituents are not put on the three bonds attached to the pyridinium ring in Problems 35–38. Each reaction in Problems 35–37 starts with the amino acid having undergone transimination with the lysine-bound coenzyme.

36.

transimination

+   Y—CH₂—CH—C

37.

38.

39.   Nonenzyme-bound FAD has a greater oxidation potential than NAD^+. Consequently, FAD oxidizes $NADH$ to NAD^+. When the enzyme binds FAD, it changes its oxidation potential, causing it to become less than that of NAD^+. Consequently, NAD^+ can oxidize enzyme-bound $FADH_2$ to FAD. In addition, there will be much more NAD^+ relative to enzyme-bound $FADH_2$ in the mitochondria (where the oxidation of $FADH_2$ by NAD^+ takes place). Therefore, the equilibrium favors the oxidation of $FADH_2$ by NAD^+ (Le Châtelier's principle).

$$E\text{-}FADH_2 + \underset{\text{excess}}{NAD^+} \rightleftharpoons E\text{-}FAD + NADH + H^+$$

40.

Chapter 25 Practice Test

1.   What two coenzymes put carboxyl groups on their substrates?

2.   What are the three one-carbon groups that tetrahydrofolate coenzymes put on their substrates?

3.   Show the mechanism for NADH reducing its substrate.

4.   Draw the structure of the compound obtained when the following amino acid undergoes transamination:

$$CH_3CH_2\overset{\overset{\displaystyle CH_3}{|}}{\underset{\underset{\displaystyle {}^+NH_3}{|}}{CH}}CH\overset{\overset{\displaystyle O}{\|}}{C}O^-$$

5.   What is the first step in the reaction of the substrate with coenzyme B_{12} in an enzyme-catalyzed reaction that requires coenzyme B_{12}?

6.   Give the coenzymes that would be required for each of the following enzyme-catalyzed reactions:

$$CH_3CH_2\overset{\overset{\displaystyle O}{\|}}{C}SCoA \xrightarrow{enzyme} CH_3\underset{\underset{\displaystyle COO^-}{|}}{CH}\overset{\overset{\displaystyle O}{\|}}{C}SCoA \xrightarrow{enzyme} \underset{\underset{\displaystyle COO^-}{|}}{CH_2}CH_2\overset{\overset{\displaystyle O}{\|}}{C}SCoA$$

7.   Show the enzyme-catalyzed reaction that requires vitamin KH_2 as a coenzyme.

8.   Give the product of the enzyme-catalyzed reaction that requires biotin and whose substrate is acetyl-CoA.

9.   Other than the substrate, enzyme, and coenzyme, what three additional reagents are needed by a reaction that requires biotin as a coenzyme?

10.   What is the function of FAD in the pyruvate dehydrogenase system?

11.   Give the structures of the two products obtained from the following transamination reaction:

$$HO-\text{⟨benzene ring⟩}-CH_2\underset{\underset{\displaystyle {}^+NH_3}{|}}{CH}\overset{\overset{\displaystyle O}{\|}}{C}O^- + {}^-O\overset{\overset{\displaystyle O}{\|}}{C}CH_2CH_2\underset{\underset{\displaystyle O}{\|}}{\overset{\overset{\displaystyle O}{\|}}{C}}O^- \xrightarrow{transamination}$$

12.   Indicate whether each of the following statements is true or false:

    a. Vitamin B_1 is the only water-insoluble vitamin that has a coenzyme function.   T   F

    b. $FADH_2$ is a reducing agent.   T   F

    c. Thiamine pyrophospate is vitamin B_6.   T   F

    d. Cofactors that are organic molecules are called coenzymes.   T   F

    e. Vitamin K is a water-soluble vitamin.   T   F

    f. Lipoic acid is covalently bound to its enzyme by an amide linkage.   T   F

CHAPTER 26
The Organic Chemistry of the Metabolic Pathways

Important Terms

acyl adenylate

$$\underset{R}{\overset{\displaystyle O}{\underset{\displaystyle ||}{C}}}\; O \;\underset{\displaystyle O^-}{\overset{\displaystyle O}{\underset{\displaystyle ||}{P}}}\; O \text{—adenosine}$$

acyl phosphate

$$\underset{R}{\overset{\displaystyle O}{\underset{\displaystyle ||}{C}}}\; O \;\underset{\displaystyle O^-}{\overset{\displaystyle O}{\underset{\displaystyle ||}{P}}}\; O^-$$

acyl pyrophosphate

$$\underset{R}{\overset{\displaystyle O}{\underset{\displaystyle ||}{C}}}\; O \;\underset{\displaystyle O^-}{\overset{\displaystyle O}{\underset{\displaystyle ||}{P}}}\; O \;\underset{\displaystyle O^-}{\overset{\displaystyle O}{\underset{\displaystyle ||}{P}}}\; O^-$$

anabolism — the reactions living organisms carry out that result in the synthesis of complex biomolecules from simple precursor molecules.

catabolism — the reactions living organisms carry out to provide energy and simple precursor molecules for synthesis.

citric acid cycle — a series of reactions that converts the acetyl group of acetyl-CoA into two molecules of CO_2 and a molecule of CoASH.

glycolysis — the series of reactions that converts glucose into two molecules of pyruvate.

high-energy bond — a bond that releases a great deal of energy when it is broken.

metabolism — reactions living organisms carry out in order to obtain the energy they need and to synthesize the compounds they require.

β-oxidation — a repeating series of four reactions that converts a fatty acyl-CoA molecule into molecules of acetyl-CoA.

oxidative phosphorylation — the fourth stage of catabolism in which NADH and $FADH_2$ are oxidized back to NAD^+ and FAD: for each NADH that is oxidized, three ATPs are formed; for each $FADH_2$ that is oxidized, two ATPs are formed.

phosphoanhydride bond — the bond that holds two phosphoric acid molecules together.

phosphoryl transfer reaction — the transfer of a phosphate group from one compound to another.

Solutions to Problems

1. Phosphoric acid has three pK_a's: the first OH group to lose a proton has a pK_a of 1.9, the second OH group has a pK_a of 6.7, and the third OH group a pK_a of 12.4.

$pK_a = 1.9$ $pK_a = 12.4$

$pK_a = 6.7$

phosphoric acid

bond linking the β-phosphorus
to the γ-phosphorus

ATP

bond linking the β-phosphorus
to the α-phosphorus

When the bond linking the β-phosphorus to the α-phosphorus breaks, the phosphorus leaving group attains its second negative charge which means that the pK_a of the conjugate acid of the leaving group is ~6.7. When the bond linking the β-phosphorus to the γ-phosphorus breaks, the phosphorus leaving group attains its third negative charge, which means that the pK_a of the conjugate acid of the leaving group is ~12.4. Consequently, the bond to the γ-phosphorus never breaks because the leaving group is a much stronger base than the leaving group that leaves when the bond to the α-phosphorus breaks.

2. **a.** Solved in the text.

b. Because pH 7.3 is much more basic than the pK_a values of the first two ionizations of ADP, these two groups will be in their basic forms at that pH, giving ADP two negative charges. We can determine the fraction of the group with a pK_a of 6.8 that will be in its basic form at pH 7.3 using the method shown in part **a**.

$$\text{fraction of group present in basic form} = \frac{K_a}{K_a + [\text{H}^+]} = \frac{1.6 \times 10^{-7}}{1.6 \times 10^{-7} + 5.0 \times 10^{-8}}$$

$$= \frac{1.6 \times 10^{-7}}{1.6 \times 10^{-7} + 0.5 \times 10^{-7}} = 0.8$$

total negative charge on ADP $= 2.0 + 0.8 = 2.8$

c. At pH 7.3 the OH group with a pK_a of 1.9 will account for one negative charge and the OH group with a pK_a of 12.4 will have no negative charge. We need to calculate the fraction of the group with a pK_a value of 6.7 that will be negatively charged at pH 7.3.

$$\text{fraction of group present in basic form} = \frac{K_a}{K_a + [\text{H}^+]} = \frac{2.0 \times 10^{-7}}{2.0 \times 10^{-7} + 0.5 \times 10^{-7}} = 0.8$$

total negative charge on phosphate $= 1.0 + 0.8 = 1.8$

3.

glycerol   glycerol-3-phosphate   ADP

4.

(R)-glycerol-3-phosphate

5. The resonance contributor on the right shows that the β-carbon of the α,β-unsaturated carbonyl compound has a partial positive charge. The nucleophilic OH group, therefore, is attracted to the β-carbon.

6. Because palmitic acid has 16 carbons and the acyl group of acetyl-CoA has 2 carbons, 8 molecules of acetyl-CoA are formed from 1 molecule of palmitic acid.

7. seven: one mole of NADH will be obtained from each of the following rounds of β-oxidation.

8.

fructose-6-phosphate   ATP   fructose-1,6-bisphosphate   ADP

9.   **a.** conversion of glucose to glucose-6-phosphate
         conversion of fructose-6-phosphate to fructose-1,6-bisphosphate

     b. conversion of 1,3-bisphosphoglycerate to 3-phosphoglycerate
         conversion of phosphoenolpyruvate to pyruvate

10. Two; each mole of D-glucose is converted to two moles of glyceraldehyde-3-phosphate, and each mole of glyceraldehyde-3-phosphate requires one mole of NAD^+ for it to be converted to one mole of pyruvate.

11. an aldehyde

12. acetaldehyde reductase

13. a ketone

14. thiamine pyrophosphate

15.

16. pyridoxal phosphate

17.

18. a secondary alcohol

19. citrate and isocitrate (and the alkene intermediate generated during the conversion of citrate to isocitrate)

20.

21. succinyl-CoA synthetase: this enzyme catalyzes a reaction of succinyl-CoA; the reverse reaction would be the synthesis of succinyl-CoA.

22. **a.** catabolic   **b.** catabolic

23.

D-galactose + ATP → D-galactose-1-phosphate + ADP

24. The hydrogen on the α-carbon.

pyruvate → lactate

25. The conversion of citrate to isocitrate.
The conversion of fumarate to (*S*)-malate.

26. The label will be on the phosphate group that is attached to the enzyme (phosphoglycerate mutase) that catalyzes the isomerization of 3-phosphoglycerate to 2-phosphoglycerate.

27. If you examine the mechanism for the isomerism of glucose-6-phosphate to fructose-6-phosphate on page 1049 of the text, you can see that C-1 in D-glucose is also C-1 in D-fructose, when the sugars are drawn in their straight chain form. Now, if you examine the mechanism for the aldolase-catalyzed cleavage of fructose-1,6-bisphosphate to form glyceraldehyde-3-phosphate on page 1051 of the text, you can see which carbons in D-glucose correspond to the carbons in dihydroxyacetone phosphate and D-glyceraldehyde-3-phosphate. Then, if you look at page 1104, you will see how the carbons in dihydroxyacetone phosphate correspond to the carbons in D-glyceraldehyde-3-phosphate.

D-glucose-6-phosphate → D-fructose-1,6-bisphosphate → dihydroxyacetone phosphate + D-glyceraldehyde-3-phosphate → D-glyceraldehyde-3-phosphate

carbons	pyruvate
3 and 4	COO^-
2 and 5	$C{=}O$
1 and 6	CH_3

Finally we see how the carbons in D-glyceraldehyde-3-phosphate correspond to the carbons in pyruvate. Therefore, C-3 and C-4 of glucose each becomes a carboxyl group in pyruvate.

28. Pyruvate loses its carboxyl group when it is converted to ethanol. Since the carboxyl group is C-3 or C-4 of glucose, half of the ethanol molecules contain C-1 and C-2 of glucose and the other half contain C-5 and C-6 of glucose.

29. The β-oxidation of a molecule of a 16-carbon fatty acyl-CoA will form 8 molecules of acetyl-CoA.

30. Each molecule of acetyl-CoA forms 2 molecules of CO_2. Therefore, the 8 molecules of acetyl-CoA obtained from a molecule of a 16-carbon fatty acyl-CoA will form 16 molecules of CO_2.

31. No ATP is formed from β-oxidation.

32. Each molecule of acetyl-CoA that is cleaved from the 16-carbon fatty acyl-CoA forms 1 molecule of $FADH_2$ and 1 molecule of NADH. Since a 16-carbon fatty acyl-CoA undergoes 7 cleavages, 7 molecules of $FADH_2$ and 7 molecules of NADH are formed from the 16-carbon fatty acyl-CoA.

33. Since each NADH forms 3 molecules of ATP and each $FADH_2$ forms 2 molecules of ATP in oxidative phosphorylation, the 7 molecules of NADH form 21 molecules of ATP and the 7 molecules of $FADH_2$ form 14 molecules of ATP. Therefore, 35 molecules of ATP are formed.

34. We have seen that each molecule of acetyl-CoA that enters the citric acid cycle forms 12 molecules of ATP (Section 26.11). A molecule of a 16-carbon fatty acid will form 8 molecules of acetyl-CoA. These will form 96 molecules of ATP. When these are added to the number of ATP molecules formed from the NADH and $FADH_2$ generated in β-oxidation, we see that 131 molecules of ATP are formed from complete metabolism of a 16-carbon saturated fatty acyl-CoA.

35. Each molecule of glucose, while being converted to 2 molecules of pyruvate, forms 2 molecules of ATP and 2 molecules of NADH.
The 2 molecules of pyruvate form 2 molecules of NADH while being converted to 2 molecules of acetyl-CoA.
Each molecule of acetyl-CoA that enters the citric acid cycle forms 3 molecules of NADH, 1 molecule of $FADH_2$, and 1 molecule of ATP. Therefore the 2 molecules of acetyl-CoA obtained from glucose form 6 molecules of NADH, 2 molecules of $FADH_2$, and 2 molecules of ATP.
Therefore, each molecule of glucose forms 4 molecules of ATP, 10 molecules of NADH (2 + 2 + 6), and 2 molecules of $FADH_2$.
Since each NADH forms 3 molecules of ATP and each $FADH_2$ forms 2 molecules of ATP, 1 molecule of glucose forms $4 + (10 \times 3) + (2 \times 2)$ molecules of ATP. That is, each molecule of glucose forms 38 molecules of ATP.

36. Attack of water on fumarate forms (S)-malate. (S)-Malate is formed when water attacks the Re face. The Re face is the face with priorities that decrease (1 > 2 > 3) in a clockwise direction (Section 18.14).

37. The conversion of propionyl-CoA to methylmalonyl-CoA requires biotin.
The conversion of methylmalonyl-CoA to succinyl-CoA requires coenzyme B_{12}.

38.   In Problem 27 we saw how the carbons in D-glyceraldehyde-3-phosphate correspond to the carbons in pyruvate.

$$\overset{3}{HC}=O \qquad \overset{4}{HC}=O \qquad 3 \text{ and } 4 \quad COO^-$$

$$H-\overset{2}{\underset{|}{C}}-OH \qquad H-\overset{5}{\underset{|}{C}}-OH \qquad 2 \text{ and } 5 \quad C=O$$

$$1\, CH_2OPO_3^{2-} \qquad 6\, CH_2OPO_3^{2-} \qquad 1 \text{ and } 6 \quad CH_3$$

D-glyceraldehyde-3-phosphate   pyruvate

Now we can answer the questions. The label in pyruvate is indicated by *.

a.
$$COO^-$$
$$C=O$$
$$*CH_3$$

c.
$$*COO^-$$
$$C=O$$
$$CH_3$$

e.
$$COO^-$$
$$*C=O$$
$$CH_3$$

b.
$$COO^-$$
$$*C=O$$
$$CH_3$$

d.
$$*COO^-$$
$$C=O$$
$$CH_3$$

f.
$$COO^-$$
$$C=O$$
$$*CH_3$$

39.   NAD$^+$ oxidizes the alcohol group to a ketone. This is followed by loss of CO_2. The metal ion complexes with two oxygen atoms of the reactant, making the carbonyl group more susceptible to receiving the electrons left behind when CO_2 is removed.

isocitrate

tautomerization

α-ketoglutarate

40.   A Claisen condensation between two molecules of acetyl-CoA forms acetoacetyl-CoA that, when hydrolyzed, forms acetoacetate.

Acetoacetate can undergo decarboxylation to form acetone, or it can be reduced to 3-hydroxybutyrate.

41.

42.   **a.**   UDP-Galactose and UDP-glucose are C-4 epimers. NAD$^+$ oxidizes the C-4 OH group of UDP-galactose to a ketone. When NADH reduces the ketone back to an OH, it attacks the sp^2 carbon from above the plane, forming the C-4 epimer of the starting material.

UDP-galactose

UDP-gulucose

b.   The enzyme is called an epimerase because it converts a compound into an epimer (in this case, a C-4 epimer).

43.   Because the compound that would react in the second step with the activated carboxylic acid group is excluded from the incubation mixture, the reaction between the carboxylate ion and ATP will come to equilibrium.

If radioactively labeled pyrophosphate is put into the incubation mixture, ATP will become radioactive if the mechanism involves attack on the α-phosphorus, because pyruvate is a reactant in the reverse reaction that forms ATP.

ATP will not become radioactive if the mechanism involves attack on the β-phosphorus, because pyruvate is not a reactant in the reverse reaction that forms ATP. (In other words, because pyrophosphate is not a product of the reaction, it cannot become incorporated into ATP in the reverse reaction.)

attack on the α-phosphorus

pyrophosphate

attack on the β-phosphorus

AMP

44.   If radioactive AMP is added to the reaction mixture, the results will be opposite. If the mechanism involves attack on the α-phosphorus, ATP will not become radioactive, because AMP is not a reactant in the reverse reaction that forms ATP. If the mechanism involves attack on the β-phosphorus, ATP will become radioactive, because AMP is a reactant in the reverse reaction that forms ATP.

Chapter 26 Practice Test

1. Give a structure for each of the following:

    a. the intermediate formed when a nucleophile (RO⁻) attacks the β-phosphorus of ATP.

    b. an acyl pyrophosphate.

    c. pyrophosphate.

2. Fill in the six blanks in the following scheme:

3. Which of the following are not citric acid cycle intermediates: fumarate, acetate, citrate?

4. Which provide energy to the cell, anabolic reactions or catabolic reactions?

5. What compounds are formed when proteins undergo the first stage of catabolism?

6. What compound is formed when fatty acids undergo the second stage of catabolism?

7. Indicate whether each of the following statements is true or false:

    a. Each molecule of $FADH_2$ forms 3 molecules of ATP in the fourth stage of catabolism.   T   F

    b. $FADH_2$ is oxidized to FAD.   T   F

    c. NAD^+ is oxidized to NADH.   T   F

    d. Acetyl-CoA is a citric acid cycle intermediate.   T   F

CHAPTER 27
The Organic Chemistry of Lipids

Important Terms

angular methyl group	a methyl substituent at the 10- or 13-position of a steroid ring system.
carotenoid	a class of compound (a tetraterpene) responsible for the red and orange colors of fruits, vegetables, and fall leaves.
cholesterol	a steroid that is the precursor of all other steroids.
dimethylallyl pyrophosphate	needed for the biosynthesis of terpenes and biosynthesized from isopentenyl pyrophosphate.
diterpene	a terpene that contains 20 carbons.
fat	a triester of glycerol that exists as a solid at room temperature.
fatty acid	a carboxylic acid with a long hydrocarbon side chain.
hormone	a chemical messenger synthesized in a gland and delivered by the bloodstream to its target tissue.
isopentenyl pyrophosphate	the starting material for the biosynthesis of terpenes.
lipid	a water-insoluble compound found in a living system.
lipid bilayer	two layers of phosphoacylglycerols arranged so that their polar heads are on the outside and their nonpolar fatty acid chains are on the inside.
membrane	the material that surrounds the cell in order to isolate its contents.
mixed triacylglycerol	a triacylglycerol in which the fatty acid components are different.
monoterpene	a terpene that contains 10 carbons.
oil	a triester of glycerol that exists as a liquid at room temperature.
phosphatidic acid	a phosphoacylglycerol in which only one of the OH groups of phosphate is in an ester linkage.
phosphoacylglycerol (phosphoglyceride)	formed when two OH groups of glycerol form esters with fatty acids and the terminal OH group is part of a phosphodiester.
phospholipid	a lipid that contains a phosphate group.
polyunsaturated fatty acid	a fatty acid with more than one double bond.

prostaglandin

a carboxylic acid, derived from arachidonic acid, that is responsible for a variety of physiological functions.

protein prenylation

the process of attaching farnesyl and geranylgeranyl groups to proteins to allow them to become anchored to membranes.

sesquiterpene

a terpene that contains 15 carbons.

simple triacylglycerol

a triacylglycerol in which the fatty acid components are the same.

sphingolipid

a lipid that contains sphingosine.

squalene

a triterpene that is a precursor of steroid molecules.

steroid

a class of compounds that contains a steroid ring system.

steroid ring system

α-substituent

a substituent on the opposite side of a steroid ring system from the angular methyl groups (the bottom face as typically drawn).

β-substituent

a substituent on the same side of a steroid ring system as the angular methyl groups (the top face as typically drawn).

terpene

a lipid isolated from a plant that contains carbon atoms in multiples of five.

terpenoid

a terpene that contains oxygen.

tetraterpene

a terpene that contains 40 carbons.

trans fused

two rings fused together such that if one ring is considered to be two substituents of the other ring, the substituents would be on opposite sides of the first ring.

triacylglycerol

the compound formed when the three OH groups of glycerol are esterified with fatty acids.

triterpene

a terpene that contains 30 carbons.

wax

an ester formed from a long straight-chain carboxylic acid and a long straight-chain alcohol.

Solutions to Problems

1. **a.** Stearic acid has the higher melting point, because it has two more methylene groups (giving it a greater surface area and therefore stronger intermolecular van der Waals interactions) than palmitic acid.

    b. Palmitic acid has the higher melting point, because it does not have any carbon-carbon double bonds, whereas palmitoleic acid has a cis double bond that prevents the molecules from packing closely together.

    c. Oleic acid has the higher melting point, because it has one cis double bond, whereas linoleic acid has two cis double bonds, which give greater interference to close packing of the molecules.

2. $CH_3CH_2CH_2CH_2CH_2CH=CHCH_2CH=CHCH_2CH=CHCH_2CH=CHCH_2CH_2CH_2COOH$

    1. excess O_3
    2. H_2O_2

3. Glyceryl tripalmitate has a higher melting point, because the carboxylic acid components are saturated and can therefore pack more closely together than the cis-unsaturated carboxylic acid components of glyceryl tripalmitoleate.

4. To be optically inactive, the fat must have a plane of symmetry. In other words, the fatty acids at C-1 and C-3 must be identical. Therefore, stearic acid must be at C-1 and C-3.

5. To be optically active, the fat must not have a plane of symmetry. Therefore, the two stearic acid groups must be attached to adjacent alcohol groups.

6.

7.   Solved in the text.

8.   No, the identities of R^1 and R^2 have no effect on the configuration of the asymmetric center at C-2 because:
the group with priority #1 will always be the carboxyl group on C-2;
the group with priority #2 will always be the group that contains the phosphorus;
the group with priority #3 will always be the group that contains the other carboxyl group;
the group with priority #4 will always be the H on C-2.

9.   Because the interior of a membrane is nonpolar and the surface of a membrane is polar, integral proteins will have a higher percentage of nonpolar amino acids.

10.   The bacteria could synthesize phosphoacylglycerols with more saturated fatty acids. These triacylglycerols would pack more tightly in the lipid bilayer and, therefore, would have higher melting points and be less fluid.

11.   a.   The sphingomyelins can differ in the fatty acid component of the amide and have either choline or ethanolamine attached to the phosphate group.

b.

12. Membranes must be kept in a semifluid state in order to allow transport across them. Cells closer to the hoof of an animal are going to be in a colder average environment than cells closer to the body. Therefore, the cells closer to the hoof have a higher degree of unsaturation to give them a lower melting point so the membranes will not solidify at the colder temperature.

13.

14.

menthol   zingerbene   β-selinene

squalene

15. The fact that the tail-to-tail linkage occurs in the exact center of the molecule suggests that the two halves are synthesized (in a head-to-tail fashion) and then joined together in a tail-to-tail linkage.

tail-to-tail linkage

16. Squalene, lycopene, and β-carotene are all synthesized in the same way. In each case, two halves are synthesized (in a head-to-tail fashion) and then joined together in a tail-to-tail linkage.

lycopene

β-carotene

17.

first step of the Claisen condensation

second step of the Claisen condensation

aldol addition

H_2O
hydrolysis of the thioester

18.   One equivalent of NADPH is required to reduce the thioester to an aldehyde. The second equivalent of NADPH is needed to reduce the aldehyde to an alcohol.

19. Solved in the text.

20.

geranyl
pyrophosphate

α-terpineol

21.

***E* isomer**

rotate about
single bond

***Z* isomer**

22. It tells you that the reaction is an S_N1 reaction, because the fluoro-substituted carbocation is less stable than the non-fluoro-substituted carbocation (due to the strongly electron-withdrawing fluoro substituent), so it would form more slowly.

less stable   more stable

If the reaction had been an S_N2 reaction, the fluoro-substituted compound would have reacted more rapidly than the non-fluoro-substituted carbocation, because the electron-withdrawing fluoro substituent would make the compound more susceptible to nucleophilic attack.

23.

farnesyl pyrophosphate

+ H$_3$O$^+$

24.   Solved in the text.

25.   Biotin is the coenzyme required to carboxylate an α-carbon.

26.   A β-**hydrogen** at C-5 means that the A and B rings are **cis** fused; an α-**hydrogen** at C-5 means that they are **trans** fused.

27.   Because the OH substituent is on the **same side** of the steroid ring system as the angular methyl groups, it is a β-**substituent**.

28.   The hemiacetal is formed by reaction of the primary alcohol with the aldehyde.

29.

30. Notice that since the methyl and hydrogen at the juncture of the A and B rings are on the same side, we know that the A and B rings are cis fused. The three OH groups are all in axial positions.

31. There are two hydride shifts and two methyl shifts. The last step is elimination of a proton.

32.   If stearic acid were at C-1 and C-3, the fat would not be optically active. Therefore, stearic acid must be attached to adjacent OH groups.

$$
\begin{array}{l}
CH_2-O-\overset{\overset{\displaystyle O}{\|}}{C}-(CH_2)_{16}CH_3 \\
CH-O-\overset{\overset{\displaystyle O}{\|}}{C}-(CH_2)_{16}CH_3 \\
CH_2-O-\overset{\overset{\displaystyle O}{\|}}{C}-(CH_2)_{14}CH_3
\end{array}
$$

33.   All triacylglycerols do not have the same number of asymmetric centers. If the carboxylic acid components at C-1 and C-3 of glycerol are not identical, the triacylglycerol has one asymmetric center (C-2). If the carboxylic acid components at C-1 and C-3 of glycerol are identical, the triacylglycerol has no asymmetric centers.

34.   **a.**   There are three triacylglycerols in which one of the fatty acid components is lauric acid and two are myristic acid. Myristic acid can be at C-1 and C-3 of glycerol, in which case the triacylglycerol does not have any asymmetric centers. If myristic acid is at C-1 and C-2 of glycerol, C-2 is an asymmetric center, and consequently, the compound can have a pair of enantiomers.

   b.   There are six triacylglycerols in which one of the fatty acid components is lauric acid, one is myristic acid, and one is palmitic acid. The three possible arrangements are shown below (with the fatty acid components abbreviated as L, M, and P). Since each has an asymmetric center, each can exist as a pair of enantiomers for a total of six triacylglycerols.

$$
\begin{array}{lll}
CH_2-O-L & CH_2-O-L & CH_2-O-M \\
{}^*CH-O-M & {}^*CH-O-P & {}^*CH-O-L \\
CH_2-O-P & CH_2-O-M & CH_2-O-P
\end{array}
$$

35.

$$
\begin{array}{l}
\overset{\overset{\displaystyle O}{\|}}{\underset{R^1}{C}}\diagdown O^- \quad \overset{\overset{\displaystyle O}{\|}}{\underset{R^3}{C}}\diagdown O^- \\
\\
\overset{\overset{\displaystyle O}{\|}}{\underset{R^2}{C}}\diagdown O^- \quad \overset{\overset{\displaystyle O}{\|}}{\underset{R^4}{C}}\diagdown O^-
\end{array}
\qquad
3\;
\begin{array}{l}
CH_2-OH \\
CH-OH \\
CH_2-OH
\end{array}
\qquad
2\;
\overset{\overset{\displaystyle O}{\|}}{\underset{\underset{\displaystyle O^-}{|}}{\underset{HO}{P}}}-O^-
$$

36.

The structure to the left has a molecular formula = $C_9H_{14}O_6$ and a molecular weight = 218.

Subtracting 218 from the total molecular weight gives the molecular weight contribution from the methylene (CH_2) groups in the triacylglycerol.

$$722 - 218 = 504$$

Dividing 504 by the molecular weight contribution from a methylene group (14) will give the number of methylene groups.

$$\frac{504}{14} = 36$$

Since there are 36 methylene groups, each fatty acid in the triacylglycerol has 12 methylene groups.

nutmeg

37.

a.

c.

b.

d.

e.

f.

38.   **a.**   Starting with mevalonyl pyrophosphate, you can trace the location of the label in the compounds that lead to the formation of geranyl pyrophosphate, and geranyl pyrophosphate is converted to citronellal.

b.   The label is lost from sample B when mevalonic pyrophosphate loses CO_2 to form isopentenyl pyrophosphate, so none of the carbons will be labeled in citronellal.

c. Because the methyl groups are equivalent in the carbocation that is formed as an intermediate when isopentenyl pyrophosphate is converted to dimethylallyl pyrophosphate, either of the methyl groups can be labeled in dimethylallyl pyrophosphate. This means that either of the two methyl groups can be labeled in geranyl pyrophosphate and in citronellal.

mevalonyl pyrophosphate

isopentenyl pyrophosphate

dimethylallyl pyrophosphate

geranyl pyrophosphate

citronellal

39.   There are 14 possible structures for compound **A**, and 4 possible structures for compound **B**.

cis and trans isomers
are possible

cis and trans isomers
are possible

A
one asymmetric
center, therefore
2 stereosiomers

A
two asymmetric
centers, therefore
4 stereosiomers

A
one asymmetric
center, therefore
2 stereosiomers

A
two asymmetric
centers, therefore
4 stereosiomers

A
one asymmetric
center, therefore
2 stereosiomers

40. Because acetyl-CoA is converted into malonyl-CoA (see Section 26.8), mevalonyl pyrophosphate will contain three labeled carbons, which means that the α-terpineol (juniper oil) will contain six labeled carbons.

41.   **a.**

b. It has 30 carbons, so it is a triterpene.

42. First draw the starting material (geranyl pyrophosphate) so that its structure is as similar as possible as that of the target molecule.

geranyl pyrophosphate

α-pinene

α-pinene

43.

44.   The OH groups will react only if they are in equatorial positions, because introduction of bulky axial substituents would decrease the stability of the molecule.

In the case of 5α-cholestane-$3\beta,7\beta$-diol, the two OH groups are on the same side of the ring system as the angular methyl group, which means that they are in equatorial positions. Both OH groups react with ethyl chloroformate.

In the case of 5α-cholestane-$3\beta,7\alpha$-diol, only one of the OH groups is on the same side of the ring system as the angular methyl group. The other is on the opposite side of the ring, which means that it is in an axial position. Only the OH group that is in the equatorial position reacts with ethyl chloroformate.

5α-cholestane-$3\beta,7\beta$-diol 5α-cholestane-$3\beta,7\alpha$-diol

45. The mechanism for the conversion of isopentenyl pyrophosphate to geranyl pyrophospate to farnesyl pyrophosphate is shown in the text on page 1138. Now we need to show how farnesyl pyrophosphate is converted to eudesmol.

farnesyl pyrophosphate

eudesmol   HB⁺

46.

47.

estradiol

DES

48.   a.   They are both sesquiterpenes.

b.

cedrol

patchouli alcohol

49.   A sesquiterpene is synthesized from farnesyl pyrophosphate. The carbons that would be labeled in farnesyl pyrophosphate, if it is synthesized from acetate with a ^{14}C-labeled carbonyl carbon, are indicated by an asterisk. (See Problem 24.)

farnesyl pyrophosphate

Chapter 27 Practice Test

1.   Mark off the isoprene units in squalene.

squalene

2.   Give the mechanism for the following reaction:

(PP$_i$ = pyrophosphate)

3.   Give structures for each of the following:

a.   a phospholipid that contains ethanolamine

b.   a wax

4.   Indicate whether each of the following statements is true or false:

		T	F
a.	Cholesterol is the precursor of all other steroids.	T	F
b.	The double bonds in unsaturated fats are conjugated.	T	F
c.	Fats have a higher percentage of saturated fatty acids than do oils.	T	F
d.	A saturated fatty acid has a lower melting point than an unsaturated fatty acid with the same number of carbons.	T	F
e.	Lipids are insoluble in water.	T	F
f.	A diterpene contains 20 carbons.	T	F
g.	Vitamin A is not a coenzyme.	T	F

5.   What is the first step in the reaction that causes fatty acids to become rancid?

6.   How many carbon atoms are there in a triterpene?

7.   What size is the ring in a prostaglandin?

8.   What is the starting material for the synthesis of prostaglandins?

CHAPTER 28
The Chemistry of the Nucleic Acids

Important Terms

anticodon	the three bases at the bottom of the middle loop in a tRNA.
base	a nitrogen-containing heterocyclic compound (a purine or a pyrimidine) found in DNA and RNA.
codon	a sequence of three bases in mRNA that specifies the amino acid to be incorporated into a protein.
deamination	a hydrolysis reaction that results in removal of ammonia.
deoxyribonucleic acid (DNA)	a polymer of deoxyribonucleotides.
deoxyribonucleotide	a nucleotide where the sugar component is D-2-deoxyribose.
dinucleotide	two nucleotides linked by a phosphodiester bond.
double helix	the term used to describe the secondary structure of DNA.
exon	a stretch of bases in DNA that are a portion of a gene.
gene	a segment of DNA.
gene therapy	a technique that inserts a synthetic gene into the DNA of an organism defective in that gene.
genetic code	the amino acid specified by each three-base sequence of mRNA.
genetic engineering	recombinant DNA technology, where DNA molecules are attached to DNA in a host cell and allowed to replicate.
human genome	the total DNA of a human cell.
intron	a stretch of bases in DNA that contain no genetic information.
major groove	the wider and deeper of the two alternating grooves in DNA.
minor groove	the narrower and more shallow of the two alternating grooves in DNA.
nucleic acid	a chain of five-membered ring sugars linked by phosphodiester groups with each sugar bearing a heterocyclic amine at the anomeric carbon in the β-position. The two kinds of nucleic acids are DNA and RNA.
nucleoside	a heterocyclic base (purine or pyrimidine) bonded to the anomeric carbon of a sugar (D-ribose or D-2-deoxyribose) in the β-position.
nucleotide	a nucleoside with one of its OH groups bonded to a phosphate group via an ester linkage.

oligonucleotide	three to ten nucleotides linked by phosphodiester bonds.
phosphodiester	two of the OH groups of phosphoric acid that have been converted to OR groups.
polymerase chain reaction (PCR)	a method that amplifies segments of DNA.
polynucleotide	many nucleotides linked by phosphodiester bonds.
primary structure	the sequence of bases in the nucleic acid.
pyrosequencing	an automated technique used to sequence DNA; it detects the identity of the base that adds to the DNA primer.
recombinant DNA	DNA that has been incorporated into a host cell.
replication	the synthesis of identical copies of DNA.
replication fork	the position on DNA where replication begins.
restriction endonuclease	an enzyme that cleaves DNA at a specific base sequence.
restriction fragment	a fragment that is formed when DNA is cleaved by a restriction endonuclease.
ribonucleic acid (RNA)	a polymer of ribonucleotides.
ribonucleotide	a nucleotide where the sugar component is D-ribose.
ribosomal RNA (rRNA)	the structural component of ribosomes, the particles on which protein synthesis takes place.
RNA splicing	the step in RNA processing that cuts out nonsense bases and splices informational pieces together.
semiconservative replication	the mode of replication that results in a daughter molecule of DNA having one of the original DNA strands plus a newly synthesized strand.
sense strand	the strand in DNA that is not read during transcription; it has the same sequence of bases as the synthesized mRNA strand, except that the mRNA has U's in place of the T's in DNA.
stacking interactions	van der Waals interactions between the mutually induced dipoles of adjacent pairs of bases in DNA.
stop codon	a codon that says "stop protein synthesis here."
template strand	the strand in DNA that is read during transcription.
transcription	the synthesis of mRNA from a DNA blueprint.
transfer RNA (tRNA)	a single-stranded RNA molecule that carries an amino acid to be incorporated into a protein.
translation	the synthesis of a protein from an mRNA blueprint.

Solutions to Problems

1.   The ring is protonated at its most basic position. In the case of a purine, this is the 7-position. In the next step, the bond between the heterocyclic base and the sugar breaks, with the anomeric carbocation being stabilized by the ring oxygen's nonbonding electrons.

The mechanism is exactly the same for pyrimidines except that the initial protonation takes place at the 3-position.

2.   **a.**   **b.**

dCDP   dTTP

c.

dUMP

e.

guanosine 5′-triphosphate
GTP

d.

UDP

f.

adenosine 3′-monophosphat
AMP

3.   The hydrolysis of cyclic AMP forms adenosine 5′-phosphate and adenosine 3′-phosphate.

$$\xrightarrow[\text{H}_2\text{O}]{\text{HCl}}$$

adenosine 5′-phosphate

+

adenosine 3′-phosphate

4.   A hydrogen bond acceptor is indicated by **A**; a hydrogen bond donor is indicated by **D**. The **A** and **D** designations show that the maximum number of hydrogen bonds that can form are two between thymine and adenine and three between cytosine and guanine. Notice that uracil and thymine have the same **A** and **D** designations.

uracil   **thymine**   **adenine**

cytosine   **guanine**

5.   If the bases existed in the enol form, no hydrogen bonds could form between the bases unless one of the bases were shifted vertically.

thymine   **adenine**

cytosine   **guanine**

6.   **a.**   3'—C—C—T—G—T—T—A—G—A—C—G—5'

b.   guanine

7.   Notice that in the first step of the mechanism shown on page 1155 of the text, when the nucleophile attacks the phosphorus of the diester, the π bond breaks. However, when a nucleophile attacks the phosphorus of the anhydride intermediate, an anhydride bond breaks in preference to the π bond.

8.

9. Thymine and uracil differ only in that thymine has a methyl substituent that uracil does not have (thymine is 5-methyluracil). Because thymine and uracil both have the same groups in the same positions that can participate in hydrogen bonding, they will both call for the incorporation of the same purine. Because thymine and uracil form one hydrogen bond with guanine and two with adenine, they will both incorporate adenine in order to maximize hydrogen bonding.

10. Because methionine is known to be the first base incorporated into the heptapeptide, the mRNA sequence is read beginning at AUG, since that is the only codon that codes for methionine.

Met-Asp-Pro-Val-Ile-Lys-His

11. Met-Asp-Pro-Leu-Leu-Asn

12. It does not cause protein synthesis to stop, because the sequence UAA does not occur within a triplet. The reading frame causes the triplets to be AUU and **AAA**. In other words, the U is at the end of a triplet and the next triplet starts with AA.

13. The sequence of bases in the template strand of DNA specify the sequence of bases in mRNA, so the bases in the template strand and the bases in mRNA are complementary. Therefore, the sequence of bases in the sense strand of DNA are identical to the sequence of bases in mRNA, except wherever there is a U in mRNA, there is a T in the sense strand of DNA.

5′——G-C-A-T-G-G-A-C-C-C-C-G-T-T-A-T-T-A-A-A-C-A-C——3′

14.

	Met	Asp	Pro	Val	Ile	Lys	His
codons	AUG	GAU	CCU	GUU	AUU	AAA	CAU
		GAC	CCC	GUC	AUC	AAG	CAC
			CCA	GUA	AUA		
			CCG	GUG			

anticodons

Note that the anticodons are stated in the 5′ → 3′ direction. For example, the anticodon of AUG is stated as CAU (not UAC).

codons   5′ A U G 3′
| | |
anticodons   3′ U A C 5′

CAU	AUC	AGG	AAC	AAU	UUU	AUG
	GUC	GGG	GAC	GAU	CUU	GUG
		UGG	UAC	UAU		
		CGG	CAC			

15.

16. Deamination involves hydrolyzing an imine linkage to a carbonyl group and ammonia, so cytosine can be deaminated to uracil.

Thymine does not have an amino substituent on the ring, which means that it cannot form an imine and therefore it cannot be deaminated.

17. **a** is the only sequence that has a chance of being recognized by a restriction endonuclease, because it is the only one that has the same sequence of bases in the $5' \rightarrow 3'$ direction that the complementary strand has in the $5' \rightarrow 3'$ direction.

$$^{5'}\text{A-C-G-C-G-T}^{3'}$$
$$^{3'}\text{T-G-C-G-C-A}^{5'}$$

18.

19. **a.** guanosine 3′-monophosphate
      b. cytidine 5′-diphosphate

c. 2′-deoxyadenosine 5′-monophosphate
d. 2′-deoxythymidine

20. Lys-Val-Gly-Tyr-Pro-Gly-Met-Val-Val

21. 5′—GAC—CAC—CAT—TCC—GGG—GTA—GCC—AAC—TTT—3′

22. 5′—AAA—GTT—GGC—TAC—CCC—GGA—ATG—GTG—GTC—3′

23.   **a.** Ile   **b.** Asp   **c.** Val   **d.** Val

24.   The third base in each codon has some variability.

mRNA 5′-GG(UCA or G)UC(UCA or G)CG(UCA or G)GU(UCA or G)CA(U or C)GA(A or G)-3′

or AG(U or C)   AG(A or G)

DNA
template   3′-CC(AGT or C)AG(AGT or C)GC(AGT or C)CA(AGT or C)GT(A or G)CT(T or C)-5′

or TC(A or G)   TC(T or C)

sense   5′-GG(TCA or G)TC(TCA or G)CG(TCA or G)GT(TCA or G)CA(T or C)GA(A or G)-3′
or AG(T or C)   AG(A or G)

Notice that Ser and Arg are two of three amino acids that can be specified by six different codons.

25.

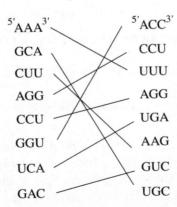

26.

27.

mRNA	CC(UCA or G)GC(UCA or G)CU(UCA or G)CG(UCA or G)
	or UU(A or G) AG(A or G)
DNA (sense strand)	CC(TCA or G)GC(TCA or G)CT(TCA or G)CG(TCA or G)
	or TT(A or G) AG(A or G)

Note that because mRNA is complementary to the template strand of DNA, which is complementary to the sense strand, mRNA and the sense strand of DNA have the same sequence of bases (expect DNA has a T where RNA has a U). Also note that Leu and Arg are each specified by six codons.

28.   **a.**   CC and GG   **c.**   CA and TG

CA and TG are formed in equal amounts, since A pairs with T and C pairs with G. (Remember that the dinucleotides are written in the $5' \rightarrow 3'$ direction. For example, part **f** is not a correct answer because it has A pairing with A and T pairing with T.)

$$5' \ \ C \ A \ \ 3'$$
$$ | \ \ \ | $$
$$3' \ \ G \ T \ \ 5'$$

29.   AZT is incorporated into DNA when the 3'-OH group of the last nucleotide incorporated into the growing chain of DNA attacks the α-phosphorus of AZT-triphosphate instead of a normal nucleotide. When AZT is thus incorporated into DNA, DNA synthesis stops because AZT does not have a 3'-OH group that can react with another nucleotide triphosphate.

30.   The number of different possible codons using four nucleotides is $(4)^n$, where n is the number of letters (nucleotides) in the code.

for a two-letter code: $(4)^2 = 16$
for a three-letter code: $(4)^3 = 64$
for a four-letter code: $(4)^4 = 256$

Since there are 20 amino acids that must be specified, a two-letter code would not provide an adequate number of codons.
A three-letter code provides enough codes for all the amino acids and also provides the necessary stop codons.
A four-letter code provides many more codes than would be needed.

31.   In the first step of the reaction, the imidazole ring of one histidine acts as a general-base catalyst, removing a proton from the 2-OH group to make it a better nucleophile. The imidazole ring of the other histidine acts as a general-acid catalyst, protonating the leaving group to make it a weaker base and therefore a better leaving group. In the second step of the reaction the roles of the two imidazole rings are reversed.

32. The normal and mutant peptides would have the following base sequence in their mRNA:

normal: CA(AG) UA(UC) GG(UCAG) AC(UCAG) CG(UCAG) UA(UC) GU(UCAG)

mutant: CA(AG) UC(UCAG) GA(AG) CC(UCGA) GG(UCGA) AC(UCAG)

a. The middle nucleotide (A) in the second triplet was deleted. This means that an A was deleted in the sense strand of DNA or a T was deleted in the template strand of DNA.

b. The mRNA for the mutant peptide has an unused 3′-terminal two-letter code, U(UCGA). The last amino acid in the octapeptide of the normal fragment is leucine, so its last triplet is UU(AG) or CU(UCAG).

This means that the triplet for the last amino acid in the mutant is U(UCGA)(UC) and that the last amino acid in the mutant is one of the following: Phe, Ser, Tyr, or Cys.

33. If deamination does not occur, the mRNA sequence will be:

AUG-UCG-CUA-AUC, which will code for the following tetrapeptide:
Met—Ser—Leu—Ile

Deamination of a cytosine results in a uracil.
If the cytosines are deaminated, the mRNA sequence will be:

AUG-UUG-UUA-AUU, which will code for the following tetrapeptide:
Met—Leu—Leu—Ile

The only cytosine that will change the particular amino acid that is incorporated into the peptide is the first one. Therefore, this is the cytosine that could cause the most damage to an organism if it were deaminated.

34.   5-Bromouracil is incorporated into DNA in place of thymine because of their similar size. Thymine pairs with adenine via two hydrogen bonds. 5-Bromouracil exists primarily in the enol form. The enol cannot form any hydrogen bonds with adenine, but it can form two hydrogen bonds with guanine. Therefore, 5-bromouracil pairs with guanine. Because 5-bromouracil causes guanine to be incorporated instead of adenine into newly synthesized DNA strands, it causes mutations.

thymine   **adenine**   **5-bromouracil** H **guanine**

35.   In an acidic environment, nitrite ion is protonated to nitrous acid. We have seen that the nitrosonium ion is formed from nitrous acid (Section 16.12).

The nitrosonium ion reacts with a primary amino group to form a diazonium ion, which can be displaced by water (Section 16.10).

cytosine   uracil

36. Ca^{2+} decreases the pK_a of water, forming calcium-bound hydroxide ion, which is a better nucleophile than water. The positively-charged nitrogen (of the guanidinium group) of arginine stabilizes the negatively charged oxygen formed when the P=O π bond breaks, and therefore makes it easier to form. Glutamic acid protonates the OR oxygen, thereby making it a weaker base and therefore a better leaving group when the P=O π bond reforms.

$$\text{Ca}^{+}\overset{\delta-}{---}\text{OH}$$

$$\underset{\text{Glu}-\overset{\overset{\displaystyle O}{\|}}{C}-\text{OH}}{}$$

37. The ribosome, the particle on which protein synthesis occurs, has a binding site for the growing peptide chain and a binding site for the next amino acid to be incorporated into the chain.

peptide binding site   amino acid binding site

In protein synthesis, all peptide bonds are formed by the reaction of an amino acid with a peptide, except formation of the first peptide bond, which has to be formed by the reaction of two amino acids. Therefore, in the synthesis of the first peptide bond, the first (N-terminal) amino acid has to have a peptide bond that will fit into the peptide binding site.

The formyl group of *N*-formylmethionine will provide the peptide group that will be recognized by the peptide binding site, and the second amino acid will be bound in the amino acid binding site.

N-formylmethionine

38. It requires energy to break the hydrogen bonds that hold the two chains together, so an enormous amount of energy would be required to unravel the chain completely. However, as the new nucleotides that are incorporated into the growing chain form hydrogen bonds with the parent chain, energy is released, and this energy can be used to unwind the next part of the double helix.

Chapter 28 Practice Test

1. Is the following compound dTMP, UMP, dUMP, or dUTP?

2. If one of the strands of DNA has the following sequence of bases running in the 5′ → 3′ direction, what is the sequence of bases in the complementary strand?

$$5'-A-C-T-T-G-C-A-T-3'$$

3. What base is closest to the 5′-end in the complementary strand?

4. Indicate whether each of the following statements is true or false:

a.	Guanine and cytosine are purines.	T	F
b.	The 3′-OH group allows RNA to be easily cleaved.	T	F
c.	The number of A's in DNA is equal to the number of T's.	T	F
d.	rRNA carries the amino acid that will be incorporated into a protein.	T	F
e.	The template strand of DNA is the one transcribed to form RNA.	T	F
f.	The 5′-end of DNA has a free OH group.	T	F
g.	The synthesis of proteins from an RNA blueprint is called transcription.	T	F
h.	A nucleotide consists of a base and a sugar.	T	F
i.	RNA contains T's and DNA contains U's.	T	F

5. Which of the following base sequences would most likely be recognized by a restriction endonuclease?

1. ACGCGT   **3.** ACGGCA   **5.** ACATCGT

2. ACGGGT   **4.** ACACGT   **6.** CCAACC

6. What would be the sequence of bases in the mRNA obtained from the following segment of DNA?

sense strand

5′—G-C-A-T-G-G-A-C-C-C-C-G-T—3′
3′—C-G-T-A-C-C-T-G-G-G-G-C-A—5′

template strand

7. Why is NADPH needed in order to convert uridines into thymidines?

8. Which of the following pairs of dinucleotides will occur in equal amounts in DNA? (Remember nucleotides are always written in the 5′ → 3′ direction.)

CA and GT   CG and AT
CG and GG   CA and TG

CHAPTER 29
Synthetic Polymers

Important Terms

addition polymer (**chain-growth polymer**)	made by adding monomers to the growing end of a chain.
alpha olefin	a monosubstituted ethylene.
alternating copolymer	a copolymer in which two monomers alternate.
anionic polymerization	chain-growth polymerization where the initiator is a nucleophile; the propagation site, therefore, is an anion.
aramide	an aromatic polyamide.
atactic polymer	a polymer in which the substituents are randomly oriented on the extended carbon chain.
biodegradable polymer	a polymer that can be degraded by microorganisms.
biopolymer	a polymer that is synthesized in nature.
block copolymer	a copolymer in which there are blocks of each kind of monomer within the polymer chain.
cationic polymerization	chain-growth polymerization where the initiator is an electrophile; the propagation site, therefore, is a cation.
chain-growth polymer (**addition polymer**)	made by adding monomers to the growing end of a chain.
chain transfer	a growing polymer chain reacts with a molecule XY in a manner that allows X· to terminate the chain, leaving behind Y· to initiate a new chain.
condensation polymer (**step-growth polymer**)	made by combining two molecules while removing a small molecule (usually water or an alcohol).
conducting polymer	a polymer that can conduct electricity down its backbone.
copolymer	a polymer formed using two or more different monomers.
cross-linking	connecting polymer chains by bond formation.
crystallites	regions of a polymer in which the chains are highly ordered.
elastomer	a polymer that can stretch and then revert back to its original shape.

epoxy resin   formed by mixing a low molecular weight prepolymer with a compound that forms a cross-linked polymer.

graft copolymer   a copolymer that contains branches of a polymer of one monomer grafted onto the backbone of a polymer made from another monomer.

head-to-tail addition the head of one molecule is added to the tail of another molecule.

homopolymer   a polymer that contains only one kind of monomer.

isotactic polymer   a polymer in which all the substituents are on the same side of the fully extended carbon chain.

living polymer   a nonterminated chain-growth polymer that remains active. Therefore, the polymerization reaction can continue upon addition of more monomer.

materials science   the science of creating new materials that will have practical applications.

monomer   a repeating unit in a polymer.

oriented polymer   a polymer obtained by stretching out polymer chains and putting them back together in a parallel fashion.

plasticizer   an organic molecule that dissolves in a polymer and allows the polymer chains to slide by each other.

polyamide   a polymer in which monomers are connected by amide groups.

polycarbonate   a step-growth polymer in which the monomers are connected by carbonate groups.

polyester   a polymer in which monomers are connected by ester groups.

polymer   a large molecule made by linking monomers together.

polymer chemistry   the field of chemistry that deals with synthetic polymers; part of the larger discipline known as materials science.

polymerization   the process of linking up monomers to form a polymer.

polyurethane   a polymer in which monomers are connected by urethane groups.

propagating site   the reactive end of a chain-growth polymer.

radical polymerization chain-growth polymerization where the initiator is a radical; the propagation site, therefore, is a radical.

random copolymer   a copolymer with a random distribution of monomers.

ring-opening polymerization   a chain-growth polymerization that involves opening the ring of the monomer.

step-growth polymer (condensation polymer)   made by combining two molecules while removing a small molecule (usually water or an alcohol).

syndiotactic polymer   a polymer in which the substituents regularly alternate on both sides of the fully extended carbon chain.

synthetic polymer   a polymer that is not synthesized in nature.

thermoplastic polymer   a polymer that has both ordered crystalline regions and amorphous non-crystalline regions.

thermosetting polymer   cross-linked polymers that, after they are hardened, cannot be re-melted by heating.

urethane (carbamate)   a compound with a carbonyl group that is both an amide and an ester.

vinyl polymer   a polymer in which the monomer is ethylene or a substituted ethylene.

vulcanization   increasing the flexibility of rubber by heating it with sulfur.

Ziegler-Natta catalyst   an aluminum-titanium initiator that controls the stereochemistry of a polymer.

Solutions to Problems

1.   a. $CH_2=CHCl$ **b.** $CH_2=CCH_3$ **c.** $CF_2=CF_2$

with the structure for b having a group:
$$C=O$$
$$OCH_3$$

2. Poly(vinyl chloride) would be more apt to contain head-to-head linkages, because a chloro substituent is less able (compared with a phenyl substituent) to stabilize the growing end of the polymer chain by electron delocalization.

3.

$-CH_2CH-CHCH_2CH_2CH-CHCH_2CHCH_2CH_2CH-CHCH_2CH_2CH-CH_2CHCH_2CHCH_2CHCH_2CH-$

(each CH bearing a phenyl group)

4.

$HO{-}OH \longrightarrow 2\ HO\cdot$

$HO\cdot\ +\ CH_2{=}CH \longrightarrow HO-CH_2CH\cdot$
with Cl substituents

$HO-CH_2CH\cdot\ +\ CH_2{=}CH \longrightarrow HO-CH_2CHCH_2CH\cdot$
with Cl substituents

$HO-CH_2CHCH_2CH\cdot\ +\ CH_2{=}CH \longrightarrow HO-CH_2CHCH_2CHCH_2CH\cdot$
with Cl substituents

5. Since branching increases the flexibility of the polymer, beach balls are made from more highly branched polyethylene.

6.

$$CH-\bigcirc$$
$$CH_2$$
$$CH-\bigcirc$$
$$CH_2$$
$$-CH_2CHCH_2CHCH_2CCH_2CHCH_2CH-$$
(with phenyl groups)

7. Decreasing ability to undergo cationic polymerization is in the same order as decreasing stability of the carbocation intermediate. (Electron donation increases the stability of the carbocation.)

donates electrons by resonance   **withdraws electrons by resonance**

$$CH_2=CH-OCH_3 \; > \; CH_2=CHCH_3 \; > \; CH_2=CH-\overset{\overset{\displaystyle O}{\|}}{C}CH_3$$

donates electrons by resonance   **withdraws electrons by resonance**

A tertiary benzylic carbocation is more stable than a secondary benzylic carbocation.

8. Decreasing ability to undergo anionic polymerization is in the same order as decreasing stability of the carbanion intermediate. (Electron withdrawal increases the stability of the carbanion.)

a.

withdraws electrons by resonance   **donates electrons by resonance**

b. $CH_2=CHC\equiv N \; > \; CH_2=CHCl \; > \; CH_2=CHCH_3$

withdraws electrons by resonance

9. Methyl methacrylate does not undergo cationic polymerization, because the carbocation propagating site would be very unstable since the ester group is strongly electron withdrawing.

a strongly electron-withdrawing group

10. In anionic polymerization, nucleophilic attack occurs at the less-substituted carbon, because it is less sterically hindered; in cationic polymerization, nucleophilic attack occurs at the more-substituted carbon, because the ring opens to give the more stable partial carbocation (Section 10.7).

position of nucleophilic attack in cationic polymerization

position of nucleophilic attack in anionic polymerization

11. a. RÖ:⁻ + (epoxide with two CH₃) ⟶ RO—CH₂CO⁻ (with two CH₃)

RO—CH₂CÖ:⁻ (with CH₃) + epoxide ⟶ RO—CH₂COCH₂CO⁻ (with CH₃ groups)

RO—CH₂COCH₂CÖ:⁻ + epoxide ⟶ RO—CH₂COCH₂COCH₂CO⁻

b. Ö: (epoxide with two CH₃) + H—B⁺ ⟶ HO⁺: (with two CH₃)

HO⁺: + Ö: (epoxide) ⟶ HOCH₂C—Ö⁺ (with CH₃ groups)

HOCH₂C—Ö⁺ + Ö: (epoxide) ⟶ HOCH₂COCH₂C—Ö⁺

12. a. CH₂=CCH₃ with CH₃ + BF₃ + H₂O

b. CH₂=CH with N-pyrrolidinone (2-pyrrolidinone) + BF₃ + H₂O

c. (epoxide) + CH₃O⁻

d. CH₂=CH with COCH₃ (=O) + BuLi

13.   3,3-Dimethyloxacyclobutane undergoes cationic polymerization by the following mechanism:

14.

15.

$$-CH_2CH-CH_2CH-CH_2CH-CH_2CH-CH_2CH-CH_2CH-$$
$$\quad\ |\qquad\ |\qquad\ |\qquad\ |\qquad\ |\qquad\ |$$
$$\quad CH\qquad CH\qquad CH\qquad CH\qquad CH\qquad CH$$
$$\quad ||\qquad ||\qquad ||\qquad ||\qquad ||\qquad ||$$
$$\quad CH_2\qquad CH_2\qquad CH_2\qquad CH_2\qquad CH_2\qquad CH_2$$

16.   **a.** $-NHCH_2CH_2CH_2\overset{\overset{O}{||}}{C}NHCH_2CH_2CH_2\overset{\overset{O}{||}}{C}-$

   b. $-NH(CH_2)_4NH\overset{\overset{O}{||}}{C}CH_2CH_2\overset{\overset{O}{||}}{C}NH(CH_2)_4NH\overset{\overset{O}{||}}{C}CH_2CH_2\overset{\overset{O}{||}}{C}-$

17.

$$-\overset{\overset{O}{||}}{C}(CH_2)_4\overset{\overset{O}{||}}{C}\!\left[\!NH(CH_2)_6NH-\overset{\overset{O}{||}}{C}(CH_2)_4\overset{\overset{O}{||}}{C}\!\right]_{\!n}\!NH(CH_2)_6NH-$$

$$\downarrow H_2SO_4$$

$$n\ HO\overset{\overset{O}{||}}{C}(CH_2)_4\overset{\overset{O}{||}}{C}OH\quad +\quad n\ H_3\overset{+}{N}(CH_2)_6\overset{+}{N}H_3$$

18.

19.   They hydrolyze to give salts of dicarboxylic acids and diols.

Kodel

↓ NaOH

n Na$^+$ $^-$OC—⬡—CO$^-$ Na$^+$ + n HOCH$_2$—⬡—CH$_2$OH

Dacron

↓ NaOH

n Na$^+$ $^-$OC—⬡—CO$^-$ Na$^+$ + n HOCH$_2$CH$_2$OH

20.   **a.**

b.

21.   The polymer is stiffer because glycerol cross-links the polymer chains.

22.   Formation of a protonated imine between formaldehyde and one amino group, followed by reaction with a second amino group, accounts for formation of the linkage that holds the monomers together.

a dimer of Melmac

23.

monomer used in the polymerization

Bakelite

24.   a.

b.

25.   **a.**   —CH$_2$CHCH$_2$CHCH$_2$CH—   chain-growth polymer
                 | | |
                 F   F   F

   b.   —CH$_2$CHCH$_2$CHCH$_2$CH—   chain-growth polymer
                 | | |
              CO$_2$H   CO$_2$H   CO$_2$H

   c.   —O(CH$_2$)$_5$CO(CH$_2$)$_5$CO(CH$_2$)$_5$C—   step-growth polymer
                   ‖   ‖   ‖
                   O   O   O

   d.   —NH(CH$_2$)$_5$NHC(CH$_2$)$_5$CNH(CH$_2$)$_5$NHC(CH$_2$)$_5$C—   step-growth polymer
                         ‖   ‖   ‖   ‖
                         O   O   O   O

   e.   —OCNH⟨ ⟩NHCOCH$_2$CH$_2$OCNH⟨ ⟩NHCO—   step-growth polymer
              ‖   ‖   ‖   ‖
              O   O   O   O

26.   **a.**   —CH$_2$CH$_2$OCH$_2$CH$_2$N⟨ ⟩N—

   b.   —⟨ ⟩—OCH$_2$CH$_2$CH$_2$O—⟨ ⟩—N=CHCH=N—

   c.   —CH=⟨ ⟩=CH—⟨ ⟩—

27.   **a.**   CH$_2$=CHCH$_2$CH$_3$

   b.   ◁O
              CH$_3$

   c.   ClSO$_2$—⟨ ⟩—SO$_2$Cl + H$_2$N(CH$_2$)$_6$NH$_2$

   d.   CH$_2$=CH—⟨ N ⟩

   e.   CH$_2$=CCH=CH$_2$
                 |
                 CH$_3$

   f.   HO(CH$_2$)$_5$COH
                       ‖
                       O

   g.   CH$_2$=CCH$_3$
              ⟨ ⟩

   h.   HOC—⟨ ⟩—COH + HOCH$_2$CH$_2$OH
              ‖   ‖
              O   O

a, **b**, **d**, **e**, and **g** are chain-growth polymers.

c, **f**, and **h** are step-growth polymers.

28. Whether a polymer is isotactic, syndiotactic, or atactic depends on whether the substituents are all on one side of the carbon chain, alternate on both sides of the chain, or are random with respect to the chain. Because a polymer of isobutylene has two identical substituents at each carbon in the chain, different configurations are not possible.

$$\begin{array}{ccccc} & CH_3 & CH_3 & CH_3 & CH_3 \\ & | & | & | & | \\ -CH_2C & CH_2C & CH_2C & CH_2C- \\ & | & | & | & | \\ & CH_3 & CH_3 & CH_3 & CH_3 \end{array}$$

polymer of isobutylene

29. **a.** $CH_3OCH_2CHOCH_2CHOCH_2CHOCH_2CHO-$
$$\begin{array}{cccc} | & | & | & | \\ CH_3 & CH_3 & CH_3 & CH_3 \end{array}$$

b. $$\begin{array}{cccc} CH_3 & CH_3 & CH_3 & CH_3 \\ | & | & | & | \\ -CH_2C-CH_2C-CH_2C-CH_2C- \\ | & | & | & | \\ CHCH_3 & CHCH_3 & CHCH_3 & CHCH_3 \end{array}$$

c. $$\begin{array}{cccc} -CH_2CH-CH_2CH-CH_2CH-CH_2CH- \\ | & | & | & | \\ COCH_3 & COCH_3 & COCH_3 & COCH_3 \\ || & || & || & || \\ O & O & O & O \end{array}$$

d. $$\begin{array}{cccc} -CH_2CH-CH_2CH-CH_2CH-CH_2CH- \\ | & | & | & | \\ OCH_3 & OCH_3 & OCH_3 & OCH_3 \end{array}$$

30.   **a.**   Because it is a polyamide (and not a polyester), it is a nylon.

b.   H$_2$N—⬡—CH$_2$—⬡—NH$_2$ and $\overset{O}{\overset{\|}{HOC}}(CH_2)_6\overset{O}{\overset{\|}{COH}}$

31.   A copolymer is a polymer composed of more than one kind of monomer. Because the initially formed carbocation can rearrange, two different monomers are involved in formation of the polymer.

$$\underset{\text{unrearranged monomer}}{-CH_2\overset{+}{C}HC\overset{\overset{\displaystyle CH_3}{|}}{\underset{\underset{\displaystyle CH_3}{|}}{C}}H_3} \xrightarrow{\text{1,2-methyl shift}} \underset{\text{rearranged monomer}}{-CH_2CH\overset{+}{C}\overset{\overset{\displaystyle CH_3}{|}}{\underset{\underset{\displaystyle CH_3}{|}}{C}}H_3}$$

32.   The polymer in the flask that contained a high molecular weight polymer and little material of intermediate molecular weight was formed by a chain-growth mechanism, whereas the polymer in the flask that contained mainly material of intermediate molecular weight was formed by a step-growth mechanism.

In a chain-growth mechanism, monomers are added to the growing end of a chain. This means that at any one time there will be polymeric chains and monomers.

Step-growth polymerization is not a chain reaction; any two monomers can react. Therefore, high molecular weight material will not be formed until the end of the reaction when pieces of intermediate molecular weight can combine.

33.   **a.**   Vinyl alcohol is unstable; it tautomerizes to acetaldehyde.

$$\underset{\text{vinyl alchol}}{CH_2{=}\overset{\overset{\displaystyle OH}{|}}{CH}} \longrightarrow \underset{\text{acetaldehyde}}{CH_3\overset{O}{\overset{\|}{CH}}}$$

b.   It is not a true polyester. It has ester groups as substituents **on** the backbone of the chain so it does have "polyester" groups, but it does not have ester groups **within** the backbone of the polymer chain. A true polyester has ester groups within the backbone of the polymer chain.

34.   Each of the five carbocations shown below can add the growing end of the polymer chain.

$$\text{—CH}_2\overset{+}{\text{CH}}$$
$$|$$
$$\text{CH}_2$$
$$|$$
$$\text{CHCH}_3$$
$$|$$
$$\text{CH}_3$$

$\xrightarrow{\text{1,2-hydride shift}}$

$$\text{—CH}_2\text{CH}_2\overset{+}{\text{CH}}$$
$$|$$
$$\text{CHCH}_3$$
$$|$$
$$\text{CH}_3$$

$\xrightarrow{\text{1,2-hydride shift}}$

$$\overset{\text{CH}_3}{\overset{|}{\text{—CH}_2\text{CH}_2\text{CH}_2\text{C}+}}$$
$$|$$
$$\text{CH}_3$$

$\downarrow$ 1, 2-methyl shift

$$\overset{\text{CH}_3}{\overset{|}{\text{—CH}_2\text{CH}_2\text{CH}\overset{+}{\text{CH}}}}$$
$$|$$
$$\text{CH}_3$$

$\xrightarrow{\text{1,2-hydride shift}}$

$$\overset{\text{CH}_3}{\overset{|}{\text{—CH}_2\text{CH}_2\text{C}\text{CH}_2\text{CH}_3}}$$
$$\underset{+}{}$$

35.   Because 1,4-divinylbenzene has substituents on both ends of the benzene ring that can engage in polymerization, the polymer chains can become cross-linked, which increases the rigidity of the polymer.

1,4-divinylbenzene

36.   The presence of three OH groups in glycerol allows for extensive cross-linking during polymerization. Glyptal gets its strength from this cross-linking.

$$
\text{HO}-\overset{\overset{\displaystyle O}{\|}}{C}-\underset{\text{terephthalic acid}}{\bigcirc}-\overset{\overset{\displaystyle O}{\|}}{C}-\text{OH} \quad + \quad \underset{\underset{\displaystyle \text{OH}}{|}}{\underset{\text{glycerol}}{\text{HOCH}_2\text{CHCH}_2\text{OH}}}
$$

37.   The five-membered ring and the four-membered ring react to form dimers. The dimers then react as monomers and polymerize to form the alternating copolymer.

38.   Both compounds can form esters via intramolecular or intermolecular reactions. The product of the intramolecular reaction is a lactone; the intermolecular reaction leads to a polymer.

5-Hydroxypentanoic acid reacts intramolecularly to form a six-membered-ring lactone, whereas 6-hydroxyhexanoic acid reacts intramolecularly to form a seven-membered-ring lactone.

5-hydroxypentanoic acid

6-hydroxyhexanoic acid

The compound that forms the most polymer will be the one that forms the least lactone, because the two reactions compete with one another.

The six-membered-ring lactone is more stable and so is the transition state for its formation, compared to a seven-membered-ring lactone. Since it is easier for 5-hydroxypentanoic acid to form the six-membered-ring lactone, than for 6-hydroxyhexanoic acid to form the seven-membered-ring lactone, **6-hydroxyhexanoic acid** will form more polymer.

39. Rubber contains cis double bonds. Ozone, which is present in the air, oxidizes double bonds to carbonyl groups, which destroys the polymer chain. Polyethylene does not contain double bonds, so it is not air-oxidized.

40.

41. The plasticizer that keeps vinyl soft and pliable can vaporize over time, causing the polymer to become brittle. For this reason, high boiling materials are preferred over low boiling materials as plasticizers.

42. **a.** Because the negative charge on the propagation site can be delocalized onto the carbonyl oxygen, the polymer is best prepared by anionic polymerization.

b. The carboxyl substituent is in position to remove a proton from water, making water a better nucleophile.

43. Hydrolysis converts the ester substituents into alcohol substituents, which can react with ethylene oxide to graft a polymer of ethylene oxide onto the backbone of the alternating copolymer of styrene and vinyl acetate.

44.   The desired polymer is an alternating copolymer of ethylene and 1,2-dibromoethylene.

$$CH_2{=}CH_2 \qquad \underset{\underset{Br \quad Br}{|\qquad|}}{CH{=}CH}$$

45.   **a.**

b.   Delrin is a chain-growth polymer.

CHAPTER 30
Pericyclic Reactions

Important Terms

antarafacial bond formation	in a pericyclic reaction, where formation and/or cleavage of two σ bonds occurs on opposite sides of the π system of the reactant.
antarafacial rearrangement	a rearrangement where the migrating group moves to the opposite face of the π system.
antibonding π molecular orbital	a molecular orbital that results when two parallel atomic orbitals with opposite phases interact. Electrons in an antibonding orbital decrease bond strength.
antisymmetric molecular orbital	a molecular orbital in which the left half is not a mirror image of the right half but would be if one half of the MO were turned upside down.
bonding π molecular orbital	a molecular orbital that results when two parallel atomic orbitals with the same phase interact. Electrons in a bonding orbital increase bond strength.
Claisen rearrangement	a [3,3] sigmatropic rearrangement of an allyl vinyl ether.
conrotatory ring closure	achieves head-to-head overlap of p orbitals by rotating the orbitals in the same direction.
conservation of orbital symmetry theory	a theory that explains the relationship between the structure and stereochemistry of the reactant, the conditions under which a pericyclic reaction takes place, and the stereochemistry of the product.
Cope rearrangement	a [3,3] sigmatropic rearrangement of a 1,5-diene.
cycloaddition reaction	a reaction in which two π-bond-containing molecules react to form a cyclic compound.
disrotatory ring closure	achieves head-to-head overlap of p orbitals by rotating the orbitals in opposite directions.
electrocyclic reaction	an intramolecular reaction in which a new σ bond is formed between the ends of a conjugated system.
excited state	a description of which orbitals the electrons of an atom or molecule occupy when an electron in the ground state has been moved to a higher energy orbital.
frontier orbital analysis	determining the outcome of a pericyclic reaction using frontier orbitals.
frontier orbitals	the HOMO and the LUMO of the two reacting species in a pericyclic reaction.
frontier orbital theory	a theory that, like the conservation of orbital symmetry, explains the relationship between reactant, product, and reaction conditions in a pericyclic reaction.

ground state   a description of which orbitals the electrons of an atom or molecule occupy when they are all in their lowest energy orbitals.

highest occupied molecular orbital (HOMO)   the molecular orbital of highest energy that contains an electron.

linear combination of atomic orbitals (LCAO)   the combination of atomic orbitals to produce a molecular orbital.

lowest unoccupied molecular orbital (LUMO)   the molecular orbital of lowest energy that does not contain an electron.

molecular orbital (MO) theory   describes a model in which the electrons occupy orbitals as they do in atoms but with the orbitals extending over the entire molecule.

pericyclic reaction   a concerted reaction that occurs as a result of a cyclic reorganization of electrons.

photochemical reaction   a reaction that takes place when a reactant absorbs light.

polar reaction   the reaction between a nucleophile and an electrophile.

radical reaction   a reaction in which a new bond is formed using one electron from one reactant and one electron from another reactant.

selection rules   the rules that determine the outcome of a pericyclic reaction.

sigmatropic rearrangement   a reaction in which a σ bond is broken in the reactant, a new σ bond is formed in the product, and the π bonds rearrange.

suprafacial bond formation   in a pericyclic reaction, where formation and/or cleavage of two σ bonds occurs on the same side of the π system of the reactant.

suprafacial rearrangement   a rearrangement where the migrating group remains on the same face of the π system.

symmetric molecular orbital   a molecular orbital in which the left half is a mirror image of the right half.

symmetry-allowed pathway   a pathway that leads to overlap of in-phase orbitals.

symmetry-forbidden pathway   a pathway that leads to overlap of out-of-phase orbitals.

thermal reaction   a reaction that takes place without the reactant having to absorb light.

Woodward-Hoffmann rules   a series of selection rules for pericyclic reactions.

Solutions to Problems

1. **a.** electrocyclic reaction   **c.** cycloaddition reaction

    b. sigmatropic rearrangement   **d.** cycloaddition reaction

2. **a.** bonding orbitals = ψ_1, ψ_2, ψ_3; antibonding orbitals = ψ_4, ψ_5, ψ_6

    b. ground-state HOMO = ψ_3; ground-state LUMO = ψ_4

    c. excited-state HOMO = ψ_4; excited-state LUMO = ψ_5

    d. symmetric orbitals = ψ_1, ψ_3, ψ_5; antisymmetric orbitals = ψ_2, ψ_4, ψ_6

    e. The HOMO and LUMO have opposite symmetries.

3. **a.** 8 π molecular orbitals   **b.** ψ_4 **c.** 7 nodes (There is also a node that passes through the nuclei, that is, through the centers of the p orbitals.)

4. **a.** 1,3-Pentadiene has two conjugated π bonds, so it has the same π molecular orbital description as 1,3-butadiene, a compound that also has two conjugated π bonds. (See Figure 30.2 on page 1214 of the text.)

    b. The π bonds in 1,4-pentadiene are isolated, so its π molecular orbital description is the same as ethene, a compound with an isolated π bond. (See Figure 30.1 on page 1213 of the text.)

    c. 1,3,5-Heptatriene has three conjugated π bonds, so it has the same π molecular orbital description as 1,3,5-hexatriene, a compound that also has three conjugated π bonds. (See Figure 30.0 on page 1215 of the text.)

    d. 1,3,5,8-Nonatetraene has three conjugated π bonds and an isolated π bond. The three conjugated π bonds are described by Figure 30.3 and the isolated π bond by Figure 30.1.

5. **a.** 2, 4, or 6 conjugated double bonds   3, 5, or 7 conjugated double bonds

    b. Under thermal reaction conditions, electrocyclic ring closure involves the HOMO of the polyene (a compound with several double bonds). If the polyene has an even number of double bonds, conrotatory ring closure will result in in-phase overlap of the terminal p orbitals in the HOMO. If the polyene has an odd number of double bonds, disrotatory ring closure will result in in-phase overlap of the terminal p orbitals in the HOMO.

    Under photochemical reaction conditions, electrocyclic ring closure involves the LUMO of the polyene, which will have the opposite symmetry of the HOMO and so will require the opposite mode of ring closure to that seen under thermal conditions.

6. **a.** (2*E*,4*Z*,6*Z*,8*E*)-Decatetraene has an even number of conjugated π bonds (4). Therefore, under thermal conditions, ring closure will be conrotatory.

    b. The substituents point in opposite directions, and conrotatory ring closure of such substituents will cause them to be trans in the ring-closed product.

    c. Under photochemical conditions, a compound with four conjugated π bonds will undergo disrotatory ring closure.

    d. Because ring closure is disrotatory and the substituents point in opposite directions, the product will have the cis configuration.

7.   **a.** correct   **b.** correct   **c.** correct

8.   **1.** **a.** conrotatory   **b.** trans   **2.** **a.** disrotatory   **b.** cis

9.   The reaction of maleic anhydride with 1,3-butadiene involves three π bonds in the reacting system, and such a reaction under thermal conditions involves suprafacial ring closure.

1,3-butadiene

The reaction of maleic anhydride with ethylene involves two π bonds in the reacting system, and such a reaction under thermal conditions involves antarafacial ring closure, which cannot occur with a four-membered ring.

ethene

10.   Solved in the text.

11.   Because we are dealing with formation of a small ring, ring closure will have to be suprafacial. Under photochemical conditions, suprafacial ring closure requires an even number of π bonds in the reacting system. Thus, a concerted reaction will occur but it will use only one of the π bonds of 1,3-butadiene.

12.   **a.** **1.** [1,7] sigmatropic rearrangement   **3.** [5,5] sigmatropic rearrangement

          2. [1,5] sigmatropic rearrangement   **4.** [3,3] sigmatropic rearrangement

    b. **1.**

          2.

3.

4.

13.   **a.** and **b.**

14.   If a nondeuterated reactant had been used, the product would have been identical to the reactant. Therefore, the rearrangement would not have been detectable.

15.   A suprafacial rearrangement can take place under photochemical conditions if there are an even number of electrons in the reacting system. Therefore, a 1,3-hydrogen shift occurs involving four electrons.

a [1,3] sigmatropic
migration of deuterium

A suprafacial rearrangement can take place under thermal conditions if there are an odd number of electrons in the reacting system. Therefore, a 1,5-hydrogen shift occurs involving six electrons.

a [1,5] sigmatropic
migration of deuterium

16.   Solved in the text.

17.   [1,3] Sigmatropic migrations of hydrogen cannot occur under thermal conditions, because the four-membered transition state does not allow the required antarafacial rearrangement.

[1,3] Sigmatropic migrations of carbon can occur under thermal conditions because carbon can achieve the required antarafacial rearrangement by using both of its *p*-orbital lobes when it migrates.

18.   **a.**   Because 1,3-migration of carbon requires carbon to migrate using both of its lobes (it involves an even number of pairs of electrons, so it takes place by an antarafacial pathway), migration will occur with inversion of configuration.

b.   Because 1,5-migration of carbon requires carbon to migrate using only one of its lobes (it involves an odd number of pairs of electrons, so it takes place by a suprafacial pathway), migration will occur with retention of configuration.

19.   Because the [1,7] sigmatropic rearrangement takes place under thermal conditions and involves an even number (4) of pairs of electrons, migration of hydrogen involves antarafacial rearrangement. Since the cyclic transition state involves eight ring atoms, antarafacial rearrangement is possible.

20.   Because the reactant (provitamin D_3) has an odd number (3) of conjugated π bonds and reacts under photochemical conditions, ring closure is conrotatory. The methyl and hydrogen substituents point in opposite directions in provitamin D_3. Conrotatory ring closure will cause substituents that point in opposite directions in the reactant to be trans in the product.

21.   Chorismate mutase catalyzes a [3,3] sigmatropic Claisen rearrangement.

22.   Were you able to convince yourself that TE-AC is valid?

23.   **a.**

b.

c.

d.

e.

f.

g. *hv*

h.

24. Because the compound has an odd number of π bonds, it will undergo disrotatory ring closure under thermal conditions and conrotatory ring closure under photochemical conditions.

In the compounds in which the two methyl substituents point in opposite directions, they will be cis in the ring-closed product when ring closure is disrotatory and trans in the ring-closed product when ring closure is conrotatory.

In the compounds in which the two methyl substituents point in the same direction, they will be trans in the ring-closed product when ring closure is disrotatory and cis in the ring-closed product when ring closure is conrotatory.

a.   **c.**

b.   **d.**

25. The hydrogens that end up at the ring juncture in the first reaction point in opposite directions in the reactant. Because ring closure is disrotatory (odd number of π bonds, thermal conditions), the hydrogens in the ring-closed product are cis. (See Table 30.2 on page 1222 of the text.)

In the second example, the hydrogens that end up at the ring juncture point in the same direction in the reactant. Ring closure is still disrotatory, so the hydrogens in the ring-closed product are trans.

26.

27.   **1.**   Because the compound has an even number of π bonds, it will undergo conrotatory ring closure under thermal conditions and disrotatory ring closure under photochemical conditions. Because the two methyl substituents point in opposite directions, they will be trans in the ring-closed product when ring closure is conrotatory and cis in the ring-closed product when ring closure is disrotatory.

a.

b.

2.   Because the compound has an even number of π bonds, it will undergo conrotatory ring closure under thermal conditions and disrotatory ring closure under photochemical conditions. Because the two methyl substituents point in the same direction, they will be cis in the ring-closed product when ring closure is conrotatory and trans in the ring-closed product when ring closure is disrotatory.

a.

b.

28.   **a.**

b.

c.

tautomerization

d.

e.

tautomerization

29.   **B** is the product. Because the reaction is a [1,3] sigmatropic rearrangement, antarafacial ring closure is required. Carbon, therefore, must migrate using both of its lobes. This means that the configuration of the migrating carbon will undergo inversion. The configuration of the migrating carbon has been inverted in **B** (the H's are cis to one another but were trans to one another in the reactant) and retained in **A**.

the H's are trans   retention   inversion
to one another
                              A   **B**
                         the H's are trans   the H's are cis
                         to one another   to one another

30.   At first glance it is surprising that the isomerization of Dewar benzene (a highly strained and unstable molecule) to benzene (a stable aromatic compound) is so slow. However, the isomerization requires conrotatory ring-opening, which is symmetry-forbidden under thermal conditions. The reaction, therefore, cannot take place by a concerted pathway and has to take place by a much slower stepwise process.

31.   Hydrogen cannot undergo a [1,3] sigmatropic rearrangement, because it cannot migrate by an antarafacial pathway that is required for a sigmatropic rearrangement involving an even number of pairs of electrons under thermal conditions. Carbon can undergo a [1,3] sigmatropic rearrangement because it can migrate by a suprafacial pathway if it uses both of its lobes. Therefore, the first compound can undergo only a 1,3-methyl group migration.

The second compound can undergo the 1,3-methyl group migration that the first compound undergoes, and the *sec*-butyl group can also undergo a [1,3] sigmatropic rearrangement. The migrating *sec*-butyl group will have its configuration inverted due to the antarafacial nature of the rearrangement.

1,3-*sec*-butyl   1,3-methyl
group migration   group migration

32. An infrared absorption band is indicative of a carbonyl group. A [3,3] sigmatropic rearrangement of the reactant leads to a compound with two enolic groups. Tautomerization of the enols results in keto groups. The ketone carbonyl groups give the absorbance at 1715 cm^{-1}.

33. The reaction is a [1,7] sigmatropic rearrangement. Since the reaction involves four pairs of electrons, antarafacial rearrangement occurs. Thus, when H migrates, because it is above the plane of the reactant molecule, it ends up below the plane of the product molecule. When D migrates, because it is below the plane of the reactant molecule, it ends up above the plane of the product molecule.

34. a.

b.

35. Disrotatory ring closure of (2E,4Z,6Z)-octatriene leads to the trans isomer, which can exist as a pair of enantiomers. One enantiomer is formed if the "top lobes" of the p orbitals rotate toward one other, and the other enantiomer is formed if the "bottom lobes" of the p orbitals rotate toward one other.

(2E,4Z,6Z)-octatriene

In contrast, disrotatory ring closure of (2E,4Z,6E)-octatriene leads to the cis isomer, which is a meso compound and, consequently, does not have a nonsuperimposable mirror image. Therefore, the same compound is formed from the "top lobes" of the *p* orbitals rotating toward one other and from the "bottom lobes" of the *p* orbitals rotating toward one other.

(2E,4Z,6E)-octatriene

36.

a.

b.

c.

d.

37. Under thermal conditions a compound with two π bonds undergoes conrotatory ring closure. Conrotatory ring closure that results in a ring-closed compound with the substituents cis to one another requires that the substituents point in the same direction in the reactant. Therefore, the product with the methyl substituents pointing in the same direction is obtained in 99% yield.

1%

Methyl groups point in opposite directions.

99%

Methyl groups point in the same directions.

38. Because the reactant has two π bonds, electrocyclic ring closure is conrotatory. Two different compounds, **X** and **Y**, can be formed because conrotatory ring closure can occur in either a clockwise or a counterclockwise direction.

Each of the compounds (**X** and **Y**) can undergo a conrotatory ring-opening reaction in either a clockwise or a counterclockwise direction to form either **A** or **B**. **A** and **B** are the only isomers that can be formed; formation of **C** and **D** would require disrotatory ring closure.

39.

40.   Because the compounds that undergo ring closure to give **A** and **B** have two π bonds, ring opening of **A** and **B** under thermal conditions will be conrotatory. Because the hydrogens in **A** and **B** are cis, they must point in the same direction in the ring-opened product. To have the two hydrogens pointing in the same direction, one of the double bonds in the ring-opened compound must be cis and the other must be trans.

An eight-membered ring is too small to accommodate conjugated double bonds with one cis and the other trans, so **A** will not be able to undergo a ring-opening reaction under thermal conditions.
A ten-membered ring can accommodate a trans double bond, so **B** will be able to undergo a ring-opening reaction under thermal conditions.

Both double bonds are cis.   **A**

cis double bond

trans double bond   **B**

41. The compound undergoes a 1,5-hydrogen shift of D or a 1,5-hydrogen shift of H. In each case, an unstable nonaromatic intermediate is formed that undergoes a subsequent 1,5-hydrogen shift to form an aromatic product.

42. Because the ring-opened compound formed in the first step has three conjugated π bonds involved in an electrocyclic reaction, conrotatory ring opening of the reactant will occur under photochemical conditions, and the trans hydrogens in the reactant require that the hydrogens point in the opposite direction in the ring-opened compound. Thermal electrocyclic ring closure of a three π bond system is disrotatory in step two, and disrotatory ring closure of a compound with hydrogens that point in opposite directions will cause those hydrogens to be cis in the ring-closed product.

43. A Diels-Alder reaction is followed by a reverse Diels-Alder reaction that eliminates CO_2. Loss of a stable gas molecule (CO_2) and formation of a stable aromatic product provide the driving force for the second step.

44.

CHAPTER 31
The Organic Chemistry of Drugs: Discovery and Design

Important Terms

antiviral drug	a drug that interferes with DNA or RNA synthesis in order to prevent a virus from replicating.
bactericidal drug	a drug that kills bacteria.
bacteriostatic drug	a drug that inhibits the further growth of bacteria.
biological activity	the activity of a compound in a biological organism.
blind screen (random screen)	the search for a pharmacologically active compound without any information about what chemical structures might show activity.
brand name	identifies a commercial product and distinguishes it from other products. It can be used only by the company that holds the patent.
combinatorial organic synthesis	the synthesis of a library of compounds by covalently connecting sets of building blocks of varying structure.
distribution coefficient	the ratio of the amount of a compound dissolving in each of two immiscible solvents.
drug	a compound that reacts with a biological molecule, triggering a physiological effect.
drug resistance	resistance to a particular drug.
drug synergism	when the effect of two drugs used in combination is greater than the sum of the effects obtained when administered individually.
generic name	a name for a drug that anyone can manufacture.
lead compound	the prototype in a search for other biologically active compounds.
molecular modeling	computer-assisted design of a compound.
molecular modification	changing the structure of a lead compound.
orphan drug	a drug for a disease or condition that affects fewer than 200,000 people.
prodrug	a compound that does not become an effective drug until it undergoes a reaction in the body.

quantitative structure-reactivity relationship (QSAR)   the relation between a particular property of a series of compounds and their biological activity.

random screen (blind screen)   the search for a pharmacologically active compound without any information about what chemical structures might show activity.

rational drug design   designing drugs with a particular structure to achieve a specific purpose.

receptor   the site where a drug binds in order to exert its physiological effect.

suicide inhibitor   a compound that inactivates an enzyme by undergoing part of the normal catalytic mechanism.

therapeutic index   the ratio of the lethal dose of a drug to the therapeutic dose.

Solutions to Problems

1. **a.** ethyl *para*-aminobenzoate   **b.** 2-(*N*,*N*-diethylamino)ethyl *para*-aminobenzoate

c. 2′-deoxy-5-iodouridine

2. There are many possibilities. The following are just a few of them.

3. The compound on the left has an electron-donating substituent, whereas the compound on the right has an electron-withdrawing substituent in the same position. Because the compounds known to be effective tranquilizers have an electron-withdrawing substituent in that position (Cl or NO_2), the compound on the right is more likely to show activity as a tranquilizer.

electron-donating substituent

electron-withdrawing substituent

4. Anesthetics have been found to have similar distribution coefficients, which means that they have similar polarities. Since diethyl ether is a known anesthetic, ethyl methyl ether with a polarity similar to that of diethyl ether is more apt to be a general anesthetic than propanol, which is considerably more polar.

5.

6.   The electron-withdrawing SO_2 group is weakly basic and, therefore, an excellent leaving group. The weak carbon-sulfur bond aids imine formation.

7.   **a.**   The chloro-substituted compound would be expected to be a more potent inhibitor because potency increases with the ability of the substituents to donate electrons into the π system.

$>$

b.   The compound with the pentyl substituent would be expected to be a more potent inhibitor because potency increases with increasing hydrophobicity of one of the substituents.

$>$

8.   Tetrahydrocannabinol is a safer drug because it has a higher therapeutic index.

$$\text{therapeutic index} = \frac{\text{lethal dose}}{\text{therapeutic dose}}$$

$$\text{tetrahydrocannabinol} = \frac{2.0\,\text{g/kg}}{20\,\text{mg/kg}} = \frac{2000\,\text{mg/kg}}{20\,\text{mg/kg}} = 100$$

$$\text{sodium pentothal} = \frac{100\,\text{mg/kg}}{30\,\text{mg/kg}} = 3.3$$

9. Hydrolysis frees the coenzyme and forms a reactive unsaturated amino acid that can react with the coenzyme bound to its enzyme. This reaction causes the imine linkage to be converted into an amine linkage that cannot be hydrolyzed to release the enzyme. In this way the enzyme is deactivated.

10. Aclovir differs from guanosine in that it has a hydroxy-ether substituent rather than a ribose.

Cytosar differs from cytidine in that it has the 2′-OH group in the β-position rather than in the α-position.

Viramid has an unusual heterocyclic base attached to ribose.

Herplex differs from 2′-deoxyuridine in that it has an iodo substituent in the 5-position.

11.

12.

13.

Tagamet + CH$_3$SH

14. The first reaction is the reaction of the substituted aniline with the more reactive acyl chloride to form an amide. In a second step, the less reactive alkyl halide undergoes an S_N2 reaction to form lidocaine.

Answers to Chapter 1 Practice Test

1.   **a.** HBr   **b.** NH_3 **c.** a carbon-fluorine bond   **d.** CCl_4

2. $^+CH_3$ $^-CH_3$ $\cdot CH_3$ **3.** $H\!:\!\ddot{\underset{\displaystyle}{O}}\!:\!\overset{\ddot{\ddot{O}}}{C}\!:\!\ddot{O}\!:^-$

      sp^2 sp^3 sp^2

4. CH_3COO^- CH_3CH_2OH CH_3OH $CH_3CH_2\overset{+}{N}H_3$

5. $^+NH_4$ **6.** $^-NH_2$

7. $CH_3CH_2CH_2CH\!=\!CH_2$ **or** $CH_3CH_2CH\!=\!CHCH_3$ **or** $CH_3\underset{\underset{\displaystyle CH_3}{|}}{C}HCH\!=\!CH_2$

8.   **a.** $CH_3OH \;+\; \overset{+}{N}H_4 \;\rightleftharpoons\; CH_3\underset{\underset{\displaystyle H}{|}}{\overset{+}{O}}H \;+\; NH_3$ **b.**   reactants

9.   **a.** $1s^2\,2s^2\,2p_x\,2p_y$ **b.** $1s^2\,2s\,2p_x\,2p_y\,2p_z$ **10.** $CH_3CH_2\underset{\underset{\displaystyle Cl}{|}}{C}HCOOH$

11. $\underset{sp}{O\!=\!C\!=\!O}$ $\underset{sp^2}{H\overset{\overset{\displaystyle O}{\|}}{C}OH}$ $\underset{sp}{HC\!\equiv\!N}$ $\underset{sp^3}{CH_3OCH_3}$ $\underset{sp^3}{CH_3CH\!=\!CH_2}$

12.

a.	A pi bond is stronger than a sigma bond.	F
b.	A triple bond is shorter than a double bond.	T
c.	The oxygen-hydrogen bonds in water are formed by the overlap of an sp^2 orbital of oxygen with an s orbital of hydrogen.	F
d.	HO^- is a stronger base than $^-NH_2$.	F
e.	A double bond is stronger than a single bond.	T
f.	A tetrahedral carbon has bond angles of $107.5°$.	F
g.	A Lewis acid is a compound that accepts a share in a pair of electrons.	T
h.	CH_3CH_3 is more acidic than $H_2C\!=\!CH_2$.	F

Answers to Chapter 2 Practice Test

1.   3-methyloctane

2.   **a.**   **b.**   **c.**

3. **a.** *sec*-butyl chloride, 2-chlorobutane   **c.** cyclopentyl bromide, bromocyclopentane

b. isohexyl alcohol, 4-methyl-1-pentanol

4. **a.** $CH_3CH_2CH_2CH_2CH_2Br$ $CH_3CH_2CH_2Br$ $CH_3CH_2CH_2CH_2Br$
$\qquad\qquad$ **1** $\qquad\qquad\qquad\qquad\qquad$ **3** $\qquad\qquad\qquad\qquad$ **2**

b. $CH_3CH_2CH_2CH_2CH_3$ $CH_3CH_2CH_2CH_2OH$ $CH_3CH_2CH_2CH_2Cl$
$\qquad\qquad\quad$ **3** $\qquad\qquad\qquad\qquad\quad$ **1** $\qquad\qquad\qquad\qquad\quad$ **2**

c.
$$\underset{\underset{\displaystyle \underset{\textbf{3}}{CH_3\ \ CH_3}}{|\qquad|}}{\overset{\overset{\displaystyle CH_3\ \ CH_3}{|\qquad|}}{CH_3C-CCH_3}}\qquad \underset{\textbf{1}}{CH_3CH_2CH_2CH_2CH_2CH_2CH_2CH_3}\qquad \underset{\underset{\textbf{2}}{CH_3}}{\overset{|}{CH_3CHCH_2CH_2CH_2CH_2CH_3}}$$

5. **a.** 6-methyl-3-heptanol   **c.** 1-bromo-3-methylcyclopentane

b. 3-ethoxyheptane   **d.** 1,4-dichloro-5-methylheptane

6.

7.

8. **a.** butyl alcohol   **c.** hexane   **e.** ethyl alcohol
b. 1-butanol   **d.** pentylamine

9. **a.** 1-bromo-3-methylbutane   **b.** 3-methyl-1-butanol   **c.** 3-methyl-1-butanamine
$\qquad\qquad$ isopentyl bromide $\qquad\qquad\qquad$ isopentyl alcohol $\qquad\qquad\qquad$ isopentylamine

10. **a.** a staggered conformer

b. the chair conformer of methylcyclohexane with the methyl group in the equatorial position

c. cyclohexane

11. **a.** $\underset{\underset{Br}{|}}{CH_3CHCH_3}$ **d.** $\underset{\underset{CH_3}{|}}{CH_3CHCH_3}$ **e.** $CH_3CH_2CH_2OH$ $\underset{\underset{CH_3}{|}}{CH_3CHOH}$ $CH_3CH_2OCH_3$

b. $CH_3CH_2NHCH_3$

c. $\underset{\underset{CH_3}{|}}{CH_3CHCH_3}$ **or** $\underset{\underset{CH_3}{|}}{\overset{\overset{CH_3}{|}}{CH_3CCH_3}}$ **or** CH_4 **or** $\underset{\underset{CH_3\ \ CH_3}{|\qquad|}}{\overset{\overset{CH_3\ \ CH_3}{|\qquad|}}{CH_3C-CCH_3}}$

Answers to Chapter 3 Practice Test

1. **a.** 4-methyl-1-hexene   **c.** 7-methyl-3-nonene

   b. 4-bromocyclopentene   **d.** 4-chloro-3-methylcyclohexene

2.

$$\underset{\textbf{2}}{-\overset{\overset{\text{O}}{\|}}{\text{C}}\text{CH}_3} \qquad \underset{\textbf{4}}{-\text{CH}=\text{CH}_2} \qquad \underset{\textbf{1}}{-\text{Cl}} \qquad \underset{\textbf{3}}{-\text{C}\equiv\text{N}}$$

3. **a.** 2-pentene   **c.** 3-methyl-2-pentene

   b. 3-methyl-1-hexene   **d.** 1-methylcyclohexene

4. **a.** Increasing the energy of activation, increases the rate of the reaction.   F
   b. Decreasing the entropy of the products compared to the entropy of the reactants
        makes the equilibrium constant more favorable.   F
   c. An exergonic reaction is one with a $-\Delta G°$.   T
   d. An alkene is an electrophile.   F
   e. The higher the energy of activation, the more slowly the reaction will take place.   T
   f. Another name for *trans*-2-butene is *Z*-2-butene.   F
   g. *trans*-2-Butene has a dipole moment of zero.   T
   h. A reaction with a negative $\Delta G°$ has an equilibrium constant greater than one.   T
   i. Increasing the energy of the reactants, increases the rate of the reaction.   T
   j. Increasing the energy of the products, increases the rate of the reaction.   F
   k. The magnitude of a rate constant is not dependent on the concentration of the reactants.   T

5. **a.** *Z*   **b.** *Z*

6. **a.** $CH_2=CHCH_2OH$ **c.**

$$\underset{\text{H}}{\overset{\text{CH}_3\text{CH}_2}{\diagdown}}\text{C}=\text{C}\underset{\text{H}}{\overset{\text{CH}_2\text{CH}_2\text{CH}_3}{\diagup}}$$

          7. 5

   b. (cyclohexene with CH₃ substituent)

   d. $CH_2=CHBr$

8. $CH_3CH=CH_2 \; + \; H\overset{\curvearrowleft}{-}\overset{..}{\underset{..}{Cl}}: \; \rightleftharpoons \; CH_3\overset{+}{CH}-CH_3 \; + \; :\overset{..}{\underset{..}{Cl}}:^- \; \longrightarrow \; CH_3CH-CH_3$

          $\underset{:\overset{..}{\underset{..}{Cl}}:}{|}$

9. b and c   **10.** **a.** 4 kcal/mol   **b.** 35 °C   **c.** one reactant forms two products

Answers to Chapter 4 Practice Test

1. **a.** $CH_3\overset{+}{\underset{|}{C}}CH_3$ with CH_3 above

   b. $CH_3CH_2\overset{+}{C}HCH_3$

   c. $CH_3CH_2\overset{+}{C}H_2$

   d. (cyclohexene ring with CH_3)

   e. (alkene: H_3C and H on left carbon, H and CH_3 on right carbon)

2. $CH_3CH_2CH_2CH{=}CH_2$ **3.** **B**

4. **a.** $CH_2{=}\overset{CH_3}{\underset{|}{C}}CH_2CH_3$ + HBr $\longrightarrow$ $CH_3\overset{CH_3}{\underset{|}{\underset{|}{C}}}CH_2CH_3$ with Br below

   b. $CH_3\overset{CH_3}{\underset{|}{C}}HCH{=}CH_2$ + HCl $\longrightarrow$ $CH_3\overset{CH_3}{\underset{|}{\underset{|}{C}}}CH_2CH_3$ with Cl below

   c. $CH_3CH_2CH{=}CH_2$ + Cl_2 $\xrightarrow{H_2O}$ $CH_3CH_2\underset{\underset{OH}{|}}{C}HCH_2Cl$

   d. $CH_3\underset{\underset{CH_3}{|}}{\overset{\overset{CH_3}{|}}{C}}CH{=}CH_2$ + HBr $\longrightarrow$ $CH_3\underset{\underset{Br}{|}}{C}{-}\underset{\underset{CH_3}{|}}{C}HCH_3$ with CH_3 above left carbon

5. **a.** $CH_3\overset{CH_3}{\underset{|}{C}}H\underset{\underset{OH}{|}}{C}HCH_3$

   b. (cyclopentane ring with CH_2CH_3 and OH)

6. **a.** $CH_3\underset{\underset{CH_3}{|}}{\overset{\overset{CH_3}{|}}{C}}CH{=}CH_2$ $\xrightarrow[Pd/C]{H_2}$ $CH_3\underset{\underset{CH_3}{|}}{\overset{\overset{CH_3}{|}}{C}}CH_2CH_3$

   b. (cyclopentene) $\xrightarrow[H_2O]{Cl_2}$ (cyclopentane with OH and Cl)

   c. (cyclohexene with CH_3) $\xrightarrow{HBr}$ (cyclohexane with CH_3 and Br)

   d. (cyclohexene with CH_3) $\xrightarrow[\substack{2.\ HO^-,\ H_2O_2,\\ H_2O}]{1.\ BH_3}$ (cyclohexane with CH_3 and OH)

7.   CH₃CH₂ĊHCHCH₃   (with CH₃ above) + cyclohexane with CH₃ +

8.
   a.   The addition of Br₂ to 1-butene to form 1,2-dibromobutane is a concerted reaction.   F
   b.   The reaction of 1-butene with HCl will form 1-chlorobutane as the major product.   F
   c.   2,3-Dimethyl-2-pentene is more stable than 3,4-dimethyl-2-pentene.   T
   d.   The reaction of HBr with 3-methylcyclohexene is more highly regioselective than the reaction of HBr with 1-methylcyclohexene.   F
   e.   The reaction of an alkene with a carboxylic acid forms an epoxide.   F
   f.   A catalyst increases the equilibrium constant of a reaction.   F

Answers to Chapter 5 Practice Test

1.   a.   a pair of enantiomers   b.   a pair of enantiomers   **2.**   −3.0

3.   (Fischer projection: CH₃ top; H—Cl; H—Cl; CH₃ bottom)

4.   CH₃CH₂CH₂CH₂Cl   CH₃CH₂CHCH₃ (with Cl)   CH₃CHCH₂Cl (with CH₃)   CH₃CCH₃ (with CH₃ and Cl)

5.
   a.   (CH₃ top; H—Br; H—Br; CH₃) + (CH₃; H—Br; Br—H; CH₃) + (CH₃; Br—H; H—Br; CH₃)

   b.   (CH₃; H—Br; H—Br; CH₂CH₃) + (CH₃; Br—H; Br—H; CH₂CH₃) + (CH₃; H—Br; Br—H; CH₂CH₃) + (CH₃; Br—H; H—Br; CH₂CH₃)

6.
   a.   CH₃CH₂—CH₃ (with Cl top, H bottom) + CH₃—CH₂CH₃ (with Cl top, H bottom)   c.   (CH₂CH₃; H—Br; H—Br; CH₂CH₃)

   b.   CH₃CH₂CH₂—CH₃ (with Br top, H bottom) + CH₃—CH₂CH₂CH₃ (with Br top, H bottom) + CH₃CH₂CHCH₂CH₃ (with Br)   d.   (CH₂CH₃; H—Br; H—Br; CH₂CH₂CH₃) + (CH₂CH₃; Br—H; Br—H; CH₂CH₂CH₃)

7.   (−)-2-Methylbutanoic acid has the *S* configuration.

8.   **a.**   **+**   **c.**

b.   **+**   **d.**   **+**

9.

10.   **or**

11.   **a.**   Diastereomers have the same melting points.   F
         b.   The addition of HBr to 3-methyl-2-pentene is stereospecific.   F
         c.   The addition of HBr to 3-methyl-2-pentene is stereoselective.   F
         d.   The addition of HBr to 3-methyl-2-pentene is regioselective.   T
         e.   Meso compounds do not rotate polarized light.   T
         f.   2,3-Dichloropentane has a stereoisomer that is a meso compound.   F
         g.   All compounds with the *R* configuration are dextrorotatory.   F
         h.   A compound with three asymmetric centers can have a maximum of nine stereoisomers.   F

Answers to Chapter 6 Practice Test

1.   **a.**   **1.** BH_3/THF **2.** HO^-, H_2O_2, H_2O

b. H_2/Lindlar catalyst

2.   **3.**   **a.** $CH_3CH_2CHC \equiv CCH_2CHCH_3$ **b.**
                                             | |
                                            CH_3 CH_3

4.   **a.**   A terminal alkyne is more stable than an internal alkyne.   F
         b.   Propyne is more reactive than propene toward reaction with HBr.   F
         c.   1-Butyne is more acidic than 1-butene.   T
         d.   An sp^2 hybridized carbon is more electronegative than an sp^3 hybridized carbon.   T

5. **a.** 1-bromo-5-methyl-3-hexyne   **b.** 5-methyl-3-hexyn-1-ol

6. $CH_3CH_2CH_2C\equiv CH$ **7.** NH_3 $CH_3\equiv CH$ CH_3CH_3 H_2O $CH_3CH=CH_2$
                                                  3   2   5   1   4

8.

$$CH_3CH_2CH_2\overset{\overset{\displaystyle O}{\|}}{C}CH_2CH_3$$

9.   **a.** $CH_3CH_2C\equiv CH$

$\downarrow$ $^-NH_2$

$CH_3CH_2C\equiv C^-$ $\xrightarrow{CH_3CH_2Br}$ $CH_3CH_2C\equiv CCH_2CH_3$

$\uparrow$ H_2/Pt

$CH_3CH_2CH_2CH_2CH_2CH_3$

b. $CH_3CH_2C\equiv CH$

$\downarrow$ 1. $^-NH_2$ 2. CH_3CH_2Br

$CH_3CH_2C\equiv CCH_2CH_3$ $\xrightarrow[\text{Lindlar catalyst}]{H_2}$ $CH_3CH_2CH=CHCH_2CH_3$

$\uparrow$ HBr

$CH_3CH_2CH_2\overset{\overset{\displaystyle}{|}}{C}HCH_2CH_3$ with Br

$CH_3CH_2CH_2CHCH_2CH_3$
                          |
                          Br

c. $CH_3CH_2C\equiv CH$

$\downarrow$ $^-NH_2$

$CH_3CH_2C\equiv C^-$ $\xrightarrow{CH_3CH_2Br}$ $CH_3CH_2C\equiv CCH_2CH_3$

$\uparrow$ H_2O | H_2SO_4

$$CH_3CH_2\overset{\overset{\displaystyle O}{\|}}{C}CH_2CH_2CH_3$$

Answers to Chapter 7 Practice Test

1.   **a.**   **c.** $CH_3\overset{\overset{\displaystyle O}{\|}}{\underset{}{C}}\!\!\overset{-}{CH}CH_3$ **e.**

b. $CH_3\overset{-}{C}HC\equiv CH$ **d.** $CH_2=CH\overset{+}{C}H_2$

2.   **a.** $CH_3CH=CH-\overset{\displaystyle ..}{\underset{\displaystyle ..}{O}}CH_3$ $\longleftrightarrow$ $CH_3\overset{-}{C}H-CH=\overset{+}{O}CH_3$

b. $CH_3CH=CH-CH=CH-\overset{+}{C}H_2$ $\longleftrightarrow$ $CH_3CH=CH-\overset{+}{C}H-CH=CH_2$

$\updownarrow$

$CH_3\overset{+}{C}H_2-CH=CH-CH=CH_2$

c. $^-CH_2-CH=CH-\overset{\displaystyle O}{\overset{\|}{CH}} \longleftrightarrow CH_2=CH-\overset{\cdot\cdot}{CH}-\overset{\displaystyle O}{\overset{\|}{CH}} \longleftrightarrow CH_2=CH-CH=\overset{\overset{\displaystyle O^-}{|}}{CH}$

3. $\underset{+}{CH_3}\overset{\overset{\displaystyle CH_3}{|}}{CCH_2}CH=CH_2 \qquad CH_2=CHCH_2CH=CH_2 \qquad CH_3CH_2NHCH_2CH=CHCH_3$

4. **a.**

b.

5. $\overset{\displaystyle O}{\overset{\|}{CH_3C}}OH \quad$ and $\quad CH_3\overset{\overset{\displaystyle O^-}{|}}{C}=\overset{+}{O}H$

6.

7. **a.**   **b.**

8. **a.**

b.

c.

9. **a.**   **b.**

10. **a.** A compound with four conjugated double bonds will have four molecular orbitals.   F
b. ψ_1 and ψ_2 are symmetric molecular orbitals.   F
c. If ψ_3 is the HOMO in the ground state, ψ_4 will be the HOMO in the excited state.   T
d. If ψ_3 is the LUMO, ψ_4 will be the HOMO.   F
e. If the ground-state HOMO is symmetric, the ground-state LUMO will be antisymmetric.   T
f. A conjugated diene is more stable than an isomeric isolated diene.   T
g. A single bond formed by an sp^2—sp^2 overlap is longer than a single bond formed by an sp^2—sp^3 overlap.   F
h. The thermodynamically controlled product is the major product obtained when the reaction is carried out under mild conditions.   F
i. 1,3-Hexadiene is more stable than 1,4-hexadiene.   T

11.

12.

13.

14. **a.**   **b.**

15.   **16.**

17. **a.**

$$CH_2{=}CH{-}\underset{\underset{Cl}{|}}{\overset{\overset{CH_3}{|}}{C}}{-}CH_3$$

product of kinetic control

$$CH_2{-}CH_2{=}\underset{\underset{Cl}{|}}{\overset{\overset{CH_3}{|}}{C}}{-}CH_3$$

product of thermodynamic control

b.

product of kinetic control

product of thermodynamic control

Answers to Chapter 8 Practice Test

1. **a.** $CH_3CH_2CH_2CH_2CHBr$ (with CH_3 on the CHBr carbon)

b. $CH_3CH{=}CHCHCH_3$ (with Br on the CH carbon)

2. **a.** $CH_3CH_2\underset{\underset{}{}}{\overset{\overset{CH_3}{|}}{C}}HBr$

b.

$-CH_2Br$

3. $CH_3\underset{}{\overset{\overset{CH_3}{|}}{C}}H\underset{\underset{Br}{|}}{C}HCH_3$

4. **a.** Increasing the concentration of the nucleophile favors an S_N1 reaction over an S_N2 reaction.   F

b. Ethyl iodide is more reactive than ethyl chloride in an S_N2 reaction.   T

c. In an S_N1 reaction, the product with the retained configuration is obtained in greater yield.   F

d. The rate of a substitution reaction in which none of the reactants is charged will increase if the polarity of the solvent is increased.   T

e. An S_N2 reaction is a two-step reaction.   F

f. The pK_a of a carboxylic acid is greater in water than it is in a mixture of dioxane and water.   F

g. 4-Bromo-1-butanol will form a cyclic ether faster than will 3-bromo-1-propanol.   T

5. **a.** CH_3O^- **b.** CH_3S^-

6. **a.** $CH_3CH_2CH_2Cl + HO^-$ **c.** $CH_3CH_2CH_2Br + HO^-$ **e.** $BrCH_2CH_2CH_2CH_2NHCH_3$

b. $CH_3CH_2CH_2I + HO^-$ **d.** $CH_3\underset{\underset{Br}{|}}{C}HCH_3 \xrightarrow[CH_3OH]{CH_3O^-}$

7. All are aprotic solvents except ethanol.

8. **a.** The rate of the reaction would increase.   **d.** The pK_a would decrease.

b. The rate of the reaction would increase.   **e.** The pK_a would decrease.

c. The rate of the reaction would decrease.

Answers to Chapter 9 Practice Test

1. **a.** $-CH_2CHCH_3$ (with Br below)   **b.** $CH_2\!=\!CHCH_2CHCH_3$ (with Br below)   **2.** CH_3CHBr (with CH_3 above)

3. CH_3CH_2—C—CH_3 (with OCH_3 above, H below) CH_3—C—CH_2CH_3 (with OCH_3 above, H below)

major   minor

4. **a.** $CH_3CH_2CO^- + CH_3CH_2CH_2Br$ (with CH_3 above and CH_3 below the C)   **b.** $-Br +$ $-O^-$ **c.** $-O^- + CH_3Br$

5. **a.** $CH_3CHCH_3 + HO^-$ (with Cl below)   **c.** $CH_3CH_2CH_2Br + HO^-$ **e.** $CH_3CCH_3 + HO^-$ (with CH_3 above, Br below)

b. $CH_3CH_2CH_2I + HO^-$ **d.** $CH_3CHCH_3 \xrightarrow[\text{CH}_3\text{OH}]{\text{CH}_3\text{O}^-}$ (with Br below)

6. **a.**   **b.**   **c.**   **d.**

7. **a.** *cis*-1-bromo-2-methylcyclohexane

b. The cis and trans isomers react at about the same rate.

Answers to Chapter 10 Practice Test

1.   HBr   2. $SOCl_2$

3.   **a.** $HOCH_2\overset{\underset{\displaystyle OCH_3}{|}}{\overset{\displaystyle \overset{CH_2CH_3}{|}}{C}}CH_2CH_3$ **b.** $CH_3OCH_2\overset{\underset{\displaystyle OH}{|}}{\overset{\displaystyle \overset{CH_2CH_3}{|}}{C}}CH_2CH_3$

4.   **a.** $CH_3\overset{\displaystyle \overset{CH_3}{|}}{C}H\,CH_2\overset{\underset{\displaystyle CH_3}{|}}{\overset{\displaystyle \overset{CH_3}{|}}{N}}$ + $CH_2{=}CHCH_2CH_3$ **b.**   **c.**

5.   **a.** $CH_3CH_2\overset{\underset{\displaystyle CH_3}{|}}{\overset{\displaystyle \overset{CH_3}{|}}{C}}{=}CCH_3$ **b.** $CH_3CH_2CH_2\overset{\underset{\displaystyle CH_3}{|}}{\overset{\displaystyle \overset{CH_3}{|}}{C}}{=}CCH_3$ **c.** $CH_3CH_2CH{=}CHCH_3$ **d.**

6.   **a.**   Tertiary alcohols are easier to dehydrate than secondary alcohols.   T
    b.   Alcohols are more acidic than thiols.   F
    c.   Alcohols have higher boiling points than thiols.   T
    d.   The acid-catalyzed dehydration of a primary alcohol is an S_N1 reaction.   F
    e.   The Hofmann elimination reaction is an E2 reaction.   T

7.   **a.** $CH_3CH_2\overset{\underset{\displaystyle CH_3}{|}}{\overset{\displaystyle \overset{CH_3}{|}}{C}}{-}I$ + CH_3OH **b.**

8.   **a.** $CH_3CH_2CH_2NHCH_2CH_3$ **c.** $CH_3CH_2CH_2\overset{\overset{\displaystyle O}{\|}}{C}H$

    b. $CH_3CH_2CH_2Cl$ **d.** $CH_3CH_2CH_2\overset{\overset{\displaystyle O}{\|}}{C}OH$

Answers to Chapter 11 Practice Test

1. an organocadmium reagent

2. Br⌇ and Br⌇

3. **a.** **b.** **c.**

4. **a.** **b.** **c.**

5. **a.** CH_2=CH_2 +

   b. CH_2=CH_2 +

   c. CH_2=CH_2 +

6. **a.**

$$\underset{H}{\overset{H}{>}}C=C\underset{Br}{\overset{CH_3}{<}} \xrightarrow[Et_2O]{(CH_3CH_2)_2CuLi} \underset{H}{\overset{H}{>}}C=C\underset{CH_2CH_3}{\overset{CH_3}{<}}$$

   b. $-CH_2OH$ $\xrightarrow{PBr_3}$ $-CH_2Br$ $\xrightarrow[Et_2O]{Mg}$ $-CH_2MgBr$ $\xrightarrow[\text{2. HCl}]{\text{1. } H_2C\!-\!CH_2}$ $-(CH_2)_3OH$

7. **a.** $CH_3CH_2CH_2CH_2CH_2OH$ **b.** $-CH_2CH_2OH$

Answers to Chapter 12 Practice Test

1. 3

2. CH_3CH_3 + ·Cl $\longrightarrow$ $CH_3\dot{C}H_2$ + HCl

3.

$$
\begin{array}{c}
\text{CH}_3 \\
\text{Br}-\!\!-\text{H} \\
\text{Br}-\!\!-\text{H} \\
\text{CH}_2\text{CH}_3
\end{array}
\qquad
\begin{array}{c}
\text{CH}_3 \\
\text{Br}-\!\!-\text{H} \\
\text{H}-\!\!-\text{Br} \\
\text{CH}_2\text{CH}_3
\end{array}
$$

4. $\text{CH}_3\text{CH}_3 + \text{Cl}\cdot \longrightarrow \text{CH}_3\dot{\text{C}}\text{H}_2 + \text{HCl}$ $\Delta H° = 101 - 103 = -2$

    $\text{CH}_3\dot{\text{C}}\text{H}_2 + \text{Cl}_2 \longrightarrow \text{CH}_3\text{CH}_2\text{Cl} + \text{Cl}\cdot$ $\Delta H° = 58 - 85 = -27$

5. $\dot{\text{C}}\text{H}_2\text{CH}_2\text{CH}_2\text{CH}\!=\!\text{CH}$ $\text{CH}_3\text{CH}_2\dot{\text{C}}\text{HCH}\!=\!\text{CH}_2$
              3   **1**

    $\text{CH}_2\text{CH}_2\text{CH}_2\text{CH}\!=\!\dot{\text{C}}\text{H}$ $\text{CH}_3\dot{\text{C}}\text{HCH}_2\text{CH}\!=\!\text{CH}_2$
              4   **2**

6. **a.**

    b. $\text{CH}_3\text{CHCH}\!=\!\text{CH}_2$ + $\text{CH}_3\text{CH}\!=\!\text{CHCH}_2\text{Br}$
                      |
                     Br

7. **a.**

$$
\begin{array}{c}
\qquad\text{CH}_3\quad\text{CH}_3 \\
\text{CH}_3\overset{|}{\text{C}}\text{CH}_2\text{CH}_2\overset{|}{\text{C}}\text{HCH}_3 \\
\overset{|}{\text{CH}_2\text{Cl}}
\end{array}
\quad
\begin{array}{c}
\quad\text{CH}_3\qquad\quad\text{CH}_3 \\
\text{CH}_3\overset{|}{\text{C}}\!-\!\overset{}{\text{C}}\text{HCH}_2\overset{|}{\text{C}}\text{HCH}_3 \\
\overset{|}{\text{CH}_3}\;\overset{|}{\text{Cl}}
\end{array}
\quad
\begin{array}{c}
\quad\text{CH}_3\quad\text{CH}_3 \\
\text{CH}_3\overset{|}{\text{C}}\text{CH}_2\overset{}{\text{C}}\text{HCHCH}_3 \\
\overset{|}{\text{CH}_3}\;\overset{|}{\text{Cl}}
\end{array}
$$

$$
\begin{array}{c}
\quad\text{CH}_3\qquad\quad\text{CH}_3 \\
\text{CH}_3\overset{|}{\text{C}}\text{CH}_2\text{CH}_2\overset{|}{\text{C}}\text{CH}_3 \\
\overset{|}{\text{CH}_3}\quad\overset{|}{\text{Cl}}
\end{array}
\qquad
\begin{array}{c}
\quad\text{CH}_3\qquad\text{CH}_3 \\
\text{CH}_3\overset{|}{\text{C}}\text{CH}_2\text{CH}_2\overset{|}{\text{C}}\text{HCH}_2\text{Cl} \\
\overset{|}{\text{CH}_3}
\end{array}
$$

    b.
$$
\begin{array}{c}
\quad\text{CH}_3\quad\text{CH}_3 \\
\text{CH}_3\overset{|}{\text{C}}\text{CH}_2\text{CH}_2\overset{|}{\text{C}}\text{HCH}_3 \\
\overset{|}{\text{CH}_2\text{Cl}}
\end{array}
$$

    c.
$$
\begin{array}{c}
\qquad\text{CH}_3\qquad\text{CH}_3 \\
\text{CH}_3\overset{|}{\text{C}}\text{CH}_2\text{CH}_2\overset{|}{\text{C}}\text{CH}_3 \\
\quad\overset{|}{\text{CH}_3}\qquad\overset{|}{\text{Br}}
\end{array}
$$

8.   23%

9. **a.**
$$
\begin{array}{c}
\qquad\quad\text{CH}_3 \\
\text{CH}_3\text{CH}_2\overset{|}{\text{C}}\text{CH}_2\text{CH}_3 \\
\qquad\quad\overset{|}{\text{Br}}
\end{array}
$$

    c.
$$
\begin{array}{c}
\qquad\quad\text{CH}_3 \\
\text{CH}_3\text{CH}_2\overset{|}{\text{C}}\text{CH}_2\text{CH}_3 \\
\qquad\quad\overset{|}{\text{Cl}}
\end{array}
$$

    b.
$$
\begin{array}{c}
\qquad\text{CH}_3 \\
\text{CH}_3\text{CH}\overset{}{\text{C}}\text{HCH}_2\text{CH}_3 \\
\quad\overset{|}{\text{Br}}
\end{array}
$$

    d.
$$
\begin{array}{c}
\qquad\quad\text{CH}_3 \\
\text{CH}_3\text{CH}_2\overset{|}{\text{C}}\text{CH}_2\text{CH}_3 \\
\qquad\quad\overset{|}{\text{Cl}}
\end{array}
$$

10. **b**

Answers to Chapter 13 Practice Test

1. **a.**

~2770 cm^{-1}

e.

OCH$_3$
~1050 or ~1250 cm^{-1}

b.

NH$_2$
~3300 cm^{-1}

f. $CH_3CH_2CH{=}CHCH_3$ $CH_3CH_2C{\equiv}CCH_3$
~1600 cm^{-1} ~2100 cm^{-1}
~3100 cm^{-1}

c. $CH_3CH_2CH_2CH_2OH$
~3600–3200 cm^{-1}

g. $CH_3CH_2C{\equiv}CH$
~3300 cm^{-1}

d.

—CHCH$_3$ ~1380 cm^{-1}
 |
 OH

2. **a.** The O—H stretch of a concentrated solution of an alcohol occurs at a higher
         frequency than the O—H stretch of a dilute solution.   F
    b. Light of 2 μm is of higher energy than light of 3 μm.   T
    c. It takes more energy for a bending vibration than for a stretching vibration.   F
    d. Propyne will not have an absorption band at 3100 cm^{-1}, because there is no
         change in the dipole moment.   F
    e. Light of 8 μm has the same energy as light of 1250 cm^{-1}.   T
    f. The M + 2 peak of an alkyl chloride is half the height of the M peak.   F
    g. A chromophore that exhibits both $n \rightarrow \pi^*$ and $\pi \rightarrow \pi^*$ transitions will have
         the $n \rightarrow \pi^*$ transition at a longer wavelength.   T

3.

         CH$_3$
          |
$CH_3CH_2CCH_2CH_2CH_3$
          |
         OH

4.   Absorbance = molar absorbtivity × concentration × length of light path (in cm)
                0.72 = molar absorbtivity × 3.8 × 10^{-4} × 1
      molar absorbtivity = 1900 M^{-1} cm^{-1}

5. **a.**

OH
OH
~3100 cm^{-1}
~1500 cm^{-1}

b.

H

~2700 cm^{-1} ~2900 cm^{-1}
~1380 cm^{-1}

c. ~3300–2500 cm⁻¹ · ~2900 cm⁻¹ / ~1380 cm⁻¹ / ~1050 cm⁻¹

e. ~3100 cm⁻¹ · ~2900 cm⁻¹ / ~1380 cm⁻¹

d. ~1700 cm⁻¹ · ~1050 cm⁻¹

6. a.   b.   c.

7.   Absorbance = molar absorbtivity × concentration × length of light path (in cm)
       0.76 = 1200 × concentration
   concentration = 6.3 × 10⁻⁴ M

Answers to Chapter 14 Practice Test

3.

O=C(CH₃)(OCH₂CH₃)   O=C(CH₃CH₂)(OCH₃)   O=C(H)(OCH₂CH₂CH₃)

3 signals   3 signals   4 signals

The signal at the highest frequency (farthest downfield) is a quartet.

The signal at the highest frequency (farthest downfield) is a singlet.

4.
- **a.** The peaks on the right of an NMR spectrum are deshielded compared to the peaks on the left.   F
- **b.** Dimethyl ketone has the same number of signals in its ¹H NMR spectrum as in its ¹³C NMR spectrum.   F
- **c.** In the ¹H NMR spectrum of the compound shown below, the lowest frequency signal (the one farthest upfield) is a singlet and the highest frequency signal (the one farthest downfield) is a doublet.   T

O_2N—⟨benzene ring⟩—CH_3

- **d.** The greater the frequency of the signal, the greater its chemical shift in ppm.   T

5.
- **a.** $CH_3CH_2CH_2Cl$

  triplet

  3 signals

- **b.** $CH_3CH_2COCH_3$ (with C=O)

  singlet

  3 signals

- **c.** CH_3CHCH_3 with Br — septet

  Br

  2 signals

6.
- **a.** $CH_3CH_2CH_2Cl$

  3 signals

  triplet

- **b.** $CH_3CH_2COCH_3$ (with C=O)

  4 signals

  singlet

- **c.** CH_3CHCH_3 with Br

  2 signals

  doublet

Answers to Chapter 15 Practice Test

1.

2.

3.

4.

5. **a.** [structure: benzenesulfonic acid with SO₂OH group]   **b.** [structure: cyclopentadiene with CH₂ (H H) at bottom]

6. [structure: propanoic anhydride] CH₃CH₂—C(=O)—O—C(=O)—CH₂CH₃

7. **a.** [structure: cyclohexene with ⁺CHCH₃]   **b.** CH₃⁺CHCH=CH₂   **c.** CH₃ĊHC(=O)CH₃

8. HÖ—NO₂ + H—OSO₃H ⇌ HÖ—⁺N(H)O₂ ⇌ H₂O + ⁺NO₂

Answers to Chapter 16 Practice Test

1. **a.** *meta*-nitrotoluene
3-nitrotoluene

c. *ortho*-ethylbenzoic acid
2-ethylbenzoic acid

b. 1,2,4-tribromobenzene

d. *para*-chlorophenol
4-chlorophenol

2. [structure: bromobenzene] **4**   [structure: acetanilide] **1**   [structure: ethylbenzene] **2**   [structure: N-methylbenzamide] **5**   [structure: benzene] **3**

3. **a.** [structure: 4-chlorobenzoic acid, COOH / Cl]   **b.** [structure: 4-nitrophenol, OH / NO₂]   **c.** [structure: 4-methylanilinium, ⁺NH₃ / CH₃]   **d.** [structure: benzoic acid, COOH]

4. **a.** *para*-bromonitrobenzene   **b.** *para*-bromoethylbenzene

5. **a.** [structure: benzene ring with NO₂ and SO₃H]

c. [structure: benzoic acid, benzene ring with COOH]

e. [structure: benzene ring with Cl, NO₂, OCH₃]

b. [structure: benzene ring with OCH₃ and CH₃] + [structure: benzene ring with OCH₃ and CH₃]

d. [structure: benzene ring with OCH₃ and NO₂]

f. [structure: benzene ring with COCH₃ and Cl]

6. **a.** Benzoic acid is more reactive than benzene toward electrophilic aromatic substitution.   F
   b. *para*-Chlorobenzoic acid is more acidic than *para*-methoxybenzoic acid.   T
   c. A CH=CH₂ group is a meta director.   F
   d. *para*-Nitroaniline is more basic than *para*-chloroaniline.   F

Answers to Chapter 17 Practice Test

1. **a.** CH_3—C(=O)—OCH_3 **b.** CH_3—C(=O)—O—[phenyl]   **c.** CH_3—C(=O)—O—[phenyl]—NO_2 **d.** CH_3—C(=O)—Cl

2. **a.** *N*-ethylpentanamide   **c.** methyl 4-phenylbutanoate

   b. 3-methylpentanoic acid   **d.** ethanoic propanoic anhydride

3. **a.** CH_3CH_2—C(=O)—O—C(=O)—CH_2CH_3

   b. CH_3—C(=O)—Cl + H_2O ⟶ CH_3—C(=O)—OH + HCl

   Any reaction in which a reactant is cleaved as a result of reaction with water.

   c. CH_3—C(=O)—OCH_3 + CH_3CH_2OH $\overset{HCl}{\rightleftharpoons}$ CH_3—C(=O)—OCH_2CH_3 + CH_3OH

4. **a.** CH_3—$\overset{\overset{\displaystyle O}{\|}}{C}$—$OH$ **b.** CH_3—$\overset{\overset{\displaystyle O}{\|}}{C}$—$OCH_3$ **c.** CH_3—$\overset{\overset{\displaystyle O}{\|}}{C}$—$NH_2$ **d.** CH_3—$\overset{\overset{\displaystyle O}{\|}}{C}$—$OH$

5. **a.** CH_3CH_2—$\overset{\overset{\displaystyle O}{\|}}{C}$—$OH$ + $\overset{+}{N}H_4$

e. (cyclohexane)—$\overset{\overset{\displaystyle O}{\|}}{C}$—$OCH_2CH_3$ + CH_3OH

b. CH_3CH_2—$\overset{\overset{\displaystyle O}{\|}}{C}$—$O^-$ + $CH_3CH_2CH_2OH$

f. $CH_3CH_2CH_2$—$\overset{\overset{\displaystyle O}{\|}}{C}$—$OCH_2CH_2CH_3$

c. $CH_3CH_2CH_2$—$\overset{\overset{\displaystyle O}{\|}}{C}$—$OH$ + CH_3CH_2—$\overset{\overset{\displaystyle O}{\|}}{C}$—$OH$

g. CH_3CH_2—$\overset{\overset{\displaystyle O}{\|}}{C}$—$OH$ + $CH_3\overset{+}{N}H_3$

d. CH_3CH_2—$\overset{\overset{\displaystyle O}{\|}}{C}$—$NHCH_2CH_3$ + Cl^-

h. CH_3CH_2—$\overset{\overset{\displaystyle O}{\|}}{C}$—$OH$ + HO—(phenyl)

Answers to Chapter 18 Practice Test

1. **a.** (phenyl)—$\overset{\overset{\displaystyle NOH}{\|}}{C}CH_2CH_3$ + H_2O

d. CH_3CH_2O, OCH_2CH_3 (on cyclohexane) + H_2O

g. $CH_3CH_2\overset{\overset{\displaystyle OH}{|}}{\underset{\underset{\displaystyle N}{\|}}{\underset{\displaystyle C}{|}}}CH_2CH_3$

b. (piperidine ring with N attached to cyclopentene) + H_2O

e. $CH_3CH_2\overset{\overset{\displaystyle OH}{|}}{\underset{\underset{\displaystyle CH_3}{|}}{C}}CH_2CH_3$

h. $CH_3\overset{\overset{}{}}{\underset{\underset{\displaystyle SCH_3}{|}}{C}}HCH_2$—$\overset{\overset{\displaystyle O}{\|}}{C}$—$CH_3$

c. $CH_3CH_2CH_2$—$\overset{\overset{\displaystyle O}{\|}}{C}$—$OH$

f. (cyclohexane)—NH_2

i. (phenyl)—$\overset{\overset{\displaystyle OH}{|}}{\underset{\underset{\displaystyle CH_2CH_2CH_3}{|}}{C}}CH_2CH_2CH_3$

2.
$$CH_3CH_2\overset{\overset{\displaystyle OH}{|}}{C}CH_2CH_2CH_3$$
$$\underset{\displaystyle CH_3}{|}$$

3.

4. **a.**

c.

e.

b.
$$\underset{\overset{\displaystyle |}{OCH_3}}{CH_3CH_2\overset{\overset{\displaystyle OCH_3}{|}}{CH}}$$

d.
$$CH_3CH_2\overset{\overset{\displaystyle OH}{|}}{\underset{\overset{\displaystyle |}{OCH_3}}{CH}}$$

5. **a.** butanal **b.** 3-pentanone

6.

Answers to Chapter 19 Practice Test

1.

2. **a.**

b.

3.

a. CH₃CH₂—C(=O)—CH₃ + CO₂

b. CH₃CH₂C(Br)(Br)—CH(=O)H + 2 Br⁻

c. CH₃CH₂CH₂—C(=O)—CH(CH₂CH₃)—C(=O)—OCH₃

d. CH₃—C(=O)—CH(—C(=O)—CH₃)(—CH₂CH₂—C(=O)—CH₃)—C(=O)—...

e. cyclopent-2-enone

f. 2-chlorocyclohexanone

g. CH₃CH₂CH₂CH₂—C(=O)—OH

4.

a. 2 CH₃CH₂—CHO $\rightleftharpoons$ CH₃CH₂CH(OH)CH(CH₃)—CHO

b. 2 CH₃CH₂—CHO $\xrightarrow{\text{HO}^-}$ CH₃CH₂CH(OH)CH(CH₃)—CHO $\xrightarrow[\Delta]{\text{H}_2\text{SO}_4}$ CH₃CH₂CH=C(CH₃)—CHO

c. 2 CH₃CH₂—C(=O)—OCH₃ $\xrightarrow[\text{2. H}_3\text{O}^+]{\text{1. CH}_3\text{O}^-}$ CH₃CH₂—C(=O)—CH(CH₃)—C(=O)—OCH₃ + CH₃OH

d. CH₃O—C(=O)—CH₂CH₂CH₂CH₂CH₂—C(=O)—OCH₃ $\xrightarrow[\text{2. H}_3\text{O}^+]{\text{1. CH}_3\text{O}^-}$ 2-(methoxycarbonyl)cyclohexanone

e. CH₃CH₂O—C(=O)—CH₂—C(=O)—OCH₂CH₃ $\xrightarrow[\text{3. H}^+, \text{H}_2\text{O}, \Delta]{\substack{\text{1. CH}_3\text{CH}_2\text{O}^- \\ \text{2. CH}_3\text{CH}_2\text{Br}}}$ CH₃CH₂CH₂—C(=O)—OH

f.

5.

6.   **a.**   **b.**

Answers to Chapter 20 Practice Test

1.   **a.** reduction   **b.** oxidation

2.   **a.**   **e.**

b. no reaction   **f.**

c. $CH_3CH_2CH_2NHCH_2CH_3$ **g.**

d.   **h.**

3. **1.** Ag$_2$O, NH$_3$ **4.**

$$\underset{CH_3CH_2}{\overset{H_3C}{>}}C=C\underset{CH_2CH_3}{\overset{CH_3}{<}} \quad \text{and} \quad \underset{CH_3CH_2}{\overset{H_3C}{>}}C=C\underset{CH_3}{\overset{CH_2CH_3}{<}}$$

2. H$_3$O$^+$

5. **a.** NaBH$_4$ is a weaker reducing agent than LiAlH$_4$.   T
   b. Esters are easier to reduce than ketones.   F
   c. In an oxidation-reduction reaction the oxidizing agent is oxidized.   F
   d. Ketones are reduced to primary alcohols.   F
   e. Aldehydes are oxidized to carboxylic acids.   T
   f. Acyl halides are oxidized to aldehydes.   F
   g. Alkenes cannot be reduced with NaBH$_4$.   T

6. CH$_3$CH$_2$CH$_2$CH$_3$ $\xrightarrow[hv]{Br_2}$ CH$_3$CH$_2$$\underset{\overset{|}{Br}}{C}HCH_3$ $\xrightarrow{HO^-}$ CH$_3$CH$_2$$\underset{}{\overset{\overset{OH}{|}}{C}}HCH_3$ $\xrightarrow{H_2CrO_4}$ CH$_3$CH$_2$$\overset{\overset{O}{\|}}{C}CH_3$

7. **c.** 1. O$_3$ 2. Zn, H$_2$O   or   1. O$_3$ 2. (CH$_3$)$_2$S

Answers to Chapter 21 Practice Test

1. **a.** 2,4-dimethyl pyrrolidine   **c.** 3-ethyltetrahydrofuran
       2,4-dimethylazacyclopentane   3-ethyloxacyclopentane

   b. *N*-methylpiperidine   **d.** 3-bromopiperidine
       N-azacyclohexane   3-bromoazacyclohexane

2. **a.**

$$\underset{CH_3}{\overset{\overset{O}{\|}}{C}}\underset{NHCH_2CH_3}{} \quad + \quad CH_3CH_2\overset{+}{N}H_3 \ Cl^-$$

   d.

   b. (cyclohexenyl-N-piperidine) + H$_2$O

   e. (furan with Cl at 2-position)

   c. (3-bromopyridine)

3. **a.** (morpholine, protonated NH$_2^+$)   **b.** (pyridinium)   **c.** (pyrimidinium)   **d.** (pyrrole, NH)

4. **a.** 4-Chloropyridine is more reactive toward nucleophilic aromatic substitution than is 3-chloropyrrole.   T

  b. Pyrrole is more reactive toward electrophilic aromatic substitution than is furan.   T

  c. Pyrrole is more reactive toward electrophilic aromatic substitution than is benzene.   T

  d. Pyridine is more reactive toward electrophilic aromatic substitution than is benzene.   F

Answers to Chapter 22 Practice Test

1. **a.**

HO—C=O
H——OH
HO——H
H——OH
H——OH
COOH

  c.

H—C=O
H——OH
HO——H
HO——H
H——OH
CH₂OH

+

H—C=O
HO——H
HO——H
HO——H
H——OH
CH₂OH

  b.

OH
CH₂OH—O
HO——OH——OCH₃

+

OH
CH₂OH—O
HO——OH
OCH₃

  d.

HO—C=O
H——OH
H——OH
H——OH
CH₂OH

2. **a.** Glycogen contains α-1,4′ and β-1,6′-glycosidic linkages.   F

  b. D-Mannose is a C-1 epimer of D-glucose.   F

  c. D-Glucose and L-glucose are anomers.   F

  d. D-Erythrose and D-threose are diastereomers.   T

  e. Wohl degradations of D-glucose and D-gulose form the same aldotetrose.   F

3.

H—C=O
HO——H
HO——H
H——OH
H——OH
CH₂OH

H—C=O
H——OH
HO——H
H——OH
H——OH
CH₂OH

4. D-mannose and D-glucose   **5.** D-tagatose   **6.** D-altrose

7.   Amylose has α-1,4'-glycosidic linkages, whereas cellulose has β-1,4'-glycosidic linkages.

8.   D-gulose and D-idose   **9.**   D-allose

10.

Answers to Chapter 23 Practice Test

1.

a.

b.

c.

d.

e.

2.

a.

b.

c.

d.

3.   **a.**   Alanine, because it is farther away from its pI.   **c.**   leucine and isoleucine

        b.   glycine   **d.**   aspartic acid

4.   The electron-withdrawing protonated amino group causes the carboxyl group of alanine to have a lower pK_a.

5.   **a.**   A cigar shaped protein has a greater percentage of polar residues than a spherical protein.   T
b.   Naturally occurring amino acids have the L-configuration.   T
c.   There is free rotation about a peptide bond.   F

6.

7.   **a.**   The sequence of the amino acids and the location of the disulfide bonds in the protein.

b.   The three-dimensional arrangement of all the atoms in the protein.

c.   A description of the way the subunits of an oligomer are arranged in space.

8.

electrophoresis pH = 5

⊖

Lys

Ser

Tyr

Asp

Phe

Ile

⊕

chromatography →

9.   **a.** $\dfrac{2.16 + 9.18}{2} = \dfrac{11.34}{2} = 5.67$ **b.** $\dfrac{9.04 + 12.48}{2} = \dfrac{21.52}{2} = 10.76$

10.   Ala   Ser   Arg   Gly   Arg   Met   His   Phe   Lys   Ile

Answers to Chapter 24 Practice Test

1.   **a.**   A catalyst increases the equilibrium constant of a reaction.   F
b.   An acid catalyst donates a proton to the substrate and a base catalyst removes a proton from the substrate.   T
c.   The reactant of an enzyme-catalyzed reaction is called a substrate.   T
d.   Complexing with a metal ion increases the pK_a of water.   F

2. It protonates the leaving group to make it a better leaving group.

3.   **a.**   **b.**

4.

first step

    a. base catalyst

    b. nucleophilic catalyst

    c. it stabilizes the positive charge on histidine

Answers to Chapter 25 Practice Test

1. biotin and vitamin KH_2

2. methyl (CH_3), methylene (CH_2), formyl ($HC{=}O$)

3.

4.

5.   The first step is removing the hydrogen atom that is going to change places with a group on an adjacent carbon.

6.   first coenzyme = biotin
second coenzyme = coenzyme B_{12}

7.

8.

9.   ATP, Mg^{2+}, HCO_3^-

10.   FAD oxidizes dihydrolipoate to lipoate.

11.

12.   **a.**   Vitamin B_1 is the only water-insoluble vitamin that has a coenzyme function.   F
　　　b. $FADH_2$ is a reducing agent.   T
　　　c.   Thiamine pyrophosphate is vitamin B_6.   F
　　　d.   Cofactors that are organic molecules are called coenzymes.   T
　　　e.   Vitamin K is a water-soluble vitamin.   F
　　　f.   Lipoic acid is covalently bound to its enzyme by an amide linkage.   T

Answers to Chapter 26 Practice Test

1.   **a.**

$$RO-\overset{\overset{\displaystyle O}{\|}}{\underset{\underset{\displaystyle O^-}{|}}{P}}-O-\overset{\overset{\displaystyle O}{\|}}{\underset{\underset{\displaystyle O^-}{|}}{P}}-O^-$$

b.

$$R-\overset{\overset{\displaystyle O}{\|}}{C}-O-\overset{\overset{\displaystyle O}{\|}}{\underset{\underset{\displaystyle O^-}{|}}{P}}-O-\overset{\overset{\displaystyle O}{\|}}{\underset{\underset{\displaystyle O^-}{|}}{P}}-O^-$$

c.

$$^-O-\overset{\overset{\displaystyle O}{\|}}{\underset{\underset{\displaystyle O^-}{|}}{P}}-O-\overset{\overset{\displaystyle O}{\|}}{\underset{\underset{\displaystyle O^-}{|}}{P}}-O^-$$

2.

$$RCH_2CH_2CH_2CH_2-\overset{\overset{\displaystyle O}{\|}}{C}-SCoA \xrightarrow[\quad]{FAD \quad FADH_2} RCH_2CH_2CH=CH-\overset{\overset{\displaystyle O}{\|}}{C}-SCoA$$

$$\xrightarrow{H_2O}$$

$$RCH_2CH_2\underset{\underset{\displaystyle OH}{|}}{CH}-CH_2-\overset{\overset{\displaystyle O}{\|}}{C}-SCoA$$

$$\xrightarrow[\quad NADH, H^+]{NAD^+}$$

$$RCH_2CH_2-\overset{\overset{\displaystyle O}{\|}}{C}-CH_2-\overset{\overset{\displaystyle O}{\|}}{C}-SCoA$$

$$\xrightarrow{CoASH} RCH_2CH_2-\overset{\overset{\displaystyle O}{\|}}{C}-SCoA \; + \; CH_3-\overset{\overset{\displaystyle O}{\|}}{C}-SCoA$$

3.   acetate

4.   catabolic reactions

5.   amino acids

6.   acetyl-CoA

7.   **a.**   Each molecule of FADH$_2$ forms 3 molecules of ATP in the fourth stage of catabolism.   F
  b.   FADH$_2$ is oxidized to FAD.   T
  c.   NAD$^+$ is oxidized to NADH.   F
  d.   Acetyl-CoA is a citric acid cycle intermediate.   F

Answers to Chapter 27 Practice Test

1.

squalene

2.

3.   **a.**

b.

x = an even number equal to or greater than 14
y = an odd number equal to or greater than 15

4.   **a.** Cholesterol is the precursor of all other steroids.   T
        b. The double bonds in unsaturated fatty acids are conjugated.   F
        c. Fats have a higher percentage of saturated fatty acids than do oils.   T
        d. A saturated fatty acid has a lower melting point than an unsaturated fatty acid with the
            same number of carbons.   F

e.	Lipids are insoluble in water.	T
f.	A diterpene contains 20 carbons.	T
g.	Vitamin A is not a coenzyme.	T

5. The first step is removal of a hydrogen atom from a methylene carbon that is flanked by two sp^2 carbons.

$$RCH = CH - CH - CH = CH -$$

$$X \cdot \quad \overset{|}{H}$$

6. 30   **7.** five-membered ring   **8.** arachidonic acid

Answers to Chapter 28 Practice Test

1. dUMP

2. 5′—A—T—G—C—A—A—G—T—3′

3. A

4.

a.	Guanine and cytosine are purines.	F
b.	The 3′-OH group allows RNA to be easily cleaved.	F
c.	The number of A's in DNA is equal to the number of T's.	T
d.	rRNA carries the amino acid that will be incorporated into a protein.	F
e.	The template strand of DNA is the one transcribed to form RNA.	T
f.	The 5′-end of DNA has a free OH group.	F
g.	The synthesis of proteins from an RNA blueprint is called transcription.	F
h.	A nucleotide consists of a base and a sugar.	F
i.	RNA contains T's and DNA contains U's.	F

5. only #1

6. 5′—G-C-A-U-G-G-A-C-C-C-C-G-U—3′

7. In the process of converting uridines into thymidines, tetrahydrofolate is oxidized to dihydrofolate. NADPH is used to reduce dihydrofolate back to tetrahydrofolate.

8. CA and TG